STATISTICAL METHODS IN BIOLOGY

Design and Analysis of
Experiments and Regression

STATISTICAL METHODS IN BIOLOGY

Design and Analysis of Experiments and Regression

S. J. Welham
Rothamsted Research, Harpenden, UK

S. A. Gezan
University of Florida, USA
(formerly Rothamsted Research, Harpenden, UK)

S. J. Clark
Rothamsted Research, Harpenden, UK

A. Mead
Rothamsted Research, Harpenden, UK
(formerly Horticulture Research International, Wellesbourne,
UK & University of Warwick, UK)

CRC Press
Taylor & Francis Group
Boca Raton London New York

CRC Press is an imprint of the
Taylor & Francis Group, an **informa** business

A CHAPMAN & HALL BOOK

CRC Press
Taylor & Francis Group
6000 Broken Sound Parkway NW, Suite 300
Boca Raton, FL 33487-2742

© 2015 by Taylor & Francis Group, LLC
CRC Press is an imprint of Taylor & Francis Group, an Informa business

No claim to original U.S. Government works

Version Date: 20140703

ISBN 13: 978-1-4398-0878-8 (hbk)

Library of Congress Cataloging-in-Publication Data

Welham, S. J. (Suzanne Jane), author.
  Statistical methods in biology : design and analysis of experiments and regression / S.J. Welham , S.A. Gezan, S.J. Clark, A. Mead.
      pages cm
  Includes bibliographical references and index.
  ISBN 978-1-4398-0878-8 (hardback : acid-free paper) 1. Biometry. 2. Regression analysis. 3. Experimental design. I. Title.

QH323.5.W45 2014
570.1'5195--dc23   2014016839

Visit the Taylor & Francis Web site at
http://www.taylorandfrancis.com

and the CRC Press Web site at
http://www.crcpress.com

To my parents and Simon, with love and thanks

SJW

To Diablita, Psycha and Luna for all their love and unconditional support

SAG

For Mum, Dad and Tony, with love. For Joe, Mike, Moira and Sue, with thanks

SJC

To Sara, Tom and my parents, with love and thanks for your continuing support

AM

Contents

Preface

This book provides an introductory, practical and illustrative guide to the design of experiments and data analysis in the biological and agricultural plant sciences. It is aimed both at research scientists and at students (from final year undergraduate level through taught masters to PhD students) who either need to design their own experiments and perform their own analyses or can consult with a professional applied statistician and want to have a clear understanding of the methods that they are using. The material is based on courses developed at two British research institutes (Rothamsted Research and Horticulture Research International [HRI – then Warwick HRI, and now the School of Life Sciences, University of Warwick]) to train research scientists and post-graduate students in these key areas of statistics. Our overall approach is intended to be practical and intuitive rather than overly theoretical, with mathematical formulae presented only to formalize the methods where appropriate and necessary. Our intention is to present statistical ideas in the context of the biological and agricultural sciences to which they are being applied, drawing on relevant examples from our own experiences as consultant applied statisticians at research institutes, to encourage best practice in design and data analysis.

The first two chapters of this book provide introductory, review and background material. In Chapter 1, we introduce types of data and statistical models, together with an overview of the basic statistical concepts and the terminology used throughout. The training courses on which this book is based are intended to follow preliminary courses that introduce the basic ideas of summary statistics, simple statistical distributions (Normal, Poisson, Binomial), confidence intervals, and simple statistical tests (including the t-test and F-test). Whilst a brief review of such material is covered in Chapter 2, the reader will need to be comfortable with these ideas to reap the greatest benefit from reading the rest of the book. Some readers may feel that their knowledge of basic statistics is sufficiently comprehensive that they can skip this review chapter. However, we recommend you browse through it to familiarize yourself with the statistical terminology that we use.

The main body of the book follows. Chapters 3 to 11 introduce statistical approaches to the design of experiments and the analysis of data from such designed experiments. We start from basic design principles, introduce some simple designs, and then extend to more complex ones including factorial treatment structures, treatment contrasts and blocking structures. We describe the use of analysis of variance (ANOVA) to summarize the data, including the use of the multi-stratum ANOVA to account for the physical structure of the experimental material or blocking imposed by the experimenter, introduce simple diagnostic methods, and discuss potential transformations of the response. We explain the analysis of standard designs, including the randomized complete block, Latin square, split-plot and balanced incomplete block designs in some detail. We also explore the issues of sample size estimation and the power of a design. Finally, we look at the analysis of unbalanced or non-orthogonal designs. Chapters 12 to 18 first introduce the idea of simple linear regression to relate a response variable to a single explanatory variable, and then consider extensions and modifications of this approach to cope with more complex data sets and relationships. These include multiple linear regression, simple linear regression with groups, linear mixed models and models for curved relationships. We also extend related themes from the earlier chapters, including diagnostic methods specific to regression. We emphasize throughout that the same type of models and principles are used for both designed experiments and regression modelling. We complete the main body of the book with a discussion of generalized linear models, which are appropriate for certain types of non-Normal data.

We conclude with a guide to practical design and data analysis (Chapter 19), which focuses on the selection of the most appropriate design or analysis approach for individual scientific problems and on the interpretation and presentation of the results of the analysis.

Most chapters include exercises which we hope will help to consolidate the ideas introduced in the chapter. In running the training courses from which this book has been developed, we often find that it is only when students perform the analyses themselves that they fully appreciate the statistical concepts and, most importantly, understand how to interpret the results of the analyses. We therefore encourage you to work through at least some of the exercises for each chapter before moving to the next one. There are fewer exercises in the earlier chapters and the required analyses build in complexity, so we expect you to apply knowledge gained throughout the book when doing exercises from the later chapters. All of the data sets and solutions to selected exercises are available online. Some of the solutions include further discussion of the relevant statistical issues.

We have set up a website to accompany this book (www.stats4biol.info) where we show how to do the analyses described in the book using GenStat®, R and SAS®, three commonly used statistical packages. Whilst users familiar with any of these packages might not refer to this material, others are encouraged to review it and work through the examples and exercises for at least one of the packages. Any errors found after publication will also be recorded on this website.

By the time you reach the end of the book (and online material) we intend that you will have gained

- A clear appreciation of the importance of a statistical approach to the design of your experiments,

- A sound understanding of the statistical methods used to analyse data obtained from designed experiments and of the regression approaches used to construct simple models to describe the observed response as a function of explanatory variables,

- Sufficient knowledge of how to use one or more statistical packages to analyse data using the approaches that we describe, and most importantly,

- An appreciation of how to interpret the results of these statistical analyses in the context of the biological or agricultural science within which you are working.

By doing so, you will be better able both to interact with a consultant statistician, should you have access to one, and to identify suitable statistical approaches to add value to your scientific research.

This book relies heavily on the use of real data sets and material from the original courses and we are hence indebted to many people for their input. Particular thanks go to Stephen Powers and Rodger White (Rothamsted Research) and John Fenlon, Gail Kingswell and Julie Jones (HRI) for their contributions to the original courses; also to Alan Todd (Rothamsted Research) for providing many valuable suggestions for suitable data sets. The majority of real data sets used arose from projects (including PhDs) at Rothamsted Research, many in collaboration with other institutes and funded from many sources; we thank Rothamsted Research for giving us general permission to use these data. We also thank, in alphabetical order, R. Alarcon-Reverte, S. Amoah, J. Baverstock, P. Brookes,

J. Chapman, R. Curtis, I. Denholm, N. Evans, A. Ferguson, S. Foster, M. Glendining, K. Hammond-Kosack, R. Harrington, Y. Huang, R. Hull, J. Jenkyn, H.-C. Jing, A.E. Johnston, A. Karp, J. Logan, J. Lucas, P. Lutman, A. Macdonald, S. McGrath, T. Miller, S. Moss, J. Pell, R. Plumb, P. Poulton, A. Salisbury, T. Scott, I. Shield, C. Shortall, L. Smart, M. Torrance, P. Wells, M. Wilkinson and E. Wright, for specific permission to use data from their own projects or from those undertaken within their group or department at Rothamsted. Rothamsted Research receives grant-aided support from the Biotechnology and Biological Sciences Research Council of the United Kingdom. We thank various colleagues, past and present, at Horticulture Research International, Warwick HRI and the School of Life Sciences, University of Warwick, for permission to use data from their research projects, particularly Rosemary Collier and John Clarkson. We thank M. Heard (Centre for Ecology and Hydrology), A. Ortega Z. (Universidad Austral de Chile) and R. Webster for permission to use data. Examples and exercises marked '*' use simulated data inspired by experiments carried out at Rothamsted Research or HRI. The small remainder of original examples and exercises (also marked '*') were invented by the authors but are typical of the type of experiments we are regularly asked to design and the data we analyse as part of our consultancy work. In the few cases where we have not been able to find examples from our own work we have drawn on data from published sources. We would like to thank Simon Harding for technical help in setting up a repository for our work and our website and Richard Webster, Alice Milne, Nick Galwey, James Bell and Kathy Ruggeiro and an anonymous referee for reading draft chapters and providing many helpful comments and suggestions.

Finally, we would like to make some individual acknowledgements. SJW, SJC and SAG thank Rothamsted Research, and in particular Chris Rawlings, for support and encouragement to pursue this project. AM thanks his colleagues at Horticulture Research International and the University of Warwick, particularly John Fenlon, for support in the development of the original training courses, and hence the development of this project, and his co-authors for the invitation to join this project. SJW thanks Simon Harding for his support, help and long-term forbearance. SAG thanks Emma Weeks for her encouragement, and the other co-authors for their patience and the fruitful discussions we had on this project. SJC thanks Tony Scott for his patience and support, Elisa Allen for her contribution to the presentation of our courses and useful comments on some chapters, and past students for their enthusiasm and constructive feedback which led to improvements in our courses and ultimately this book. AM also thanks his family, Sara and Tom, for their continuing support and understanding.

S J Welham
Welwyn Garden City, UK

S A Gezan
Harpenden, UK and Gainesville, Florida, USA

S J Clark
Harpenden, UK

A Mead
Leamington Spa, UK

Authors

Suzanne Jane Welham obtained an MSc in statistical sciences from University College London in 1987 and worked as an applied statistician at Rothamsted Research from 1987 to 2000, collaborating with scientists and developing statistical software. She pursued a PhD from 2000 to 2003 at the London School of Hygiene and Tropical Medicine and then returned to Rothamsted, during which time she coauthored the in-house statistics courses that motivated the writing of this book. She is a coauthor of about 60 published papers and currently works for VSN International Ltd on the development of statistical software for analysis of linear mixed models and presents training courses on their use in R and GenStat.

Salvador Alejandro Gezan, PhD, is an assistant professor at the School of Forest Resources and Conservation at the University of Florida since 2011. Salvador obtained his bachelor's from the Universidad of Chile in forestry and his PhD from the University of Florida in statistics-genetics. He then worked as an applied statistician at Rothamsted Research, collaborating on the production and development of the in-house courses that formed the basis for this book. Currently, he teaches courses in linear and mixed model effects, quantitative genetics and forest mensuration. He carries out research and consulting in statistical application to biological sciences with emphasis on genetic improvement of plants and animals. Salvador is a long-time user of SAS, which he combines with GenStat, R and MATLAB® as required.

Suzanne Jane Clark has worked at Rothamsted Research as an applied statistician since 1981. She primarily collaborates with ecologists and entomologists at Rothamsted, providing and implementing advice on statistical issues ranging from planning and design of experiments through to data analysis and presentation of results, and has coauthored over 130 scientific papers. Suzanne coauthored and presents several of the in-house statistics courses for scientists and research students, which inspired the writing of this book. An experienced and long-term GenStat user, Suzanne has also written several procedures for the GenStat Procedure Library and uses GenStat daily for the analyses of biological data using a wide range of statistical techniques, including those covered in this book.

Andrew Mead obtained a BSc in statistics at the University of Bath and an MSc in biometry at the University of Reading, where he spent over 16 years working as a consultant and research biometrician at the Institute of Horticultural Research and Horticulture Research International at Wellesbourne, Warwickshire, UK. During this time, he developed and taught a series of statistics training courses for staff and students at the institute, producing some of the material on which this book is based. For 10 years from 2004 he worked as a research biometrician and teaching fellow at the University of Warwick, developing and leading the teaching of statistics for both postgraduate and undergraduate students across a range of life sciences. In 2014 he was appointed as Head of Applied Statistics at Rothamsted Research. Throughout his career he has had a strong association with the International Biometric Society, serving as International President and Vice

President from 2007 to 2010 inclusive, having been the first recipient of the 'Award for Outstanding Contribution to the Development of the International Biometric Society' in 2006, serving as a Regional Secretary of the British and Irish Region from 2000 to 2007 and on the International Council from 2002 to 2010. He is a (co)author of over 80 papers, and coauthor of *Statistical Principles for the Design of Experiments: Applications to Real Experiments* published in 2012.

1

Introduction

This book is about the design of experiments and the analysis of data arising in biological and agricultural sciences, using the statistical techniques of analysis of variance (ANOVA) and regression modelling. These techniques are appropriate for analysis of many (although not all) scientific studies and form an important basic component of the statistician's toolbox. Although we provide some of the mathematical formulae associated with these techniques, we have also tried to interpret the equations in words and to give insight into the underlying principles. We hope that this will make these useful statistical methods more accessible.

This chapter presents an introduction to the different types of data and statistical models that are considered in this book, together with an overview of the basic statistical concepts and terminology which will be used throughout. In particular, we discuss

- Types of scientific study
- Populations and samples
- Mathematical and statistical models used to describe biological processes
- The linear model – which underlies all the models and methods introduced in this book
- Parameter estimation and statistical inference
- ANOVA – the major statistical tool used to evaluate and summarize linear models

At the end of this chapter, we preview the contents of the remaining chapters.

1.1 Different Types of Scientific Study

We shall be concerned with data arising from both experimental and observational studies. Although they have many common features, there are some subtle differences that influence the conclusions that can be drawn from the analyses of data from these two types of study.

An **experimental study** is a scientific test (or a series of tests) conducted with the objective of studying the relationship between one or more outcome variables and one or more condition variables that are intentionally manipulated to observe how changing these conditions affects the results. The outcome of a study will also depend on the wider environment, and the scientist will endeavour to control other variables that may affect the outcomes, although there is always the possibility that uncontrolled, perhaps unexpected, variables also influence the outcome. Adequate planning is therefore crucial to experimental success. There are a few key elements that need to be clearly specified and considered for an experimental study (Kuehl, 2000). These are the

- aims of the experiment – usually expressed as questions or hypotheses
- physical structure of the study materials
- subjects or entities to be used
- set of conditions to be investigated
- other variables that might affect the outcome
- outcome variables to be measured
- protocols that define how the measurements are taken
- available resources (e.g. money, time, personnel, equipment, materials)

The aims of an experimental study need to be clearly specified, often in the form of hypotheses to be tested or a set of questions to be answered; this is a vital part of the planning process. The physical structure and subjects to be used should be chosen so that the results of the experiment can be related to a wider context (see Section 1.2). In addition, the set of conditions to be investigated must be chosen to answer directly the scientific questions. Other variables likely to affect the outcome should be identified and evaluated so that they can be controlled, as far as possible, and therefore do not interfere with the measured outcome. If they cannot be controlled then they should be measured. Consideration of the variables to be measured is often overlooked at the planning stage, but is important because it may affect both the statistical analysis and the efficiency of the design. As discussed later (Chapter 18), the analysis required for binary data (e.g. absence or presence of disease) or count data (e.g. numbers of insects or weeds present) may be different from that for a continuous variable (e.g. shoot length). A full defini-tion of measurement protocols is good practice and should reduce differences in proce-dure between scientists working on the same experiment, and improve repeatability of the results. Finally, the resources available will usually limit the size and scope of the experiment.

Design of experiments is a process that brings together all the elements above to pro-duce an experiment that efficiently answers the questions of interest and aims to obtain the maximum amount of information for the resources available, or to minimize the resources needed to obtain the information desired. The main statistical principles used in constructing a good design are replication, randomization and blocking. These concepts are discussed in detail in Chapter 3.

An **observational study** differs from an experimental study in that the application of conditions that affect the outcome is not directly controlled by the scientist. However, all the elements listed above for experimental studies should still be considered when you plan an observational study, although opportunities for the random allocation of conditions to subjects will be limited and sometimes non-existent. In observational stud-ies, the set of conditions to be investigated is first defined, and then subjects with these characteristics are sought and measurements made. Observational studies are often used in ecology where it is difficult to set up an experiment whilst retaining natural habitats. For example, a study might aim to determine the difference in beetle populations using selected field margins as the subjects under two conditions: with and without hedges. In this context, it is harder than in experimental studies to control other variables that may affect the outcome. For example, the set of hedges available may be composed of several plant types, which might in turn affect the species and abundance of beetles present. In addition, hedges are already in place, and fields with hedges may differ systematically in other characteristics from fields without hedges – in an extreme case they might be on different farms, with different farming methods used. The scientist should therefore consider that differences between conditions in an observational study might be caused by other unrecorded, or possibly unobserved, variables. In experimental studies, where we have greater control over conditions, this can still be true, but we can use randomization to guard against such unknown differences between subjects. But where there are potential uncontrolled sources of variability, the scientist should be wary of inferring direct causal relationships. Hill (1965) gave criteria that should be satisfied by a causative relationship in the context of epidemiology, and many of these criteria can be applied more widely and may be helpful in deciding whether a causal relationship is plausible for any observational study.

The separation between experimental and observational studies is not complete, as some studies may have both experimental and observational components. However, both types of study incorporate structure, and we should take account of this structure in the planning, design, statistical analysis and interpretation of such studies.

1.2 Relating Sample Results to More General Populations

For most scientific studies there is an implicit assumption that the results obtained can be applied to a population of subjects wider than those included in the study, i.e. that the conclusions will apply more generally (although usually with caveats) to the real world. For example, in a field trial to investigate disease control it will generally not be possible to have very large plots, nor to assess visually every plant in a plot, and so a random sample of plants is selected from each plot. It is assumed that the sampled plants are representative of all the plants in the plot and so the results from the sample are inferred to apply to the whole plot. In turn, we should usually have several plots within the trial with the same treatment applied and hope to infer the results from this sample of plots to the whole field. However, it is well-known that field experiment results vary markedly over years and locations, so the trial would ideally be performed at several locations across several years to provide a representative sample of environments. The combined results from the whole set of trials can then be claimed to apply to the region where they were carried out, rather than to a single field in a single year.

In planning any scientific study, it is therefore important to consider the frame of reference when experimental subjects are selected. The scientist should identify the **population** (wider group of subjects) to which they hope the experimental results will apply. Ideally, the subjects should then consist of a **sample**, or subset, drawn from this population. If the process of selecting a sample, known as **sampling**, is made at random, then it is reasonable to assume that the sample will have similar properties to the whole population, and we can use it to make statistical inferences about the population. Generally, as the number of observations in the sample increases, the inferences made about the population become more secure. If a sample is not taken at random, then this sense of the sample being representative of the population may be lost.

1.3 Constructing Models to Represent Reality

A **model** is an abstract representation of a hypothesized process that underpins a biological or physical phenomenon, that is, a way of describing a real system in words, diagrams, mathematical functions, or as a physical representation. In biology, models usually correspond to a simplification of the real process, as no existing model can represent reality in all details. However, this does not mean that models cannot be useful. A good model summarizes the major factors affecting a process to give a representation that provides the level of detail required for the objective of a particular study.

Mathematical models use mathematical notation and expressions to describe a process. A **statistical model** is a mathematical model that allows for variability in the process that may arise from sampling variation, biological variation between individuals, inaccuracies in measurement or influential variables being omitted (knowingly or not) from the model. Therefore, any statistical model has a measure of uncertainty associated with it.

Models are additionally often classified as either process (or mechanistic) models or empirical models. A **process model** purports to give a description of the real underlying process. This type of model can be useful in testing our knowledge: if a model can be built to reproduce the behaviour of the system accurately, then our knowledge of the process (theory) is at least consistent with reality. Conversely, and arguably more usefully, failure of a process model may indicate gaps in knowledge that can be pursued by further experimentation. Process models are often complex, with many parameters, but can sometimes be fitted using statistical principles (see e.g. Brown and Rothery, 1993, Chapter 10).

Statistical models usually fall under the category of **empirical models**, which use the principle of correlation to construct a simple model to describe an observed response in terms of one or more explanatory variables. Empirical models use the correlation between the explanatory (input) variable(s) and the measured response (output) variable to build a model without explicit reference to the true underlying process (although knowledge of this process may be used both to select suitable input variables and to identify the appropriate form of the relationship). This can be useful to identify variables that are influential where no detailed knowledge of the process exists, although some care should be taken with interpretation as there may be no direct causative relationship between the input and output variables; instead they may both be driven by some other hidden (latent) or unmeasured variable.

We shall consider statistical models of the general form

$$\text{response} = \text{systematic component} + \text{random component}.$$

This model can exist in abstract form, but we usually relate it to a set of measurements that have been made. The **response**, or response variable, relates to one type of numerical outcome from the study, sometimes also called the **set of observations**. The **systematic component** is a mathematical function of one or more explanatory variables that provide a representation of the experimental conditions. The systematic component describes the relationship between the response and these explanatory variables and hence between the response and the experimental conditions. Where the conditions have a direct numerical evaluation, such as count, weight or height, the explanatory variable is termed **quantitative**. We refer to quantitative variables as **variates**. Where the conditions are classified into groups or categories the explanatory variable is termed **qualitative**. In this case, the explanatory variable indicates the group to which each subject belongs. We shall refer to qualitative variables as **factors** and identify the distinct groups in the factor as the **factor levels**. For example, sex would be a factor with two levels: male and female. Note that it is sometimes convenient to group a quantitative variable into categories so as to treat it as a qualitative variable, for example, heights can be classified as short, medium or tall. However, this change cannot always be made in reverse; some explanatory variables, such as sex, are inherently qualitative. Similarly, if a scientist had compared three types of fertilizer, or one fertilizer across three different plant varieties, then the levels of the explanatory variable (fertilizer type or plant variety) cannot be translated into meaningful numbers. In the context of experimental studies, the conditions imposed by the experimenter are usually represented as factors and referred to as **treatments**. We also use this term more generally to describe the set of conditions present in observational studies when represented by factors. In some contexts, where it is more natural, we use the alternative term **groups** instead of treatments.

In general, the systematic component of the statistical models that we consider can be partitioned further into explanatory and structural components as

systematic component = explanatory component + structural component.

The **explanatory component** corresponds to the conditions of interest, or treatments, in the study. The **structural component** is used to account for the structure of the study, such as sub-sampling within an observational study or blocking within a designed experiment. The structural component is not always present: it may be omitted in the (rare) case that the experimental material consists of an unstructured sample. This partition facilitates the accurate specification of the whole model, as it encourages us to consider the two components separately: the explanatory component relates to our hypothesis (or hypotheses) of interest, and the structural component relates to the structure of the experimental material.

The **random component**, also known as error or noise, corresponds to variation in the response that is not explained by the systematic component. This component may have several sources, such as inherent between-subject variability, measurement errors and background variation within the environment of the study. Mathematically, we usually describe the random component in terms of some appropriate probability distribution (see Chapters 2 and 4).

The systematic component is used to predict the response for any set of experimental conditions, and the random component is used to estimate the uncertainty in those predictions. Here are two simple examples of statistical models.

EXAMPLE 1.1: QUALITATIVE EXPLANATORY VARIABLE

Consider an experiment to investigate nutrient feeding strategies for plants grown in pots. A scientist has obtained a new liquid feed and wishes to evaluate its effect on plant growth. The instructions provided by the manufacturer suggest three feeding regimes labelled A, B and C. The scientist decides to grow 12 plants of a single plant variety, each one in a separate pot, and to allocate four plants at random to each of the three suggested regimes. After six weeks, the height of each plant is measured. Here, the response variable is plant height and the only explanatory variable is the feeding regime, which is a qualitative variable with three levels.

We might hypothesize that the plant height for a given feeding regime can be expressed symbolically as

height = overall mean + effect of feeding regime + deviation.

This is a simple (empirical) statistical model with height as the response. For an unstructured sample of 12 pots, there is no need for a structural component. So here the systematic part of the model (i.e. overall mean + effect of feeding regime) relates only to the explanatory component, with plant height modelled as an overall mean modified by some specific amount for each feeding regime. The random part (labelled deviation) allows for the deviation of individual observations from the feeding regime value given by the systematic component. Using mathematical notation (see also Section 2.1) we can write this model as

$$y_{jk} = \mu + \tau_j + e_{jk} \ . \tag{1.1}$$

Here, we have identified each plant by labelling it by the treatment applied ($j = 1, 2, 3$ for regimes A, B, C, respectively) and then we number the plants within each treatment group (using $k = 1, 2, 3, 4$). Hence, y_{jk} represents the height of the kth plant with the jth feeding regime. We use μ to represent the population mean height (the 'overall mean'), and τ_j represents the difference in response for the jth feeding regime relative to the overall mean (the 'effect of the feeding regime'). Finally, e_{jk} is the deviation associated with the kth replicate plant under the jth feeding regime.

The symbols μ and τ_1, τ_2, τ_3 (usually written as τ_j, $j = 1 \ldots 3$) are unknown population parameters that have to be estimated from the observed sample from the experiment. This model represents the height of a plant under the jth regime using the systematic component $\mu + \tau_j$, so a different value pertains to each regime, as shown in Figure 1.1a.

In Example 1.1, the explanatory variable (feeding regime) is a qualitative variable, or factor, with three levels (A, B and C). Without further information we cannot infer relationships between these factor levels and so we model the response by fitting a separate effect for each level. However, if the different feeding regimes correspond to different application rates for the liquid feed, then the scientist could evaluate the quantities corresponding to each feed rate and turn them into quantitative values (numbers). We can then consider other models for these data as shown in Example 1.2.

EXAMPLE 1.2: QUANTITATIVE EXPLANATORY VARIABLE

Suppose now that the scientist in Example 1.1 has evaluated the volumes (or doses) for feeding regimes A, B and C as 20, 40 and 60 mL per plant, respectively. The explanatory variable now corresponds to a quantitative variable (i.e. dose) with numeric values, and we can reasonably consider the response as a function of this continuous variable, expressed symbolically as

$$height = f(dose) + deviation,$$

where f(dose) indicates some mathematical function of dose. Here, we assume the simplest case, namely that the function is a straight line relationship (see Figure 1.1b). We can formally write this simple model as

$$y_{jk} = \alpha + \beta x_j + e_{jk} \ . \tag{1.2}$$

We again label each plant by the treatment applied (here $j = 1, 2, 3$ for doses 20, 40 and 60 mL, respectively) and then number plants within each treatment group (using $k = 1, 2, 3, 4$) so y_{jk} is the height of the kth replicate plant with the jth dose. Now, x_j is the numerical quantity of the jth dose ($x_1 = 20$, $x_2 = 40$, $x_3 = 60$), α is the plant height at zero dose (the intercept of the line in Figure 1.1b with the y-axis at $x = 0$), β is the linear response to increasing the dose by 1 mL (the slope of the line in Figure 1.1b), and e_{jk} is the deviation

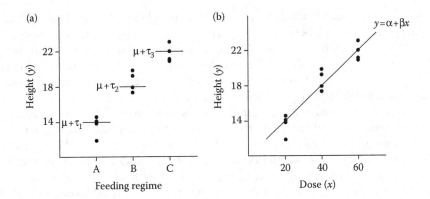

FIGURE 1.1

Two linear models with observed (•) and population (—) responses (heights) for the plant growth experiment for (a) a qualitative explanatory variable representing three feeding regimes (A, B and C, Example 1.1), and (b) a quantitative explanatory variable representing three doses (20, 40 and 60 mL, Example 1.2).

from the linear trend for the kth replicate plant with the jth dose. The symbols α and β are unknown population parameters that have to be estimated from the observed sample.

The model represented by Equation 1.2 differs from the model represented by Equation 1.1 in several important respects, even though it could arise from the same experiment. In Example 1.1, feeding regime was considered to be a qualitative variable (and so here we call Equation 1.1 the qualitative model), and a separate effect was allowed for each level. In Example 1.2, we used additional information, that is the numeric values of dose, to fit height as a linear function of dose (and so here we call Equation 1.2 the quantitative model). The qualitative model might be considered more flexible, as it does not make any assumption about the shape of the relationship. However, the quantitative model has the advantage that it is more **parsimonious**, i.e. that it uses fewer parameters to describe the pattern. It has the further advantage that we can also make predictions at intermediate doses (e.g. 50 mL) using the fitted model (under the assumption that the straight line model is appropriate).

1.4 Using Linear Models

Equations 1.1 and 1.2 are simple examples of **linear models**, an important sub-class of the statistical models introduced in Section 1.3. In this context, the response variable is sometimes called the dependent variable and the explanatory variables are sometimes called independent or predictor variables. The explanatory and structural components of a linear model each consist of a set of terms added together (an **additive structure**) and each term consists of either a single unknown parameter (such as τ_j in Equation 1.1), or an unknown parameter multiplied by a known variable (such as βx_j in Equation 1.2) – this is the **linear structure**. The random component, or deviation, is added to the systematic component to give the full model. In general, linear models might contain terms for several qualitative or quantitative explanatory variables or both. It is important, but slightly confusing, to note that the output from a complex linear model will generally not be a straight line (e.g. Equation 1.1), although the straight line relationship between a response variable and a single explanatory variable (e.g. Equation 1.2) is the simplest example of a linear model. The class of linear models is a large and flexible one and, although the models themselves are usually approximations, they can adequately represent many real-life situations. The most common uses for linear models are model specification, parameter estimation and prediction.

The main objective in **model specification** is to determine what form of statistical model best describes the relationship between the response and explanatory variable(s). There will often be a biological hypothesis behind a study that suggests a suitable form of model and the explanatory variables that should be included in the model. For example, in Example 1.1 the scientist wanted to investigate whether the different feeding regimes had detectable effects on plant growth. The process of statistical **hypothesis testing** can be used to refine the model by determining whether there is any evidence in the data that the proposed explanatory variables explain patterns in the response. Often several competing models might be compared. If many potential explanatory variables have been measured, **variable screening** may be used to select the variables that best explain the variation in the response. For example, in field studies on insect abundance, many climatic and environmental variables can be measured, and those that are most highly related to insect counts then identified.

Once an appropriate model has been determined, **parameter estimation** (see Section 1.5) is required to interpret the model and, potentially, the underlying biological process. Associated with each parameter estimate is a measure of uncertainty, known as the **standard error**.

The fitted model can be derived by substitution of estimates in place of the unknown parameter values in the model, and uncertainty in the fitted model is derived from the parameter standard errors. **Prediction** involves the use of the fitted model to estimate functions of the explanatory variable(s) – for example, the prediction of a treatment mean together with some measure of its precision. Again, uncertainty in predictions is derived from uncertainty in the parameter estimates.

1.5 Estimating the Parameters of Linear Models

Any linear model has an associated set of unknown parameters for which we want to obtain estimates. For example, in fitting the models represented by Equations 1.1 and 1.2 to the observed data, our aim is to find the 'best' estimates of the parameters μ, τ_1, τ_2 and τ_3, or α and β, respectively. In Chapters 4 (qualitative model) and 12 (quantitative model) we present detailed descriptions of how to obtain estimates of these parameters; here, we outline the basic principles. Before we consider the estimation process, some basic notation is required. In general, we represent estimated parameter values by placing a 'hat' (^) over the parameter symbol, for example, $\hat{\mu}$ indicates an estimate of μ, the population mean. Then, the fitted value for an observation y_{jk}, denoted $\hat{y}_{jk}$, consists of the systematic component of the model with all parameters replaced by their estimates. So, in the qualitative model represented by Equation 1.1, the fitted value for the kth plant with the jth feeding regime is

$$\hat{y}_{jk} = \hat{\mu} + \hat{\tau}_j \,, \tag{1.3}$$

which is an estimate of the population mean for plants with the *j*th feeding regime. For the quantitative model in Equation 1.2, the corresponding fitted value is

$$\hat{y}_{jk} = \hat{\alpha} + \hat{\beta} x_j \, . \tag{1.4}$$

For all linear models, parameters are estimated with the **principle of least squares**. This method finds the 'best-fit' model in the sense that it finds estimates for the parameters that minimize the sum, across all observations, of the squares of the differences between the observed data and fitted values. For example, for the qualitative model (Equation 1.1) we minimize

$$\sum_{j=1}^{3} \sum_{k=1}^{4} (y_{jk} - \hat{y}_{jk})^2 \, ,$$

where $\hat{y}_{jk}$ was defined in Equation 1.3. For the quantitative model (Equation 1.2), the quantity minimized has the same generic form, but now Equation 1.4 is used to define the fitted values. The symbol Σ is used to indicate the sum across the specified index (see Section 2.1 for more details). Note that these summations are over all combinations of the three factor levels ($j = 1, 2, 3$) and the four replications ($k = 1, 2, 3, 4$), and hence over the full set of 12 observations. Having found the best-fit model for our observed data, we can calculate fitted values based on the parameter estimates. We can then obtain estimates of the deviations, called **residuals**, from the discrepancy between the observed and fitted values, as

$$\hat{e}_{jk} = y_{jk} - \hat{y}_{jk} \, .$$

If the residuals are relatively small, then our model gives a good description of the data. These residuals can be examined to assess the validity of our model (to diagnose any lack of fit of the model to the data) and the assumptions made in fitting the model to the data (Chapters 4 and 12). One such assumption concerns an underlying probability distribution for the deviations (see Chapter 4), and the estimated variance of this distribution is used to calculate the parameter standard errors. This variance, often called the residual variance, provides a measure of uncertainty which can also be used in hypothesis testing and to form confidence intervals for predictions.

1.6 Summarizing the Importance of Model Terms

The main tool we use for the statistical analysis of any linear model, with either qualitative (factor) or quantitative (variate) explanatory variables, or both, is the **analysis of variance**, usually abbreviated as ANOVA. As the name suggests, the principle behind ANOVA is the separation and comparison of different sources of variation. In its simplest form, ANOVA quantifies variation in the response associated with the systematic component of the model (systematic variation) and compares it with the variation associated with the random component of the model (often called noise or background variation). Informally, if the ratio of systematic variation to background variation is large then we can conclude that the proposed model accounts for much of the variation in the response, and that the explanatory variables provide a good explanation of the observed response. However, if the ratio of systematic variation to background variation is small, then it does not necessarily indicate that the response is not related to the explanatory variables – it may just be that the background variation is too large to clearly detect any relationship. We can use ANOVA to assess whether the variation associated with different levels, or groups of levels, of a qualitative explanatory variable (factor) is larger than the background variation, which would give evidence that the explanatory variable is associated with substantive changes in the response. Similarly, we can assess whether there is substantive variation in the response associated with some trend in a quantitative explanatory variable (variate). We can often also partition variation associated with different explanatory variables to assess their relative importance, and a well-designed experiment can make this easier. We use ANOVA to summarize model fitting in two related contexts.

We first consider the use of ANOVA in structured scientific studies where we include the experimental conditions as factors, and wish to relate variation in the response to variation in the conditions. For example, consider a traditional field trial to assess the yield response of a set of plant varieties to different levels of fertilizer application. Here, the experimental conditions are combinations of plant variety and fertilizer application, with both considered to be qualitative variables. In a basic analysis, we are interested in identifying whether differences between plant varieties or fertilizer application levels, or particular combinations of these factors, provide an explanation for the observed differences in yield response. Within this context we can then generalize this basic analysis in several different ways: to take account of the physical structure of the experimental units (e.g. to allow for the blocking of experimental units); to take account of any quantitative scale underlying the factor levels (e.g. the nitrogen content of the fertilizer applications); and, in a limited way, to account for other explanatory variables that may have been measured (e.g. perhaps soil pH varies across the field and affects yield). This is the traditional framework for ANOVA and most statistical packages have algorithms tailored to the analysis of data within this framework (e.g. the ANOVA command in GenStat, the aov() function in R and the proc glm procedure in SAS).

We then consider the use of ANOVA in scientific studies where the main aim is to model the response as a mathematical function of one or more quantitative explanatory variables. This context is usually called **regression modelling** or **regression analysis**, and we emphasize the particular case of linear regression, where only linear functions of one or more continuous explanatory variables are permitted. For example, suppose a forester wishes to build a model to predict timber volume from easily measured field variables such as tree diameter and height. In a basic analysis, having measured both the field variables and the actual timber volume for a number of trees, we are interested in determining which field variables (or combinations of field variables) explain the observed differences in timber volume. Again, within this context we can generalize the basic analysis to take account of any grouping of observations, such as tree variety or location. Within regression modelling, ANOVA is the main statistical tool used for assessment of the importance of different explanatory variables. Statistical software packages usually contain more general algorithms for regression analyses (e.g. FIT in GenStat, the lm() function in R and the proc reg procedure in SAS).

It should be clear that there is much overlap between these two contexts. For example, both the qualitative model of Example 1.1 and the quantitative model of Example 1.2 could be analysed by either type of algorithm. However, using different algorithms to analyse the same data set can be confusing, because even when the methods are equivalent, the results may appear to differ if different conventions are used for their presentation. One of the main aims of this book is to explain the rationale behind these different conventions, and so to eliminate this confusion.

1.7 The Scope of This Book

We follow this chapter with a review chapter. Although we minimize the use of mathematical formulae, some are essential, and so we provide a review of mathematical notation in Chapter 2, along with the basic statistical concepts and methods used elsewhere in the book. Many readers will be familiar with these concepts and might treat this chapter as optional.

The early chapters of the book (Chapters 3 to 11) focus on the design of experiments and the analysis of data from designed experiments. In Chapter 3, we concentrate on the essential statistical principles of design: replication, randomization and blocking. We consider the structure of an experiment and describe some common designs. In Chapters 4 to 7, we consider analysis of simple designs. In Chapter 4, we consider in detail the analysis of data from the simplest design – the completely randomized design – to explain the concepts of ANOVA. We explain how the ANOVA table is formed, how it relates to a model for the data and how to interpret it. In Chapter 5, we explore the assumptions underlying the model and analysis and describe the diagnostic tools we can use to check them. We consider how these assumptions might be violated and the possible consequences, and ways to remedy these problems. In Chapter 6, we discuss transformations of the response variable as one remedy for failure to satisfy the model assumptions. In Chapter 7, we extend the analysis to the simplest design that includes blocking, the randomized complete block design, and introduce the concept of strata, or different structural levels, within a design and its analysis. In Chapters 8 to 11, we consider more advanced issues in the analysis of designed experiments. In Chapter 8, we consider how best to extract answers about our experimental hypotheses from our analysis. The advantages of factorial treatment structures, used to test the effects of several treatment factors simultaneously, will be explained. We describe the use of crossed and nested models for factorial structures, and how to make pairwise comparisons of treatments. In Chapter 9, we describe the analysis of some designs with somewhat more complex blocking structures, namely the Latin square, split-plot and balanced incomplete block designs. Then in Chapter 10, we consider how to calculate the replication required to obtain a specified precision for treatment comparisons in simple designs, and we introduce the concept of statistical power. We also discuss the case of equivalence testing, where the interest is in detecting equivalence rather than differences between treatments. Finally, Chapter 11 examines the issues that arise for non-orthogonal designs, where an unambiguous analysis can no longer be obtained.

In the later chapters of the book (Chapters 12 to 18) we turn our attention to regression modelling. In Chapter 12, after a brief general introduction, we concentrate first on simple linear regression, relating the response to a linear function of a single explanatory variate. The diagnostic tools introduced in Chapter 5 can be used for regression modelling, but additional diagnostic tools are available to check the validity of a regression analysis, and these are introduced in Chapter 13. In Chapters 14 and 15, we then extend regression models. In Chapter 14, we introduce multiple linear regression, extending the simple linear regression model to include several explanatory variates and considering problems of collinearity and variable selection. In Chapter 15, we show how to investigate the best form of a regression model when observations arise from different groups, how to incorporate simple designs into regression models and discuss analysis of covariance. We then move beyond linear regression. In Chapter 16, we introduce linear mixed models for the analysis of unbalanced studies where structure is present. In Chapter 17, we first use functions of explanatory variables to model curved relationships with linear models and then give a brief introduction to non-linear models. This concept is extended in Chapter 18 to the case of the generalized linear model, which can be used to model responses with certain types of non-Normal errors. We introduce two special, but commonly used, cases – the logit model for Binomial (proportion) data, and the log-linear model for Poisson (count) data.

Finally, the concluding chapter (Chapter 19) provides an overview of the full process of design and statistical analysis by way of real examples.

Our website (www.stats4biol.info) provides an overview and basic introduction to three commonly used statistical packages: GenStat, R and SAS. All of the examples are analysed with each of these packages, together with answers to selected exercises. Our personal preference is for the GenStat statistical software, because of its excellent implementation of algorithms for the analysis of designed experiments, and the provision of menus to make analyses easily accessible to all. The R package provides functions for all the standard analysis approaches introduced in this book, and has the benefits and drawbacks associated with being free, open-source software. We include SAS because of its wide user base and general availability. Most results presented in the book can be obtained with any of these packages; we comment where results may differ between packages and output has been obtained from a specific package.

2

A Review of Basic Statistics

This chapter briefly reviews some basic mathematical and statistical concepts that are fundamental to the material that comes later. Readers familiar with the mathematical notation commonly used to define summary statistics and with simple statistical tests, such as the t-test, can treat this chapter as optional revision or as reference material. We first introduce two commonly used statistics, the sample mean and sample variance, and in doing so define the mathematical notation we use throughout the rest of the book (Section 2.1). We then review random variables and probability distributions with particular reference to the Binomial distribution for discrete variables and the Normal distribution for continuous variables (Section 2.2). Later, we discuss statistical inference (Section 2.3), review one- and two-sample t-tests (Section 2.4), and discuss the concept of correlation (and covariance) between two variables (Section 2.5). To complete this chapter, we describe our conventions for presentation of calculations and numerical results (Section 2.6).

2.1 Summary Statistics and Notation for Sample Data

As discussed in Section 1.2, when we obtain data from a study we regard them as a sample from the broader set of results that we might obtain if we repeated the experiment many times, and use statistical techniques to make inferences from our sample to this wider population. The first step in any analysis is to summarize the data. In defining the tools we use to do this, and later to analyse data, we often express the mathematical or statistical concepts algebraically using some standard mathematical notation. Symbols with predefined meanings, often Greek letters (e.g. μ and σ), and various shorthand expressions are commonly used. For example, in a laboratory experiment where several treatments are to be compared, we can use the letter N to represent the total number of observations made (e.g. $N = 20$), the letter t to represent the number of treatments or groups (e.g. $t = 2$), and, if they are equally replicated, the letter n to represent the number of replicates of each experimental treatment (e.g. $n = 10$). If the treatments are unequally replicated then the notation is extended by the use of subscripts to denote the replication for any given treatment, for example, $n_1 = 12$ and $n_2 = 8$ indicates that treatment 1 has 12 replicates and treatment 2 has eight. An individual response (datum or measurement) is often represented by a lower case italic letter (usually y) with an index (usually a subscript) to identify it uniquely. For example, y_i might be used to denote the response from the ith observation. To specify a set of N responses we write y_i, $i = 1 \ldots N$ (where $1 \ldots N$ denotes all integers from 1 to N). Such notation is useful as it allows us to write down general expressions or formulae applicable to any statistical analysis. For a particular data set, we then substitute the actual numerical values recorded in place of the symbols. Note that whilst there are some generally accepted conventions, notation often differs between books and subject areas. In this section we define notation to be used throughout this book. Occasionally, we have used the same symbol to represent different quantities in different contexts. We have tried to minimize this practice, as it is potentially confusing, and try to explain all of our notation as it is introduced.

Many statistical formulae are written as sums of several components. The Greek letter Σ (capital sigma) is commonly used to denote the sum of a set of values defined by their index numbers, over a range with the lower limit specified by a subscript and the upper limit specified by a superscript, so $\sum_{i=1}^{N}$ or $\Sigma_{i=1}^{N}$ specifies a sum over the index numbers from 1 up to N. For example, the sum (or total) of a set of N responses, labelled as y_i, $i = 1 \ldots N$, would be written as

$$y_1 + y_2 + \ldots + y_{N-1} + y_N = \sum_{i=1}^{N} y_i \, .$$

For brevity, we sometimes write Σ_i to indicate summation over all available values of index i.

An important summary statistic, the **sample mean** (or sample grand mean), defined as the arithmetic mean of the N data values and denoted $\bar{y}$, would then be written

$$\bar{y} = \frac{y_1 + y_2 + \ldots + y_{N-1} + y_N}{N} = \frac{1}{N} \sum_{i=1}^{N} y_i \, . \tag{2.1}$$

The use of the summation symbol Σ has simplified and generalized the expression for the sample mean, which is the sum of all responses from label 1 to label N, divided by the number of observations, N.

Sometimes it is useful to label observations within treatment groups using two (or more) subscripts. So in an experiment with $t = 2$ treatments and $n = 10$ replicates of each treatment, the resulting set of responses might be concisely represented as y_{jk}, $j = 1, 2$, $k = 1 \ldots 10$, where the index j indicates the treatment applied and the index k labels the replicates within treatments. Formulae may then be simplified by using 'double sums', for example, the expression

$$\sum_{j=1}^{2} \sum_{k=1}^{10} y_{jk}$$

represents summation over 20 responses. The two indices are summed over in turn, the 'inner' (or rightmost) sum being executed first (here for values of index k from 1 to 10), to give

$$\sum_{j=1}^{2} \sum_{k=1}^{10} y_{jk} = \sum_{j=1}^{2} (y_{j1} + y_{j2} + \ldots + y_{j10}) \, .$$

Then the 'outer' (or leftmost) sum is executed (here across values of index j from 1 to 2) to give a sum across all combinations of the two indices and hence the full set of observations. We adapt and extend these basic forms of notation as necessary throughout this book.

When data are identified by more than one index, it is common to express totals and means using the 'dot notation'. Suppose that for t treatment groups, y_{jk} identifies the kth replicate response belonging to the jth treatment group. Then $y_{j\cdot}$ and $\bar{y}_{j\cdot}$ represent the group total and group mean response, respectively, for observations on the jth treatment. If n_j is used to represent the number of observations in that jth group, then these expressions can be written algebraically as

$$y_{j\cdot} = \sum_{k=1}^{n_j} y_{jk}; \quad \bar{y}_{j\cdot} = \frac{1}{n_j} \sum_{k=1}^{n_j} y_{jk} .$$

Here, the dot symbol '.' in the position of index k indicates summation across the observations for all possible values of that index, i.e. for $k = 1 \ldots n_j$, with the other index (or, in general, indices) kept constant. The bar symbol '$-$' over the y indicates that the mean is taken by dividing the total by the number of observations included in the summation. Note that within this system of notation the overall sample mean should strictly be denoted $\bar{y}_{\cdot\cdot}$ but, for simplicity, the dots are generally omitted here; therefore we use $\bar{y}$ to represent the sample grand mean. The overall and group means are conventionally used as summaries of the location (average response) for the whole sample and particular treatment groups, respectively, and as estimates for the corresponding values in the population from which the sample has been taken.

The **sample variance** quantifies the amount of variation, or spread, in the sample about its mean. For responses y_i, $i = 1 \ldots N$, the sample variance can be expressed algebraically as

$$\frac{1}{N} \sum_{i=1}^{N} (y_i - \bar{y})^2 . \tag{2.2}$$

In Equation 2.2, the N deviations of the individual responses about the overall sample mean are squared and then added together, and the result is divided by the number of observations, N. The process of subtracting the sample mean from all of the responses is known as **centering**, so the variance is proportional to the sum of the squared centered responses. We might think of using this quantity to estimate the variance of the population from which the sample has been taken, but that estimator is biased and tends to underestimate the population variance. We therefore use a scaled version of the sample variance to estimate the population variance, known as the **unbiased sample variance** and here denoted s^2, written as

$$s^2 = \frac{1}{(N-1)} \sum_{i=1}^{N} (y_i - \bar{y})^2 . \tag{2.3}$$

These sample variance statistics are not on the same measurement scale as the original responses. However, their square roots do have the same units as the response, and we usually choose to work with the **unbiased sample standard deviation**, s. The **coefficient of variation** (%CV) expresses the unbiased sample standard deviation as a percentage of the sample mean, calculated as

$$\%\mathrm{CV} = 100 \times \frac{s}{\bar{y}} .$$

This quantity is sometimes used as a measure of the relative precision of an experiment, particularly in the context of field experiments. However, %CV provides a sensible summary statistic only for variables measured relative to an absolute (as opposed to arbitrary) zero that forms a lower limit for observed values.

EXAMPLE 2.1A: WHEAT YIELDS*

A total of $N = 7$ measurements of the yield of a commercial variety of wheat were obtained from a field trial. The samples were converted into equivalent yields per hectare as: 7, 9, 6, 12, 4, 6 and 9 t/ha. The sample mean is calculated (as in Equation 2.1) as

$$\bar{y} = \frac{7 + 9 + 6 + 12 + 4 + 6 + 9}{7} = \frac{53}{7} = 7.5714 \, ,$$

or 7.57 when rounded to two decimal places. Using Equation 2.3, we calculate the unbiased sample variance as

$$s^2 = \frac{(7 - 7.57)^2 + (9 - 7.57)^2 + \dots + (9 - 7.57)^2}{(7 - 1)} = \frac{41.71}{6} = 6.95 \, .$$

Note that although we have shown values rounded to two decimal places within this calculation, we actually perform this (and all following) calculations using full accuracy, as explained in Section 2.6. The unbiased sample standard deviation is then calculated directly as $s = \sqrt{6.95} = 2.64$ t/ha, with %CV $= 100 \times 2.64/7.57 = 34.82$. This is larger than is usual for a well-managed agricultural trial, but is based on a small number of values and so may be poorly estimated.

To make statistical inferences on samples, we usually make some assumptions about the probability distribution underlying the data, and we introduce this concept in the next section.

2.2 Statistical Distributions for Populations

Before discussing probability distributions, we want you to understand the concept of a random variable. A **random variable** represents the possible outcomes of a stochastic process, i.e. a process that is not deterministic, but includes some unpredictable variation. Conventionally, random variables are represented by upper case symbols, with realizations of the variable represented by lower case symbols. For example, we might denote yield from a field plot as a random variable Y, with the realized yield denoted y. When defining models in later chapters, we will often not make this distinction, and simply use the lower case symbols.

We use probability distributions to help us make inferences from data. If we can realistically assume that the population of possible outcomes from an experiment behave like a sample of a random variable from a certain probability distribution, then we can use known properties of that distribution to derive inferences for the population from our observations.

The mathematical theory underlying probability distributions requires a distinction to be made between discrete and continuous random variables. A **discrete random variable** is one that can take only a certain pre-specified set of possible values, such as integer counts. A **continuous random variable** may take any real value within its defined range.

In this book, we are primarily concerned with continuous random variables that follow a Normal distribution, except in Chapter 18 where we consider discrete random variables following either Binomial or Poisson distributions. However, as it is easier to understand the concepts associated with probability distribution functions by working with discrete distributions, we start with definitions for this type of random variable, using the Binomial distribution as an illustration.

2.2.1 Discrete Data

As stated above, a discrete random variable Y can take only a certain pre-specified set of possible values, which we denote by S. The **probability distribution** associated with Y is a function that gives the probability of observing a particular value, y, called a **point probability**, denoted as $P_Y(y) = \text{Prob}(Y = y)$ ('the probability that variable Y takes value y'). It is also often useful to consider the **cumulative distribution function**, denoted F_Y, defined as the probability that the random variable is less than or equal to a certain value y, written as

$$F_Y(y) = \text{Prob}(Y \le y) = \sum_{v \le y} P_Y(v) ,$$

i.e. the sum of point probabilities over all values v in the set S that are less than or equal to the target value y. This function takes values between zero (for values of y less than the minimum value of Y) and one (for values of y equal to or greater than the maximum value of Y).

The Binomial distribution is an example of a discrete probability distribution. It is usually derived as the distribution of the number of successes out of a series of m independent binary trials (i.e. trials with only two possible outcomes: success or failure), where each trial has an equal probability of success, denoted p. For example, the number of heads (successes) obtained after the tossing of two separate coins ($m = 2$) can take the values $y = 0$, 1 or 2, and follows a Binomial distribution with success probability $p = 0.5$ (for fair coins). The **Binomial probability distribution** takes the form

$$P_Y(y; m, p) = \frac{m!}{y!(m - y)!} p^y (1 - p)^{m-y} . \tag{2.4}$$

The calculated probability of y successes is then a function of the number of successes, y, the number of tests, m, which is known, and the probability of success p, which is often unknown. Note that $p^0 = (1 - p)^0 = 1$, and the factorial function $x!$ is defined for any positive integer x as the product of all integer values less than or equal to x, i.e.

$$x! = x \times (x - 1) \times (x - 2) \times \ldots \times 1 ,$$

so that $1! = 1$, $2! = 2$, $3! = 6$ and so on. A value is also needed for $x = 0$, and by convention $0!$ is defined to be equal to 1.

EXAMPLE 2.2A: PLANT INFECTION*

Consider an experiment in which three plants in a pot are inoculated with a virus, where each plant has a 40% chance of becoming infected. The number of plants (0–3) that show symptoms several days after inoculation can be considered to follow a Binomial distribution. The possible values that can be observed (the set S) are the integers 0, 1, 2, 3. The number of tests here is the number of plants, so $m = 3$, and the probability of success is the probability of infection, so $p = 0.4$. We can calculate the probability of each outcome using Equation 2.4 as follows:

$$\text{Prob}(Y = 0; m = 3, p = 0.4) = 1 \times 0.6^3 = 0.216$$

$$\text{Prob}(Y = 1; m = 3, p = 0.4) = 3 \times 0.4 \times 0.6^2 = 0.432$$

$$\text{Prob}(Y = 2; m = 3, p = 0.4) = 3 \times 0.4^2 \times 0.6 = 0.288$$

$$\text{Prob}(Y = 3; m = 3, p = 0.4) = 1 \times 0.4^3 = 0.064 \ .$$

The cumulative distribution can be derived directly as

$$\text{Prob}(Y \leq 0; m = 3, p = 0.4) = \text{Prob}(Y = 0) = 0.216$$

$$\text{Prob}(Y \leq 1; m = 3, p = 0.4) = \text{Prob}(Y \leq 0) + \text{Prob}(Y = 1) = 0.216 + 0.432 = 0.648$$

$$\text{Prob}(Y \leq 2; m = 3, p = 0.4) = \text{Prob}(Y \leq 1) + \text{Prob}(Y = 2) = 0.648 + 0.288 = 0.936$$

$$\text{Prob}(Y \leq 3; m = 3, p = 0.4) = \text{Prob}(Y \leq 2) + \text{Prob}(Y = 3) = 0.936 + 0.064 = 1.000 \ .$$

This cumulative distribution function is shown in Figure 2.1a. It is a discontinuous function, defined on the range 0–3, with jumps at the values in S. The function values are shown using solid lines and filled circles; the open circles and dashed lines are used to join the discontinuous segments.

The inverse of the cumulative distribution function is known as the **quantile function**. The **quantiles** of a distribution divide its range into intervals such that each interval contains an equal proportion of the distribution. Special cases include the **median** (which divides the distribution into two parts) and the **quartiles** (four parts). The **interquartile range** (first to third quartile, or central part of the distribution) gives a measure of the spread of a distribution. **Percentiles** are often used and divide the distribution into 100 parts. Hence, the median and the first and third quartiles can alternatively be termed

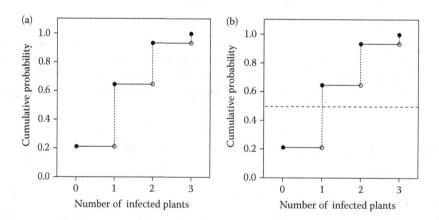

FIGURE 2.1
(a) Cumulative distribution function for plant infection data (Example 2.2A) and (b) with 0.5 quantile marked (Example 2.2B).

the 50th, 25th and 75th percentiles, respectively. Quantiles for common distributions are widely available in statistical software and in books of statistical tables. Formally, the q quantile (for $0 \leq q \leq 1$) can be defined as the value v satisfying

$$\min\{v \in S : F_Y(v) \geq q\} .$$

The symbol $\in$ means 'is an element of', so the q quantile is the smallest value in the set S that has cumulative distribution function value greater than or equal to q.

EXAMPLE 2.2B: PLANT INFECTION*

We can find any quantile for the distribution underlying the plant infection experiment from the cumulative distribution function shown in Figure 2.1a. For example, suppose we wish to find the median, i.e. quantile $q = 0.5$. In Figure 2.1b, we draw a horizontal line at height 0.5, and find that the smallest valid value (i.e. in the set 0, 1, 2, 3) with cumulative probability greater than this value is 1; hence, 1 is the median value for this distribution.

The mean, or **expected value**, of a discrete random variable Y is calculated as

$$E(Y) = \sum_{y \in S} y \, P_Y(y) . \tag{2.5}$$

This equation is interpreted as 'the sum, over all possible values of Y (i.e. for $y \in S$), of the values multiplied by their point probabilities'. This is a measure of the location (average or mean value) of the distribution. Similarly, the spread of the distribution is measured by its variance, which can be expressed as

$$\mathrm{Var}(Y) = \sum_{y \in S} [y - E(Y)]^2 \, P_Y(y) . \tag{2.6}$$

This expression (Equation 2.6), writes the variance as the sum, over all the possible values of Y, of the squared deviation of each value from the mean, multiplied by its point probability. We can interpret these quantities as the mean and variance of a population that follows the given probability distribution. Unsurprisingly, the expression for the variance of the random variable in Equation 2.6 has a similar structure to that for the variance of a sample (Equation 2.2) and we explore this connection further below.

The expected value (mean) of the Binomial distribution takes the form

$$E(Y) = \sum_{y=0}^{m} y \, P_Y(y; m, p) = \sum_{y=0}^{m} y \left(\frac{m!}{y!(m-y)!} p^y (1-p)^{m-y} \right) = mp ,$$

with variance

$$\mathrm{Var}(Y) = \sum_{y=0}^{m} (y - mp)^2 \left(\frac{m!}{y!(m-y)!} p^y (1-p)^{m-y} \right) = mp(1-p) .$$

Obtaining the simplified forms requires mathematical manipulations outside the scope of this book (see for example Wackerly et al., 2007). Note that both the mean and the variance are functions of the population parameters m and p.

EXAMPLE 2.2C: PLANT INFECTION*

We can now use the probability distribution obtained in Example 2.2A to calculate the mean and variance of the distribution of the number of infected plants. The distribution mean is calculated as

$$E(Y) = (0 \times 0.216) + (1 \times 0.432) + (2 \times 0.288) + (3 \times 0.064) = 1.2,$$

and we can verify directly that $E(Y) = 1.2 = 3 \times 0.4 = mp$. Similarly, we can calculate the distribution variance as

$$Var(Y) = [(0 - 1.2)^2 \times 0.216] + [(1 - 1.2)^2 \times 0.432] + [(2 - 1.2)^2 \times 0.288]$$

$$+ [(3 - 1.2)^2 \times 0.064]$$

$$= 0.3110 + 0.0173 + 0.1843 + 0.2074$$

$$= 0.72,$$

and again we can verify directly that $Var(Y) = 0.72 = 3 \times 0.4 \times 0.6 = mp(1 - p)$.

In practice, the true probability distribution of any sample is usually unknown. If a data set can be considered as a set of samples from the same underlying distribution, then the **empirical probability distribution** of the sample, i.e. the relative frequency of each value within the sample, gives information on the form of that underlying probability distribution. The relative frequency is defined as the frequency of each value as a proportion of the total number of values and gives an estimate of each point probability, and can be graphically represented by using a bar chart. The sample mean (Equation 2.1) can then be calculated from the empirical probability distribution using the formula for the expected value (Equation 2.5) after substitution of the relative frequencies for the unknown point probabilities. The sample variance (Equation 2.2) can similarly be calculated using the formula for the variance (Equation 2.6). In this sense, the sample mean and variance can be seen as estimates of the true mean and variance of the underlying random variable, although in practice we usually use the unbiased sample variance (Equation 2.3) to get an unbiased estimate of the variance of the random variable.

There are various different definitions of the sample quantiles, and we use one of the simpler (but common) definitions. For a sample $y_1 \ldots y_N$, the kth sample percentile is found in several steps. First, the sample is put into order of increasing size of its values. Then the index number, j, within the ordered set is calculated as

$$j = (N + 1) \times k/100 .$$

If j is an integer value then the kth sample percentile is the jth value in the ordered set. If j is not an integer value, then let l denote the largest integer smaller than j (i.e. the next smallest integer). The kth sample percentile is then defined as the average of the lth and $(l + 1)$th values in the ordered set.

EXAMPLE 2.2D: PLANT INFECTION*

Suppose that our plant infection experiment is now carried out with 20 pots, each containing three plants, with the following numbers of plants per pot becoming infected: 0, 2, 1, 1, 0, 1, 1, 1, 2, 1, 3, 1, 3, 1, 0, 0, 0, 2, 1 and 1, i.e. no plants infected in five pots, one plant infected in 10 pots, two plants infected in three pots, and three plants infected in two pots. The empirical probability distribution is thus

$$\text{Prob}(Y = 0) = 5/20 = 0.25$$

$$\text{Prob}(Y = 1) = 10/20 = 0.50$$

$$\text{Prob}(Y = 2) = 3/20 = 0.15$$

$$\text{Prob}(Y = 3) = 2/20 = 0.10 .$$

This empirical distribution is shown as a bar chart in Figure 2.2. The sample mean can be calculated either directly from the data values (i.e. as 22/20 = 1.1), or via the empirical probability distribution as

$$E(Y) = (0 \times 0.25) + (1 \times 0.50) + (2 \times 0.15) + (3 \times 0.10) = 1.1 ,$$

which is a slight underestimate of the true population mean (obtained as 1.2 in Example 2.2C). Similarly, the sample variance can be either calculated directly, or via the empirical distribution as

$$\text{Var}(Y) = \{[(0 - 1.1)^2 \times 0.25] + [(1 - 1.1)^2 \times 0.50] + [(2 - 1.1)^2 \times 0.15]$$

$$+ [(3 - 1.1)^2 \times 0.10]\}$$

$$= 0.3025 + 0.0050 + 0.1215 + 0.3610$$

$$= 0.79 .$$

We can convert this into the unbiased sample variance by multiplying by $N/(N - 1)$, giving $0.79 \times 20/19 = 0.83$ as an estimate of the variance of the underlying random variable. For this sample, this estimate is larger than the true value of the population variance (obtained as 0.72 in Example 2.2C).

To calculate sample quantiles, we list the observations in order as 0, 0, 0, 0, 0, 1, 1, 1, 1, 1, 1, 1, 1, 1, 2, 2, 2, 3, 3. The sample median ($k = 50$, so $k/100 = 1/2$) requires $j = 21 \times 1/2 = 10.5$. The 10th and 11th values in the ordered set are both 1 and hence

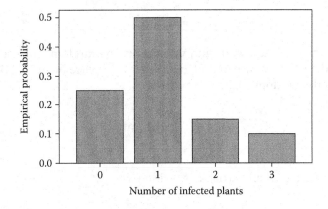

FIGURE 2.2
Bar chart showing the empirical probability distribution of the number of infected plants in the plant infection trial. Three plants were tested in each of 20 pots (Example 2.2D).

the sample median is 1. The sample lower quartile ($k = 25$, $k/100 = 1/4$) requires $j = 21 \times 1/4 = 5.25$, with estimate 0.5 (the average of the 5th value, 0, and the 6th value, 1, from the ordered set) and the sample upper quartile ($k = 75$, $k/100 = 3/4$) requires $j = 21 \times 3/4 = 15.75$, with estimate 1.5 (the average of the 15th value, 1, and the 16th value, 2, from the ordered set).

   These sample statistics deviate from those associated with the theoretical distribution calculated in Examples 2.2B and C because of variations inherent in the sampling process – if the experiment was repeated, then somewhat different results would be obtained each time.

2.2.2 Continuous Data

A continuous random variable can take any real value within a defined range. For example, plant heights can take any value greater than zero. Theoretically, therefore, there are infinitely many possible values, each with negligible probability (because there are so many possibilities), and the formulae for discrete random variables have to be adapted to take this into account. In this context, we refer to density functions rather than distribution functions and we work with integrals rather than sums.

   It is helpful in this case to start with the **cumulative density function** (CDF), defined as in the discrete case as $F_Y(y) = \text{Prob}(Y \leq y)$, which again takes values between zero (for values at or below the minimum value of Y) and one (for values at or above the maximum value of Y). The **probability density function** (PDF), $f_Y(y)$, can be interpreted as the probability of Y falling in the range $y \leq Y \leq y + \delta$, divided by δ, as δ decreases to zero, which is the derivative of the CDF. In mathematical terms, the CDF is written in terms of the PDF as

$$F_Y(y) \;=\; \int_{v=-\infty}^{v=y} f_Y(v)\,dv \,.$$

Informally, this can be interpreted as meaning that the integral ($\int$) sums the probabilities $f_Y(v)$ across all the possible values of v between the lower limit (here, minus infinity) and the upper limit, the target value y. As such, this is directly analogous to the formula in the discrete case. The quantile function is now defined straightforwardly in terms of the inverse CDF with quantile q ($0 \leq q \leq 1$) defined as the value v such that

$$v \;=\; F_Y^{-1}(q) \,.$$

The formulae for the expected value and variance of the random variable are also analogous to those in the discrete case, but using integrals in place of summations. Hence, the expected value of the random variable is written as

$$E(Y) \;=\; \int_{y=-\infty}^{y=\infty} y\, f_Y(y)\,dy \,,$$

and the variance of the random variable is written as

$$\text{Var}(Y) \;=\; \int_{y=-\infty}^{y=\infty} (y - E(Y))^2 f_Y(y)\,dy \,.$$

A histogram, the continuous analogue of the bar chart, can be used to give information on the shape of the empirical PDF. For continuous variables, data values have to be grouped into contiguous intervals. In the simple case where all intervals have equal width, then the relative frequency of observations in each interval is plotted. If intervals are of unequal widths then, for each interval, the relative frequency is divided by the interval width, so that the area under the histogram in each interval is equal to its relative frequency. As in the discrete case, the sample mean and variance can be considered as estimates of the expected value and variance of the random variable, although again we usually use the unbiased sample variance to get an unbiased estimate of the variance of the random variable.

EXAMPLE 2.3A: WILLOW BEETLE MEASUREMENTS

A sample of 50 willow beetles (*Phratora vulgatissima*) was taken from a willow crop located close to Bristol, UK, and various characteristics were measured, including the total length and width of each beetle (Peacock et al., 2003). The data are presented in Table 2.1 and can be found in file WILLOW.DAT.

The length measurements ranged from 4.10 to 4.95 mm. The sample mean for length was 4.552 mm with unbiased sample variance 0.0260 mm^2 and standard deviation 0.1611 mm. The empirical probability distribution is illustrated using a histogram in Figure 2.3. The histogram uses 10 intervals of length 0.1 mm, starting at 4 mm length. The relative frequencies plotted are calculated as the number of observations in each interval divided by $N = 50$. In this case, the sample median is the average of the 25th and 26th values in the ordered set of observations (smallest first), which is 4.55 mm. The index number for the lower quartile is $j = 12.75$, so the lower quartile is the average of the 12th and 13th values in the ordered set, here 4.45 mm, slightly smaller than the mean. Similarly, the index number for the upper quartile is $j = 38.25$, so the upper quartile is the average of the 38th and 39th values in the ordered set, here 4.65 mm. The inter-quartile range is thus 0.20 mm, which is slightly larger than the standard deviation.

TABLE 2.1

Length and Width (mm) of 50 Willow Beetles (*Phratora vulgatissima*) Sampled from a Willow Crop Located Close to Bristol, UK (Example 2.3A and File WILLOW.DAT)

Beetle	Length	Width	Beetle	Length	Width	Beetle	Length	Width
1	4.60	1.50	18	4.55	1.50	35	4.60	1.60
2	4.70	1.65	19	4.60	1.70	36	4.55	1.65
3	4.50	1.55	20	4.55	1.60	37	4.775	1.55
4	4.55	1.65	21	4.35	1.60	38	4.60	1.60
5	4.75	1.65	22	4.45	1.60	39	4.45	1.475
6	4.40	1.50	23	4.55	1.55	40	4.60	1.60
7	4.20	1.70	24	4.35	1.55	41	4.65	1.625
8	4.70	1.55	25	4.65	1.65	42	4.725	1.65
9	4.55	1.60	26	4.50	1.55	43	4.95	1.725
10	4.70	1.65	27	4.45	1.50	44	4.65	1.65
11	4.65	1.55	28	4.45	1.60	45	4.60	1.625
12	4.50	1.55	29	4.10	1.40	46	4.45	1.55
13	4.30	1.50	30	4.50	1.50	47	4.30	1.525
14	4.65	1.65	31	4.60	1.60	48	4.75	1.60
15	4.75	1.65	32	4.75	1.575	49	4.525	1.55
16	4.65	1.60	33	4.70	1.65	50	4.35	1.50
17	4.45	1.60	34	4.35	1.55			

Source: Data from Rothamsted Research (A. Karp).

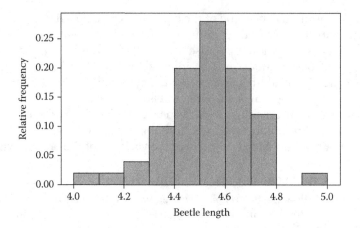

FIGURE 2.3
Histogram of relative frequencies for lengths (mm) of willow beetles from a sample of size 50 (Example 2.3A).

2.2.3 The Normal Distribution

In this book we assume in most instances that random variables follow a **Normal distribution** (sometimes called the Gaussian distribution), which approximately describes many types of continuous measurements, such as lengths, weights and so forth. The PDF for the Normal distribution is a bell-shaped symmetric curve, taking its maximum value at the mean (Figure 2.4a). As for all symmetric distributions, the median of this distribution is equal to its mean. The Normal distribution is defined by two parameters, the mean, μ, and the variance, σ^2, and its PDF takes the form

$$f_Y(y) = \frac{1}{\sqrt{2\pi\sigma^2}} \exp\left[-\frac{(y - \mu)^2}{2\sigma^2} \right],$$

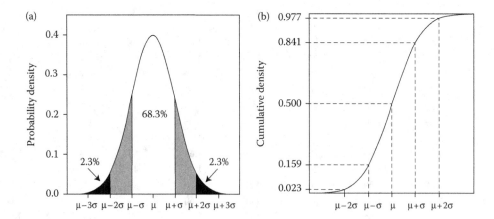

FIGURE 2.4
(a) PDF and (b) CDF of a Normal random variable with mean μ and standard deviation σ.

where exp() denotes the exponential function. Where a random variable, Y say, is assumed to follow a Normal distribution, it is conventional to write $Y \sim \text{Normal}(\mu, \sigma^2)$ (or 'Y follows a Normal distribution with mean μ and variance σ^2'). It is useful to remember that approximately 68% of the distribution lies within one standard deviation (σ) of the mean, i.e. in the range from $(\mu - \sigma)$ to $(\mu + \sigma)$. The inter-quartile range (middle 50% of the distribution) is therefore smaller than twice the standard deviation. In addition, approximately 95% of the distribution lies within two standard deviations of the mean, i.e. in the range from $(\mu - 2\sigma)$ to $(\mu + 2\sigma)$, and almost all of the distribution (more than 99.7%) lies within three standard deviations of the mean. Figure 2.4 shows these properties in terms of both the PDF (Figure 2.4a) and its integral, the CDF (Figure 2.4b). So, for example, the 2.3% of the distribution lying above $y = \mu + 2\sigma$ corresponds to a CDF value of

$$\text{Prob}(Y \le \mu + 2\sigma) = 1 - 0.023 = 0.977 \,.$$

The Normal distribution has the useful property that any linear function of a Normal random variable also has a Normal distribution. So if $Y \sim \text{Normal}(\mu, \sigma^2)$, then for known constants a and b, the random variable $Z = aY + b$ has a Normal distribution with mean $a\mu + b$ and variance $a^2\sigma^2$, i.e. $Z \sim \text{Normal}(a\mu + b, a^2\sigma^2)$. Z is conventionally used to represent the **standard Normal distribution** with mean 0 and variance 1, obtained by setting $a = 1/\sigma$ and $b = -\mu/\sigma$ to centre and standardize any Normal distribution, i.e.

$$Z = \frac{Y - \mu}{\sigma} \sim \text{Normal}(0, 1) \,.$$

Quantiles of the standard Normal distribution are widely available in both books of statistical tables and statistical packages.

The sum of a set of Normal random variables is also a Normal random variable. In particular, for a set of N independent (uncorrelated) Normal random variables $Y_1 \ldots Y_N$, with common mean μ and variance σ^2 then

$$\sum_{i=1}^{N} Y_i \sim \text{Normal}(N\mu, N\sigma^2) \,,$$

i.e. the sum of the variables has a Normal distribution with mean $N\mu$ and variance $N\sigma^2$. The mean of these variables ($\overline{Y}$) is also Normally distributed with

$$\overline{Y} = \frac{1}{N} \sum_{i=1}^{N} Y_i \sim \text{Normal}\left(\mu, \frac{\sigma^2}{N}\right) , \tag{2.7}$$

i.e. the mean of the variables has a Normal distribution with the same expected value, μ, as the original variables, and with a smaller variance than the original variables, equal to σ^2/N. The square root of this latter quantity is known as the standard error of the mean, which is used often in statistical tests. As we might expect, as the number of random variables (or sample size), N, increases then the uncertainty associated with their mean, $\bar{Y}$, decreases.

In fact, the Central Limit Theorem (see Casella and Berger, 2002) states that the distribution of the mean of any set of independent and identically distributed random variables will tend towards a Normal distribution, and this approximation becomes more accurate as the number of random variables contributing to the mean increases. This theorem holds even if the distributions of the individual random variables are not Normal. For example, suppose we have samples of 100 bean seeds, and we assess each seed for weevil infestation. The mean rate of infestation in each sample may well have an approximate Normal distribution, although this would certainly not hold for the observations on the individual seeds, or for the means of small samples. This property means that in practice we frequently encounter observations with a distribution that is either Normal or approximately so, and hence we will usually make (and verify) this assumption.

2.2.4 Distributions Derived from Functions of Normal Random Variables

Once we have made the assumption of a Normal distribution for our random variables, then several other distributions, derived from functions of these variables, become useful. We merely state some results here, in their simplest form, in order to give context to their use later. Full details and derivations of the distributions introduced below can be found in standard statistical texts such as Hoel (1984) or Wackerly et al. (2007).

The chi-squared distribution is associated with sums of squared Normal random variables. For a set of N independent random variables $Z_1 \ldots Z_N$ with a standard Normal distribution, the sum of the squares of these variables has a chi-squared distribution with N degrees of freedom, written as

$$\sum_{i=1}^{N} Z_i^2 \sim \chi_N^2 \,.$$

The symbol χ_k^2 indicates a **chi-squared distribution** on k **degrees of freedom** (df) for $k > 0$. The df determines the mean (equal to k) and variance (equal to $2k$) of the distribution, which is defined for positive values only and is right-skewed (has a long tail on the right-hand side of the distribution). In this context, the df are related to the number of independent variables contributing to the sum. Now suppose we have a set of N independent Normal random variables $Y_1 \ldots Y_N$ with common mean μ and variance σ^2. The unbiased sample variance of this set, denoted as random variable S^2, has a scaled chi-squared distribution with $N-1$ df, i.e.

$$S^2 = \frac{1}{(N-1)} \sum_{i=1}^{N} (Y_i - \bar{Y})^2 \sim \frac{\sigma^2}{(N-1)} \chi_{N-1}^2 \,.$$

The variables in this sum are no longer independent, due to centering by the sample mean, and so the df of the distribution are reduced by one. We can rescale S^2 by factor $(N-1)/\sigma^2$ to obtain a variable (Q^2) with an unscaled chi-squared distribution, as

$$Q^2 = \frac{(N-1)}{\sigma^2} S^2 = \sum_{i=1}^{N} \frac{(Y_i - \bar{Y})^2}{\sigma^2} \sim \chi_{N-1}^2 \,.$$

Student's t-distribution is associated with test statistics calculated as the ratio of an estimate of location to its standard error. In an abstract context, if a random variable Z has a standard Normal distribution, i.e. $Z \sim \text{Normal}(0,1)$, and V is an independent random variable with a chi-squared distribution on v df, i.e. $V \sim \chi^2_v$, then

$$T = \frac{Z}{\sqrt{V/v}} \sim t_v \, ,$$

where t_v denotes a **Student's t-distribution** on v df. The t-distribution is a bell-shaped symmetric distribution, with mean zero, but with fatter tails and a flatter peak than the Normal distribution (Figure 2.5). As the number of df becomes large, the t-distribution converges towards the standard Normal distribution.

We usually meet this distribution in the context of a set of independent Normal random variables $Y_1 \ldots Y_N$, with common mean μ and variance σ^2. Consider the following statistic:

$$\frac{(\bar{Y} - \mu)}{\sqrt{S^2/N}} = \frac{\sqrt{N}(\bar{Y} - \mu)}{\sqrt{\sigma^2 Q^2/(N - 1)}} = \frac{\sqrt{N}(\bar{Y} - \mu)}{\sigma} \times \frac{1}{\sqrt{Q^2/(N - 1)}} \, .$$

In the first step above, we have rewritten the divisor of the denominator (N) as a multiplier of the numerator, and rewritten S^2 in terms of Q^2. In the second step, we associated the square root of σ^2 with the numerator. From Equation 2.7 and results given above, we can deduce that $\sqrt{N}(\bar{Y} - \mu)/\sigma$ has a standard Normal distribution, and we know $Q^2 \sim \chi^2_{N-1}$, so, as these quantities are independent, it follows that

$$\frac{(\bar{Y} - \mu)}{\sqrt{S^2/N}} \sim t_{N-1} \, . \tag{2.8}$$

Here, the df are associated with the denominator, i.e. the estimated standard error of the mean, $\sqrt{(S^2/N)}$. We use this result frequently, for example Section 2.4.1. For calculation of

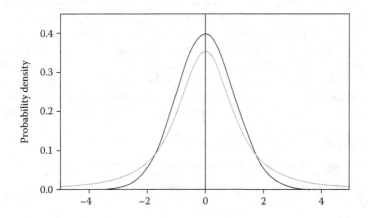

FIGURE 2.5
PDF for standard Normal distribution ($\mu = 0$, $\sigma = 1$, black line) and t-distribution with 2 df (grey line).

power (Chapter 10), we also use the result that for any constant c, $(Z + c)/\sqrt{V/v} \sim t_v(c)$, where $t_v(c)$ denotes a **non-central t-distribution** with non-centrality parameter c.

The F-distribution is associated with test statistics calculated as the ratio of two sums of squares. In abstract, if U_1 and U_2 are independent chi-squared random variables with d_1 and d_2 df, respectively, then

$$\frac{U_1/d_1}{U_2/d_2} \sim F_{d_1,d_2} \; ,$$

where F_{d_1,d_2} denotes an **F-distribution** with d_1 and d_2 df. This distribution is right-skewed and is defined only for positive values.

2.3 From Sample Data to Conclusions about the Population

An important role of statistics is to provide information on a population based on data obtained from a sample. This is what is commonly known as **statistical inference**. Two types of inference are described below: point and interval estimation and hypothesis testing.

2.3.1 Estimating Population Parameters Using Summary Statistics

We have already seen that we can interpret the sample mean and unbiased sample variance as estimates of the mean and variance of the underlying distribution of the population. **Point estimation** corresponds to this process of calculating (or estimating) a single summary value (or **statistic**) from the sample data that constitutes our 'best guess' of a population parameter. By convention, Greek letters are used to denote population parameters and sample statistics are denoted with 'equivalent' lower case Roman letters. The parameters for the population mean and variance, and their unbiased sample estimates (defined in Section 2.1), are usually labelled as shown in Table 2.2.

There is always uncertainty in the parameter estimates because of variability in the sampling process, and frequently because measurements themselves are imprecise. This uncertainty should also be estimated and reported by a **standard error** (SE). For example, we estimate the population mean using the sample mean, and we write this as $\hat{\mu} = \bar{y}$. Equation 2.7 states that the distribution of the mean of a sample from a Normal population is itself Normal, with mean equal to the population mean and variance equal to σ^2/N. This distribution can be interpreted as the set of outcomes that can be achieved by taking independent samples from the underlying population. The standard error of the estimate

TABLE 2.2

Notation for Population Parameters and Their Sample Statistics

Sample Statistics	Population	Sample
Mean	μ	$\bar{y}$
Variance	σ^2	s^2
Standard deviation	σ	s

is then the standard deviation of this distribution, written as $SE(\hat{\mu}) = \sigma/\sqrt{N}$. When σ is unknown, as is usually the case, we substitute its estimate, s, to get an estimated SE which we write as

$$\widehat{SE}(\hat{\mu}) = s/\sqrt{N} \, .$$

The hat over SE emphasizes the fact that the SE is itself estimated. This quantity is known as the **estimated standard error of the mean**, which we denote SEM.

Whereas point estimation provides a single estimate of a population parameter, **interval estimation** provides a range of estimates within which the parameter is likely to occur. **Confidence intervals** (CIs) are the most common example of interval estimates. For example, consider a 95% CI for the mean of a Normal random variable Y, with known variance σ^2. Using a sample of size N, $y_1 \ldots y_N$, with mean $\bar{y}$, a 95% CI takes the form

$$\left[\; \bar{y} - \left(1.960 \times \frac{\sigma}{\sqrt{N}} \right), \; \bar{y} + \left(1.960 \times \frac{\sigma}{\sqrt{N}} \right) \; \right] , \tag{2.9}$$

where 1.960 corresponds to the 97.5th percentile of the Normal distribution. The left-hand value is the lower limit and the right-hand value is the upper limit of the CI. This interval is derived from the property of the random variable that

$$\text{Prob}\left(-1.960 \le \frac{\bar{Y} - \mu}{\sigma} \le 1.960 \right) = 0.95 \, ,$$

where $\bar{Y}$ is the mean of a hypothetical sample from Y of size N. Note that we obtain the 95% coverage property by excluding 2.5% of the distribution in each tail. Unfortunately, since an actual sample mean is a fixed quantity, we cannot make this same probability statement about the CI in Equation 2.9 – the calculated CI either does or does not contain the population mean. The probabilistic interpretation of this CI is that, if we repeat the study many times, then 95% of our calculated CIs will contain the population mean. More information about the derivation of CIs can be found in standard statistical texts, for example Wackerly et al. (2007).

2.3.2 Asking Questions about the Data: Hypothesis Testing

Hypothesis testing is a form of inference where pairs of **hypotheses** for a population parameter are compared using the information from a random sample of observations. The two hypotheses are defined as the **null** (denoted H_0) and the **alternative** (denoted H_1 or, sometimes, H_a). The null hypothesis H_0 represents the *status quo*, and we accept (cannot reject) this unless we obtain strong evidence from our sample that it is false. The hypothesis H_1 represents an alternative state that contradicts the null hypothesis and may be one-sided or two-sided. With a **two-sided** H_1 we do not specify a direction for the alternative hypothesis – we are just interested in detecting whether the *status quo* is implausible. With a **one-sided** H_1 we are interested in detecting deviations in a particular direction. For example, consider the situation where a scientist has obtained samples of aphids from two adjacent fields, one sprayed and one unsprayed, and wants to know if there are differences in resistance to the applied pesticide. The null hypothesis assumes no difference in resistance between the two fields, and the alternative hypothesis might be that the population from the sprayed field shows greater resistance than the other (one-sided) or simply that resistances in the two fields are not the same (two-sided).

To test the null hypothesis against the alternative, we first need to identify an appropriate **test statistic** which has a known statistical distribution when the null hypothesis is true. The actual statistic used depends on the characteristics of the problem, but you may already be familiar with some common statistical tests, such as the t-test and the F-test, and the test statistics associated with them. Having constructed a test statistic, the final step in the hypothesis testing process involves assessing the consistency of the observed test statistic with the null hypothesis. Under the null hypothesis (i.e. on the assumption that the null hypothesis is true), if the probability (P) of obtaining a test statistic as extreme as the observed value is small then we have statistical evidence against the null hypothesis. The test statistic thus assesses how well the data support the null hypothesis. The strength of this evidence is quantified by the observed significance level of the test, denoted above as P and sometimes called the **P-value**. It is good practice to predetermine the level of significance required to reject the null hypothesis (denoted α_s); this is often chosen as 5% ($\alpha_s = 0.05$). This is known as the **Type I error**, and represents the probability of rejecting the null hypothesis when in fact it is true. More details of hypothesis testing, including the concepts of Type II error and statistical power, are discussed in Chapter 10. The calculations associated with the test statistics used in hypothesis tests can also be used to derive CIs (Section 2.3.1) for population parameters, i.e. an indication of the likely range of values for the quantity of interest, taking account of the uncertainty associated with the estimate of the population parameter. These concepts are illustrated in more detail in the examples below.

2.4 Simple Tests for Population Means

In the following sections we describe the one- and two-sample t-tests, which are used extensively later in the contexts of regression modelling and treatment comparisons.

2.4.1 Assessing the Mean Response: The One-Sample t-Test

When we collect a single sample of observations we often want to make inferences about the value of the unknown mean of the population from which we have drawn the sample. We assume that we have a sample of N independent observations from a single population, $y_1 \ldots y_N$. Suppose we wish to test the null hypothesis that the population mean is equal to a given value, i.e. $H_0: \mu = c$, against a general two-sided alternative hypothesis, i.e. $H_1: \mu \neq c$, where c is some pre-determined constant value (referred to as a two-sided test, see Section 2.3.2).

We can estimate the population mean, μ, by the sample mean, $\bar{y}$ (Equation 2.1). Usually, the population variance is also unknown and is estimated by the unbiased sample variance, s^2 (Equation 2.3). The null hypothesis is then evaluated using a **one-sample t-test**, with test statistic, t, computed as

$$ t = \frac{\hat{\mu} - c}{\widehat{SE}(\hat{\mu} - c)} = \frac{\bar{y} - c}{s/\sqrt{N}} , $$

i.e. as the ratio of the difference between the estimated mean and the constant c and the estimated standard error of that difference. As c is fixed and known, it has no uncertainty and so, in this case, the standard error of the difference between $\bar{y}$ and c is simply equal to the standard error of $\bar{y}$ and estimated by $SEM = s/\sqrt{N}$, as in Section 2.3.1. If the null hypothesis is true then the population mean is equal to c and the t-statistic should be close to zero. If it is not then this gives evidence against the null hypothesis. For a given significance level, α_s, we can determine whether the data provide sufficient evidence against the null hypothesis by comparing our test statistic with a critical value from the appropriate distribution.

If H_0 is true and the observations are Normally distributed (or at least approximately so) then the observed test statistic t follows a Student's t-distribution with $N - 1$ df, as in Equation 2.8. The **critical value** defines a threshold such that test statistics more extreme than this value occur with probability α_s when the null hypothesis is true. Because we are considering a two-sided test, both large positive and large negative values of the test statistic are unlikely under H_0, so we consider the probabilities associated with extreme values in both tails of the t-distribution. We denote $t_{N-1}^{[\alpha_s/2]}$ to be the $100(1 - \alpha_s/2)$th percentile of the t-distribution with $N - 1$ df, and use this as our critical value. For example, if we choose $\alpha_s = 0.05$, then $t_{N-1}^{[0.025]}$, the 97.5th percentile of this distribution, is the critical value. From the definition of percentiles (Section 2.2.1) and the symmetry of the t-distribution about zero it follows that, for a random variable t_{N-1} with a t-distribution on $N - 1$ df,

$$\text{Prob}(t_{N-1} \leq -t_{N-1}^{[\alpha_s/2]}) = \text{Prob}(t_{N-1} \geq t_{N-1}^{[\alpha_s/2]}) = \alpha_s/2 .$$

So $\text{Prob}(|t_{N-1}| \geq t_{N-1}^{[\alpha_s/2]}) = \alpha_s$, i.e. the probability of equalling or exceeding the critical value is α_s, as required. We reject the null hypothesis if the absolute value of our test statistic, denoted $|t|$, meets or exceeds this value. This situation is illustrated in Figure 2.6.

The observed significance level of this test is calculated as $P = \text{Prob}(|t_{N-1}| \geq t)$, i.e. the probability under H_0 of obtaining a result more extreme (larger positive or negative) than that observed. If the observed significance level is less than the pre-determined significance level α_s then the null hypothesis is rejected.

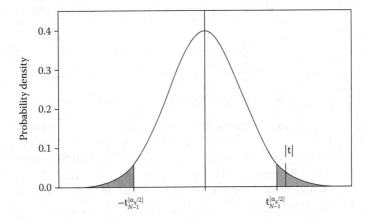

FIGURE 2.6
Critical regions for a two-sided one-sample t-test with probability level α_s. Shaded area covers $100\alpha_s\%$ of distribution containing the most extreme values. $|t|$ is the absolute value of an observed t-statistic greater than the critical value at significance level α_s.

The test can also be used to calculate a $100(1 - \alpha_s)$% CI (Section 2.3.1) for the population mean μ as the range

$$\left[\ \bar{y} - \left(t_{N-1}^{[\alpha_s/2]} \times \frac{s}{\sqrt{N}} \right) \ , \ \bar{y} + \left(t_{N-1}^{[\alpha_s/2]} \times \frac{s}{\sqrt{N}} \right) \ \right].$$

Note that if the t-test is not significant at level α_s, i.e. if the data are consistent with the null hypothesis, then this $100(1 - \alpha_s)$% CI will contain the hypothesized value, c, as a plausible value for the population mean. Conversely, if the null hypothesis is rejected, then this CI will not contain c, as it is then an unlikely value for the population mean.

EXAMPLE 2.1B: WHEAT YIELDS*

Consider the set of seven yield measurements presented in Example 2.1A. Historical records indicate that for these field plots, the expected yield should be close to 9 tonnes per hectare (t/ha). Because the mean yield from this harvest was 7.6 there is concern that this year is atypical. To evaluate this, we perform a one-sample t-test with null hypothesis H_0: $\mu = 9$, and alternative hypothesis H_1: $\mu \neq 9$. Recall that the unbiased sample variance was 6.95. The t-statistic is calculated by substituting the (unrounded) sample statistics into the t-test formula as

$$t = \frac{7.57 - 9}{\sqrt{6.95/7}} = \frac{-1.43}{0.997} = -1.433 \ .$$

This statistic has $N - 1 = 6$ df. The critical value is the 97.5th percentile value of the t-distribution with 6 df, equal to 2.447. Because the absolute value of the test statistic ($|t| = 1.433$) is smaller than the critical value, we fail to reject the null hypothesis at the 5% significance level and conclude that there is not enough evidence to indicate that this year is atypical.

A 95% CI for the population mean can be calculated as

$$\left[\ 7.57 - (2.447 \times \sqrt{6.95/7}) \ , 7.57 + (2.447 \times \sqrt{6.95/7}) \ \right] = (5.1, 10.0) \ .$$

As expected from the results of the significance test, this CI contains the expected yield of 9 t/ha.

2.4.2 Comparing Mean Responses: The Two-Sample t-Test

We now consider the case in which we have two treatment groups and wish to test for differences between the population means for the two treatments. In this case the null hypothesis (H_0) is that the population means for the two treatment groups are equal, and the two-sided alternative hypothesis (H_1) is that the two population means are different. If we label the population means for the two treatments as μ_1 and μ_2, respectively, these hypotheses can be written as

$$H_0: \mu_1 = \mu_2 \quad \text{and} \quad H_1: \mu_1 \neq \mu_2 \ .$$

We have again specified a two-sided test, but we could specify a one-sided alternative hypothesis if that were appropriate. The population means for the two treatment groups are estimated by their sample means. The test compares the difference between these two sample means with an estimate of the uncertainty in this difference due to background variation. If the difference in sample means is large compared to the background variation then there is evidence for a difference between the two population means.

Suppose we have a sample of n_1 observations for treatment group 1 (denoted y_{1k}, $k = 1 \ldots n_1$), and a sample of n_2 observations for group 2 (denoted y_{2k}, $k = 1 \ldots n_2$). The estimates for the group means are calculated as

$$\hat{\mu}_1 = \bar{y}_{1.} = \frac{1}{n_1} \sum_{k=1}^{n_1} y_{1k} \quad \text{and} \quad \hat{\mu}_2 = \bar{y}_{2.} = \frac{1}{n_2} \sum_{k=1}^{n_2} y_{2k} ,$$

with the dot notation as defined in Section 2.1. If we can reasonably assume that the background variation is the same for the two groups, then the background variability is estimated by a weighted sum of the unbiased within-group sample variances, also known as a **pooled estimate of variance**

$$s_{\text{pooled}}^2 = \frac{(n_1 - 1) \times s_1^2 + (n_2 - 1) \times s_2^2}{n_1 + n_2 - 2} ,$$

where the unbiased within-group sample variances are calculated as usual (see Equation 2.3) as

$$s_1^2 = \frac{1}{(n_1 - 1)} \sum_{k=1}^{n_1} (y_{1k} - \bar{y}_{1.})^2 \quad \text{and} \quad s_2^2 = \frac{1}{(n_2 - 1)} \sum_{k=1}^{n_2} (y_{2k} - \bar{y}_{2.})^2 .$$

If the treatment variances cannot be assumed equal, then a modified statistic must be used (see for example Wackerly et al., 2007). The df associated with the pooled estimate of variance are equal to the total number of elements in the two summations, $N = n_1 + n_2$, minus the number of treatment means estimated, here two; hence, the df are $n_1 + n_2 - 2 = N - 2$. The estimated **standard error of the difference** between the two sample means, denoted SED, can then be calculated as

$$\text{SED} = \widehat{\text{SE}}(\hat{\mu}_1 - \hat{\mu}_2) = \sqrt{s_{\text{pooled}}^2 \times \left(\frac{1}{n_1} + \frac{1}{n_2} \right)} .$$

The null hypothesis is then evaluated using a **two-sample t-test** with the test statistic, t, computed as

$$t = \frac{\hat{\mu}_1 - \hat{\mu}_2}{\widehat{\text{SE}}(\hat{\mu}_1 - \hat{\mu}_2)} = \frac{\bar{y}_{1.} - \bar{y}_{2.}}{\text{SED}} , \tag{2.10}$$

i.e. as the difference between the estimated population means, divided by the estimated uncertainty in that difference due to the background variability. If the data follow a Normal distribution and the observations are independent then under the null hypothesis this test statistic has a t-distribution with $N - 2$ df. This distribution can be used to obtain critical values, as for the one-sample t-test, and to obtain the observed significance level. Alternatively (and equivalently), the squared value of the test statistic has an F-distribution on 1 and $N - 2$ df. This distribution is useful later when we have more than two groups to compare (Section 4.3).

As for the one-sample case, the t-distribution can be used to construct a CI, but now for the difference in population means, $\mu_1 - \mu_2$. A $100(1 - \alpha_s)\%$ CI for this difference can be calculated as

$$\left[(\bar{y}_1. - \bar{y}_2.) - (t_{N-2}^{[\alpha_s/2]} \times \text{SED}) , \ (\bar{y}_1. - \bar{y}_2.) + (t_{N-2}^{[\alpha_s/2]} \times \text{SED}) \right] .$$

If the t-test is not significant at level α_s, i.e. if the data are consistent with the null hypothesis, then this $100(1 - \alpha_s)\%$ CI will contain zero as a plausible value for the difference in population means. Conversely, if the null hypothesis is rejected, then this CI will not contain zero, as it is then an unlikely value for the difference between the two population means.

EXAMPLE 2.1C: WHEAT YIELDS*

A standard commercial and a new 'improved' wheat variety are to be compared using yield measurements obtained from 14 plots in a field trial. The objective of the study is to determine whether the varieties produce different average yields. For each variety, yields were obtained from $n = 7$ small plots, and converted into tonnes per hectare (t/ha). The data for the commercial variety were analysed in Examples 2.1A and B, and the complete data set can be found in Table 2.3 and in file WHEAT.DAT.

The hypotheses can be stated as

H$_0$: both varieties give the same mean yield
H$_1$: the varieties give different mean yields.

In mathematical terms this can be written as

H$_0$: $\mu_1 = \mu_2$ (the population means are equal)
H$_1$: $\mu_1 \neq \mu_2$ (the population means are not equal)

TABLE 2.3

Yield Measurements (in t/ha) from a Standard Commercial and an Improved Wheat Variety Obtained from Plots in a Field Trial (Example 2.1C and file WHEAT.DAT)

Plot	Variety	Yield	Plot	Variety	Yield
1	Commercial	7	8	Improved	12
2	Commercial	9	9	Improved	8
3	Commercial	6	10	Improved	12
4	Commercial	12	11	Improved	9
5	Commercial	4	12	Improved	8
6	Commercial	6	13	Improved	16
7	Commercial	9	14	Improved	7

Here, μ_1 corresponds to the population mean of the commercial variety, and μ_2 to that of the improved variety. To test the null hypothesis we substitute the summary statistics for these data into the t-test formula (Equation 2.10). The sample means for each variety are $\bar{y}_{1\cdot} = 7.57$ (from Example 2.1A) and

$$\bar{y}_{2\cdot} = \frac{12 + 8 + 12 + 9 + 8 + 16 + 7}{7} = 10.29 .$$

The unbiased sample variances are $s_1^2 = 6.952$ and $s_2^2 = 10.238$, with pooled variance

$$s^2_{\text{pooled}} = \frac{(7-1) \times 6.952 + (7-1) \times 10.238}{(7+7-2)} = 8.595 ,$$

giving

$$\text{SED} = \sqrt{s^2_{\text{pooled}} \times \left(\frac{1}{n_1} + \frac{1}{n_2} \right)} = \sqrt{8.595 \times \frac{2}{7}} = 1.567 .$$

Finally, the observed test statistic is

$$t = \frac{\bar{y}_{1\cdot} - \bar{y}_{2\cdot}}{\text{SED}} = \frac{7.57 - 10.29}{1.567} = \frac{-2.71}{1.567} = -1.732 .$$

The absolute value of the test statistic ($|t| = 1.732$) is compared with a critical value of the t-distribution with $7 + 7 - 2 = 12$ df. If we set $\alpha_s = 0.05$ (i.e. use a 5% significance level) then the critical value for this two-sided test is $t_{12}^{[0.025]} = 2.179$ (the 97.5th percentile of the t-distribution with 12 df). As 1.732 is less than 2.179, we cannot reject H_0. We might report that there is insufficient statistical evidence to conclude that the mean yields of the commercial and improved varieties are different. Alternatively, we might report the observed significance level for this test statistic as $P = 0.109$, obtained from $t_{12}^{[0.0545]} = 1.732$. Again, because this value is larger than our chosen significance level of $\alpha_s = 0.05$, we cannot reject H_0. A 95% CI for the difference in population means ($\mu_1 - \mu_2$) is

$$(-2.714 - (2.179 \times 1.567), -2.714 + (2.179 \times 1.567)) = (-6.1, 0.7) .$$

It follows that zero is a plausible value for the difference in population means as it is contained in the CI, agreeing with the result of the hypothesis test.

Note that to have achieved significance at the 5% level the test statistic, t, would have needed to satisfy $|t| > 2.179$. If the unbiased sample variance is assumed equal to the pooled estimate, s^2_{pooled}, then this requires

$$\frac{|\bar{y}_{1\cdot} - \bar{y}_{2\cdot}|}{1.567} > 2.179 ,$$

so that the difference between the two variety means would have needed to be at least as large as 3.42 t/ha ($3.42 = 1.567 \times 2.179$). This large difference is required because the pooled variance is relatively large, and the total number of observations is small, and hence there is considerable uncertainty in the estimates of the population means for the two varieties.

2.5 Assessing the Association between Variables

When two variables have been measured on the same material, it is often of interest to know whether the variables are independent or whether they show some association. For example, the diameter and weight of seeds would be expected to show a strong positive association such that seeds with larger diameters also have larger weights. Covariance and correlation are both measures of the strength and direction of a relationship between two variables. Correlation is a more useful measure, because it uses a standardized scale and is independent of the scale of measurement. However, correlation is often defined in terms of covariance, so we discuss the latter first.

Suppose we have observations of two variables, here labelled x and y, measured on the same units so the data consist of N pairs (x_i, y_i). The **unbiased sample covariance** between the two variables, s_{xy}, is defined as

$$s_{xy} = \frac{1}{(N-1)} \sum_{i=1}^{N} (x_i - \bar{x})(y_i - \bar{y}) , \tag{2.11}$$

i.e. we centre both variables, calculate the product of the centered variables for each unit, and then sum the resulting values over all units and divide by $N-1$. This is an unbiased estimate of the population covariance between these two variables, denoted σ_{xy}. If both variables tend to be large on the same units, then the covariance will be large and positive. If one variable tends to be large when the other is small then the covariance will be large and negative. If the variables are completely unrelated then the covariance will be close to zero. The formula for the sample covariance in Equation 2.11 has a similar form to that for the sample variance in Equation 2.3, and it is easy to verify that the covariance of a variable with itself is simply its unbiased sample variance. Analogous to Equation 2.2, the **sample covariance** would use the divisor N in Equation 2.11 in place of $N-1$, but gives a biased estimate of the population covariance.

Where we have several random variables measured on the same units, we can assemble their variances and covariances into a single structure, called the **variance–covariance matrix**. This is a symmetric matrix with rows and columns indexed by the variables. The variances are held on the diagonal, and the covariances are held in the off-diagonal positions. For example, for three variables, x, y and z, the variance–covariance matrix takes the form

$$\begin{pmatrix} \sigma_x^2 & \sigma_{xy} & \sigma_{xz} \\ \sigma_{yx} & \sigma_y^2 & \sigma_{yz} \\ \sigma_{zx} & \sigma_{zy} & \sigma_z^2 \end{pmatrix} . \tag{2.12}$$

The matrix is symmetric because covariances are invariant to the order in which the variables are specified, so that $\sigma_{xy} = \sigma_{yx}$, and so on. For this reason, it is sufficient to present only the values on or below the diagonal (the lower triangle). The unbiased sample variance–covariance matrix replaces the population values by the unbiased sample variances and covariances.

Unfortunately, covariance is strongly dependent on scale – if you convert a set of measurements from centimetres to inches then the covariance will also change, although relationships between the variables clearly do not change. **Correlation** is a standardized measure of the strength and direction of a relationship between two variables that is independent of the scale of measurement. In general statistical usage, correlation quantifies the departure of two variables from independence, and many correlation coefficients have been defined.

We use **Pearson's product–moment correlation coefficient**, which is derived directly from the covariance, and is appropriate for estimating correlation between two variables with an underlying bivariate Normal distribution defined by the population means and standard deviations of each variable and the population correlation coefficient, ρ. This coefficient measures the strength of a *linear* relationship between the variables and can be estimated by the **sample correlation coefficient**, r, calculated as the unbiased sample covariance divided by the product of the unbiased sample standard deviations. If we write s_y and s_x to be the unbiased sample standard deviations of variables y and x, respectively, the sample correlation coefficient is defined as

$$r = \frac{s_{xy}}{s_x \times s_y}.$$

The value of r has no units and lies between –1 and +1. A few examples are illustrated in Figure 2.7. In Figure 2.7a, a strong positive correlation is shown ($r = 0.96$). If the observations lie exactly on a straight line with a positive slope, then x and y are perfectly correlated ($r = 1$), a concept that is known as **collinearity** where one variable can be completely determined from the other. In Figure 2.7b, a weaker negative correlation is pictured ($r = -0.57$). When $r = -1$ the paired observations also lie exactly on a straight line but with a negative slope. Both Figures 2.7c and d show a weak sample correlation between the variables (r close to 0). For Figure 2.7c the points show a random scatter, whilst for Figure 2.7d there is a clear relationship between x and y, but one that is not linear. When two variables are independent, so that a value of y does not depend in any way on the value of x, then their sample correlation coefficient will be close to zero; however, the converse is not necessarily true (as shown in Figure 2.7d) because the correlation coefficient detects only linear dependencies between two variables. Hence, a scatter plot of the variables should always be considered alongside the summary value of r.

For more than two variables, it is conventional to present the set of pairwise correlations in matrix form, similar to the variance–covariance matrix in Equation 2.12 but with value 1 on the diagonal and correlation coefficients for pairs of variables in the off-diagonal positions.

It is important to understand that strong correlation does not necessarily imply causation. If x and y are correlated, then there are four possibilities to consider: (1) x causes y; (2) y causes x; (3) a third variable, z, influences both x and y; or (4) there is no relationship but by chance an atypical joint sample has been produced. For this reason, causal conclusions should not be drawn from correlations without further information or experimentation, or both.

It is sometimes of interest to investigate whether there is any statistical evidence of correlation between two variables. Here, the null hypothesis states that the population correlation coefficient is equal to zero, i.e. H_0: $\rho = 0$, and is tested against a two-sided alternative hypothesis that the correlation is non-zero, i.e. H_1: $\rho \neq 0$. The null hypothesis is evaluated using a t-statistic based on the sample correlation, calculated as

$$t = r\sqrt{\frac{N-2}{1-r^2}}.$$

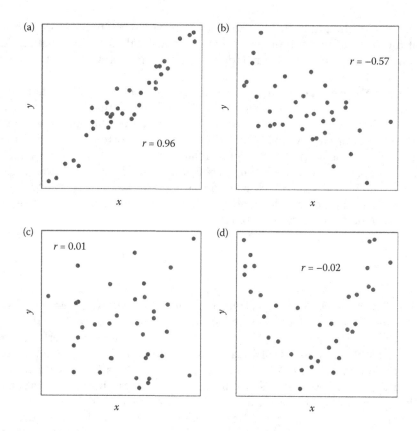

FIGURE 2.7
Scatter plots illustrating correlation patterns between two variables: (a) strong positive correlation; (b) moderate negative correlation; (c) uncorrelated and unrelated variables; (d) uncorrelated but related variables.

If the two variables have a Normal distribution, then under the null hypothesis this statistic has a t-distribution with $N - 2$ df. The significance level of the test depends on both the sample correlation and the sample size: a given value of r becomes more significant as N increases. This test can also be used for variables with a non-Normal distribution, but the distribution of the test statistic is then approximate and may be inaccurate unless the sample size is large.

EXAMPLE 2.3B: WILLOW BEETLE MEASUREMENTS

The lengths and widths (mm) of the sample of willow beetles described in Example 2.3A are plotted in Figure 2.8. Some points represent multiple observations, with the area of each plotted symbol proportional to the number of observations represented. It seems clear that there is a positive relationship between the two variables, although there is one beetle much wider than would be expected from its length (point in bottom right of plot). The unbiased sample variance–covariance matrix takes the form

$$\begin{pmatrix} s_L^2 & \\ s_{LW} & s_W^2 \end{pmatrix} = \begin{pmatrix} 0.0260 & \\ 0.0059 & 0.0043 \end{pmatrix},$$

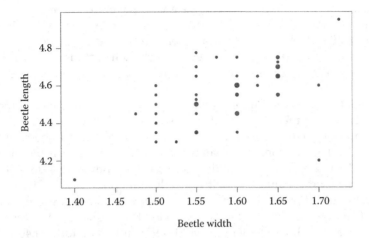

FIGURE 2.8
Length (mm) plotted against width (mm) for 50 willow beetles (Example 2.3B). Area of points is proportional to the number of observations at that position.

where we use the subscript L to indicate length and W to indicate width measurements. The sample correlation between length and width is therefore calculated as

$$r = \frac{s_{LW}}{s_L \times s_W} = \frac{0.00586}{\sqrt{0.02595} \times \sqrt{0.00434}} = 0.5528 \; .$$

Hence, we observe a positive correlation between these traits of 0.55. The distribution of these variables is consistent with a Normal distribution (e.g. see Figure 2.3), and so we formally test whether this result is consistent with an uncorrelated population. We calculate the t-statistic as

$$t = r\sqrt{\frac{N-2}{1-r^2}} = 0.5528 \times \sqrt{\frac{48}{0.6944}} = 4.596 \; .$$

The value 4.596 is then evaluated against a t-distribution with 48 df, giving observed significance level $P < 0.001$. Hence, we have strong evidence that the population correlation coefficient between length and width is not zero.

2.6 Presenting Numerical Results

The presentation of numerical results is fraught with difficulties because various, somewhat arbitrary, choices have to be made on rounding and the number of significant figures shown. In this section we describe the conventions that we try to follow in this book. This should make our written calculations easier to follow, as well as suggesting general guidelines for use in other contexts, such as scientific publications.

Where we show calculations in the text we have necessarily rounded the numbers presented, and this includes any intermediate results. However, to get answers that match statistical software, we have not actually implemented this rounding in our calculations; we have always retained full accuracy. This means that there will be small differences (usually in the last decimal place) between the answers we present and those achieved by the calculations shown. Similarly, we round quantities presented in ANOVA tables (see Chapter 4) but do all of our calculations based on unrounded quantities.

In presenting the results of hypothesis tests, we show test statistics, critical values and observed significance levels to three decimal places. This gives sufficient accuracy for most situations.

The definition of a suitable scale for data-dependent quantities is more difficult. For any variate, we first identify the scale at which we see variation within it, and we call this its natural granularity. This may be related to the size of the numbers, but this is not always the case, and so we consider several examples, starting with the yields recorded in Example 2.1 (Table 2.3). These are all integer values in the range 4–16. All of the information is captured by these numbers with zero decimal places and so we say that this is the natural granularity for these observations. Exactly the same argument holds for the integer count data of Example 2.2. However, the situation is slightly different for Example 2.3 (Table 2.1). Here both the beetle lengths (4.10–4.95 mm) and widths (1.40–1.725 mm) are measured to the nearest 0.025 mm. Accurate representation of the measurements requires three decimal places, but this implies much more accuracy than is actually present, and so we denote the natural granularity as two decimal places (0.01) even though this still gives slightly greater accuracy than is present in the measurements.

A different situation holds for many machine-calculated measurements, for example where small-plot yields have been converted into acre or hectare yields, or observations have been transformed to or from the logarithmic scale. In these cases, the conversion between scales can introduce many superfluous decimal places. All of the decimal places (up to rounding error) should be retained prior to analysis, so that no accuracy is lost if any further transformation is required (e.g. as described in Chapter 6). At the point of analysis, we suggest that the natural granularity is decided from the range of the data, calculated as the minimum value subtracted from the maximum value, and by use of the first three significant figures of this range to define the natural granularity for the presentation of results. For example, data on 0–100 (range 100) has natural granularity of 0 (zero decimal places); data on 1000–1010 (range 10) has its natural granularity defined as 0.1 (one decimal place). If the natural granularity is greater than 1, we do not generally advocate rounding values, but it may be worth rescaling the data (dividing by the natural granularity) to avoid the appearance of spurious accuracy. Similarly, if there is no variation in the leading significant figures, it may make sense to remove those digits. Occasionally these guidelines may fail and, in that case, common sense should be applied.

Once the natural granularity of a data set has been defined, it can be used to define a sensible scale for reporting statistics. We should report sample means and estimates of other location parameters to one decimal place more than the natural granularity. We should report sums of squares, mean squares, and estimates of variances and standard deviations or errors to two decimal places more than the natural granularity. We use greater accuracy for standard deviations and errors because multiples of these quantities are often used to compare differences between location estimates (e.g. group means). A suitable scale for a regression coefficient is harder to define as it depends on the scale of its explanatory variate. Here, we suggest using sufficient decimal places for such coefficients to ensure that we can report the fitted values to one decimal place more than the natural granularity, with additional precision on coefficient standard errors. We have often found ourselves breaking these rules for convenience of presentation in this book, but nevertheless recommend that you think carefully about the appropriate level of precision for your own circumstances.

EXERCISES

2.1* A plant ecologist is interested in the distribution of one species of grass within a field. She investigates this by throwing a 0.1 m² quadrat to 20 random positions in the field and counting the number of plants of the species in the quadrat at each position. The counts for the 20 quadrats were: 15, 12, 6, 7, 4, 2, 10, 14, 3, 6, 9, 9, 2, 11, 10, 3, 2, 11, 9 and 10. File GRASS.DAT contains the unit number (variate *Quadrat*) and plant count (variate *Count*) for each quadrat. Consider whether these data should be considered as continuous or discrete, and draw a bar chart or histogram (as appropriate). Obtain the sample mean, median and inter-quartile range. What can you say about the distribution of these data?

2.2 Obtain a histogram of the beetle widths (mm) given in Table 2.1 (and variate *Width* in file WILLOW.DAT). Do these data seem consistent with a Normal distribution, as asserted in Example 2.3B?

2.3* The one-sample t-test is rarely used in analysis of experimental data, except in the context of regression, but it can be useful for analysis of paired samples from a set of subjects. In this scenario, the two sample t-test is not valid because two samples from a single subject are not independent. However, if we analyse the differences between the samples from each subject, we can use a one-sample t-test to test the null hypothesis of no difference between samples.

An experiment made measurements of Rubisco protein (on a relative scale) in 12 grass plants before and after a drought stress period of five days. File PROTEIN.DAT contains the unit number (*DPlant*) and Rubisco measurements (variates *Before* and *After*) for each plant. Calculate the change in amount of Rubisco protein in each plant and analyse this change using a two-sided one-sample t-test. Write down the null and alternative hypotheses for this test and interpret them in the context of this experiment. Is there any evidence that the amount of Rubisco has changed after five days of drought stress?

2.4* A soil scientist sampled two fields to get background measurements of carbon biomass (measured as mg C per kg of soil) prior to a field experiment. Six samples were taken from each field: the samples from the first field gave 910, 1058, 929, 1103, 1056, 1022 mg C kg⁻¹; the samples from the second field gave 1255, 1121, 1111, 1192, 1074, 1415 mg C kg⁻¹. File CARBON.DAT contains the unit number (*Sample*), field number (factor *Field*) and carbon biomass measurement (variate *Carbon*) for each sample. Use a two-sided two-sample t-test to test whether there is any difference in average biomass between the two fields, and calculate a 95% CI for the difference.

2.5 In Example 12.1 (Tables 12.1 and A.1), we describe an experiment in which several morphological traits were measured on 190 seeds from a line of diploid wheat. Two of the traits measured on each seed were length (mm) and weight (mg). The unit numbers (*DSeed*) and length and weight measurements (variates *Length* and *Weight*) can be found in file TRITICUM.DAT. Produce a scatter plot of these two traits and calculate the unbiased sample variances and covariances between them. Derive their sample correlation coefficient, *r*. Is there evidence of association between these two variables?

3

Principles for Designing Experiments

This chapter presents the basic concepts that are required to construct designs to address directly and efficiently the aims of a biological experiment. We first discuss the choice of treatments and materials; treatments should be determined by the aims and the materials should be chosen according to the frame of reference for the experiment (Section 3.1). These two components must then be combined to produce an appropriate design. A good design takes proper consideration of three statistical principles: replication (Section 3.1.1), randomization (Section 3.1.2) and blocking (Section 3.1.3), to reduce bias and maximize the precision of treatment comparisons. We describe the structure of a design with respect to underlying factors using a symbolic form (Section 3.2). Many experiments will use one of the wide and flexible family of standard designs, such as the completely randomized design (CRD) (Section 3.3.1), the randomized complete block design (RCBD) (Section 3.3.2), the Latin square (LS) design (Section 3.3.3), the split-plot (SP) design (Section 3.3.4) or the balanced incomplete block design (BIBD) (Section 3.3.5). Once the structure of the design has been determined, a properly randomized layout can be generated with statistical software (Section 3.3.6).

3.1 Key Principles

As described in Chapter 1, an experimental study investigates the relationship between an outcome and one or more conditions that are manipulated by the researcher. Before considering the appropriate design for any experiment, it is important to be clear about its aims, which are usually associated with one or more scientific questions or hypotheses to be tested. Examples of such questions might be

- Is any reduction in disease infection achieved with a new 'resistant' variety compared with a standard 'control' variety?
- How do plant metabolites respond to increasing drought stress at different stages of development?
- Which chemicals, of several under study, show insecticidal activity?
- How is yield related to plant spacing, and does this relationship vary between varieties?

The aims of the experiment should be well defined to make it easy to assess whether the chosen treatments are sufficient to achieve them. In this context, the term **treatments** is used to describe the set of different experimental conditions to be tested, for example, varieties, nitrogen rates, or chemical compounds, or, more usually, combinations of several such classifying variables. Control treatments – either positive or negative controls – may be used to provide a baseline, or to verify that the experiment has worked as expected. A negative control usually corresponds to a 'null' treatment, and a positive control usually corresponds to a standard treatment with a known effect. This is discussed further in Section 8.5. In addition to defining the experimental treatments, the experimental units must be chosen. The **experimental unit** for a treatment is defined as 'the smallest division of the experimental material such that any two units may receive different treatments in the actual experiment' (Cox, 1992). For some treatments, this may be larger than the size of unit on which individual observations are recorded (sometimes called the **observational** or **measurement unit**), and may occur at a range of scales, as in the following examples.

- *An area of land on a farm.* A field trial typically has numerous small plots, and experimental treatments are applied to the individual plots. The experimental unit is the plot, and the measurement unit may be either the plot (e.g. yield) or sub-samples from the plot area (e.g. individual plant measurements).

- *Individual soil samples taken from a field.* In the context of a field trial with treatments applied to plots, if a single soil sample is taken from each field plot for processing in the lab, then the soil sample becomes the experimental unit. If multiple soil samples are taken from each plot, then the experimental units are the sets of samples from each plot.

- *Pots, each containing three plants.* If experimental treatments, such as soil nutrient content, are applied to whole pots, then the pot is the experimental unit. The measurement unit may be either the whole pot (e.g. combined biomass) or individual plants.

- *Different leaves from an individual plant.* In the investigation of the response of plants to aphid attack, clip cages with or without aphids might be attached to individual leaves within a plant. The experimental unit is then the individual leaf.

- *Samples of RNA extracted from different plants.* Investigation of gene expression often involves the application of different treatments to individual plants, followed by extraction of RNA from each plant. The experimental unit for further study is then the RNA taken from an individual plant.

- *A batch of 10 insects in a Petri dish.* Experiments on small insects are often done on groups of insects kept together in dishes (or cages), with treatments applied to the dishes. The experimental unit is then the dish. The measurement unit may be the dish, via a summary of insect behaviour such as percent survival, or the individual insects.

Recall from Section 1.2 that the experimental units are considered to consist of a sample from a wider population for which inferences can be made, and that this population should be identified according to the frame of reference for the experiment. For example, if RNA samples for microarray work are taken from only a single plant, then conclusions regarding gene expression in the plant population cannot safely be made without further experimentation, because variation between different plants would be expected. If samples are taken from several randomly selected plants, then variation between plants can be accounted for, and inferences can be applied to the wider population. A similar situation occurs when an experiment is established in a single site, or in a single year, or on a single variety, as there can be no certainty that results can be safely extrapolated to wider circumstances. This issue is especially relevant to field trials, where the chance peculiarities of a single environment can produce anomalous results; for this reason, many journals will not publish the results of field trials that have not been repeated over several sites or seasons or both. It is therefore important to recognize the frame of reference implied by the choice of experimental units, so that appropriate conclusions can be drawn from the results.

Although we wish to have a representative sample of experimental units, we can get more precise estimates of differences between treatments by making comparisons across similar units. We can deal with this apparent contradiction by using sets of reasonably homogeneous experimental units, and then repeating the comparison across a wider range of circumstances. In some cases, the experimental units may have some intrinsic structure, introducing some heterogeneity between groups of more homogeneous units. This structure should be incorporated in both the choice of experimental units for treatment application and in the statistical analysis. For example, experimental materials may be arranged as plants within pots within trays, giving three structural levels, and we expect different levels of variation within each of these levels. Depending on practical considerations and the aims of the experiment, it may be appropriate to apply treatments at any of these levels and a statistical analysis should account for this structure.

The choice of experimental units and treatments should be made separately: units are chosen according to the appropriate frame of reference for the experiment, and treatments are chosen to enable the hypotheses to be tested. A good design then matches the treatments with the units so that the treatment differences can be estimated without bias (i.e. without systematic over- or under-estimation) and as precisely as possible (i.e. to minimize uncertainty in the results). Our main tool to avoid experimental bias is randomization, i.e. the random allocation of treatments to experimental units, and experimental precision can be improved by the use of proper replication and blocking (terms we discuss in more detail below).

First, it is helpful to recall the role of the underlying unit-to-unit variation in biological experimentation. It is well known that biological individuals vary in any given characteristic or response. The amount of variation may depend on several factors, such as differing genetic backgrounds and environmental effects, but some variation is always present. This natural variation may be inflated by uncertainty introduced by the measurement process in cases where exact measurement is not possible (also known as measurement error). This combined background variation is a potential cause of both bias and uncertainty in experimental results. For example, if two treatments are each applied to one plant only, it is not possible to assess whether any difference in the measured response is due to treatment differences or natural plant-to-plant variation. Statistical design and analysis aim to distinguish, quantify and, subsequently, compare variation between treatments (signal) with background variation (noise). A large signal:noise ratio indicates that substantive treatment differences are present. A small signal:noise ratio indicates that any apparent treatment differences could be explained by the background variation in the system, and therefore cannot confidently be attributed to treatment effects. Proper identification and control of background variation is thus an essential aim of any statistical design.

A good design considers each of the three basic principles: replication, randomization and blocking. **Replication** is the process of applying each treatment to more than one experimental unit, so the number of replicates of a treatment is the number of independent experimental units to which each treatment is applied. **Randomization** means the random allocation of treatments to experimental units and is used to ensure the fair assessment of treatments without bias. For this reason, it can be regarded as an insurance against potential unknown differences between units, and it should be used whenever possible. In some circumstances, it may be possible to identify or construct groups of experimental units expected to have similar responses in the absence of any treatment effects. This process is known as **blocking**. A **block** is a subset of the experimental material within which experimental units are expected to be homogeneous, with more heterogeneity expected between experimental units in different blocks. In the analysis of experimental data, variation due to blocks can be separated from background variation and, if there are differences among blocks, this separation will increase the precision of treatment comparisons by reducing the estimate of the unit-to-unit background variability. An appropriate use of these three design principles will give confidence that any treatment differences observed are real and not due to some chance combination of circumstances, and will also enable the maximum amount of information to be obtained from the available resources. We now discuss each of these principles in more detail.

3.1.1 Replication

The natural background variation among experimental units means that it is necessary to replicate the application of each treatment to several experimental units. This replication serves two important purposes. First, by repeating each treatment on several experimental units, we get a more reliable estimate of the effect of each treatment. Second, and possibly more importantly, the replicated observations provide an estimate of the background variation between units, which we can use to assess whether treatments differ and to indicate the precision associated with the estimated treatment effect. Usually each treatment will be replicated an equal number of times, but in circumstances where particular treatments are of greater interest, it may be advantageous to have increased replication for those treatments. Conversely, reduced replication may be used where resources for particular treatments are either scarce or expensive, for example, seed for a new breeding line.

To illustrate some issues regarding replication, consider an experiment to compare two pesticide treatments (a standard and a new formulation) applied to six insect-net cages, each cage containing 10 aphids. The new formulation is applied to three cages selected at random, with the standard formulation applied to the remaining three cages. The replication of each treatment in this experiment is only three, even though 30 aphids have been treated with each pesticide and even if the measurement unit is the individual aphid, because each treatment is applied to a cage of aphids and so this is the experimental unit. The individual aphids here are an example of pseudo-replicates. **Pseudo-replication** describes the situation in which multiple measurements are taken from each experimental unit. This can be a very useful experimental technique, but must be properly incorporated into any statistical analysis, which may otherwise produce an incorrect estimate of the between-unit variability (usually too small), possibly leading to incorrect conclusions about the importance of treatment effects. Pseudo-replication usually causes problems when the smallest level of experimental material (i.e. the measurement unit, here, the aphid) is wrongly identified as the experimental unit in a statistical analysis, in place of the level at which treatments were actually applied (here, the cage). As a rule of thumb, replication needs to occur at the level at which the treatments have been applied to be considered 'real'. Consider the following examples of designs for experiments.

- Twelve pots, each containing four plants at the three-leaf stage, with six treatments (A–F) each applied to two of the pots with the allocation made at random (Figure 3.1). Treatments were applied to pots, and so the pot is the experimental unit and the replication of each treatment is two. Measurements from individual plants are pseudo-replicates.

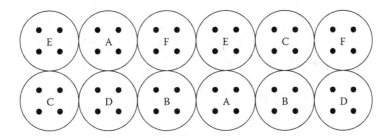

FIGURE 3.1
Design for an experiment with four plants (•) in each of 12 pots with treatments (A–F) applied to pots.

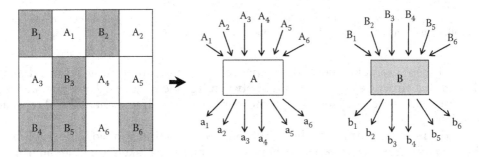

FIGURE 3.2
A two-stage design. The first stage (left) is a field trial with 12 plots and two treatments (A and B). A soil sample is taken from each plot and labelled by its treatment and replicate (A_1–A_6 and B_1–B_6). At the second stage, samples from each treatment are mixed together (bulked) then sub-sampled. Sub-samples from each bulked sample are labelled by lower-case letters and sample number and then measured.

- A field experiment consisting of 12 plots, with two treatments (A and B) each applied to six of the plots selected at random (Figure 3.2). One soil sample was taken per plot, and labelled by the treatment and replicate number, giving samples $A_1 \ldots A_6$ and $B_1 \ldots B_6$. The soil samples from the six replicate plots for each treatment (e.g. $A_1 \ldots A_6$) were bulked (combined) together and mixed thoroughly and six sub-samples taken and measured. These sub-samples were labelled as $a_1 \ldots a_6$ and $b_1 \ldots b_6$. In this case, although treatments were originally applied to plots, at the analysis stage there is only a single replicate for each treatment because samples from independent plots have been bulked and the sub-samples are not independent. The sub-samples are pseudo-replicates and give information on the homogeneity of the bulked sample rather than on the variation between plots. Ideally, samples from each plot should have been kept separate, giving six true replicates for each treatment.

- Two controlled environment (CE) cabinets, one at 10°C and one at 20°C, each containing eight seed trays, with two different watering regimes (A and B) each applied to four trays chosen at random within each cabinet (Figure 3.3). Both temperature and watering regime are considered as treatments here. The experimental units for watering regime are the seed trays and each watering regime is applied to eight independent seed trays, giving replication of eight. The experimental units for temperature are the cabinets, and the temperature treatments are unreplicated. To achieve replication of temperature, it would be necessary to use another two cabinets, or to repeat the study under the same controlled conditions with a new randomization of both factors.

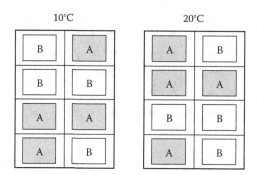

FIGURE 3.3
Design for an experiment with eight trays, with two watering regimes (A and B) applied within each of two CE rooms, with each room operating at a different temperature (10°C or 20°C).

The last of these examples shows that experimental units can occur at several different levels within a structure, and may differ between treatment factors. Therefore, one level of structure may represent pseudo-replicates for one type of treatment and real replicates for another.

It is also important to draw the distinction between technical and biological replication. **Technical replication** occurs when several measurements are taken from the same biological material, while **biological replication** occurs when measurements are taken from several independent biological subjects. The use of adequate biological replication is required to make inferences valid for the population from which the samples were obtained, rather than for a single individual. Technical replication is always pseudo-replication, but biological replication may correspond to either pseudo-replication or true replication, depending on the context. It is clearly important to recognize when measurements are pseudo-replicates that do not increase treatment replication. Note however, that technical replication can be useful in increasing precision where measurement is seriously subject to error, provided that it is properly accounted for in the statistical analysis. We give an example of an analysis accounting for pseudo-replication in Section 7.5.

3.1.2 Randomization

Randomization is required to ensure the fair allocation of treatments to units to guard against bias, and to cope with the natural variation between experimental units. In the simplest case, randomization requires that each permutation of the set of treatments has an equal probability of occurring, so that (for equal replication) every experimental unit has an equal chance of receiving any treatment. Hence, each treatment is equally likely to be applied to 'good' units as to 'bad' units. Where randomization has not taken place, there will always be a question about possible bias in the experiment.

To obtain a proper randomization for a given design, a method is required for assigning treatments to experimental units at random. We use the convention that treatments get assigned to experimental units, though the opposite approach can also be used. So, for example, for an experiment comprising two treatments each replicated six times, we might write A and B on six pieces of paper each, to represent the six replicates of each treatment, and put these in a bag and draw (without replacement and without looking) to obtain a sequence which allocates treatments to units 1 ... 12. Alternative approaches could use six tokens of each of two different colours, or six playing cards from each of two different suits, in the same way. Random number tables can also be used for allocating treatments to units, though care is needed to define the protocol where too many repeats of a treatment occur in the random sequence. However, most randomizations are now done via statistical software. The mechanics of the process are unimportant as long as the property of equal probability for each permutation of treatments is preserved – this is discussed further in Example 3.1. Randomization ensures that any (possibly unconscious) bias of the experimenter (e.g. a tendency to assign the biggest plants to their favoured treatment) is avoided and that any unknown differences between the units are unlikely to consistently favour particular treatments.

To reinforce objectivity in some areas of research, particularly medical research, trials are carried out as either single- or double-blind trials. In a single-blind trial, the subject does not know which treatment has been allocated, while in a double-blind trial, neither the subject nor the investigator knows which treatment has been applied. In plant science, the perception of the subject is not considered relevant. However, the perception of the investigator measuring or assessing the experimental material could be influenced (possibly unconsciously) by their expectation of the applied treatment. It is therefore good practice to make experimental measurements (especially subjective assessments) without knowledge of the treatment allocation as far as possible. For example, in field trials, this can sometimes be achieved by the use of a field plan with plot numbers marked but not treatments.

Randomization leads to estimates of treatment differences that are unbiased when considered across the whole set of possible randomizations. However, this does not guarantee that any individual randomization will produce unbiased results. For example, all instances of one treatment may be assigned to larger plants by chance. For this reason, experimental units should be chosen to be as homogeneous as possible while still being representative of the population of interest. Selecting homogeneous units has the added advantage of reducing the background variation or noise. Where it is not possible to select a completely homogenous set of experimental units, the units need to be grouped into sets (blocks) of more homogeneous experimental units to avoid potential bias (see Section 3.1.3). Sometimes, even where the units are thought to be homogeneous, a randomization can give cause for concern. For example, for 12 pots arranged in a line with two treatments (labelled A and B) each replicated six times, consider the following randomization

$$A\ A\ A\ A\ A\ A\ B\ B\ B\ B\ B\ B.$$

This particular randomization does not look random, but can occur (with probability 1 in 924). If we are not happy to accept this randomization, it is probably an indication that we do not consider the experimental units to be completely homogeneous, so that some sort of blocking is needed.

EXAMPLE 3.1: RANDOMIZATION

The efficacy of a new pesticide is to be tested in the field with 15 plots of size $5\,\mathrm{m} \times 10\,\mathrm{m}$ arranged in a 3×5 array. Five plots will be sprayed with the pesticide and 10 will be untreated (controls) for comparison. In this case, extra replication of the control treatment is used to obtain a good estimate of background variability and because the new pesticide is available in only small amounts. We evaluate two methods of determining a randomization for this experiment and consider whether each of the methods gives a valid randomization, i.e. with equal probability for each permutation of treatment effects.

TABLE 3.1

Experimental Layout Achieved Using Randomization by
Playing Cards for a Field Trial with 15 Plots (Numbered
1–15) and Two Treatments: A Pesticide Treatment (Labelled
P) with Five Replicates and a Control Treatment (Labelled
C) with 10 Replicates (Example 3.1)

1	2	3	4	5
P	P	C	P	C
6	7	8	9	10
C	C	P	C	P
11	12	13	14	15
C	C	C	C	C

First, we use a pack of cards. We might take 15 cards: five red cards to represent the
pesticide treatment and 10 black cards to represent the control. We shuffle the cards
(randomization), then deal them out in shuffled order to allocate treatments to plots (in
order 1–15) to get, for example, the randomization shown in Table 3.1.

Is this a valid randomization? Let us consider the process. Assuming a fair shuffle,
when we pick the first card we have a probability of 5/15 of picking the pesticide treatment for the first plot (and 10/15 of picking the control). These probabilities change as
the allocation proceeds. For example, if the first plot is allocated the pesticide treatment,
when we pick the second card we have a probability of 4/14 of the second plot also being
allocated the pesticide treatment (as we have four red cards out of 14 left), and so on. With
this method, the probability of plots 1–5 all being allocated the pesticide treatment is
1/3003 (= 5/15 × 4/14 × 3/13 × 2/12 × 1/11). This is the same, for example, as the probability of plots 11–15 all being allocated the pesticide treatment (equivalent to the probability of plots 1–10 being allocated the control, i.e. 10/15 × 9/14 × ... × 2/7 × 1/6 = 1/3003).
There are, in fact, 3003 possible permutations of five pesticide-treated plots and 10 control
plots, calculated as the factorial function of 15 (the number of plots) divided by the product of the factorial function of five (the number of pesticide-treated plots) and the factorial
function of 10 (the number of control plots):

$$\frac{15!}{5! \times 10!} = \frac{15 \times 14 \times 13 \times \ldots \times 2 \times 1}{(5 \times 4 \times 3 \times 2 \times 1) \times (10 \times 9 \times \ldots \times 2 \times 1)}$$
$$= \frac{15 \times 14 \times 13 \times 12 \times 11}{5 \times 4 \times 3 \times 2 \times 1}$$
$$= 3003 .$$

With this randomization approach, each of these permutations is equally likely.

An alternative, and perhaps at first sight simpler, approach is to toss a coin to construct
the randomization, with heads corresponding to the pesticide treatment, and tails corresponding to the control. Working through the plots one by one, we toss the coin once
for each plot, allocating the pesticide treatment to the plot if it comes up heads (subject
to a maximum of five pesticide plots), and allocating the control treatment to the plot if it
comes up tails (subject to a maximum of 10 control plots). Is this a valid randomization?
Let us again consider the process. We have a probability of 1/2 (assuming a fair coin) of
allocating pesticide to the first plot. The second coin toss takes no account of the allocation for the first plot, so again we have probability 1/2 of pesticide being allocated to the
second plot, and so on. With this method, the probability of plots 1–5 being allocated
the pesticide treatment is 1/32 (= 1/2 × 1/2 × 1/2 × 1/2 × 1/2). In contrast, the probability
of plots 11–15 being allocated the pesticide treatment (i.e. plots 1–10 being allocated the

control) is $(1/2)^{10}$ (= 1/1024). The probabilities of these different permutations are obviously not the same, and the probability of any particular permutation depends on how we number the plots!

It is clear that the two processes are not equivalent. The 'coin tossing' approach gives an invalid randomization, with different permutations having different probabilities. By contrast, the 'card shuffling' approach associates the same probability with each permutation, and therefore provides a valid randomization. This example illustrates some of the issues that must be considered when you derive a randomization scheme, and that are automatically accounted for by statistical software.

Some other examples of randomization are presented in Figures 3.1 to 3.3 and in Section 3.3. Note that if an experiment is repeated, then a new randomization should be generated each time the design is used; it is not statistically valid (or sensible) to generate a single randomization and then to use it repeatedly.

3.1.3 Blocking

It is desirable for the set of experimental units that are used to compare treatments to be reasonably uniform (homogeneous) in their natural response, as this decreases our estimate of the background variation, thus increasing precision and the potential for the experiment to detect small treatment differences. So, if the experimental units are intrinsically diverse (heterogeneous), then the experiment is likely to be insensitive. Further, as noted in Section 3.1.2, using a set of homogeneous units increases the chances of a fair comparison between treatments. However, it is not always possible to obtain a sufficient number of homogeneous experimental units for a whole experiment and, even if it is possible, it might not be desirable if it means restricting the frame of reference for the experiment. In such cases, it might be possible to identify groups of experimental units such that the units within each group are reasonably homogeneous, but with different underlying responses between groups. These groups of units can then be considered as blocks within the design. Blocking the units in this way potentially increases the precision of an experiment, as comparisons between treatments within blocks are made against a more uniform background. In this sense, blocking is said to be used for the control of variation, and for this reason is also known as **local control**.

The term block originated in agricultural experiments, where a block corresponded to a set of contiguous field plots; however, the specification of blocking can take more general forms, including the recognition of any physical structure present in the experiment. We often use the term 'block' as synonymous with 'structure'. Blocks may therefore be defined according to proximity of units in space (e.g. neighbouring plots), proximity of units in time (e.g. units measured in the same day or hour), units with similar physical characteristics (e.g. size of plant, age of insect), or logistical factors (e.g. machine, technician). Note that the number of units per block should ideally be determined by consideration of the uniformity or structure of experimental units and not by what is convenient in relation to the number of experimental treatments.

Consider the following examples of types and causes of heterogeneity among experimental units, which can be addressed by the use of blocking.

- *Field characteristics.* A slope, or fertility or pH trend across a field, or local pest problems (e.g. pigeons next to woodland) may be present. Blocks are usually formed from sets of contiguous plots that are expected to be similar in as many respects as possible. Occasionally, blocks may be formed from non-contiguous plots with similar properties, for example, soil pH, but, in such cases, the other spatial characteristics need to be reasonably homogeneous.

- *Glasshouse characteristics.* Differential shade or temperature due to positioning with respect to walls and doors are common in glasshouses. Blocks are usually formed from sets of trays or pots placed close together and hence in similar environmental conditions.

- *Time of measurement.* Some experiments may be processed over a lengthy period, and time of measurement may have a systematic effect on results. In the laboratory, there may be a limit to the number of samples that can be measured in one batch, and equipment may give slightly different readings on different days. In either case, a set of units processed within the same time period can be considered as a block.

- *Investigator.* For subjective measures, such as visual scores, individuals often perceive pre-determined scores differently. However, even in more objective situations, for example, an investigator following a standard protocol, the use of subtly different procedures can lead to systematic differences in results. If several different investigators are scoring material or carrying out a laboratory process, then it makes sense to regard each person as a block.

- *Batches of chemical, of plants, or of other organisms such as insects.* Again, if there is any possibility of (even small) differences between batches, then batches should be considered as blocks.

- *General structure.* There will often be a natural structure in experimental material. For example, trays of plants may be held on shelves within a CE cabinet. Conditions are more similar for plants within the same tray, for trays on the same shelf, and for shelves within the same cabinet (in the case of several cabinets), so all of these levels of structure should be considered as possible blocking factors.

In each of the examples above, information on the causes of heterogeneity is used to define blocks of reasonably homogeneous units and treatments can then be assigned at random to units within blocks. Note that each block might not be able to contain the full set of treatments (see Section 3.3.5), and that all blocks might not even contain the same numbers of experimental units. The randomization process needs to take account of the structure of the blocks, so that each treatment has the same probability of being applied to any unit within each block. If there are large differences between blocks, this also ensures a fairer allocation of treatments to units, as each treatment will occur within several (often all) blocks. For this reason, blocking can be seen as a set of restrictions on the randomization of treatments to the experimental units. We consider this in more detail for specific designs later (Section 3.3). Note that although blocking is generally intended to increase the precision of treatment comparisons where groups of heterogeneous units are present, the precision may decrease if too much blocking is used where there is no heterogeneity.

3.2 Forms of Experimental Structure

To successfully design, and later analyse, an experiment, it is necessary to identify all components of the experiment, i.e. both the treatments imposed and the structure of the units. In Section 1.3, we partitioned the systematic part of our mathematical model into two components: the explanatory component describes the treatments present and the structural component describes the blocking, or other structure, of the experimental units. We describe both components using factors which label the different groups present. Often, several factors are required to describe each component fully. For example, in a CE experiment where trays were placed on shelves within cabinets, we need factors to label each of the cabinets, shelves and trays to fully describe the structure. Similarly, a set of experimental treatments may be constructed from an underlying set of treatment factors. A factorial treatment structure consists of all possible experimental treatments constructed by taking one level from each of a set of treatment factors; this gives a particularly efficient form of design and is discussed further in Sections 8.2 and 8.3.

We write our model components using a symbolic notation similar to that commonly used in statistical software. To use this notation effectively, we first need to understand two different types of relationships between factors, namely nested and crossed structures.

Nested structures are used to describe hierarchical relationships. These most often occur within the structural component, but also occasionally within the explanatory component (e.g. see Section 8.4). A nested structure describes the situation where multiple units at one structural level are entirely contained within units at a higher level, and there is no direct relationship between units with the same label at the lower level. For example, consider an experiment with different treatments to be applied to four leaves (factor Leaf, with four levels) within each of 10 plants (factor Plant, with 10 levels). Leaves within plants are the experimental units, and we consider the Leaf factor to be nested within the Plant factor, written symbolically as Plant/Leaf. In this hierarchical structure, there is no association between leaves with the same label across plants, for example there is no association between leaf 1 on plant 1 and leaf 1 on plant 2. The / **(forward slash) operator** is used to indicate a nested relationship. In fact, this operator generates two separate model terms, as

$$Plant/Leaf = Plant + Plant.Leaf$$

The first term consists of the Plant factor alone, and labels each of the 10 individual plants. In the second term, the . **(dot) operator** generates all combinations of levels of the two factors, in this case labelling the 40 individual leaves. These two terms label the units within the two levels of the design.

Crossed structures occur when two factors are used to classify experimental units both independently and simultaneously. This type of structure occurs frequently within both the explanatory and structural components of the model. For example, consider a laboratory experiment to examine an extraction procedure in which three different filtering methods (factor Filter, with three levels) are tested with four different reagents (factor Reagent, with four levels), giving 12 experimental treatments in total. Both factors act simultaneously, and the crossed structure can be written as Filter*Reagent. In a crossed structure, there is an association between units with the same level of either factor. The * **(star) operator** indicates a crossed relationship and again generates several model terms, as

$$Filter*Reagent = Filter + Reagent + Filter.Reagent$$

The first two terms are the individual factors, and the third term labels all combinations of the two factors, here the 12 individual treatments. The interpretation of these terms is discussed further in Section 8.2.

In the structural component of the model, the terms generated describe different levels of the design at which variation may occur; these different levels are known as **strata**. For example, the crossed structure of a rectangular layout, Row*Column, generates three strata (Row, Column and Row.Column) and the nested structure Plant/Leaf generates two strata (Plant and Plant.Leaf).

In general, either of the model components may contain nested or crossed relationships or both. Examples 3.2 and 3.3 describe some specific situations in the context of the structural component; in Chapter 8, we consider examples in the context of the explanatory component.

EXAMPLE 3.2: NESTED STRUCTURAL FACTORS

An experiment is set up with two identical CE rooms with three trays, each containing six pots, within each room (see Figure 3.4), with the potential to allocate different treatments at both the tray and pot levels.

The CE rooms can be considered as the highest level of structure, with trays within rooms as the middle level, and pots within trays as the lowest level. This nested structure thus has multiple units at any one level (e.g. trays) completely contained within each unit at the level above (e.g. rooms). We can verify that this is a nested structure by noting that there is no association between tray 1 in room 1 and tray 1 in room 2, and similarly no association between pots with the same label in different trays. The structural factors can be denoted as Room (two levels), Tray (trays labelled within rooms, three levels) and Pot (pots labelled within trays, six levels) and we can write this as

Structural component: Room/Tray/Pot

which can be expanded as

Structural component: Room + Room.Tray + Room.Tray.Pot

The three terms label the three strata, or levels of the hierarchy: the individual rooms (term Room), the individual trays (term Room.Tray) and the individual pots (term Room.Tray.Pot). The appropriate experimental unit for the application of any treatment must then be decided as a separate exercise.

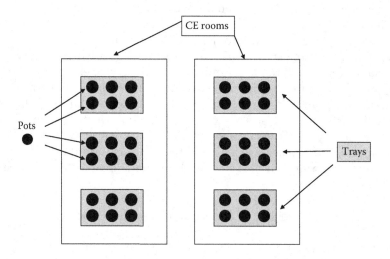

FIGURE 3.4
A nested structure with six pots per tray and three trays per CE room (Example 3.2).

EXAMPLE 3.3: CROSSED AND NESTED STRUCTURAL FACTORS

Consider an experiment where a large set of plant samples are to be processed by a machine. There are two machines that could be used (factor Machine, with two levels) and two scientists available to do the work (factor Scientist, also with two levels). We might want to allow for potential differences in results both between scientists and between machines. If we want to separate the effects of the different scientists from the effects of the two machines, then both scientists must use both machines. So an appropriate design might allocate four sets of samples to be processed by the four machine-by-scientist combinations. Each of the machine-by-scientist combinations can be considered as a block (see Figure 3.5).

Because there may be an association between samples processed either by the same machine or by the same scientist, this is a crossed relationship. The structure of the four blocks can then be written with our symbolic notation as

Machine*Scientist

which can be expanded as

Machine + Scientist + Machine.Scientist

These three terms, or strata, describe an overall effect for each machine, an overall effect for each scientist, and a combined effect for each machine-by-scientist combination. There is no association across samples processed by different machine-by-scientist combinations, and so samples can be considered to be nested within these blocks. The full structure can thus be written as

Structural component: (Machine*Scientist)/Sample

and expanded as

Structural component: Machine + Scientist + Machine.Scientist
+ Machine.Scientist.Sample

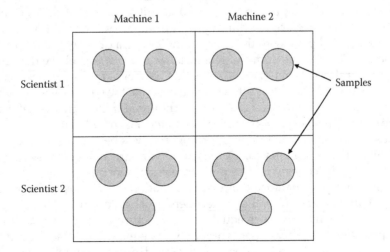

FIGURE 3.5
A crossed and nested structure with three samples within each of four machine-by-scientist combinations (Example 3.3).

So the full structure has four strata: the three described above, plus a fourth that labels the full set of individual samples. One advantage of using the crossed structure at the higher level is that it allows us to establish whether differences between scientists and machines are present, and to estimate their relative size and potential impact on the results. This information can be useful in designing future experiments although, in practice, we would require several repeats of this structure to get reliable estimates of the different sources of variation. An alternative structure for this experiment might combine (or, in statistical terminology, **confound**) the scientist and machine effects, so that each scientist uses just one of the two machines. In this case, the effects of the two scientists and the effects of the two machines would not be separable. Whether this is important depends on the information required from the experiment: this confounded blocking still achieves the main aim of separating block variation from the background variation, but does not allow us to compare the relative influence of the scientists and machines used in the process.

The symbolic notation for model formulae that we have used here was introduced by Wilkinson and Rogers (1973) and is used within GenStat. Unfortunately, conventions for model specification differ somewhat between statistical packages. For example, the R package uses the : (colon) operator where we have used the . (dot) operator. Details can be found by consulting software documentation.

To define a model fully, we need to specify in full both the explanatory and structural components. For analysis, we also need to define the response variable to be analysed. In the examples above, and throughout this book, we have included the individual units within the structural model. This is not strictly necessary, as the individual units obviously correspond to individual observations. However, we believe that retention of information on the full design structure as well as the treatment factors from an experiment is good practice. For example, unless we retain full information on the design, we cannot use all of the diagnostic procedures described in Chapter 5; we cannot plot residuals on the experimental layout if we do not know where each unit was placed. If a few pots in a corner of a glasshouse behave differently to the rest, the cause may be obvious when observations are plotted on the experimental layout but not when examined by treatment classification. Unfortunately, it is not common practice to record this information, for example, we have been unable to determine the full experimental layout for many of the examples in this book. Where this is the case, we use dummy structural factors, for example, we use factor DPot, to arbitrarily label pots, with the D prefix indicating a dummy factor. Treatment factors can sometimes be used as dummy structural factors, but we believe that this practice is confusing and prefer to avoid it. We discuss this further at the end of Section 7.3.

Finally, we mention two concepts often used to describe the structure of designs: orthogonality and balance. Two factors are said to be **orthogonal** if the estimated effects for each factor are the same regardless of whether the other term is included or not in the model. A more rigorous mathematical definition of orthogonality is beyond the scope of this book (but details can be found in Bailey, 2008). Most of the designs that we consider in this book are orthogonal, and the concept and consequences of non-orthogonality are discussed in Chapter 11. The concept of **balance** refers to information on treatment differences. In the simplest case of an unstructured set of experiment units, a design is balanced if the precision of all treatment comparisons is equal. For a structured set of units, a design is balanced if the precision of all treatment comparisons is equal within each stratum. Most of the designs that we consider in this book are balanced, and the complications introduced by unbalanced designs are discussed within Chapters 11 and 16.

3.3 Common Forms of Design for Experiments

There are many types of statistical design for experiments, which differ from one another in their complexity and in their statistical properties. In the following sections, we describe briefly some common designs illustrated using a simple treatment structure (more complex treatment structures are considered in Chapter 8).

3.3.1 The Completely Randomized Design

The **completely randomized design** (CRD) is the simplest form of design and is appropriate if the experimental units are unstructured and homogeneous, so that there is no need for any form of blocking. The random allocation of treatments to experimental units is not constrained in any way, so that each treatment is equally likely to be allocated to each unit. This is the only case in which we omit the structural component from our model, as this comprises only a single factor which indexes each observation.

> **EXAMPLE 3.4: CALCIUM POT TRIAL***
>
> An experiment was devised to investigate the effect of differences in soil calcium on the root growth of plants. The experimental material consisted of 20 pots, each containing one plant, arranged in a grid with four rows and five columns, under uniform controlled conditions. The treatments comprised four relative concentrations of calcium (A = 1, B = 5, C = 10, D = 20). Each treatment was applied to five pots selected at random to give a CRD. The layout for this design is shown in Table 3.2.

The main advantages of this design are that it is easy to set up and has a simple form of analysis. It is also flexible, as the statistical analysis is still simple if the replication varies between treatments or if data are missing for some units. The CRD also provides maximal information on the background unit-to-unit variation, as none of the between-unit information is used to assess blocking. However, this is also a weakness of the design if heterogeneity among units is present, as this heterogeneity will inflate the background variation and decrease the precision of estimates of treatment differences. The statistical analysis for this design is presented in Chapter 4.

TABLE 3.2

Randomization for the Calcium Pot Trial, with Pot Numbers (1–20), as a CRD with Four Treatments Labelled A–D, Each with Five Replicates (Example 3.4)

1	2	3	4	5
D	A	B	C	D
6	7	8	9	10
A	D	A	C	B
11	12	13	14	15
A	C	A	D	C
16	17	18	19	20
B	D	C	B	B

3.3.2 The Randomized Complete Block Design

The **randomized complete block design** (RCBD) is the simplest design that includes blocking and is probably the most frequently used design. In this design, the number of experimental units in each block must be the same as the number of treatments. Within each block, treatments are then randomly assigned to experimental units with a different randomization for each block. The design is called **complete** because all treatments occur within each block. If we use factors Block to label the blocks, and Unit to label the units within each block, then this is a nested structure with two strata, represented using our symbolic notation as

Structural component: Block/Unit

or written in expanded form as

Structural component: Block + Block.Unit

> **EXAMPLE 3.5: POTATO YIELDS***
>
> An experiment was devised to investigate the effects of four different types of fungicides (labelled F1, F2, F3, F4) on the yield of potatoes in field plots. An untreated control treatment (labelled Control) was also included to give a baseline comparison. In the field designated for the trial, heterogeneity was thought to be present at large scales, but suitable blocks of five field plots could be identified and so a RCBD with four blocks each of five plots could be used. The randomized layout is shown in Table 3.3. The structural component is written as
>
> Structural component: Block/Plot

The RCBD is popular (and useful) because it includes some blocking to deal with heterogeneity between experimental units, while still being straightforward to manage and with a simple statistical analysis. Because each treatment occurs once in each block, this design is both orthogonal (treatments are orthogonal to blocks) and balanced (all treatment comparisons are made with equal precision). Details of statistical analysis for this design are presented in Chapter 7. A weakness of the design is that the block size must be equal to the number of treatments, and so the RCBD may be inefficient if the natural block size, as determined by the experimental material, is smaller than the number of treatments. The RCBD will also be inefficient if two independent sources of background heterogeneity are present. We introduce appropriate designs for these situations (the balanced incomplete block design and the Latin square design, respectively) below.

TABLE 3.3

Randomization for the Potato Yields Trial as a RCBD with Five Treatments in Four Blocks of Five Plots (Example 3.5)

	Plot 1	Plot 2	Plot 3	Plot 4	Plot 5
Block 1	F3	Control	F2	F1	F4
Block 2	F2	Control	F3	F4	F1
Block 3	Control	F2	F3	F4	F1
Block 4	F3	F2	F1	Control	F4

3.3.3 The Latin Square Design

The **Latin square** (LS) **design** is useful where patterns of heterogeneity are associated with two crossed structural factors with the same numbers of levels. Because this design was originally used for square layouts in field trials, the structural factors are often called Row and Column, corresponding to the spatial arrangement of the rows and columns of the layout, respectively. However, these factors often correspond to non-spatial factors, such as time of day and observer. Using our symbolic notation, we write the crossed structure as

    Structural component: Row*Column

or written in expanded form as

    Structural component: Row + Column + Row.Column

This structure has three strata. The number of treatments must be equal to the numbers of rows and columns, and the treatment allocation is such that each treatment appears exactly once in each row and once in each column (see Figure 3.6a).

Construction of a Latin square is more complex than for the RCBD, as the three-way inter-relationship between rows, columns and treatments must be preserved. Tables of standard Latin squares have been published for small numbers of treatments (e.g. see Cochran and Cox, 1957, or Fisher and Yates, 1963), but statistical software can be used to obtain Latin squares for any number of treatments. To generate a randomization, one standard square of the right size is first selected at random. The order of the columns is then randomized, followed by the order of the rows (as illustrated in Figure 3.6). This randomization preserves the structure of the design while giving a very large number of possible squares, and thus avoiding bias.

EXAMPLE 3.6: LUPIN TRIAL

An experiment was devised to investigate the effects of soil type and water availability on the growth of lupins. The experiment was to be done with pots on a bench in a glasshouse, with a systematic trend running along the bench (left–right) as a result of a temperature gradient, and across the bench (up–down) because of differing light levels. The rows and columns within the array of pots were therefore considered as blocking factors with a crossed structure, and a LS design is appropriate. Four treatments, labelled CL, CH, SL and SH, representing different combinations of soil type (clay or

(a)	C1	C2	C3	C4
R1	A	B	C	D
R2	B	C	D	A
R3	C	D	A	B
R4	D	A	B	C

(b)	C4	C2	C1	C3
R1	D	B	A	C
R2	A	C	B	D
R3	B	D	C	A
R4	C	A	D	B

(c)	C4	C2	C1	C3
R2	A	C	B	D
R3	B	D	C	A
R4	C	A	D	B
R1	D	B	A	C

FIGURE 3.6
Randomization of a LS design. Rows, columns and treatments are labelled R1–R4, C1–C4 and A–D, respectively. (a) Start with a standard LS design, then (b) randomize the order of the columns and (c) finally randomize the order of the rows.

TABLE 3.4

Randomization for the Lupin Trial as a Latin Square Design with Four
Treatments Labelled CH, CL, SH and SL (Example 3.6)

	Column 1	Column 2	Column 3	Column 4
Row 1	CH	SL	CL	SH
Row 2	CL	SH	CH	SL
Row 3	SH	CH	SL	CL
Row 4	SL	CL	SH	CH

sand) and the amount of water supplied (low or high) were used. A randomized layout
for a LS design for this experiment is shown in Table 3.4. It is straightforward to verify
that each treatment can be found once in each row and once in each column and that
each row or column contains all four treatments.

The main disadvantage of the LS design is the restriction that the number of rows, columns and treatments must all be equal. This is discussed further in Section 9.1, where
some extensions of the LS design are also described.

3.3.4 The Split-Plot Design

The **split-plot** (SP) **design** has a nested structure, and is used in the case where (at least)
two treatment factors are present, with the levels of one treatment factor having to be
applied to large experimental units while the levels of another treatment factor can be
applied to smaller units. Here we consider a standard form of the SP design with two treatment factors, A and B, with a crossed structure, and a nested structural component with
three strata. The highest level of structure corresponds to complete replicates of the set of
treatments, and we denote this level using the factor Block. Each block is then divided into
a number of whole plots (factor WPlot), with levels of treatment factor A randomized to the
whole plots separately within each block. Finally, each whole plot is divided into a number
of subplots (factor Subplot), and the levels of factor B are randomized onto subplots within
each whole plot. This design can be represented in symbolic form as

Explanatory component: A*B
Structural component: Block/WPlot/Subplot

EXAMPLE 3.7: WEED COMPETITION EXPERIMENT

A field trial was set up to study the competitive effects of three different weed species
in winter wheat under different levels of water stress. Variation in water stress was
provided by the presence or absence of irrigation, which could be applied only to large
areas of land whereas the weed species could be applied to small plots. A SP design was
therefore deemed suitable, with the two irrigation treatments (factor Irrigation, with two
levels) applied to whole plots (factor WholePlot, with two levels). Each whole plot was
split into four subplots (factor Subplot, with four levels), and a pre-determined population of each weed species (*Alopecurus myosuroides* (black-grass), *Galium aparine* (cleavers)
and *Stellaria media* (chickweed), abbreviated as Am, Ga and Sm, respectively) was sown
in one of these four subplots. The remaining subplot within each whole plot had no
weed seeds added, and it was used as a control. Factor Species was used to label the

TABLE 3.5

Randomization for the Weed Competition Experiment as a Split-Plot Design with Two Whole-Plot Treatments (Irrigated, Highlighted in Grey, and Non-Irrigated) and Four Subplot Treatments (Weed Species Am, Sm, Ga or Control, –) (Example 3.7)

	Block 1		Block 2		Block 3		Block 4	
Whole plot 1	1	2	1	2	1	2	1	2
	–	Am	–	Ga	Sm	Ga	Am	–
	3	4	3	4	3	4	3	4
	Sm	Ga	Sm	Am	–	Am	Ga	Sm
Whole plot 2	1	2	1	2	1	2	1	2
	–	Sm	Am	Ga	Am	Sm	Ga	Sm
	3	4	3	4	3	4	3	4
	Ga	Am	–	Sm	Ga	–	Am	–

four weed treatments, i.e. the three added populations and control. This structure was repeated another three times, giving four blocks (factor **Block**, with four levels), with a different randomization in each block, as shown in Table 3.5. The model for this design can be written in symbolic form as

Explanatory component:	Irrigation*Species
Structural component:	Block/WholePlot/SubPlot

The statistical analysis for, drawbacks of and variations on this design are discussed in Section 9.2.

3.3.5 The Balanced Incomplete Block Design

The **balanced incomplete block design** (BIBD) can be useful when there is only one blocking factor but the number of units per block is smaller than the number of treatments. In this case, each block can contain only a subset of the treatments, and designs with this property are known as **incomplete block designs**. A BIBD has the additional characteristic of balance, which requires that all treatment comparisons have equal precision. This is achieved if the treatments have equal replication and each pair of treatments occurs together within a block exactly the same number of times over the whole experiment. If we again use factor **Block** to label the blocks, and factor **Unit** to label the units within blocks, then this design has the same nested blocking structure as the RCBD, represented as

Structural component: Block/Unit

Construction of a BIBD is more complex than for the RCBD, as the balanced inter-relationship between blocks and treatments must be preserved. Tables of standard BIBDs have been published (e.g. see Cochran and Cox, 1957, or Fisher and Yates, 1963) and can be used to generate a BIBD. These designs are also available in many statistical packages. The first step in construction is to choose a standard design with the right block size and number of treatments. The standard layout is then randomized first by randomization of the order of the blocks, and then randomization of the order of the treatments present within each block.

TABLE 3.6

Randomization for Grain Protein Content Experiment as a BIBD with Six Treatments
Labelled A–F in Six Sessions (Blocks), Each Containing Five Samples (Units) (Example 3.8)

	Sample 1	Sample 2	Sample 3	Sample 4	Sample 5
Session 1	C	F	E	B	A
Session 2	C	E	F	D	A
Session 3	E	D	F	C	B
Session 4	A	C	B	E	D
Session 5	D	B	A	C	F
Session 6	E	F	A	D	B

EXAMPLE 3.8: GRAIN PROTEIN CONTENT *

An experiment was devised to evaluate the grain protein content for six different varieties of pea (labelled A, B, C, D, E and F). Five independent samples of grain were available for each variety. Only five samples (factor Sample, five levels) could be assessed within a session (factor Session, six levels), with possible heterogeneity between sessions, so a BIBD with six blocks (corresponding to sessions), each containing five units (corresponding to samples) was used. The structural component was specified as

Structural component: Session/Sample

A randomized plan for this design is shown in Table 3.6. For this design each variety is replicated five times, as each appears in five of the six sessions, and any pair of varieties is present together in four sessions, for example, varieties E and F are both present in sessions 1, 2, 3 and 6.

One drawback of BIBDs is that the range of available designs is fairly limited: it is not always possible to construct a BIBD for a given number of treatments, number of blocks and block size (number of units per block). More details are given in Section 9.3.

3.3.6 Generating a Randomized Design

Once a design has been chosen, and the numbers of treatments and replicates have been defined, then a randomized layout or plan for the experiment can be generated. Most general statistical software (including GenStat, R and SAS) have some facilities for generating standard designs, including most of those considered in this book. For non-standard designs (including some BIBDs), more specialist software, such as CycDesigN (see http:/www.vsni.co.uk/software/cycdesign) must be used.

EXERCISES

3.1 Suppose that you are planning an experiment to investigate the impact of nutrient deprivation on plant metabolites. You have four different nutrient levels to test, obtained by applying appropriate nutrients to four sub-samples from a single bag of base compost. The resulting volume of each nutrient level is sufficient for four seed trays (i.e. 16 seed trays in total), and six plants will be grown in each seed tray. To achieve the required growing conditions, a small CE cabinet will be used. The cabinet has four shelves, and you can fit four seed trays on each shelf in a 2×2

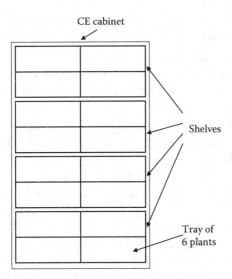

CE cabinet

Shelves

Tray of
6 plants

FIGURE 3.7
Structure of a CE cabinet to be used for an experiment to investigate the impact of nutrient deprivation on plant metabolites (Exercise 3.1).

arrangement (Figure 3.7). Although the cabinet is supposed to provide a uniform environment, a technician suggests that light levels may vary between the shelves, and that this might affect plant growth. When they reach the required growth stage, the six plants from each seed tray will be bulked and processed together to form a single sample to be read by a machine. Your colleague tells you that the machine shows some drift over time, but that readings should stay stable across a set of up to six samples.

How might you design this experiment to obtain an unbiased assessment of differences between the four nutrient levels? Consider and discuss the different factors which might affect your choice of design and produce a candidate design. You should consider both stages of the experiment and the following issues:

- What is the experimental unit for the nutrient treatments?
- What are the sources of heterogeneity in the experimental process?
- How might you deal with this heterogeneity?
- How would you allocate the treatments to the experimental units?
- What replication do you have for each treatment?
- What are the advantages/disadvantages of your design?

How would you modify your design if

a. A temperature gradient was discovered between the front and back of the shelves

b. You want to include a CO_2 treatment that can only be applied to a whole CE cabinet and you obtain sufficient resources for 32 trays (eight for each nutrient level)

3.2 Identify the experimental unit, the replication for each treatment and whether pseudo-replication is present in the following experiments.

    a. A pot experiment with 12 circular pots in a 2×6 array, in a uniform environment. Each pot contains four plants at the three-leaf stage, and each of four treatments (labelled A–D) were applied at random to one plant per pot as shown in Figure 3.8.

    b. A field experiment with 12 homogeneous rectangular plots in a 3×4 grid. Two treatments (labelled A and B) were applied at random to six plots each (Figure 3.9). At harvest, 25 plants are to be sampled per plot, and the plants from each plot will be processed as a single batch for measurement.

    c. The field experiment described in part (b) (Figure 3.9) with the height of 25 individual plants per plot measured and recorded *in situ* at 4-weekly intervals from tillering until harvest.

3.3 Four replicates of each of four treatments, labelled A–D, are to be applied at random to batches of aphids in 16 Petri dishes laid out in a 2×8 array (Figure 3.10). The environment is thought to be homogeneous. Use a pack of playing cards to determine an appropriate randomization for this experiment.

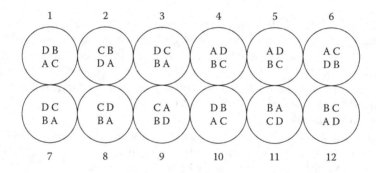

FIGURE 3.8
Experimental layout for a pot experiment with 12 pots and four treatments, labelled A–D, applied to plants within pots. Letters denote the positions of the plants and the treatment applied (Exercise 3.2a).

B	A	B	A
A	B	A	A
B	A	B	B

FIGURE 3.9
Experimental layout for a field experiment with plots in a 3×4 grid, showing the allocation of treatments (A or B) to plots (Exercises 3.2b and c).

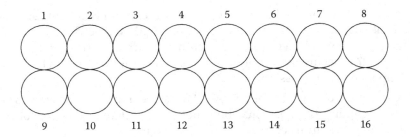

FIGURE 3.10
Layout of 16 numbered Petri dishes for an aphid experiment using four replicates of four treatments (Exercise 3.3).

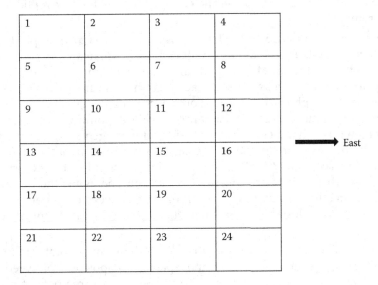

FIGURE 3.11
Layout (with numbered plots) for an experiment testing six treatments in a field with a pH gradient running from west to east (Exercise 3.4).

3.4 Four novel herbicides (labelled A–D) are to be compared with a commercial product (labelled P) and a hand-weeded control (labelled H) in a field trial, giving six treatments in total. The field available can accommodate 24 plots in an array of four columns running north to south, by six rows running west to east (Figure 3.11). The field has a known pH gradient running west to east, i.e. along rows, which may affect crop growth. Produce a RCBD which accounts for this gradient with a randomized allocation of treatments to plots using (a) a pack of playing cards, and (b) a standard six-sided die.

3.5 The efficacy of six synthetic insect pheromones is to be tested in the field. Traps are baited with a single pheromone, deployed at dusk, left out overnight, then retrieved the next morning and the insect catches recorded. There is sufficient material to bait six traps with each pheromone.

a. Consider how you might use a RCBD for this experiment if only six traps are available at any one time and all six traps will be placed in the same field, but a different field will be used each night. Are any structural factors confounded? What are the assumptions of this design? Write down the structural component for this design.

b. How would you change your design if the same trap locations were to be used each night and the positions could not be considered homogeneous? Write down the structural component for this design.

c. How might you modify your design if 18 traps are available at any one time? Under which conditions would designs based on CRD, RCBD or LS arrangements be preferable?

d. What design might you use if only four pheromones are to be tested, with six traps available at any one time, and enough material for nine replicates of each pheromone?

3.6 The effect of temperature on the transmission of a virus by five aphid species is to be investigated. Three small growth chambers are available and three temperatures will be tested. The temperature for each chamber can be set and then applies to the whole chamber, and each chamber can hold five plants in individual pots. One aphid will be placed onto each plant using a clip cage. Forty-five plants and 15 aphids of each species are available. Assuming that chambers (and positions within chambers) can be considered homogeneous, suggest a design to test the effects of temperature and aphid species. What are the experimental units for each factor? Produce a randomized design for this experiment and write down the explanatory and structural components for the design. If you suspected that there were systematic differences between chambers, how would you modify your design? Write down the structural component for this new design.

3.7 A field experiment was set up to investigate how invertebrate abundance is affected by the spatial structure and species composition of weed patches (Smith, 2007). Small weed patches were formed from three pots of plants in a tray. Species composition was varied by using different numbers of mayweed (M) or thistle (T) plants in the patch, i.e. 3M, 2M + 1T, 1M + 2T or 3T. Spatial structure was varied by changing the distance between patches (12 or 6 m). Five blocks of two whole plots were set up, with the two spacings allocated at random to whole plots within blocks. Each whole plot contained 16 patches laid out in a 4 × 4 array with the designated spacing, with patches allocated to four replicates of each of the four species compositions according to a LS design. A different randomization was used within each whole plot. Write down the explanatory and structural components for this design.

3.8 A glasshouse experiment to compare the effect of two nutrition regimes on the growth of three wheat varieties was set up as a RCBD with 12 blocks of six pots each, as shown in Figure 3.12. The treatments comprise the six combinations of nutrition regime (labelled N1, N2) and variety (labelled V1–V3). The blocks accommodate an expected temperature gradient running from the door to the

1	2	3	4	5	6 N2 V2	7	8	9	10	11	12
13	14	15	16	17	18 N1 V2	19	20	21	22	23	24
25	26	27	28	29	30 N2 V3	31	32	33	34	35	36
37	38	39	40	41	42 N1 V1	43	44	45	46	47	48
49	50	51	52	53	54 N1 V3	55	56	57	58	59	60
61	62	63	64	65	66 N2 V1	67	68	69	70	71	72

Far end (warmer)   Door (cooler)

FIGURE 3.12

Layout of pots (labelled 1–72) as a RCBD in a greenhouse experiment to compare the effects of two nutrition regimes (N1 and N2) on the growth of three wheat varieties (V1, V2 and V3). Blocks (columns) contain six pots. One block shows treatment labels in addition to pot numbers (Exercise 3.8).

far end of the glasshouse. Several characteristics of each plant, including height and number of leaves, are to be recorded every week. Suggest acceptable protocols for recording data if

a. You are the only person available to take the measurements

b. There are two people available to take the measurements

Which protocols would be unacceptable and why?

4

Models for a Single Factor

In this chapter, we present the analysis for data classified by a single explanatory factor. In the context of designed experiments, this would correspond to the case of a completely randomized design (Section 3.3.1) with a single treatment factor. Equivalently, this type of data might arise from an observational study in which the observations have been selected to conform to a single pre-defined classification, or grouping variable. In both cases, the only structure in the data is the treatment or grouping factor; there must be no other explanatory variables and no other structure, such as blocking or pseudo-replication, associated with the experimental material. If any such structure is present, then you should use a more complex analysis (see Chapters 7, 9 and 16 for details). In the case where only a single factor – representing treatments or groups – is present, the aim of the analysis is to discover if there are any differences in response between the factor levels. For brevity, here we use the term 'treatments' to cover either a set of imposed treatments or a set of observed groups. The first step in the analysis is to write down a model for the data in terms of the unknown population mean for each treatment (Section 4.1). The principle of least squares is used to estimate these treatment means (Section 4.2). The technique of ANOVA is then used to partition the variation in the data (Section 4.3). This analysis serves several purposes: we can obtain an estimate of the background variation, which in turn is used to indicate uncertainty on estimates of treatment means; we can also obtain an estimate of the amount of variation in the data accounted for by treatment differences, and compare this with the background variation. If the variation between treatments is large compared with the background variation, then we conclude that substantive differences between treatments are present in the data. This comparison is formalized in an F-test, and differences between pairs of treatment means can be compared with the standard error of the difference (SED) to identify significant differences between responses for different treatments (Section 4.4). There are several forms (parameterizations) of the ANOVA model for a single treatment factor, and we explain some of the different forms used in statistical software (Section 4.5).

4.1 Defining the Model

Here, we consider a set of observations classified by a single treatment factor. It is natural to label observations by their factor level, i.e. the treatment group to which they belong, and then to number observations within each treatment. We use a general notation that can be adapted to apply to any data set. The treatments are labelled by index j, and the number of treatments is denoted as t. We label observations within treatments using index k, allow the replication to differ between treatments, and denote the number of observations for the jth treatment as n_j. Then, y_{jk} represents the response from the kth observation on the jth treatment and the full set of responses can be denoted as y_{jk}, $j = 1 \ldots t$, $k = 1 \ldots n_j$ (see Section 2.1 for more explanation of this notation). The total number of observations is the sum of the number of replicates across all the treatments, denoted $N = n_1 + n_2 + \ldots + n_t$ or $N = \sum_{j=1}^{t} n_j$.

The only structure associated with the observations is due to the treatment groups, but there will also be variation among the responses within each treatment. We can describe this structure using a simple model, as

$$y_{jk} = \mu_j + e_{jk} , \tag{4.1}$$

where μ_j is the true (but unknown) population mean for the jth treatment, and e_{jk} is the deviation of the kth response on the jth treatment from its population mean. The set of t unknown population means, μ_j, $j = 1 \ldots t$, are the parameters of this model. Equation 4.1 essentially says that each observation consists of two parts, a contribution due to the treatment and a deviation due to the individual. These two parts correspond to the systematic and random components, respectively, of the general model described in Section 1.3. An example of this situation is shown in Figure 1.1a. The individual deviation is sometimes called random noise, residual error or measurement error. The term 'error' here just reflects the presence of variation, and hence uncertainty in ascertaining true population values: it is not intended to imply that a mistake has occurred. Hence, we prefer the alternative term **deviation**. In general, the deviation reflects natural between-unit biological variation, variation within the study environment and inaccuracies in measurement. The deviations are regarded as random, without any structure related to the experimental units and not under control of the experimenter.

Note that the labelling used here for the observations, subscripts j and k, is chosen to identify the observations associated with each treatment. Other labelling schemes, such as use of subscript i to number the units in the order of the experimental layout, are equally valid, and are sometimes preferable. For example, ordering by layout is required to check for spatial trends. We strongly recommend recording information on the full experimental layout within any data set so that the link between the treatments and units is retained.

Using the symbolic notation introduced in Chapter 3, we can represent the model in Equation 4.1 as

Response variable: Y

Explanatory component: Treatment

Here, we have a variate named Y containing the observed response and a factor called Treatment that identifies the treatment group from which each observation arises. Note that we use italic font to distinguish variate names from factor names. In this case, the explanatory component is represented by a single factor and there is no structural component.

To make inferences on the unknown parameters in Equation 4.1 (and in any linear model), we make some assumptions about the deviations. These assumptions are given here for the general case, so, for simplicity, we replace the subscript jk by the subscript i so that e_i is the deviation corresponding to the ith unit (for $i = 1 \ldots N$). In estimating the unknown parameters, we assume that e_i is a realization of a random variable with the properties

Assumption 1

$$E(e_i) = 0 \quad \text{for } i = 1 \ldots N .$$

The expected value (function E) of each deviation is assumed to be zero. This means that the population mean of the deviations is zero, which implies no systematic bias in the observations. ■

Assumption 2

$$\text{Var}(e_i) = \sigma^2 \quad \text{for } i = 1 \ldots N .$$

The variances (Var) of the deviations are the same for all units. This is also known as homoscedasticity, or homogeneity of variances. ■

Assumption 3

$$\text{Cov}(e_i, e_j) = 0 \quad \text{for all } i \neq j, \quad \text{and} \quad i, j = 1 \ldots N .$$

The covariance (Cov) between deviations for two separate observations is zero and the deviations are independent. ■

Assumption 4

$$e_i \sim \text{Normal}(0, \sigma^2) \quad \text{for } i = 1 \ldots N .$$

The deviations follow a Normal distribution with mean 0 and variance σ^2. ■

In addition, we make an assumption on the explanatory variables:

Assumption 5

The values of the explanatory variables (factors or variates) are known without error. ■

The first three assumptions require that the deviations are independent and identically distributed, i.e. arise from the same underlying probability distribution. The fourth assumption adds the requirement that this is the Normal distribution, and this is necessary to make valid statistical inferences that rely on the properties of the Normal distribution. This includes significance testing (F-test or t-test, see Sections 4.3 and 4.4) and the calculation of confidence intervals (CIs) (Section 4.4). In Chapters 5, 6 and 18, common violations of Assumptions 1–4 are discussed in detail. For the case of data with a single explanatory factor, Assumption 5 requires that each observation can be allocated to a treatment group without error, which is usually a realistic requirement. In general, when explanatory variates have been measured, possibly with error, then the assumption may become unrealistic. Consequences of violating this assumption are discussed in Chapter 13.

In some areas of biological science, it is common practice to summarize treatment responses from an experiment graphically, with bar charts showing the sample means and standard deviations (SDs) for each treatment, as in Example 4.1A. This can be a useful precursor to a formal statistical analysis as it provides an informal check as to whether the observations comply with Assumption 2, i.e. that all random deviations share a common variance. If this is the case, then the sample SDs should be roughly equal across treatments. In practice, for little or moderate replication, there may appear to be differences in sample SDs across treatments even when the assumption is true.

EXAMPLE 4.1A: CALCIUM POT TRIAL*

An experiment devised to evaluate the effect of four relative concentrations (levels) of calcium (A = 1, B = 5, C = 10, D = 20) on root growth was introduced in Example 3.4. Each treatment was applied to five individual plants growing in pots. The experiment used a CRD, as shown in Table 3.2, and measurements of total root length (cm) were made on pots 1–20 (in order) at the end of the experiment. The data set is presented in summary form in Table 4.1, and can be found in file CALCIUM.DAT in the flat file format required by statistical software. This format puts the explanatory variables and responses into parallel columns. Here, the file contains three columns: a variate (*Pot*) to uniquely identify each pot; a factor (*Calcium*, with four levels) to identify the treatment group for each pot; and the response variate (*Length*) obtained from each pot.

The model for these data can be written in symbolic notation as

Response variable: *Length*
Explanatory component: Calcium

We also use an informal version of the mathematical model given in Equation 4.1, and write this model in a more interpretable form that is relevant to the data, as

$$Length_{jk} = Calcium_j + e_{jk} ,$$

where $Length_{jk}$ represents the root length for the kth plant in the jth treatment group (with $j = 1 \ldots 4$ corresponding to A–D respectively), $Calcium_j$ represents the mean of the jth treatment group and e_{jk} is the deviation for the kth plant in the jth treatment group.

Before a formal analysis, it can be helpful to use sample statistics, such as treatment means and SDs, to summarize the data. These values are given in Table 4.1 and also

TABLE 4.1

Calcium Pot Trial Data: Total Root Length (cm) and Summary Statistics for Plants Treated with Four Relative Concentrations of Calcium (A–D) according to a CRD (Example 4.1A and File CALCIUM.DAT)

Replicate	A	B	C	D
1	58	80	49	47
2	52	68	70	49
3	74	72	72	45
4	58	74	74	48
5	79	85	71	38
Treatment mean	64.2	75.8	67.2	45.4
Treatment variance	135.20	45.20	105.70	19.30
Treatment SD	11.63	6.72	10.28	4.39

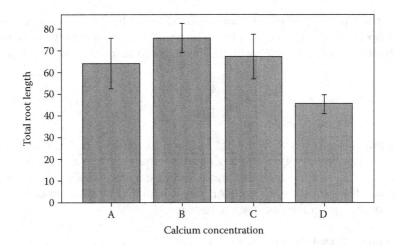

FIGURE 4.1
Summary statistics: treatment means and unbiased sample standard deviations (SD) for a CRD measuring root lengths (cm) under four calcium concentrations (A, B, C, D), each with five replicates (Example 4.1A). Vertical bars represent $\pm 1 \times$ SD for each treatment.

shown graphically in a bar chart in Figure 4.1. The sample grand mean, $\bar{y}$, is equal to 63.15 cm. Root growth appears to be greater for calcium levels B and C and the SDs are broadly similar across treatments A–C, but appear smaller for calcium level D. This plot is exploratory and we discuss better ways to present experimental results in Section 4.4.

4.2 Estimating the Model Parameters

The parameters associated with the model in Equation 4.1, namely the population means for each treatment, μ_j, $j = 1 \ldots t$, are unknown quantities to be estimated from the data. Recall (from Section 1.5) that we denoted parameter estimates by placing a hat symbol (^) above the parameter symbol. The estimated population mean for the jth treatment is thus denoted $\hat{\mu}_j$. Recall also that the fitted value for an observation y_{jk}, denoted $\hat{y}_{jk}$, consists of the systematic component of the model with parameters replaced by their estimates. Hence, here

$$\hat{y}_{jk} = \hat{\mu}_j \ .$$

The parameters are estimated by the principle of least squares, which finds the values of the parameters that minimize the sum of the squares of the differences between the observed data and their fitted values, called the residual sum of squares (ResSS), which can be written mathematically as

$$\text{ResSS} = \sum_{j=1}^{t} \sum_{k=1}^{n_j} \left(y_{jk} - \hat{y}_{jk} \right)^2 = \sum_{j=1}^{t} \sum_{k=1}^{n_j} \left(y_{jk} - \hat{\mu}_j \right)^2 . \tag{4.2}$$

The details of the minimization are not required to understand the principles, but are shown in Section C.1 for interested readers. The resulting estimate of the population mean for the jth treatment is the sample mean of the observed responses for that treatment, i.e.

$$\hat{\mu}_j = \frac{1}{n_j}\sum_{k=1}^{n_j} y_{jk} = \bar{y}_{j\cdot}\,.$$

This sample mean is written with the dot notation introduced in Section 2.1. Recall that the dot symbol, '$\cdot$', in the position of index k indicates summation across all values of that index, i.e. for $k = 1 \ldots n_j$, with the other index (or, in general, indices) held fixed. The bar symbol, '$-$', over the y indicates that the mean is calculated by division of the sum by the number of components in the summation, here n_j.

EXAMPLE 4.1B: CALCIUM POT TRIAL*

From Table 4.1, we can now estimate the population means for the four calcium treatments as

$$\overline{Calcium_1} = 64.2; \quad \overline{Calcium_2} = 75.8; \quad \overline{Calcium_3} = 67.2; \quad \overline{Calcium_4} = 45.4\,.$$

Having obtained the treatment sample means as estimates of the treatment population means, we require some measure of the uncertainty in these estimates to form standard errors and CIs. It is also helpful to be able to formally test whether the data show evidence of substantive differences among the treatment population means. The ANOVA provides a framework for estimation of the within- and between-treatment variances, and formal tests of differences among treatment means. Given that the assumption of a common variance for the set of deviations (denoted σ^2) is realistic, a pooled estimate of this variance (denoted s^2) is obtained from ANOVA of the data, and can be used to derive standard errors and CIs for single treatments or treatment comparisons.

4.3 Summarizing the Importance of Model Terms

For a set of observations classified by a single treatment factor, the primary question of interest is whether there are any differences in the responses among treatments. **Analysis of variance** (ANOVA) is a statistical technique that enables us to address this question directly. Intuitively, it is easy to see that if treatment differences exist, then variation in responses among observations for different treatments ('between treatments') will be larger than the variation in responses among observations for the same treatment ('within treatments'), with the additional variation being directly attributable to the treatment differences. ANOVA uses this principle to partition the total variation for a given response into the variation attributable to treatment differences and the variation due to random deviations (which we refer to as background variation).

In the general case, given any model for a set of observations in the form described in Section 1.3, i.e.

response = systematic component + random component,

ANOVA considers the variation in the response associated with all the parts of the systematic component of the model (systematic variation) and compares it to the background variation associated with the random component of the model. Informally, if the ratio of systematic variation to background variation is large, then we can conclude that the proposed model accounts for much of the variation in the response, and that the explanatory variable(s) provide a good explanation of the observed response. Formally, ANOVA can be used to test various hypotheses about the form of the systematic component, from the simplest case of a single treatment factor, presented in this chapter, to the more complex explanatory and structural components described in Chapters 7 to 9.

The simplest application of ANOVA, to data classified by a single factor, is usually referred to as **one-way analysis of variance**. This analysis can be regarded as an extension of the two-sample t-test to allow comparison of several treatments simultaneously (with exact equivalence of these methods in the case of only two treatments). Recall that the two-sample t-test (Section 2.4.2) is used to compare the observed difference between two sample means with the expected variation in that difference, based on a pooled estimate of the background variation. The t-test was used to evaluate a null hypothesis of equality of the population means for two treatments against an alternative hypothesis that the population means were not equal. In the case where there are several (t) treatments, it is also of interest to establish whether there is any evidence of differences among population means for the treatments. The null hypothesis is that the treatment population means are all equal. Mathematically, this is written as

$$H_0: \mu_1 = \mu_2 = \ldots = \mu_t .$$

Given the assumptions underlying the analysis (presented in Section 4.1), if this hypothesis is true, then it implies that the observations for all treatments arise from a single Normal distribution with common mean, μ (e.g. Figure 4.2a). The general alternative hypothesis, H_1, is that the treatment population means are not all equal. Taken in combination with the model assumptions, this hypothesis implies that the observations arise from a set of Normal distributions with a common variance, but with some variation within the set of population means (e.g. Figure 4.2b).

The test of this null hypothesis compares the variation among treatments with the background variation. In common with the two-sample t-test of Section 2.4.2, this test uses an

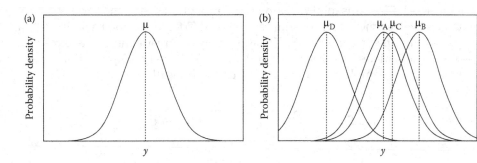

FIGURE 4.2
(a) The assumed distribution of responses for a single factor model under the null hypothesis (treatment means equal), and (b) the assumed distributions for a single factor model with four groups (A, B, C, D) in a case where the null hypothesis is not true (treatment means not equal).

estimate of the background, or unit-to-unit, variability that is pooled across all treatment groups, in accordance with the assumption of common variance underlying the analysis.

In the following sections, we give details of the ANOVA calculations for a model with a single treatment factor to demonstrate the basic principles of the approach. Nowadays, statistical packages are usually used for these calculations and so it is not necessary to give detailed formulae for all cases. We present full details for the RCBD (Chapter 7), but ANOVA for more complex structures is discussed in less mathematical detail in Chapters 8 and 9.

The calculations in this section are presented for the general case of unequal replication across treatments, but we also give the simpler formulae used for the case of equal replication (i.e. $n_j = n$ for $j = 1 \ldots t$).

4.3.1 Calculating Sums of Squares

As stated above, the aim of ANOVA is to partition the variation in the response between two or more sources. The statistics used to quantify variation are initially calculated as sums of squared deviations about means, and hence referred to as **sums of squares**. The total sum of squares (TotSS) is related to the sample variance, and we calculate it by taking the difference between each observed value and the sample grand mean, squaring these differences and then adding them together. Algebraically, this is written as

$$\text{TotSS} = \sum_{j=1}^{t} \sum_{k=1}^{n_j} \left(y_{jk} - \bar{y} \right)^2 . \tag{4.3}$$

Note that this is equal to $(N-1) \times$ the unbiased sample variance (Equation 2.3). With equal replication, the expression becomes

$$\text{TotSS} = \sum_{j=1}^{t} \sum_{k=1}^{n} \left(y_{jk} - \bar{y} \right)^2 .$$

This calculation is illustrated in Figure 4.3, for the case of two treatments ($t = 2$) each with four replicates ($n = 4$). The lengths of the vertical lines represent the differences to be squared and then added together.

For a one-way ANOVA, this total variation is then partitioned into the variation between treatments (or the treatment sum of squares, denoted TrtSS), and the background variation, which is quantified by the ResSS. Variation between treatments is calculated as the sum, over all observations, of the squared differences between the appropriate treatment sample mean and the sample grand mean, written algebraically as

$$\text{TrtSS} = \sum_{j=1}^{t} \sum_{k=1}^{n_j} \left(\bar{y}_{j.} - \bar{y} \right)^2 = \sum_{j=1}^{t} n_j \left(\bar{y}_{j.} - \bar{y} \right)^2 .$$

As the contributions from observations within the same treatment group are repeated, the expression can be simplified (as shown) to be a sum across index j (the different treatment groups) multiplied by the replication (n_j) for each treatment. This calculation is illustrated in Figure 4.4, where again the lengths of the vertical lines represent the

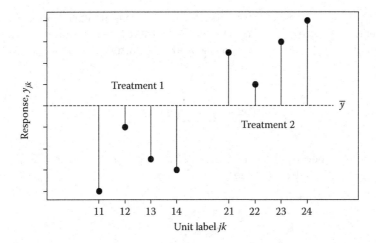

FIGURE 4.3
Calculation of the total sum of squares (TotSS) for a single factor model with two treatment groups ($j = 1, 2$) and four replicates per group ($k = 1 \ldots 4$). Each vertical line represents a difference between a response and the sample grand mean, or $y_{jk} - \bar{y}$.

differences to be squared and then added together. With equal replication of all treatments, this becomes

$$\text{TrtSS} = n \sum_{j=1}^{t} \left(\bar{y}_{j\cdot} - \bar{y} \right)^2.$$

The final step is the calculation of the ResSS. Within each treatment, background variation is represented by the variation of each response about its treatment population mean.

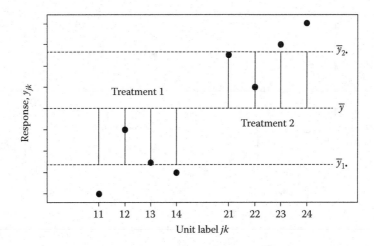

FIGURE 4.4
Calculation of the treatment sum of squares (TrtSS) for a single factor model with two treatment groups ($j = 1, 2$) and four replicates per group ($k = 1 \ldots 4$). Each vertical line represents a difference between a group sample mean and the sample grand mean, or $\bar{y}_{j\cdot} - \bar{y}$.

The treatment population means are unknown, so we assess variation around their estimates, the treatment sample means. The ResSS is therefore calculated as the sum of the squared differences between each response and its treatment sample mean, written algebraically as

$$\text{ResSS} = \sum_{j=1}^{t} \sum_{k=1}^{n_j} \left(y_{jk} - \bar{y}_{j\cdot} \right)^2 .$$

This can also be obtained by the substitution of $\bar{y}_{j\cdot} = \hat{\mu}_j$ into Equation 4.2. With equal replication of all treatments, this becomes

$$\text{ResSS} = \sum_{j=1}^{t} \sum_{k=1}^{n} \left(y_{jk} - \bar{y}_{j\cdot} \right)^2 .$$

This calculation is illustrated in Figure 4.5, where again the lengths of the vertical lines represent the differences to be squared and then added together.

The name 'residual sum of squares' arises from a connection with the model residuals, $\hat{e}_{jk}$, defined as the discrepancy between the data and the fitted systematic component as

$$\hat{e}_{jk} = y_{jk} - \hat{y}_{jk} = y_{jk} - \hat{\mu}_j = y_{jk} - \bar{y}_{j\cdot} . \tag{4.4}$$

It is immediately clear that the ResSS is simply equal to the sum of the squared residuals, i.e.

$$\text{ResSS} = \sum_{j=1}^{t} \sum_{k=1}^{n_j} \left(y_{jk} - \bar{y}_{j\cdot} \right)^2 = \sum_{j=1}^{t} \sum_{k=1}^{n_j} \hat{e}_{jk}^2 .$$

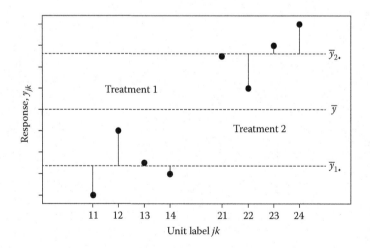

FIGURE 4.5
Calculation of the residual sum of squares (ResSS) for a single factor model with two treatment groups ($j = 1, 2$) and four replicates per group ($k = 1 \ldots 4$). Each vertical line represents a difference between a response and its group mean, or $y_{jk} - \bar{y}_{j\cdot}$.

ANOVA produces an additive partition of the total variation such that the total sum of squares is equal to the sum of the treatment and residual sums of squares, i.e.

$$\text{TotSS} = \text{TrtSS} + \text{ResSS} . \tag{4.5}$$

The relationship in Equation 4.5 means that given any two out of the total, treatment or residual sums of squares, one can calculate the third quantity directly. For example, the residual sum of squares is equal to the total sum of squares minus the treatment sum of squares, i.e. $\text{ResSS} = \text{TotSS} - \text{TrtSS}$. It is straightforward (but fiddly) to verify this relationship algebraically, i.e. by a rearrangement of the formula, and this is shown for the interested reader in Section C.2.

EXAMPLE 4.1C: CALCIUM POT TRIAL*

Calculation of sums of squares by hand is often helped by the use of structured tables. For example, to aid in the calculation of the TotSS, it is useful to draw up a table like Table 4.2. Note that the treatments here correspond to different levels of the factor Calcium.

The first three columns list the treatment groups and label the units by treatment (j) and replicate within treatment (k). The fourth column lists the responses (root lengths) and the fifth column takes the difference between the responses and the sample grand mean ($\bar{y} = 63.15$). The sixth column holds the squares of the differences from the fifth column and the total sum of squares can be obtained as the sum of the values in this final column, and here is equal to 3684.55.

TABLE 4.2

Calculation of Total Sum of Squares for Root Lengths from the Calcium Pot Trial (Example 4.1C)

Calcium	j	k	y_{jk}	$y_{jk} - \bar{y}$	$(y_{jk} - \bar{y})^2$
A	1	1	58	−5.15	26.5225
A	1	2	52	−11.15	124.3225
A	1	3	74	10.85	117.7225
A	1	4	58	−5.15	26.5225
A	1	5	79	15.85	251.2225
B	2	1	80	16.85	283.9225
B	2	2	68	4.85	23.5225
B	2	3	72	8.85	78.3225
B	2	4	74	10.85	117.7225
B	2	5	85	21.85	477.4225
C	3	1	49	−14.15	200.2225
C	3	2	70	6.85	46.9225
C	3	3	72	8.85	78.3225
C	3	4	74	10.85	117.7225
C	3	5	71	7.85	61.6225
D	4	1	47	−16.15	260.8225
D	4	2	49	−14.15	200.2225
D	4	3	45	−18.15	329.4225
D	4	4	48	−15.15	229.5225
D	4	5	38	−25.15	632.5225
Total	−	−	−	0.00	3684.5500

TABLE 4.3

Calculation of Treatment Sum of Squares for Root Lengths from the Calcium Pot Trial
(Example 4.1C)

Calcium	j	n_j	$\bar{y}_{j\cdot}$	$\bar{y}_{j\cdot} - \bar{y}$	$(\bar{y}_{j\cdot} - \bar{y})^2$	$n_j(\bar{y}_{j\cdot} - \bar{y})^2$
A	1	5	64.2	1.05	1.1025	5.5125
B	2	5	75.8	12.65	160.0225	800.1125
C	3	5	67.2	4.05	16.4025	82.0125
D	4	5	45.4	−17.75	315.0625	1575.3125
Total	–	–	–	0.00	492.5900	2462.9500

A similar table can be useful in constructing the calculations for the TrtSS (see Table 4.3). This table has a similar format, but in this case, the differences are between the treatment means and the sample grand mean. Again, the treatment sum of squares, TrtSS, is equal to the sum of the values in the final column. In this case, with equal replication of $n = 5$, we could equivalently have taken the sum of the values in the penultimate column (492.59) and multiplied it by the replication to get the same answer, i.e. TrtSS = $5 \times 492.59 = 2462.95$.

Finally, calculation of the residual sum of squares is most easily done by subtraction as

$$\text{ResSS} = \text{TotSS} - \text{TrtSS} = 3684.55 - 2462.95 = 1221.60 \ .$$

4.3.2 Calculating Degrees of Freedom and Mean Squares

To make statistical inferences about the sums of squares, we must also consider the amount of information used to form them. Each contribution to a sum of squares is a positive value (after being squared), so the sum must increase as the number of contributions increases. To compare sums of squares, it is therefore necessary to standardize them onto a common scale. We do this by considering the amount of information, or **degrees of freedom** (df), used in their construction. A rigorous mathematical definition of degrees of freedom is beyond the scope of this book but available elsewhere (e.g. Bailey, 2008) for the interested reader.

All the sums of squares we consider take the form

$$\text{SS} = \Sigma \ (\text{value} - \text{adjustment})^2 ,$$

where the summation is across all the observations, and may use several indices. Both the 'value' and the 'adjustment' arise from an estimated model. The degrees of freedom can be calculated as the (minimum) number of parameters required to calculate the model used for 'values' minus the (minimum) number of parameters required to calculate the model used for 'adjustment'. So, for example, TotSS was obtained from the deviations of the individual observations about the sample grand mean. In this case, the 'values' in the SS are the N individual observations (n_j from each of the treatments), which can only be formed from a model using N parameters. The adjustment is the sample grand mean, an estimate of a single parameter (the overall population mean). Therefore, the degrees of freedom for TotSS are

$$\text{TotDF} \ = \ \left(\sum_{j=1}^{t} n_j \right) - 1 = N - 1 .$$

This expression can be written as $(n \times t) - 1$ (or $nt - 1$) for the case of equal replication. Similarly, the treatment sum of squares (TrtSS) uses t values, i.e. the treatment sample means $\bar{y}_{j\cdot}$, again adjusted by the sample grand mean, so the treatment degrees of freedom becomes

$$\text{TrtDF} = t - 1 \, .$$

Finally, the ResSS uses all the individual observations, i.e. N values. The adjustment arises from a model in which each treatment has a separate mean, which requires t parameters, one for each treatment. The residual degrees of freedom are therefore

$$\text{ResDF} = N - t.$$

For equal replication, this can be written as $nt - t$ or $(n - 1)t$.

Note that the TotDF is partitioned between the treatment and residual degrees of freedom in a similar way to the TotSS, namely

$$\text{TotDF} = \text{TrtDF} + \text{ResDF} \, ,$$

and that the residual degrees of freedom can be calculated by subtraction as $\text{ResDF} = \text{TotDF} - \text{TrtDF}$.

Having calculated the degrees of freedom for each term, we put the treatment and residual sums of squares onto a common scale by dividing each by their degrees of freedom, producing ratios known as **mean squares**. For one-way ANOVA, these calculations are

$$\text{TrtMS} = \text{TrtSS}/\text{TrtDF} \, ,$$

$$\text{ResMS} = \text{ResSS}/\text{ResDF} \, .$$

4.3.3 Calculating Variance Ratios as Test Statistics

The **residual mean square** (ResMS), sometimes also called the **mean square error** (MSE), provides an unbiased estimate of background variation (σ^2), with this estimate usually called s^2. Intuitively, the ResMS quantifies the variation within each treatment group, which should arise from background variation alone, then combines this information across treatments, giving an estimate of background variation pooled across treatments.

If there are no differences between treatments, then contributions to the treatment mean square, i.e. differences between the treatment means and the sample grand mean, arise from background variation alone. In this case, the treatment and residual mean squares should be of similar sizes, allowing for sampling variation. We can formalize this comparison by considering the expected value of each of the respective mean squares, which are

$$E(\text{TrtMS}) = \sigma^2 + \frac{1}{(t - 1)} \sum_{j=1}^{t} n_j \left(\mu_j - \mu \right)^2 \, ,$$

$$E(\text{ResMS}) = \sigma^2 \, ,$$

where μ is the overall population mean. The expected value of the TrtMS is equal to the background variation plus a scaled sum of the squared differences between the treatment population means and the overall population mean. The expected value of the ResMS is simply equal to the background variation. If there are no differences between treatments, then the treatment means, μ_j, are equal to the overall mean, μ, and the second term in the expectation of the TrtMS becomes zero; both mean squares then have the same expected value. This is the basis of the ANOVA F-test, which tests the null hypothesis of equal treatment population means, i.e.

$$H_0: \mu_1 = \mu_2 = \ldots = \mu_t ,$$

against the general alternative of some variation within the set of treatment population means. The test statistic is obtained by dividing the treatment mean square by the residual mean square, and is known as the **variance ratio** or observed F-statistic, which we denote as F, i.e.

$$F = \text{TrtMS}/\text{ResMS} .$$

If the null hypothesis is true, we expect the value of the variance ratio to be close to one. A larger ratio implies that the variation between treatment means is greater than the background variation, and is evidence of differences among the treatment population means. More formally, if the assumptions on the deviations hold and the null hypothesis is true, such a ratio of two independent mean squares has an F-distribution, and the amount of evidence can be quantified. The F-distribution depends on two sets of degrees of freedom, one associated with the numerator in the ratio (here, TrtMS with $t-1$ df) and one associated with the denominator (here, ResMS with $N-t$ df). For clarity, we sometimes specify the observed variance ratio with its df as subscripts, for example, $F_{t-1,N-t}$. As for the two-sample t-test (Section 2.4.2), we usually have a pre-determined level of significance for testing, denoted α_s (usually $\alpha_s = 0.05$). This is a one-sided test, as only large variance ratios indicate differences among the population means. The critical value required is therefore the $100(1 - \alpha_s)$th percentile of the F-distribution, denoted $F_{t-1,N-t}^{[\alpha_s]}$, which satisfies

$$\text{Prob}(F_{t-1,N-t} \geq F_{t-1,N-t}^{[\alpha_s]}) = \alpha_s ,$$

where $F_{t-1,N-t}$ denotes a random variable with an F-distribution on $(t-1)$ and $(N-t)$ df. Variance ratios larger than the critical value give evidence (at significance level α_s) against the null hypothesis. Tables of 95th percentiles for F-distributions with a range of numerator and denominator df are provided in Appendix B, but these are also available in statistical software. Alternatively, the observed significance level can be calculated as the proportion of the F-distribution greater than the observed variance ratio, or

$$P = \text{Prob}(F_{t-1,N-t} \geq F_{t-1,N-t}) .$$

This calculation requires the quantile function of the F-distribution, which is also available in statistical software.

4.3.4 The Summary ANOVA Table

All of these calculations can be neatly summarized in the **ANOVA table** (see Table 4.4). A summary ANOVA table can be constructed for any linear model. The treatment sum of squares (TrtSS) is sometimes known as the 'between' sum of squares, because it quantifies

TABLE 4.4

Structure of the ANOVA Table for a CRD with t Treatments (Factor Treatment) and N Observations in Total

Source of Variation	df	Sum of Squares	Mean Square	Variance Ratio	P
Treatment	$t-1$	TrtSS	TrtMS = TrtSS/$(t-1)$	F = TrtMS/ResMS	Prob($F_{t-1,N-t} \geq$ F)
Residual	$N-t$	ResSS	ResMS = ResSS/$(N-t)$		
Total	$N-1$	TotSS			

variation between (among) treatment means. Similarly, the residual sum of squares (ResSS) is sometimes called the 'within' sum of squares, as it quantifies a pooled measure of variation within treatment groups. Table 4.4 shows the full form of the one-way ANOVA table. Where space is limited, we will occasionally omit one or more of the columns, and you should note that the columns holding the sums of squares and degrees of freedom are sometimes interchanged.

Prior to interpreting the output from any ANOVA, it is essential to check that the data do not violate any of the underlying assumptions made about the deviations (Section 4.1), as the validity of the F-test depends upon them. In this context, you should note that use of randomization in setting up an experiment increases the robustness of the F-test – this is discussed further in Section 5.2.4. The true values of the deviations are not known, but we have the residuals (defined in Equation 4.4) as estimates of these values. Validation methods that use residuals are known as diagnostic tools and are described in Chapter 5. Once the assumptions have been checked and found to be reasonable, within certain limits, we can then examine the results from the ANOVA and begin to draw conclusions.

EXAMPLE 4.1D: CALCIUM POT TRIAL*

The calculations in Example 4.1C can be combined to construct the ANOVA table in Table 4.5. Residual plots for these data can be seen in Section 5.2, where they are discussed in some detail. For the moment, we merely state that the residual plots are reasonably consistent with the assumptions underlying the analysis. The 5% critical value of the relevant F-distribution ($F_{3,16}^{[0.05]}$) is 3.239 (Table B.1) and the 1% critical value ($F_{3,16}^{[0.01]}$) is 5.292. Both of these values are considerably smaller than the observed variance ratio, $F_{3,16} = 10.753$. The observed significance level can be obtained as the proportion of the F-distribution with 3 and 16 df greater than the observed variance ratio. In this example, this is $P = 0.00041$ (often reported as $P < 0.001$ in statistical software). Hence, we reject the null hypothesis and conclude that there is very strong evidence that the population means are not all equal, indicating that there is some effect of calcium concentration on root growth. The next step in the analysis is to interpret this effect.

TABLE 4.5

ANOVA Table for Root Lengths from the Calcium Pot Trial with Four Treatments (Factor Calcium) (Example 4.1D)

Source of Variation	df	Sum of Squares	Mean Square	Variance Ratio	P
Calcium	3	2462.95	820.98	10.753	< 0.001
Residual	16	1221.60	76.35		
Total	19	3684.55			

Having constructed the ANOVA table, calculated the variance ratio, and compared this value to the appropriate F-distribution, we know whether there is evidence to support rejection of the null hypothesis that all treatments have the same population mean. If we have a significant result for the F-test, then we should examine the patterns in the treatment sample means. If the F-test result is not significant, then the analysis is complete and we do not have evidence to reject the null hypothesis. It is often worth considering whether a more complex treatment structure is present, requiring a more complicated form of explanatory model (see Chapter 8). If differences between treatment sample means appear large from a biological point of view but the F-test is not significant, this may indicate that the experiment was not sufficiently large or precise to be able to detect these differences as statistically significant. This relates to the power of the experiment, which can generally be increased by adding more replicates across the whole experiment. Approaches for determining the replication level required to detect a given size of treatment difference for a given amount of background variation are described in Chapter 10.

4.4 Evaluating the Response to Treatments

4.4.1 Prediction of Treatment Means

If the fitted model gives a good representation of the observations, then the best prediction of the population mean for the jth treatment comes from the parameter estimates (Section 4.2) as

$$\hat{\mu}_j = \bar{y}_{j\cdot} \, ,$$

i.e. the best prediction of the treatment population mean is the treatment sample mean. If we reject the null hypothesis, having obtained a significant F-test result, and we have checked the validity of our model assumptions (Chapter 5), then we can examine the table of sample means to identify the source(s) and size(s) of any treatment differences associated with this significant test result. It is important to realize that statistical significance is not the same as biological significance – with sufficient replication, it is possible to find statistically significant differences that are too small to have any real biological meaning – and so it is important to also consider the biological significance of any statistically significant comparison.

To make statistical inferences about the treatment population means, we need a measure of the uncertainty associated with our estimates of these values, the treatment sample means. This uncertainty is measured by the variance of these estimates, usually expressed on the SD scale (i.e. as the square root of the variance) and commonly referred to as the standard error of the mean. We use the general notation $\text{SE}(\hat{\mu}_j)$ to denote the SE of such estimates. For the jth treatment with n_j replicate observations, the variance of the estimated mean is σ^2/n_j (as introduced in Section 2.2.3) and the SE is the square root of this variance, i.e.

$$\text{SE}(\hat{\mu}_j) = \sqrt{\frac{\sigma^2}{n_j}} \, .$$

In fact, the true value of the background variation, σ^2, is unknown, and so we replace it with our best estimate, s^2 (equal to the ResMS), to get an estimate of this SE, written as

$$\text{SEM}_j \;=\; \widehat{\text{SE}}(\hat{\mu}_j) \;=\; \sqrt{\frac{s^2}{n_j}} \;.$$

In general, for simplicity, we use the notation SE to denote the estimate, $\widehat{\text{SE}}$, and use SEM_j to denote this SE for the estimate of the population mean for the jth treatment. The subscript can be dropped when all treatments are equally replicated and the SEMs are all equal.

A $100(1 - \alpha_s)$% CI for the population mean of the jth treatment can be constructed as

$$\left(\hat{\mu}_j - [t_{N-t}^{[\alpha_s/2]} \times \text{SEM}_j] \,, \quad \hat{\mu}_j + [t_{N-t}^{[\alpha_s/2]} \times \text{SEM}_j] \right) ,$$

where $t_{N-t}^{[\alpha_s/2]}$ is the $100(1 - \alpha_s/2)$th percentile of the Student's t-distribution with df equal to the residual degrees of freedom from the ANOVA (here $N - t$). The confidence limits can alternatively be expressed as

$$\hat{\mu}_j \;\pm\; [t_{N-t}^{[\alpha_s/2]} \times \text{SEM}_j] \,.$$

EXAMPLE 4.1E: CALCIUM POT TRIAL*

From the ANOVA table obtained above (Example 4.1D), we have ResMS = s^2 = 76.35. The sample means for this study were shown in Table 4.1. Each treatment has five observations, $n = 5$, so the standard errors of the predicted means are all equal to

$$\text{SEM} = \sqrt{\frac{s^2}{n}} = \sqrt{\frac{76.35}{5}} = \sqrt{15.27} = 3.91 \,.$$

The 97.5th percentile of the t-distribution with 16 df is $t_{16}^{[0.025]} = 2.120$. Using the estimate for treatment A obtained in Example 4.1B, $\widehat{Calcium}_1 = 64.2$, we calculate a 95% CI for this treatment as

$$\widehat{Calcium}_1 \pm (t_{16}^{[0.025]} \times \text{SEM}) = 64.2 \pm (2.120 \times 3.91) = 64.2 \pm 8.29 \,,$$

giving the 95% CI as (55.9, 72.5).

4.4.2 Comparison of Treatment Means

Usually, one of the aims of a study is the comparison of the population means for different treatments. The best estimate of a difference in population means is the difference in sample means, for example, for the jth and kth treatments

$$\hat{\mu}_j - \hat{\mu}_k = \bar{y}_{j\cdot} - \bar{y}_{k\cdot} \,.$$

The standard error of this difference between the predicted population mean of the jth treatment, with replication n_j, and the kth treatment, with replication n_k (also presented in Section 2.4.2) takes the form

$$\mathrm{SE}(\hat{\mu}_j - \hat{\mu}_k) = \sqrt{\sigma^2 \left(\frac{1}{n_j} + \frac{1}{n_k} \right)} .$$

Again, we estimate this SE using s^2 in place of the unknown background variance, written as

$$\mathrm{SED}_{j,k} = \widehat{\mathrm{SE}}(\hat{\mu}_j - \hat{\mu}_k) = \sqrt{s^2 \left(\frac{1}{n_j} + \frac{1}{n_k} \right)} .$$

We thus use $\mathrm{SED}_{j,k}$ to denote the SE for the estimated difference between the population means for the jth and kth treatments. The subscripts can be dropped when all treatments are equally replicated and the SEDs are all equal with

$$\mathrm{SED} = \sqrt{\frac{2s^2}{n}} .$$

Note that the SED, the standard error of a difference between predictions, is always larger than the SEM for a single prediction because the SED contains uncertainty associated with the estimation of two population means.

Under the null hypothesis that the population means for the two treatments are equal, i.e. H_0: $\mu_j = \mu_k$, the statistic

$$t_{N-t} = \frac{\hat{\mu}_j - \hat{\mu}_k}{\mathrm{SED}_{j,k}} = \frac{\bar{y}_{j\cdot} - \bar{y}_{k\cdot}}{\mathrm{SED}_{j,k}} ,$$

has a t-distribution with degrees of freedom equal to the residual df, here $N - t$. The statistic can be used to test this null hypothesis against the two-sided alternative H_1: $\mu_j \neq \mu_k$ by comparing it with critical values of that t-distribution (Section 2.4.2). The differences between several pairs of treatment means can be evaluated in this way; however, in general, the testing of many pairwise comparisons without taking precautions with regard to overall levels of significance can give misleading results – we return to this topic in Section 8.8.

From the form of this two-sample t-test, it follows that the smallest absolute difference between two treatment sample means that would result in a statistically significant two-sided t-test for the difference between the corresponding population means can be calculated as

$$\mathrm{LSD}_{j,k} = t_{N-t}^{[\alpha_s/2]} \times \mathrm{SED}_{j,k} ,$$

where $\mathrm{LSD}_{j,k}$ denotes this **least significant difference** (LSD) between the jth and kth treatment means. A $100(1 - \alpha_s)\%$ CI for the difference in population means between the jth and kth treatments can be computed in terms of this LSD as

$$\left((\hat{\mu}_j - \hat{\mu}_k) - \text{LSD}_{j,k}, \ (\hat{\mu}_j - \hat{\mu}_k) + \text{LSD}_{j,k} \right).$$

Because of the connection with the t-test, if the CI for a difference between the population means for two treatments does not include zero, then there is statistical evidence for a significant difference (at level α_s) between the means for these two treatments.

EXAMPLE 4.1F: CALCIUM POT TRIAL*

For this study, we want to know whether a small increase in calcium (treatment B) has any impact on root growth compared with the standard (treatment A). To calculate the SED between these two treatment means requires the ResMS ($s^2 = 76.35$) and residual df (ResDF = 16) from the ANOVA table (Table 4.5). Each treatment has five observations, $n = 5$, so there is also a common SED for comparing any pair of treatments, i.e.

$$\text{SED} = \sqrt{\frac{2s^2}{n}} = \sqrt{\frac{2 \times 76.35}{5}} = \sqrt{30.54} = 5.53 \,.$$

The tabulated 97.5th percentile of the t-distribution with 16 df is $t_{16}^{[0.025]} = 2.120$, so at a 5% significance level, the LSD is calculated as

$$\text{LSD} = t_{16}^{[0.025]} \times \text{SED} = 2.120 \times 5.53 = 11.72 \,.$$

Hence, any difference between treatment means greater than 11.72 is significant at the 5% level. The predicted difference between treatments B and A can be obtained from Example 4.1B as $\widehat{Calcium_2} - \widehat{Calcium_1} = 75.8 - 64.2 = 11.6$, slightly smaller than the LSD, and the 95% CI for the difference between these treatments is

$$(\widehat{Calcium_2} - \widehat{Calcium_1}) \pm \text{LSD} = (75.8 - 64.2) \pm 11.72 = 11.6 \pm 11.72 \,,$$

giving the 95% CI as (–0.1, 23.3). As expected from the comparison between the treatment means with the LSD, zero is contained in this CI, confirming the conclusion that these treatments are not statistically different at a significance level of 5%.

It is often more informative to consider the overall pattern in the treatment means, as shown in Figure 4.6, rather than to make numerous pairwise treatment comparisons. Figure 4.6 indicates an initial increase in root growth as calcium increases, then a clear decrease at the highest level. The LSD bar indicates the likely size of differences caused by background variation. These results could be used to suggest rough limits on calcium concentrations likely to be beneficial to root growth that could be verified using field studies. Given the numerical relationship between the calcium factor levels, we might also try to model the pattern by considering the calcium content as a quantitative explanatory variable (recall Examples 1.1 and 1.2), and we describe this type of model in Section 8.7.

Note that although there is a superficial similarity between Figures 4.6 and 4.1, there are important differences between the two plots. Both presentations show the treatment sample means, but they differ in their presentations of uncertainty. Figure 4.1 shows only the unbiased sample SD for each treatment, giving some indication of the within-group variability for each treatment. This is useful as a precursor to ANOVA in assessing the assumption of homogeneity of variance (although formal tests can also be used,

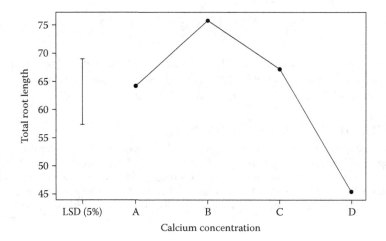

FIGURE 4.6
Estimated LSD and treatment means for the calcium pot trial (see Examples 4.1B and F).

see Section 5.3). On the other hand, Figure 4.6 shows the LSD, based on an estimate of background variation pooled across the treatment groups, which gives a direct and more appropriate measure for statistical comparison of treatment population means as a result of the ANOVA.

4.5 Alternative Forms of the Model

In the preceding sections, we have used the simplest form of the model for a single factor, i.e.

$$y_{jk} = \mu_j + e_{jk} \, ,$$

using a single parameter to represent the population mean for each treatment. Statistical packages generally use other forms, or parameterizations, of the model, and we give an introduction to these forms here. As long as the forms are equivalent, the same ANOVA table and the same conclusions will be obtained. One widely used form writes the single factor model as

$$y_{jk} = \mu + \tau_j + e_{jk} \, , \tag{4.6}$$

where μ is the overall population mean across all treatments, and τ_j is the unknown population **treatment effect** for the jth group, i.e. the difference between the population mean for the jth treatment and the overall mean, which can be positive or negative.

The explanatory component can then be written in symbolic form as

Explanatory component: [1] + Treatment

Here, the term [1] denotes a factor that takes value 1 everywhere, and is associated with the overall mean, μ.

This model has $t + 1$ parameters – one for each treatment group plus the overall mean. However, as the structure can be described with only t parameters, i.e. one for each group, the version in Equation 4.6 is called **over-parameterized**. The consequence of over-parameterization is that estimation of the parameters by the least squares principle does not result in a unique solution, and so we impose constraints to obtain a unique solution. To keep the interpretation of μ as the overall mean, the average treatment effect must be zero; hence we impose the constraint $\Sigma_j \tau_j = 0$, i.e. the sum (and hence mean) of the treatment effects is zero (sometimes called the **sum-to-zero constraint**). The parameter estimates then take the form

$$\hat{\mu} = \bar{y} ,$$
$$\hat{\tau}_j = \bar{y}_{j\cdot} - \bar{y} ,$$

so the overall population mean μ is estimated by the sample grand mean, and the population treatment effects τ_j are estimated by the differences between the treatment sample means and the sample grand mean. The fitted values then take the form

$$\hat{y}_{jk} = \hat{\mu} + \hat{\tau}_j = \bar{y}_{j\cdot} ,$$

and hence are equal to the treatment sample means, exactly as before. In this parameterization, we note that the treatment sum of squares is simply a sum of squares of the estimated treatment effects, i.e.

$$\text{TrtSS} = \sum_{j=1}^{t} n_j \left(\bar{y}_{j\cdot} - \bar{y} \right)^2 = \sum_{j=1}^{t} n_j \hat{\tau}_j^2 .$$

This parameterization is used in the GenStat ANOVA algorithm.

Other parameterizations can be used, but here we just consider variations on one of the most common, often called the **corner-point constraint** or **first-level-zero constraint**, written as

$$y_{jk} = \mu_1 + \nu_j + e_{jk} .$$

In this parameterization, the constraint $\nu_1 = 0$ is used. The term μ_1 then represents the population mean for the first treatment, and the effect ν_j represents the difference between the population mean for the jth treatment and that of the first treatment. Here, the least squares estimates take the form

$$\hat{\mu}_1 = \bar{y}_{1\cdot} ,$$
$$\hat{\nu}_j = \bar{y}_{j\cdot} - \bar{y}_{1\cdot} .$$

So the parameter μ_1 is estimated by the sample mean for the first treatment, and the effects v_j are estimated by differences between the sample means for the jth treatment and the first treatment. However, the fitted values, calculated as

$$\hat{y}_{jk} = \hat{\mu}_1 + \hat{v}_j = \bar{y}_{j\cdot} \, ,$$

are again equal to the treatment sample means, matching previous forms of the model. This parameterization is used in the GenStat regression algorithm (commands MODEL and FIT) and the lm and aov algorithms implemented in R.

Another version of this parameterization uses **last-level-zero constraints**, taking the form

$$y_{jk} = \mu_t + \omega_j + e_{jk} \, ,$$

with constraint $\omega_t = 0$, then the ω_j, $j = 1 \dots t-1$, represent differences with the last treatment. Again, although individual parameter estimates differ from previous forms, the fitted values are the same. This algorithm is used by the PROC GLM algorithm in SAS and can be used in the GenStat regression algorithm (commands MODEL and FIT) if the last factor level is chosen as the reference level.

All of the above forms are specific to the linear model with a single factor, and must be extended for more complex models; the principles are the same however. Details for more complex models are given in Chapters 7, 8 and 11. In general, although the value and interpretation of individual parameter estimates depend on the parameterization used, the fitted values and predictions for population treatment means are unchanged. For this reason, we usually make inferences for the population means, which we still denote as μ_j, $j = 1 \dots t$, rather than for the individual model parameters.

EXERCISES

4.1*   A glasshouse experiment to evaluate control of a weed species by three different chemical treatments used a CRD with seven replicates of each treatment.

    a.   What are the null and alternative hypotheses for this experiment?

    b.   Construct the ANOVA table given that TrtSS = 121.5 and ResSS = 87.4.

    c.   What is the appropriate F-distribution for the variance ratio under the null hypothesis? What is the 5% critical value from this distribution?

    d.   Would we accept or reject the null hypothesis?

4.2   A laboratory experiment investigated the effect of different treatments on grain production in wheat ears infected with *Fusarium graminearum* (Baldwin et al., 2010). Single wheat ears on 30 separate plants were inoculated with *F. graminearum*. Four treatments (labelled A–D) and a negative (untreated) control were then allocated to the inoculated ears as a CRD. The number of grains in the region above the inoculation position of each ear was counted. File GRAINS.DAT contains the unit number (*DEar*), the treatment applied (factor Treatment) and the number of grains (variate *Grains*) for each ear.*

    a.   Write down a mathematical model for the numbers of grains.

---

* Data from K. Hammond-Kosack, Rothamsted Research.

    b. Write down the null and alternative hypotheses associated with this experiment.

    c. Construct an ANOVA table by calculating the total, treatment and residual sums of squares and df and then deriving the other columns. Is there any evidence that grain production is affected by the treatments?

    d. Calculate the predicted mean for each treatment group and the SED and LSD for treatment comparisons.

    e. State your conclusions from this analysis.

(We re-visit these data in Exercises 5.1 and 5.2.)

4.3 An experiment was done to assess whether fungal infection affected aphid reproduction. Sixty adult aphids were equally divided among three treatment groups, which were either inoculated with a fungus (either *Beauveria bassiana* or *Pandora neoaphidis*) or not inoculated. One first-generation (FG) nymph was taken from each adult; these nymphs were placed individually into Petri dishes which were then arranged randomly within a controlled environment chamber. The development time for each FG aphid was observed and the number of nymphs produced by each FG aphid during a time equal to its own development time was counted. Some FG aphids died before producing nymphs and were removed from the experiment. File FUNGUS.DAT contains the unit numbers (*DAphid*), the treatments (factor Fungus) applied, and the numbers of nymphs (variate Nymphs) produced by the 33 remaining FG aphids.[*]

    a. Write down a mathematical model for the aphid counts.

    b. Write down the null and alternative hypotheses associated with this experiment.

    c. Construct an ANOVA table by calculating the total, treatment and residual sums of squares and df and then deriving the other columns. Is there any evidence that reproduction of FG progeny is affected by fungal infection of the original adult aphids?

    d. Calculate the estimated mean, with a 95% confidence interval, for each treatment group.

    e. State your conclusions from this analysis.

    f. Comment on whether omitting the FG aphids that died might bias the results – what assumptions has your analysis made?

(We re-visit these data in Exercises 5.1, 5.2 and 5.4.)

4.4 An experiment investigated the effect of conidia density on transmission of a fungus that attacks aphids. Cadavers of aphids killed by the fungus, and from which the fungus was releasing spores, were placed on bean plants at three densities (A = 1, B = 5 or C = 10 cadavers per plant) to give different doses of fungal conidia. The densities were allocated to individual bean plants as a CRD with six replicates. Twenty uninfected live aphids were placed on each plant with one ladybird which was allowed to forage to facilitate transfer of conidia between the cadavers and the live aphids. For each plant, the proportion of aphids that became infected after 7 days was recorded and transformed to the logit scale for analysis (see Chapter 6). The unit numbers (*DPlant*), treatment

---

[*] Data from J. Baverstock, Rothamsted Research.

allocations (factor Density) and transformed responses (variate *LogitP*) are in file TRANSMISSION.DAT.[*]

a. Write down the null and alternative hypotheses associated with this experiment.

b. Obtain the ANOVA table. Is there any evidence that the density of fungal conidia affects the rate of transmission of the fungus to the aphids?

c. Plot the predicted means for each density with the LSD. What does this plot suggest?

d. State your conclusions from this analysis.

(We re-visit these data in Exercise 5.2.)

4.5   A variety of maize was genetically modified, and plants were classified as homozygous, heterozygous or null according to the number of glutamine mutants present (2, 1 or 0, respectively; Haines, 2000). The dry weights (g) of single kernels from each of 10 plants of each type, sampled at random, were recorded. The unit number (*DKernel*), classification (factor Type) and dry weight (variate *Weight*) for each kernel are in file MAIZE.DAT.

a. Write down the null and alternative hypotheses associated with this experiment.

b. Obtain the ANOVA table. Is there any evidence that mean kernel weights differ among the different genetic types?

c. Plot the predicted means for each genetic type with the LSD. What genetic hypothesis does this plot suggest?

d. State your conclusions from this analysis.

(We re-visit these data in Exercise 5.2.)

---

[*] Data from J. Pell, Rothamsted Research.

5

Checking Model Assumptions

In Chapter 4, we introduced a simple additive linear model to describe the effects of a single explanatory factor. The ensuing statistical analysis is based on the assumption that this additive form of model is true, and also on the assumptions about the properties of the model deviations given in Section 4.1. Conclusions from the statistical analysis are valid only if these assumptions are consistent with the properties of the data. In this chapter, we describe some simple diagnostic tools that can be used to check the assumptions underlying analysis of variance.

The term **diagnostic tools** refers to a collection of techniques used to detect inconsistencies between the statistical model and the data. Diagnostic tools are used primarily to check that the assumptions underlying the analysis are not violated and that unusual individual data values do not unduly affect the fit of the model and hence any inferences drawn from it. If you find problems, then you can take corrective measures, such as transforming the data (see Chapter 6). Diagnostics for linear models take two main forms: analysis of the properties of the residuals and computation of influence statistics. In this chapter, we concentrate on the former; influence statistics, and related diagnostic tools more relevant to regression models, are presented in Chapter 13. However, all of the diagnostic tools discussed in this chapter are applicable to assess the assumptions for any linear model.

First, we describe two of the most commonly used forms of residual (Section 5.1). We then discuss graphical diagnostic tools (residual plots) for inspecting the residuals and checking the model assumptions (Section 5.2). We also briefly describe permutation tests, which can be used when the assumption of a Normal distribution for the deviations is not plausible (Section 5.2.4). We then describe one formal test, Bartlett's test, for checking homogeneity of variances between treatment groups (Section 5.3). We end this chapter with a short discussion of how to identify and deal with outliers (Section 5.4).

5.1 Estimating Deviations

To examine whether the assumptions made about the deviations are plausible, we would ideally like to examine them directly. As this is not possible (because they are not known), we examine estimates of the deviations, called the **residuals**. Unfortunately, even if the assumptions underlying the model are true, the statistical properties of the residuals are not exactly the same as those of the deviations. For this reason, different types of residuals have been developed to examine different aspects of the distributions of deviations. Here, we describe simple and standardized residuals; later, in Chapter 13, we introduce prediction and deletion residuals, which are particularly useful in regression analysis.

5.1.1 Simple Residuals

In Equation 4.4, a residual from the model fitted for a CRD was defined as the discrepancy between the response and the fitted systematic component. This is the definition of a **simple** (or **ordinary**) **residual**, and can be easily extended for any linear model. Using a general notation that labels observations by index i for $i = 1 \ldots N$, we define the simple residual for the ith observation as

$$\hat{e}_i = y_i - \hat{y}_i \, ,$$

where y_i is the response and $\hat{y}_i$ is the fitted value for that observation. As in previous chapters, the 'hat' notation denotes an estimate; the residuals $\hat{e}_i$ are estimates of the unknown deviations e_i. In this general notation, the subscript i refers to the ith observation, but recall that it is often convenient to relabel the units to reflect the structure of a specific data set, for example, to use $\hat{e}_{jk}$ for a CRD to represent the simple residual for the kth replicate of the jth treatment, as in Equation 4.4.

> **EXAMPLE 5.1A: CALCIUM POT TRIAL***
>
> Recall the calcium pot trial analysed in Example 4.1 (data in file CALCIUM.DAT). Figure 5.1a shows the fitted values, here the treatment means, and the observed responses plotted against treatment group. The value of the simple residual for each observation is equal to the vertical distance between the observation and its group mean, indicated for the second smallest response in treatment A by a dotted line. Observations larger than the group mean have positive residuals; those smaller than the group mean have negative residuals. Figure 5.1b shows the simple residuals plotted by treatment group. The residuals show the same pattern as the observations within each treatment, but the groups are now centered about zero rather than about the treatment means. There is a suggestion in Figure 5.1b that the variance differs between treatments; this is investigated further in Section 5.3.

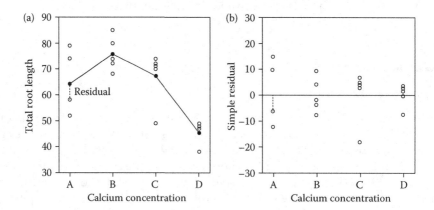

FIGURE 5.1

(a) Observed (○) and fitted (●) values (root lengths, cm) from the calcium pot trial (Example 5.1A). (b) Simple residuals for all pots in the calcium pot trial. Vertical dotted line indicates the simple residual for the second smallest response in treatment A.

For a model with a single explanatory factor, such as the CRD, the residuals within each group must sum to zero. Recall that in the CRD we label observations y_{jk} by their treatment group (index j) and replicate (index k). From Equation 4.4, we know that the residuals take the form

$$\hat{e}_{jk} = y_{jk} - \bar{y}_{j\cdot} \, ,$$

where $\bar{y}_{j\cdot}$ is the sample mean for the jth treatment group. The sum of the residuals in this group is therefore calculated as

$$\sum_{k=1}^{n_j} \hat{e}_{jk} = \sum_{k=1}^{n_j} \left(y_{jk} - \bar{y}_{j\cdot} \right) = \left(\sum_{k=1}^{n_j} y_{jk} \right) - n_j \bar{y}_{j\cdot} = y_{j\cdot} - y_{j\cdot} = 0 \, .$$

In other words, the residuals sum to zero because the sum of the observations is equal to the sum of the fitted values. Summing to zero is a constraint that means that residuals within treatment groups are not independent, even when the model deviations are truly independent. Similar constraints across treatment groups and other structures also hold, even for more complex models. The sum of the complete set of residuals across the experiment must also be equal to zero.

A general expression for the variances of residuals is given in Section 13.4.2. For the CRD, the variance of the residuals is directly related to the replication within each treatment group, as

$$\text{Var}(\hat{e}_{jk}) = \sigma^2 \left(\frac{n_j - 1}{n_j} \right) .$$

The variances of the residuals are therefore all equal for the CRD when the treatment groups have equal replication, but differ between treatment groups otherwise. As usual, we can estimate this quantity by replacing the unknown variance, σ^2, by its estimate $s^2 = \text{ResMS}$ (Section 4.3).

5.1.2 Standardized Residuals

Standardized residuals are used to deal with the problem of unequal variances within a set of simple residuals. A standardized residual is defined as the simple residual divided by an estimate of its standard error. We denote the standardized residual here as r_i, with

$$r_i = \frac{\hat{e}_i}{\widehat{\text{SE}}(\hat{e}_i)} \, ,$$

where $\widehat{\text{SE}}(\hat{e}_i)$ is the estimated standard error of the simple residual for the ith observation (see Section 13.4.2 for details). The standardized residuals have a common variance equal to one (unit variance), but are not independent as they are subject to the same constraints as the simple residuals. For reasonably large data sets, most of the standardized residuals should fall within the range ± 2, and individual points outside this band may be investigated as potential outliers (see Section 5.4).

For the CRD, the standardized residuals (labelled by treatment group j and replicate number k) are thus

$$r_{jk} \;=\; \frac{\hat{e}_{jk}}{\widehat{\text{SE}}\left(\hat{e}_{jk}\right)} \;=\; \frac{\hat{e}_{jk}}{\sqrt{s^2\left(n_j - 1\right)/n_j}} \;=\; \frac{\hat{e}_{jk}}{s}\sqrt{\frac{n_j}{\left(n_j - 1\right)}} \;,$$

where s is the square root of the residual mean square. When all groups have equal replication, so all n_j are equal to a common value n, the set of standardized residuals are simply a scaled version of the set of simple residuals.

Note that there is no common nomenclature for residuals: the standardized residuals defined here are called 'internally Studentized residuals' or 'Studentized residuals' in some statistical texts and software. You should therefore make sure you understand the definition of the residuals being used in any context.

5.2 Using Graphical Tools to Diagnose Problems

Examination of the distribution of residual values is an essential step in the fitting of any linear model. The two types of residuals described above (simple and standardized) are both useful for detection of general inadequacies in the model and violations of the assumptions. In the case of designed experiments with equal replication, the simple residuals are proportional to the standardized residuals, and either set can be used. However, it is usually more appropriate to use the standardized residuals, as these have the advantages of a standard scale and common unit variance. In this section, we concentrate on a few of the most commonly used graphical procedures (usually called **residual plots**) for checking the validity of the most important assumptions (homogeneity of variance, independence and Normality) underlying a fitted linear model. We describe several forms of residual plot, often used in combination to provide an overall picture of the validity, or otherwise, of the model.

5.2.1 Assessing Homogeneity of Variances

The assumption that all deviations have equal variance (Assumption 2 of Section 4.1) is often called the assumption of homogeneity of variances (or homoscedasticity). If the data conform to this assumption, a plot of the standardized residuals, r_i, against the fitted values, $\hat{y}_i$ (usually called a **fitted values plot**), should show approximately equal variance (indicated by the vertical spread about zero) across the range of fitted values (e.g. Figure 5.2a). A variant of the fitted values plot (called an **absolute residuals plot**) replaces the residuals with their absolute values, $|r_i|$. This plot should also show approximately equal variance across the range of fitted values (e.g. Figure 5.2b). A smooth trend line can be added to both plots to emphasize any pattern.

One common departure from constant variance occurs where the spread of the residuals increases as the fitted values get larger (e.g. Figure 5.3). This pattern is often seen for discrete data, such as counts, or continuous data, such as weights, where the data span a large range. For data in the form of counts as a percentage of a fixed total, the variance is usually

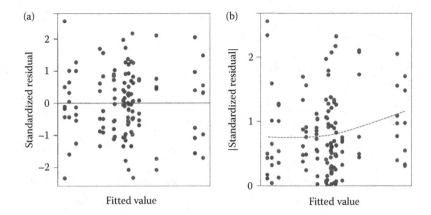

FIGURE 5.2
(a) Fitted values plot (residuals against fitted values) and (b) absolute residuals plot (absolute value of residuals against fitted values) with trend line, both showing acceptable homogeneity of variance.

small where the fitted value is close to either 0 or 100%, increasing towards a maximum at the centre of the range (50%). Occasionally, variances may differ systematically among treatment groups in a manner unrelated to their fitted values; these differences are more difficult to detect from a fitted values plot but may be tested formally by Bartlett's test, as described in Section 5.3.

The pattern in Figure 5.3 might be removed by the application of a variance-stabilizing transformation of the response, and Chapter 6 deals with this topic in detail. Alternatively, we might re-evaluate the assumption of Normality and decide that a different probability distribution would be more appropriate and use the methods presented in Chapter 18. In general, any strong pattern in the spread of the residuals, whether symmetric or asymmetric, suggests a failure to meet the assumption of homogeneous variances. In regression modelling, the fitted values plot can also be used to detect systematic deviations of the model from the pattern in the response (see Section 13.1).

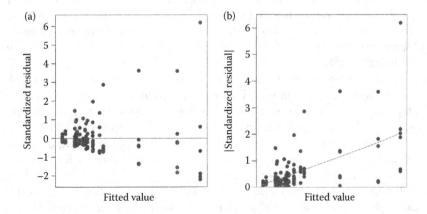

FIGURE 5.3
(a) Fitted values plot and (b) absolute residuals plot with trend line, both showing a strong pattern of larger variance for larger fitted values.

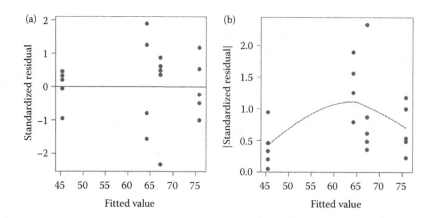

FIGURE 5.4
(a) Fitted values plot and (b) absolute residuals plot with trend line, for the calcium pot trial (Example 5.1B).

EXAMPLE 5.1B: CALCIUM POT TRIAL*

Figure 5.4 shows the fitted value and absolute residuals plots (based on the standardized residuals) for total root length. The fitted values for this model correspond to the sample means of the four treatments, and so there are four columns of points on the graph. There is no strong pattern in the spread of the residuals across the fitted values.

5.2.2 Assessing Independence

Usually, the assumption of independent deviations (Assumption 3 of Section 4.1) can safely be made given knowledge of the experimental procedure. For example, when we select a random sample of individual plants from locations in a large field and measure the height of each plant, we can reasonably assume that the deviation from the mean for any particular plant has no association with that for any other plant. However, there are situations when dependence (or correlation) among deviations arises. For example, suppose plants were sampled in pairs from locations in a field; owing to local environmental conditions, we might expect the deviations within a pair to be correlated. Similarly, if the plants are processed by a machine that shows drift in measurements over time, this drift might be observed as a positive correlation in deviations from consecutive measurements. In general, proximity in location or time of measurement can provide a mechanism for correlation among deviations.

Again, the distribution of the residuals can be used to investigate departures from this assumption of independent deviations. However, we stated above that the set of residuals are not independent, even if the deviations are independent. Since the induced correlation among residuals occurs within treatment groups, or within blocks, it should not be associated with trends in time or space (the usual sources of correlation) in a randomized experiment, and so the exploratory graphs described here should still reveal any strong serial (spatial or temporal) correlation.

The first step in detecting correlation is to order the residuals by the suspected source of correlations, usually location or time. This order will often correspond to the physical layout or processing of the experiment, and so requires full details of the design to be recorded and stored as part of the data set. We denote a variable that gives this order as the **index variable**. Note that in some contexts there may be several index variables. For

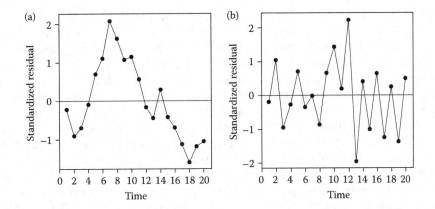

FIGURE 5.5
Index plots showing (a) positively and (b) negatively correlated standardized residuals when plotted against the time order in which the responses were obtained.

example, in a two-phase study that consists of a field experiment followed by laboratory processing, both the layout of the field plots and the order of processing in the laboratory might be used as index variables. Dependence, or correlation, of deviations may be detected graphically with an **index plot**, where the residuals ordered by the index variable are plotted against the values $1 \dots N$. Positive correlation is demonstrated by a tendency for adjacent residuals to have similar magnitude and sign (e.g. Figure 5.5a), and negative correlation is indicated by alternating signs (e.g. Figure 5.5b).

Alternatively, each residual can be plotted against the residual immediately preceding it when ordered by the index variable. Correlation in the residuals would then be revealed by evidence of a trend in the graph, with a positive slope indicating a positive correlation (e.g. Figure 5.6a) and a negative slope indicating a negative correlation (e.g. Figure 5.6b), rather than a random scatter over all four quadrants of the graph.

Sometimes correlation between observations will purposely be incorporated into a study. For example, to examine the effect of a growth regulator on the height of plants

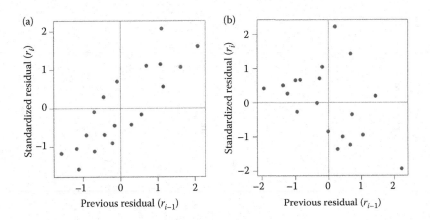

FIGURE 5.6
Standardized residuals (r_i), ordered by the index variable, plotted against the previous value in the series (r_{i-1}) for (a) positively and (b) negatively correlated residuals.

over time, we might measure the same set of plants on consecutive days. The data then contain several observations for each experimental unit which will show positive correlation; a plant that is taller than average at the first measurement is very likely to still be taller than average at the second measurement. This type of data is often called repeated measurements or longitudinal data, and is closely related to time series data (see e.g. Diggle et al., 2002). A similar situation occurs if data are related in space, for example, if measurements are made at different distances along a transect (e.g. along a river or into a field). While the index plot is useful for detecting such correlation, formal analysis of such data usually requires a more complex approach that takes account of the spatial or temporal correlation. There are several formal statistical tests for detecting serial (auto)correlation in the deviations (i.e. correlation between adjacent deviations in space or time), of which the most well known is the Durbin–Watson test. The discipline of geostatistics provides some more modern diagnostics; see Webster and Oliver (2007) or Chilès and Delfiner (2012).

If temporal or spatial correlation is expected prior to experimentation, then it should be incorporated at the design stage. For example, if the correlation is associated with machine drift, then all samples within a block should be processed together to confound differences between blocks with differences in larger-scale time effects. However, if there is still evidence of strong correlation within the set of residuals then the best solution is to model this correlation formally. This requires the use of more sophisticated techniques (e.g. linear mixed models) which we briefly introduce in Chapter 16, and which are described further in the references suggested above.

EXAMPLE 5.1C: CALCIUM POT TRIAL*

Figure 5.7a shows the index plot of standardized residuals for total root length, where the index variable is the pot number (variate *Pot*), which defines the order in which the experimental units were arranged and measured. There is no indication of strong positive or negative correlation. This is supported by Figure 5.7b where each residual is plotted against the residual for the preceding pot; there is a slight suggestion of negative correlation, but the points lie in all four quadrants with no strong trend.

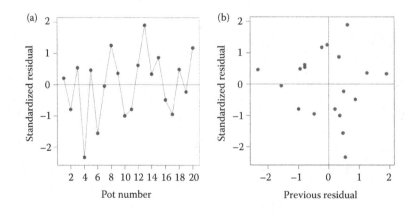

FIGURE 5.7

(a) Index plot of standardized residuals, and (b) plot of standardized residuals versus previous residual (with order defined by pot number), for the calcium pot trial (Example 5.1C).

5.2.3 Assessing Normality

The assumption that the deviations arise from a Normal distribution (Assumption 4 of Section 4.1) underlies the validity of the F-test for detecting differences between treatment means or evaluating the effect of a specific explanatory variable in a linear model. Note that this assumption is required only for tests of significance or computation of confidence intervals and not for parameter estimation, and, in fact, that the tests associated with the ANOVA are reasonably robust to departures from Normality, especially for randomized designs (see Section 5.2.4).

If the deviations arise from a Normal distribution, then the residuals inherit this distribution, so it is natural to examine the residuals in this context. To ensure that the residuals have common variance, we use standardized residuals. We consider two types of residual plot for assessing the validity of this assumption: histograms (see Section 2.2) of the residuals and Q–Q (quantile–quantile) plots. If the Normality assumption holds, then a histogram of the residuals should exhibit an approximately symmetrical, bell-shaped distribution centered around zero (e.g. Figure 5.8a). The standard **Q–Q plot** displays the ordered residuals plotted against the quantiles of the proposed probability distribution. Several standard variations are used, for example, plotting the ith smallest residual against the $100(i - 0.375)/(n + 0.25)$th percentile of the proposed distribution. As we wish to assess whether our residuals arise from a Normal distribution, we use quantiles from a standard Normal distribution (with zero mean and unit variance), and the resulting plot is also called a **Normal plot**. If the residuals are consistent with a sample from a Normal distribution, then the plot should yield an approximately straight line passing through the origin. The slope of this line is determined by the standard deviation of the residuals, and so a Normal plot of standardized residuals should lie on the 1:1 line (e.g. Figure 5.8b).

A skewed distribution of the residuals (e.g. a few very large positive residuals with a corresponding increase in the number of small negative residuals, or vice versa) results in a curved pattern, probably not passing through the origin. Distributions of residuals with fatter or thinner tails than a Normal distribution (i.e. with more or fewer large residuals, respectively), result in the relationship deviating from a straight line towards the extremes (both positive and negative).

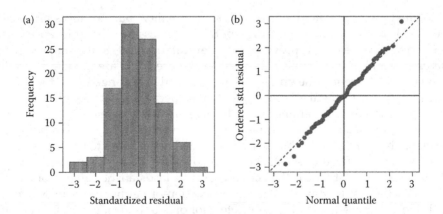

FIGURE 5.8
Checking Normality: (a) histogram of standardized residuals, and (b) Normal (or Q–Q) plot based on standardized (std) residuals with 1:1 line (– –).

The **half-Normal plot** is a variation in which the ordered set of the absolute values of the residuals are plotted against quantiles of the standard Normal distribution. In this case, the ith smallest absolute residual is plotted against the $[50 + 50\,(i - 0.375)/(n + 0.25)]$th percentile. If the residuals are consistent with a Normal distribution, the plot should again show an approximately straight line starting at the origin.

Note that formal statistical distribution tests, such as the Kolmogorov–Smirnov or Anderson–Darling tests, are not strictly appropriate here as they make an assumption of independent observations which is not obeyed by the set of residuals. We cannot therefore recommend the use of these tests in this context.

There are several approaches for dealing with non-Normality of the residuals. One is to use a non-parametric permutation test as an alternative to the F-test (see Section 5.2.4 for more details). Skewness can sometimes be corrected by an appropriate transformation of the data (see Chapter 6). In this situation, the shape of the histogram of the residuals (e.g. in terms of the amount of skewness displayed) may give clues as to possible transformations of the response. Alternatively, where the form of the data suggest that it is unlikely that a Normal distribution can be assumed (e.g. for discrete data including counts or counts as a percentage of a fixed total), more advanced techniques based on other probability distributions (e.g. Poisson or Binomial) for the response variable can be used instead. These analytical approaches are discussed briefly in Chapter 18.

EXAMPLE 5.1D: CALCIUM POT TRIAL*

It is often useful to consider a set of residual plots together, and Figure 5.9 shows a composite display for the standardized residuals, consisting of the fitted values and absolute residuals plots from Figure 5.3 with a histogram of residuals and a Normal plot.

In this case, the fitted values plot (Figure 5.9a), the absolute residuals plot (Figure 5.9b) and the histogram (Figure 5.9c), while not perfect, do not indicate serious departures from homogeneity of variances or from Normality. In the Normal plot (Figure 5.9d), the residuals follow an approximately straight line and it appears that the data are consistent with the assumptions underlying the linear model.

5.2.4 Using Permutation Tests Where Assumptions Fail

Where residual plots indicate departures from Normality (but variances are acceptably homogeneous) an alternative to the F-distribution for assessing the significance of the size of an observed variance ratio is provided by a **permutation test**. In the simplest case of an unstructured sample, for example, the CRD, the observed data are randomly re-allocated to the units (or equivalently, the unit labels are randomly rearranged with the data values remaining fixed) and the analysis is repeated on the permuted data to obtain a new value of the test statistic, here the variance ratio. If there are no treatment differences, then the test statistic should take a similar value (subject to variations due to sampling) for any permutation. This permutation procedure is repeated many times to provide an empirical reference distribution for the test statistic formed under the null hypothesis of no treatment differences, against which the actual observed test statistic can be compared. The probability for the test is computed as the proportion of permutations in which the test statistic is more extreme (defined appropriately for one- or two-sided tests) than the original observed test statistic. An exact (exhaustive) test can be made if the number of possible permutations is small, but in general the test is evaluated for a large subset of random permutations (often 999 to provide a three-digit significance level). Note that where structure is present, the permutation procedure must take it into account.

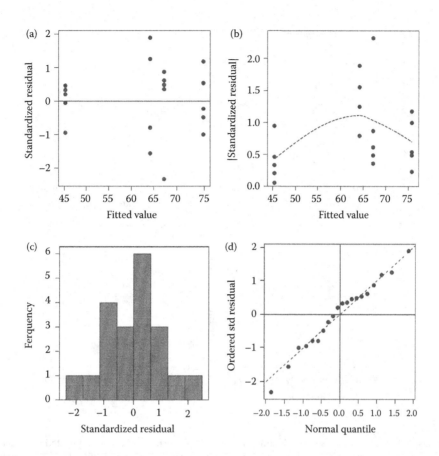

FIGURE 5.9
A composite set of residual plots based on standardized (std) residuals for the calcium pot trial (Example 5.1D). (a) Fitted values plot, (b) absolute residuals plot, (c) histogram of residuals, and (d) Normal plot.

Permutation tests are a form of non-parametric test, i.e. a test for which no probability distribution is assumed, and can be derived for most hypothesis tests. Permutation tests also have a connection with the validity of the F-tests derived from an ANOVA table. If variances are similar across groups and the residual df are large, then the F-distribution used in ANOVA gives a good approximation to the distribution of the permutation test statistic under the null hypothesis (for more details see e.g. Box, Hunter and Hunter, 1978). It follows that if treatments have been allocated to experimental units at random, then under the null hypothesis, the observed F-statistic can be considered as a random sample from the permutation distribution. The F-distribution thus approximates the correct reference distribution without the need for distributional assumptions. Note that this reasoning follows only for properly randomized designs (i.e. experimental studies), and does not apply to many observational studies.

5.2.5 The Impact of Sample Size

Interpretation of the residual plots can be somewhat subjective, and properties of the residuals are much easier to assess visually when sample sizes are large. Take care therefore when dealing with small data sets and, in these situations, a more lenient approach to interpreting residual plots is generally allowed (although remember that the assumptions

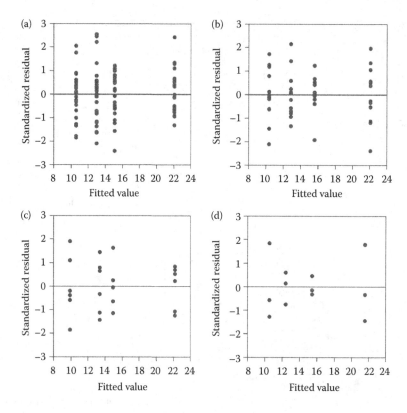

FIGURE 5.10
Fitted values plots for four data sets, each simulated as a CRD with four equally replicated treatments with Normal deviations and total number of observations equal to (a) 100, (b) 48, (c) 24 and (d) 12.

must still be met for the analysis to be valid). Conversely, a stricter approach should usually be taken for larger data sets.

To illustrate the effects of sample size, we have simulated data from a CRD with four equally replicated treatment groups, with replication $n = 25, 12, 6$ and 3, i.e. with a total of $N = 100, 48, 24$ and 12 observations, respectively. The deviations were generated to obey all the assumptions underlying the model, and the true population means for the treatment groups took values 10, 13, 15 and 22. The estimated treatment means show some variation about these true values, as expected. Figures 5.10, 5.11 and 5.12 show the fitted values plots, histograms of residuals and Normal plots, respectively, based on standardized residuals from ANOVA for each of the four simulated data sets. Although each of these data sets obeys the assumptions of homogeneity of variance, independence and Normality, the resemblance between the residual plots and the ideal patterns clearly gets worse as the sample size decreases.

5.3 Using Formal Tests to Diagnose Problems

It is often difficult to judge by eye whether variances are similar across treatment groups, either in residual plots (e.g. Figure 5.1), or in bar charts of treatment means including

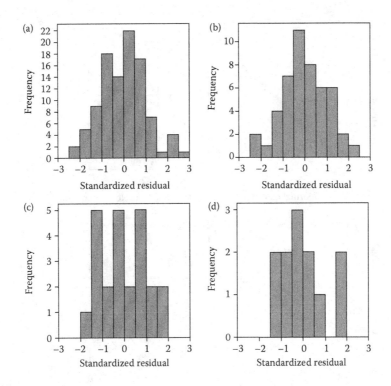

FIGURE 5.11

Histograms of residuals for four data sets, each simulated as a CRD with four equally replicated treatments with Normal deviations and total number of observations equal to (a) 100, (b) 48, (c) 24 and (d) 12.

within-group standard deviations (e.g. Figure 4.1). From both of these graphs, it appears that the variation within treatment D is smaller than that within the other treatments. However, the sampling variation in estimates of variances can be large, especially for small data sets, and so it is often sensible to test formally whether such a set of sample variances could plausibly have arisen from a population with a common variance (but allowing for different group means). Here, we give details of Bartlett's test (Bartlett, 1937; Snedecor and Cochran, 1989), but other tests (e.g. the F_{max}-test of Hartley, Sheffé–Box test, Levene test) can also be used (Sokal and Rohlf, 1995).

Bartlett's test is based on the assumption that data have arisen as a number of samples from Normally distributed populations. The number of samples corresponds to the t different treatment groups in the CRD. The null hypothesis of the test is that the population variances for the different treatment groups are all equal, and the alternative hypothesis is that these population variances are not all equal. The test statistic is based on a comparison, on the logarithmic scale, of the average of the treatment sample variances with a pooled variance estimate. To construct these sample variances algebraically, it is convenient to label observations as y_{jk}, where index j labels the treatment groups ($j = 1 \ldots t$) and k labels the replicate values within treatments ($k = 1 \ldots n_j$), i.e. the same labelling as for the CRD (Chapter 4). The unbiased sample variance for the jth treatment is then calculated (as in Section 2.1) as

$$s_j^2 = \frac{1}{(n_j - 1)} \sum_{k=1}^{n_j} \left(y_{jk} - \bar{y}_{j.} \right)^2 ,$$

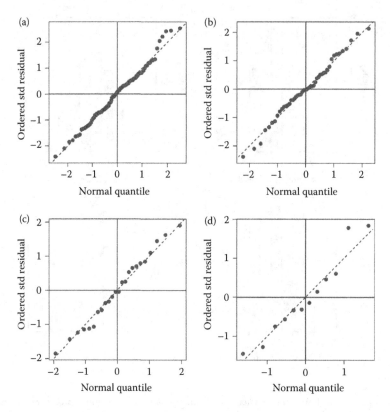

FIGURE 5.12
Normal plots based on standardized (std) residuals for four data sets, each simulated as a CRD with four equally replicated treatments with Normal deviations and total number of observations equal to (a) 100, (b) 48, (c) 24 and (d) 12.

where $\bar{y}_{j\cdot}$ is the sample mean for the jth treatment. The pooled variance estimate, denoted s^2_{pooled}, is calculated as

$$s^2_{\text{pooled}} = \frac{1}{(N-t)} \sum_{j=1}^{t} (n_j - 1)s_j^2 \, , \qquad (5.1)$$

i.e. as a weighted sum of the unbiased treatment sample variances, and is equal to the estimate of s^2 (the ResMS) from the CRD analysis (Section 4.3). The test statistic, X^2, is

$$X^2 = \frac{1}{(1+c)} \left[(N-t)\log_e(s^2_{\text{pooled}}) - \sum_{j=1}^{t} (n_j - 1)\log_e(s_j^2) \right] \, ,$$

where c is a scaling factor calculated as

$$c = \frac{1}{3(t-1)} \left[\left(\sum_{j=1}^{t} \frac{1}{(n_j - 1)} \right) - \frac{1}{N-t} \right].$$

For groups with equal replication, i.e. $n_j = n$ for $j = 1 \ldots t$, these expressions simplify to

$$X^2 = \frac{(n-1)}{(1+c)}\left[t\log_e(s^2_{\text{pooled}}) - \sum_{j=1}^{t}\log_e(s^2_j)\right] \text{ with } c = \frac{t+1}{3t(n-1)} \,.$$

Under the null hypothesis, $X^2 \sim \chi^2_{t-1}$, i.e. X^2 has an approximate Chi-squared distribution with $t - 1$ df (Section 2.2.4). Inequality of sample variances is indicated by larger values of the test statistic, so a one-sided test is appropriate. So, if X^2 is larger than the $100(1 - \alpha_s)$th percentile of this chi-squared distribution, there is evidence (at significance level α_s) that the variances differ between treatment groups.

If data show evidence of unequal variances across treatment groups, and the variances also change systematically with the treatment means (as seen in the fitted values plot), then a transformation might resolve the issue (see Chapter 6). Alternatively, the test may reflect non-Normality of the deviations, and a different probability distribution might be considered (see Chapter 18). If the deviations appear to follow a Normal distribution but the pattern in the variances is not related in a simple manner to trends in the treatment means, then a weighted analysis can be used to account for the different variances associated with different treatment groups. More details about fitting weighted linear models can be found in Rawlings et al. (1998).

EXAMPLE 5.1E: CALCIUM POT TRIAL*

The unbiased sample variances for each treatment in the calcium pot trial were given in Table 4.1 as $s^2_A = 135.20$, $s^2_B = 45.20$, $s^2_C = 105.70$, $s^2_D = 19.30$, with equal replication of $n = 5$ for all treatments. The range of variances seems large, but each estimate is based on only five observations. The pooled variance estimate is calculated with Equation 5.1 as

$$s^2_{\text{pooled}} = \frac{4}{16}(135.20 + 45.20 + 105.70 + 19.30) = \frac{305.40}{4} = 76.35\,.$$

As expected, this is equal to the ResMS from the ANOVA table in Example 4.1C. The scaling factor c is equal to

$$c = \frac{t+1}{3t(n-1)} = \frac{5}{3 \times 4 \times 4} = 0.1042\,,$$

and so

$$X^2 = \frac{16 \times \log_e(76.35) - 4 \times (\log_e(135.20) + \log_e(45.20) + \log_e(105.70) + \log_e(19.30))}{1 + 0.1042} = 3.633\,.$$

The 95th percentile of the chi-squared distribution with 3 df is 7.815 and so the test statistic is consistent with the null hypothesis of equal population variances across treatments.

5.4 Identifying Inconsistent Observations

An **outlier** is an observation that is in some way inconsistent with the rest of the data set. In the context of designed experiments with replication, an outlier is usually an observation that is inconsistent with the other observations in a treatment group. Here, we use the term outlier to describe points with large residuals from the fitted model; therefore, the identification of a point as an outlier may change with the proposed model. Residual plots can be used to identify potential outliers, and many statistical packages automatically identify observations with large residuals as potential outliers.

EXAMPLE 5.2: DISEASE PROGRESS

An experiment, designed as a CRD, investigated disease progression within leaves of oilseed rape plants. The amount of pathogen DNA extracted from leaves of inoculated plants in 11 lines of oilseed rape was measured for 12 replicate plants of each line. The $\log_{10}$-transformed DNA values were analysed. Figure 5.13 shows a composite set of residual plots based on the standardized residuals. One observation clearly stands out

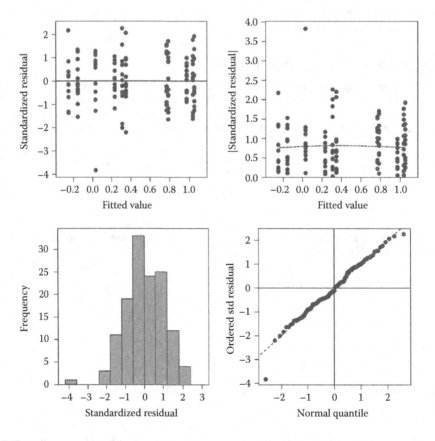

FIGURE 5.13
Composite set of residual plots based on standardized (std) residuals from an experiment to measure disease progress within leaves (Example 5.2). (Data from Y. Huang, Rothamsted Research.)

as having a much larger (negative) residual and should be investigated as a possible outlier. Apart from this one observation, the residual plots are otherwise satisfactory: the
histogram is symmetric, the variances are similar across the treatments (range of fitted
values) and the Normal plot shows a straight line.

Outliers may be ascribed to several different sources: (1) problems with the experimental
procedure; (2) errors in the recording, transcription or data input procedures; (3) an incorrect or incomplete model specification; and (4) a genuine observation that is incompatible
with the rest of the observations.

The first step in dealing with an outlier is to try to determine its origin. Laboratory
notebooks may reveal where problems with the experimental procedure were experienced
or suspected. For example, the effectiveness of the inoculation on the outlying plant in
Example 5.2 might be in question. Maintenance records can be consulted to check the
calibration of equipment. Original data records should always be cross-checked to detect
errors: common errors include the transposition of digits (e.g. 54.2 becomes 45.2) and
movement of the decimal point (e.g. 54.2 becomes 542 or 5.42). Any proven errors should
be corrected or, if they cannot be corrected, the observations should be set as missing or
removed, and any dubious observations should be flagged. Finally, you should consider
whether an observation might correspond to a different population (e.g. a different species
or subspecies) from the one of interest, in which case a different result might be expected
for that observation.

The next step is to decide what to do with the observations identified as anomalous but
where no error can be proved: you must decide whether to retain or remove these observations. It is helpful to consider at this point whether the model is compatible with the data.
Are there any important explanatory variables that have not been included in the model?
Inclusion of these variables might improve the fit of the model and reduce the number of
potential outliers. For example, if species are showing different reactions to a treatment,
then explicitly allowing for this differential response in the model might accommodate
all of the observations. If the deviations do not follow a Normal distribution with equal
variance, then data transformation might be advantageous (see Chapter 6) or another
probability distribution can be used (see Chapter 18). In Figure 5.13, the residual plots are
entirely acceptable apart from the single outlying point, so neither of these options would
be justified.

Outliers should not be discarded indiscriminately or without careful consideration, not
least because observations are often expensive to obtain, but also because this will affect
the analysis. Sometimes, particularly with large data sets (such as Example 5.2), the effect
of removing an outlier is negligible, but usually estimates of the model parameters will
change and the estimated residual variance will decrease, sometimes substantially, thus
increasing the chance of rejecting H_0. This is not necessarily desirable, as the residual variance will be underestimated when we eliminate outliers that are genuine observations,
resulting in a larger Type I error (more false-positive results) than expected. The decision
on which outliers to retain and which to eliminate must always be reported. If the results
change markedly when outliers are excluded, then it is good practice to report these differences. Remember that it may be considered fraudulent to remove 'inconvenient' data
points from an analysis without good justification. In addition, always bear in mind that
if the outlier is a genuine observation, then it might be the most important point in the
study, because it indicates unexpected behaviour in the system. A story about the hole in
the ozone layer is often quoted in this context. The traditional version of the tale relates
that NASA scientists should have been the first to discover the ozone hole over Antarctica,

but their satellite instruments used automatic outlier detection and deletion methods that removed anomalous readings, and so the seasonal ozone depletion went unnoticed until reported by Farman et al. (1985). In fact, Pukelsheim (1990) reports that the satellite observations were flagged, checked against data from a ground station, and only then dismissed because they contradicted the ground station readings. It later transpired that the ground station data were misleading due to faulty instrument settings! We believe that the full version of this story carries even more warnings than the traditional version: certainly you should not use automatic methods to delete outliers without examining them first, but when you cross-check against another method, you also need to be sure that the other method is reliable.

EXAMPLE 5.1F: CALCIUM POT TRIAL*

In the calcium pot trial, treatment group C has four observations in the range 70–74 (Table 4.1) and one much smaller observation with value 49 (pot 4). This large discrepancy suggests that this observation should be investigated as a potential outlier. This observation has a standardized residual of −2.33. However, although it is the largest residual (in absolute value), it does not appear to be inconsistent with the overall distribution of the residuals shown in Figure 5.9. It is therefore sensible to examine this observation for potential sources of error, as suggested above, but if none is found, there is no justification for removing it from the analysis.

An alternative analytical approach that allows the retention of potential outliers is the use of 'robust' statistical methods. These are designed to be less sensitive to the presence of outliers, and further details can be found in Barnett and Lewis (1994).

EXERCISES

5.1 Obtain the simple and standardized residuals from the ANOVA for the data from (a) Exercise 4.2 and (b) Exercise 4.3. Use a scatter plot to compare the simple and standardized residuals in each case. Can you explain the patterns that you see? Are there any potential outliers?

5.2 For the data sets in each of Exercises (a) 4.2, (b) 4.3, (c) 4.4 and (d) 4.5, produce a set of residual plots based on standardized residuals, including a histogram of residuals, a fitted values plot, an absolute residuals plot and a Normal plot. Give a critical assessment of whether the ANOVA assumptions are reasonable in each case. Is there any evidence of outliers?

5.3* An experiment compared the growth of tomato seedlings in eight commercial composts. Space was available in a glasshouse to place 32 small pots in a single line along the edge of one bench. The area was assumed to be homogeneous and so a CRD was used, with four pots of each type of compost. One seedling was transplanted into each pot and the heights of the young plants (cm) were measured after 2 weeks. Pots were numbered 1–32 along the bench and plants were measured in order of pot number. The pot numbers (*Pot*), composts used (factor Compost) and resulting plant heights (variate *Height*) are in file COMPOST.DAT. Analyse these data and inspect standardized residual and index plots. Are the assumptions of your model satisfied?

5.4 Compare the unbiased sample variances for each treatment group from Exercise 4.3 using Bartlett's test. Is there any evidence of variance heterogeneity?

5.5* An experiment was devised to evaluate the effect of four watering regimes on root growth using a CRD. Each regime was applied to 12 individual plants growing in pots. Measurements of total root length (cm) were made at the end of the experiment. The unit numbers (*Pot*), watering regimes (factor Regime) and total root lengths (variate *Length*) are in file WATERING.DAT. Analyse these data and inspect plots of standardized residuals. Are there any potential outliers? On inspection of the original data sheets, it was discovered that one observation had been mistyped as 47 rather than 74. Correct this data value and rerun the analysis. Comment on whether the model assumptions are reasonable for these data.

6

Transformations of the Response

In Chapter 5, we discussed graphical diagnostic tools for inspecting residuals and checking the assumptions underlying the linear model (see Section 4.1). In particular, we considered how to evaluate the assumptions that the deviations have a common variance (homogeneity of variances, Section 5.2.1) and come from a Normal distribution (Section 5.2.3). We indicated that transformations of the response variable might be used to remove skewness in the distribution of the residuals (i.e. evidence of a non-Normal distribution) or to correct systematic changes in the variances of the residuals (i.e. evidence of variance heterogeneity, as illustrated in Figure 5.3).

In this chapter, we show how different transformations can be used to deal with particular violations of the model assumptions. We start by discussing the general rationale for data transformations (Section 6.1) and then concentrate on three particularly useful transformations: the log and square root transformations for positive response variables (i.e. with values ≥ 0), and the logit transformation for proportions (between 0 and 1) or percentages (between 0 and 100, Section 6.2). It is usually desirable to relate results to the original scale of measurement, and so we also discuss the procedure of back transformation (Section 6.3). We then take a closer look at the interpretation of the log transformation in terms of a multiplicative model (Section 6.4). Finally, we review some other methods for analysing non-Normal responses, which are useful when transformation is either unsuccessful or inappropriate (Section 6.5).

6.1 Why Do We Need to Transform the Response?

The validity of the conclusions from a statistical analysis depends on the validity of the assumptions underlying that analysis. Applying a transformation to the response variable is an option that may allow the assumptions to be satisfied sufficiently so that we can continue to use the simple linear model and reach valid conclusions. A transformation, or **data transformation**, is the process of using a mathematical function to map the response variable from the original scale of measurement onto another (the transformed) scale.

In the context of simple models involving factors, the most common reason for transforming responses is to stabilize the variance of the residuals so that a pooled estimate (across all treatments) can sensibly be used, i.e. to make the assumption of homogeneity of variance, or homoscedasticity, valid. A second important use of transformation is to make the distribution of the deviations closer to a Normal distribution (recall that conclusions of ANOVA based on the t- and F-statistics rely on the deviations having an approximate Normal distribution). Finally, the use of a transformation may provide a scale on which the additive form of the linear model is more realistic. If we are fortunate, then a transformation may achieve several aims simultaneously. However, on some occasions, whilst one aim is achieved by the use of transformation, other aspects of the analysis may be made worse.

The assumption that the deviations follow a Normal distribution implies that the response should be measured on a continuous, or close-to-continuous, scale without any abrupt truncation. There are several common types of response that do not apparently comply with this description. For example, counts must be positive (i.e. $y \geq 0$) and can take only integer values; they are clearly not continuous. If the counts are small then the distribution may also be abruptly truncated at zero. These factors are especially important for small counts. Proportions ($0 \leq p \leq 1$) that are calculated with respect to small samples (e.g. the proportion of dead aphids in a sample of 10) can take only certain values (e.g. 0, 0.1, 0.2 … 1) and so again are not continuous, and may also be abruptly truncated if many proportions are close to zero or one. These are obvious cases where the Normal assumption is invalid. In addition, for these types of response the variance is almost always related to the expected response for an observation (usually termed a variance–mean relationship). For counts and proportions there are other analytical approaches based on more appropriate probability distributions (i.e. Poisson and Binomial, respectively). We briefly outline these approaches, and when they are likely to be required, in Section 6.5 but we leave a more detailed discussion until Chapter 18. Other types of response may be continuous but also show a variance–mean relationship (and possibly truncation). For example, quantities such as height or weight are usually (effectively) continuous but must be positive, and larger expected values are often associated with greater variation. Some proportions may be effectively continuous, for example, percentage area as assessed by eye or a computer, but may show greater variance around 0.5 than at the limits of the scale (0 or 1). In all of these cases, transformation often provides a good-enough approximation to the assumptions of the linear model to make the analysis valid and reliable.

After transformation (based on consideration of the residual plots introduced in Chapter 5), the analysis should be repeated for the transformed response, and residuals from the new analysis should be inspected as usual. Sometimes, it is necessary to try several different transformations until satisfactory residual plots are obtained, and sometimes it will not be possible to find a suitable transformation.

6.2 Some Useful Transformations

In this section we describe the most common transformations, concentrating on the logarithmic, square root and logit transformations.

6.2.1 Logarithms

Possibly, the most common transformation is the **logarithmic (log) transformation**, which can take several related forms. The concept of the log transformation can be most easily understood for the common logarithm (log to base 10), which maps the original response variable, y, to a new variable, z, such that y is equal to 10 raised to the power z, i.e. $y = 10^z$, and usually written as $z = \log_{10}(y)$. Values of $z = 0$, 1 or 2 on the $\log_{10}$ scale thus correspond to values of $y = 10^0 = 1$, $10^1 = 10$ or $10^2 = 100$ on the original scale. The natural logarithms (log to base e) work in a similar way but use Euler's number, $e = 2.71828\ldots$ The natural logarithm maps y to z such that y is equal to e raised to the power z, i.e. $y = e^z$, or equivalently $y = \exp(z)$, and can be written as $z = \log_e(y)$. Logarithms to other bases are similarly defined, and any logarithm can be used for a log transformation. There are several different conventions of notation for specifying logarithms. Elsewhere you may see $\log(y)$ used for either $\log_{10}(y)$ or $\log_e(y)$, and $\ln(y)$ for $\log_e(y)$. In this book we use the unsubscripted term 'log' to represent a logarithm in general discussion, but always specify the base as a subscript for specific examples.

All log transformations are defined only for $y > 0$, i.e. for strictly positive values, but the transformed variable, z, can take any value. Values greater than one on the original scale ($y > 1$) are mapped onto positive values ($z > 0$), and values less than one on the original scale ($0 < y < 1$) are mapped onto negative values ($z < 0$), with $y = 1$ always mapped onto $z = 0$. Log transformations bring larger values closer together and spread smaller values relatively further apart. They therefore provide a useful way of dealing with responses that exhibit a right-skewed distribution, i.e. with only a short tail to the left of the distribution peak and a long tail stretched out to the right (such as that illustrated in Figure 6.1a), making the distribution more symmetrical (as shown in Figure 6.1b) after log transformation. Common examples include counts of insects on plants (where a few plants are very heavily infested), measures of size such as weights or lengths (where a few individuals are particularly heavy or large), counts of colony-forming units in pathogen cultures, and concentrations of plant nutrients and trace metals in soil samples. Clearly, if the responses are already reasonably symmetric or left-skewed then a log transformation is unlikely to be helpful.

Log transformations are also useful for stabilizing the variance of responses for which the variance increases in proportion to the expected response, a feature often associated with integer counts in ecology, and usually detected via fitted value plots as in Figure 5.3. On applying a log transformation, the variation associated with larger values is decreased, hopefully achieving homogeneity of variances across the range of the variable.

Finally, a log transformation can be applied to transform a multiplicative model onto an additive scale, as required for the form of the linear model. This valuable consequence of applying a log transformation is discussed in detail in Section 6.4.

The choice of which base to use is arbitrary, though the type of response may suggest the choice of a particular base. For example, for numbers of colony-forming units, where values are often in the range from 10^4 to 10^9 across a set of observations, and for which changes of an order of magnitude are important, the obvious choice is the

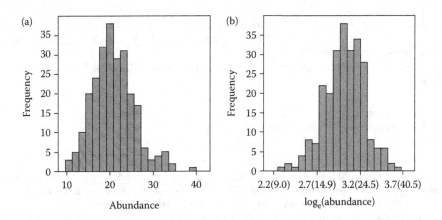

FIGURE 6.1
Distribution of a sample of insect counts: (a) original responses and (b) following $\log_e$ transformation (axis labels within parentheses indicate value on original scale).

logarithm to base 10 because, for example, a unit change on the $\log_{10}$ scale reflects a 10-fold increase in numbers. The default used by many statisticians in the life sciences is the natural logarithm; possibly because Euler's number, e, seems to be related to many natural phenomena.

An important consequence of applying a log transformation is that the influence of larger observations on treatment or group means is less than on the original scale, whilst the influence of small observations is increased. This may be inappropriate in some contexts.

A major constraint of applying a log transformation is that it is defined only for positive values, $y > 0$. However, for many types of positive response, such as integer counts, zero is a valid observation and applying a log transformation then results in an undefined value. When there are only a few zero values, it is common practice to add a small **offset**, c, to every response prior to applying the transformation function, so the transformation becomes

$$z = \log(y + c) \, .$$

The inclusion of an offset provides a degree of flexibility in the transformation process, but the choice made can affect the outcome, so the offset should be chosen with some care. For integer insect counts, it is usual to add an offset of $c = 1$, i.e. $z = \log(y + 1)$. In other cases, a simple rule of thumb is to use an offset equal to half the smallest positive value recorded. For example, if the smallest positive value for a response variable is 0.1 g, we might add an offset of $c = 0.05$. In practice, it might be necessary to try several different offsets to find a value that gives adequate residual plots.

EXAMPLE 6.1A: BEETLE MATING

An experiment was conducted to investigate the viability of interspecies mating in leaf beetles by examination of the results when females from two species of willow beetle (the brassy willow beetle, *Phratora vitellinae*, and the blue willow beetle, *Phratora vulgatissima*) were mated with males from either their own species (intraspecies mating) or the other species (interspecies mating), i.e. there were four treatments ($t = 4$) in total (for further details, see Peacock et al., 2004). The experiment was carried out as a CRD (completely randomized design) with 10 replicates of each treatment ($n = 10$). We analyse the number of eggs laid by each female; the data are presented in Table 6.1 and can be found in file BEETLES.DAT, where factor Treatment has four levels (labelled as 1 = *P. vit.* × interspecies, 2 = *P. vit.* × intraspecies, 3 = *P. vulg.* × interspecies, 4 = *P. vulg.* × intraspecies) and the response is held in variate *Eggs*.

The untransformed counts were analysed by one-way ANOVA (see Section 4.3) using model

Response variable: *Eggs*
Explanatory component: Treatment

A composite set of residual plots based on standardized residuals from this analysis is shown in Figure 6.2. In the fitted values and absolute residual plots, the variance appears greater for larger fitted values, with a suggestion of skewness in the histogram and a slight curve in the Normal plot, both showing some possible outliers.

In an attempt to remove the observed variance–mean relationship, the counts were transformed to the $\log_{10}$ scale as *logEggs* = $\log_{10}(Eggs)$, and the transformed response was analysed, using model

TABLE 6.1

Number of Eggs Laid by Females of Two Willow Beetle Species
(*P. vitellinae* and *P. vulgatissima*) Following Inter- or Intraspecies
Mating (Example 6.1A and File BEETLES.DAT)

	P. vitellinae		*P. vulgatissima*	
Replicate	Interspecies $j = 1$	Intraspecies $j = 2$	Interspecies $j = 3$	Intraspecies $j = 4$
1	57	90	82	136
2	15	80	91	117
3	40	101	66	181
4	34	59	98	41
5	42	73	82	89
6	19	51	134	106
7	43	43	51	133
8	39	57	96	98
9	36	42	52	106
10	24	66	91	79
$\bar{y}_{j.}$	34.9	66.2	84.3	108.6
$\tilde{y}_{j.}$	32.6	63.6	81.1	102.0

Source: Data from Rothamsted Research (A. Karp).
Note: $\bar{y}_{j.}$ is the arithmetic sample mean and $\tilde{y}_{j.}$ is the geometric sample mean
for the *j*th treatment.

Response variable: *logEggs*
Explanatory component: Treatment

This model can be written in mathematical form as

$$logEggs_{jk} = Treatment_j + e_{jk} ,$$

where $logEggs_{jk}$ is the $\log_{10}$-transformed number of eggs for the *k*th replicate ($k = 1 \ldots 10$)
of the *j*th treatment (numbered $1 \ldots 4$ as for the factor levels given above) with deviation
e_{jk}, and $Treatment_j$ is the population mean for the *j*th treatment on the $\log_{10}$ scale. Plots
of standardized residuals from this new model are shown in Figure 6.3. The spread of
residuals is now more consistent across the range of fitted values, although still not perfect. The histogram seems slightly skewed in the other direction, and the Normal plot
still shows some curvature, although the extreme points are now more consistent with
the overall pattern. The ANOVA table for this model is Table 6.2.

The null hypothesis is that the treatment population means on the $\log_{10}$ scale are
all equal. On the basis of the ANOVA of the transformed response ($F_{3,36} = 19.254$,
$P < 0.001$) we should reject this hypothesis and conclude that differences exist
between the treatment means on the $\log_{10}$ scale. The sample means calculated from
the logged counts for the four treatments, used to predict the corresponding population means, are listed in Table 6.3. The ResMS gives $s^2 = 0.0238$, from which we derive
the common SEM of $\sqrt{(s^2/10)} = 0.0488$, and the SED of $\sqrt{(2s^2/10)} = 0.0690$ for comparing
pairs of treatments, both with 36 df. Most eggs are laid by the intraspecies-mated
P. vulgatissima females (treatment 4) and fewest by the interspecies-mated *P. vitellinae*
females (treatment 1). In Section 6.3, we discuss how to relate these predictions back
to the original scale.

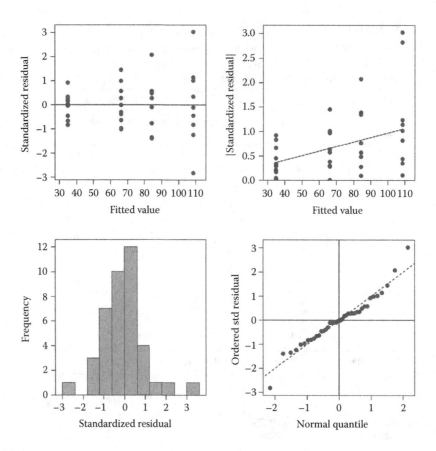

FIGURE 6.2
A composite set of residual plots based on standardized residuals for the number of eggs laid in the beetle mating experiment (Example 6.1A).

TABLE 6.2

ANOVA Table for the Log_{10}-Transformed Number of Eggs from the Beetle Mating Experiment with Four Treatments (Factor **Treatment**) (Example 6.1A)

Source of Variation	df	Sum of Squares	Mean Square	Variance Ratio	P
Treatment	3	1.3751	0.4584	19.254	< 0.001
Residual	36	0.8571	0.0238		
Total	39	2.2322			

TABLE 6.3

Predicted Treatment Means for Log_{10}-Transformed Numbers of Eggs from the Beetle Mating Experiment, with SEM = 0.0488, SED = 0.0690 on 36 df (Example 6.1A)

Treatment 1 (*P. vit.* × Inter)	Treatment 2 (*P. vit.* × Intra)	Treatment 3 (*P. vulg.* × Inter)	Treatment 4 (*P. vulg.* × Intra)
1.513	1.804	1.909	2.008

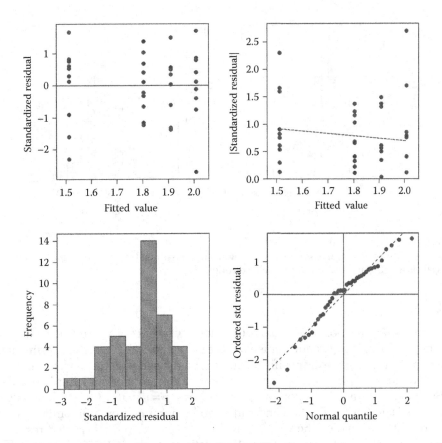

FIGURE 6.3
A composite set of residual plots based on standardized residuals for the $\log_{10}$-transformed number of eggs from the beetle mating experiment (Example 6.1A).

We return to the log transformation in Sections 6.3 and 6.4 and discuss alternatives to the log transformation for counts in Section 6.5.

6.2.2 Square Roots

The **square root transformation**, $z = \sqrt{y}$, is defined for all non-negative numbers, i.e. positive numbers and zero ($y \geq 0$), and maps onto that same range ($z \geq 0$). Hence, it is a potentially useful transformation for any response that can take only non-negative values, but it can be particularly appropriate for responses measured as areas, as their square roots can be interpreted as being proportional to an average radius or diameter. It can also be used as an alternative to a log transformation for positive responses. Like the log transformation, the square root transformation tends to bring larger values closer together and spread smaller ones relatively further apart. However, for the square root transformation this rescaling is not as strong and so it may be more successful in cases where the log transformation has over-corrected skewness or variance heterogeneity. The effect of the square root transformation in correcting skewness is shown in Figure 6.4. Because this transformation is defined for zeros ($y = 0$), it can be used without an offset when zero responses are present. However, some authors have suggested that an offset of $c = 0.5$ might still be useful in this case (see Sokal and Rohlf, 1995).

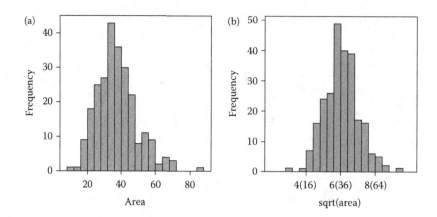

FIGURE 6.4
Distribution of a sample of area measurements: (a) original responses and (b) following square root transformation (axis labels within parentheses indicate value on original scale).

6.2.3 Logits

Data in the form of proportions usually (but not always) occur where each experimental unit contains a sample of m entities (usually the same number for each experimental unit) and the number that fall into a specific category, y, has been counted. For example, we might sample 25 weed plants from a field plot and count the number showing herbicide resistance. Such data are usually reported as proportions, $p = y/m$ (for $0 \leq p \leq 1$), or as percentages, $P = 100 \times y/m$ (for $0 \leq P \leq 100$) and can often be described by a Binomial distribution (see Section 2.2.1), for which the variance is directly related to the expected response. For this type of response, the **logit transformation**, defined as $z = \log_e(y/(m - y))$ or equivalently $z = \log_e(p/(1 - p))$ or $z = \log_e(P/(100 - P))$, is often applied to remove or reduce the variance–mean relationship and provide the homogeneity of variance assumed for a linear model.

Occasionally, proportions (or percentages) do not have a direct interpretation as $p = y/m$. For example, computer measurement of the proportion of lesion area on plant leaves works on an effectively continuous scale. However, these measurements may display similar patterns of variance heterogeneity so that application of the logit transformation in the form $z = \log_e(p/(1 - p))$ or $z = \log_e(P/(100 - P))$ is still appropriate.

Theoretically, the logit transformation maps from the range (0, 1) onto an unrestricted range, but in practice we need consider only the range (–4, 4) for most applications. The proportion 0.5 maps onto a logit of zero, with proportions less than 0.5 resulting in negative values and proportions greater than 0.5 resulting in positive values. The logit transformation tends to bring values at the centre of the range (~0.5) closer together and to spread values at the ends of the range (approaching 0 or 1) relatively further apart. Figure 6.5a shows the distribution of proportions obtained as the number of diseased plants out of samples of size 25 for a survey of a single variety with mean prevalence of 0.8. Figure 6.5b shows the logit-transformed proportions, which are all positive as the original proportions were greater than 0.5. The logit transformation has corrected the left-skewness of the untransformed responses by spreading out the larger proportions relative to those closer to 0.5. Proportions often show an increased variance for values with an expected response around 0.5, relative to those nearer to the ends of the scale (in many cases, a property inherited from the Binomial distribution), and the logit transformation tends to counteract this property and, if we are fortunate, result in homogeneity of variances across the range of the variable.

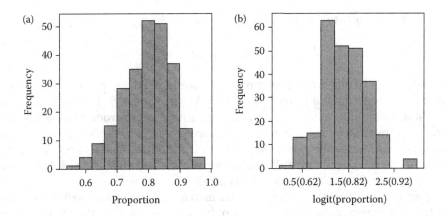

FIGURE 6.5
Distribution of a sample of proportions: (a) original responses and (b) following logit transformation (axis labels within parentheses indicate value on original scale).

The logit transformation has the constraint that it is undefined at the limits of the range, i.e. for $p = 0$ or $p = 1$ (or $P = 0$ or $P = 100$). To avoid the problem that this creates, we might compute the proportion from a sample of size m using an offset c as $p = (y + c)/(m + 2c)$. This moves the ends of the range symmetrically away from zero and one, as the minimum proportion is then $c/(m + 2c)$ (> 0) and the maximum is $(m + c)/(m + 2c)$ (< 1), with the centre of the range unmoved, as $(m/2 + c)/(m + 2c) = 0.5$. The logit transformation can then be calculated directly in terms of the adjusted proportions or percentages or as $z = \log_e[(y + c)/(m + c − y)]$. Here, the offset c is usually chosen to be equal to 0.5 or 1, but other offsets can be used. If the responses are recorded as proportions or percentages directly (so p or P is known but y and m are not), then the adjusted transformation takes the form $z = \log_e[(p + c)/(1 + c − p)]$, or $z = \log_e[(P + c)/(100 + c − P)]$, where c is chosen to be the minimum of two values: the difference between 0 and the smallest observation, and the difference between 1 (or 100 when using percentages) and the largest observation.

The quantity $p/(1 − p)$ is sometimes called the **odds**, so that the logit transformation corresponds to the logarithm of the odds, or **log-odds**. Results of an analysis with the logit transformation are therefore sometimes interpreted in terms of a change in the odds, or odds ratios. Whilst this is an interpretation that is commonly used in medical statistics and in the betting industry, it is an interpretation that is often difficult to relate to the biological background of an analysis, and so we generally avoid this form of interpretation.

We discuss alternatives to the logit transformation in Sections 6.2.4 and 6.5.

6.2.4 Other Transformations

Many other transformations have been suggested for data analysis, and these are often related to a physical interpretation of the measurement scale. For example, a cube root transformation might be considered for volumes, as the transformed response could be related to average size in one dimension. Or a reciprocal transformation might be considered for growth rates measured as mm/day, as the transformed response could be interpreted as the number of days required to grow 1 mm. In an ideal case, a transformation will give an interpretable physical representation of the response as well as enabling it to satisfy the assumptions of the analysis so that the conclusions are valid.

The family of **power transformations**, defined as

$$z = \frac{(y^\lambda - 1)}{\lambda} \quad \text{for } \lambda \neq 0 \quad \text{and} \quad z = \log_e y \quad \text{for } \lambda = 0,$$

encompasses many of the transformations commonly used for positive responses. The parameter λ is known as the power parameter. When $\lambda = 0$, 0.5, 1 and –1, the resulting transformations are equivalent to a natural logarithm, square root, simple linear transformation and reciprocal transformation, respectively. The **Box–Cox transformation** provides a method for deciding the best power transformation to use to obtain an approximate Normal distribution, but a full description of this approach is outside the scope of this book. For more details, see Sokal and Rohlf (1995).

For proportions or percentages, the most common alternative to the logit is the **arcsine**, or **angular** transformation, defined as $z = \arcsin \sqrt{p}$ or $z = \arcsin \sqrt{P/100}$. Other possibilities include the **probit** and **complementary log–log** transformations. Further details on these alternative transformations can be found in Sokal and Rohlf (1995).

6.3 Interpreting the Results after Transformation

Following the analysis of a transformed response (e.g. Example 6.1A), interpretation of the results is often aided if they can be represented on the scale of the original measurements. Unfortunately, standard errors and other measures of variability (i.e. SEMs, SEDs and LSDs) cannot be back-transformed directly because most transformations impose a non-linear re-scaling so that the size of the back-transformed error should differ according to the predicted value(s) with which it is associated. This means that the comparison of the difference between two predictions based on an appropriate SED must be made on the transformed scale. However, the limits of confidence intervals for predictions of treatment population means or differences (see Section 4.4) derived on the transformed scale can be back-transformed, along with the predicted value, for presentation and interpretation on the original scale. Note that whilst confidence intervals on the transformed scale are symmetric about the estimated value, they are usually asymmetric on the back-transformed scale.

Formulae for the transformations and back-transformations corresponding to the log, logit and square root functions are presented in Table 6.4. Because of the importance of the log transformation, we discuss the interpretation of back-transformed predictions from the log scale in detail in Section 6.4.

Note that, even when a transformation appears to have been successful, when an offset has been included then the back-transformation may lead to estimates outside the valid range of the original variable (e.g. negative values for counts, values exceeding one for proportions, etc.), which is clearly undesirable. In this case, other methods should be used (see Section 6.5).

EXAMPLE 6.1B: BEETLE MATING

Table 6.5 shows back-transformed values of the treatment sample means and 95% confidence limits calculated from the formulae in Section 4.4. For example, the 95% confidence interval for *P. vitellinae* × interspecies mating (treatment 1) calculated on the $\log_{10}$ scale requires the predicted population mean, its SEM, and the 97.5th percentile of the t-distribution with 36 df, given by

TABLE 6.4

Common Transformations and Their Inverses (Back-Transformations)

Transformation	Description	Back-Transformation
$z = \log_{10}(y)$	Common logarithm	$y = 10^z$
$z = \log_e(y)$	Natural logarithm	$y = e^z$
$z = \log_{10}(y + c)$	Common logarithm with offset c	$y = 10^z - c$
$z = \log_e(y + c)$	Natural logarithm with offset c	$y = e^z - c$
$z = \log_e(y/(m - y))$	Logit	$y = m\, e^z/(1 + e^z)$
$z = \log_e[(y + c)/(m - y + c)]$	Logit with offset c	$y = [(m + c)\, e^z - c]/(1 + e^z)$
$z = \log_e[(P + c)/(100 - P + c)]$	Logit of percentages with offset c	$P = [(100 + c)\, e^z - c]/(1 + e^z)$
$z = \sqrt{y}$	Square root	$y = z^2$
$z = \sqrt{(y + c)}$	Square root with offset c	$y = z^2 - c$

TABLE 6.5

Predicted Means (Middle Value) and Lower and Upper 95% Confidence Limits (First and Third Values, Respectively) on the Back-Transformed Scale for the Beetle Mating Experiment (Example 6.1B)

		Mating Type	
		Interspecies	Intraspecies
Species of female	*P. vit.*	25.9, 32.6, 40.9	50.7, 63.6, 79.9
	P. vulg.	64.6, 81.1, 101.8	81.2, 102.0, 128.1

$$\overline{Treatment}_1 = 1.513; \quad SEM = 0.0488; \quad t_{36}^{[0.025]} = 2.028 \;.$$

The CI is then computed as $1.513 \pm (2.028 \times 0.0488) = (1.414, 1.612)$. These limits are then back-transformed to give $(10^{1.414}, 10^{1.612}) = (25.9, 40.9)$. The predicted mean is back-transformed similarly, i.e. $10^{1.513} = 32.6$. Note that the back-transformed mean is smaller than the midpoint of the confidence interval (which is 33.4), and is slightly smaller than the treatment mean calculated from the original data ($\overline{y}_1. = 34.9$, Table 6.1); this is discussed further in Section 6.4.

It is possible to obtain an approximation of the standard errors (SEMs or SEDs) for the means on the back-transformed scale. These can be obtained by the **delta method**, which is outside the scope of this book, but is commonly used for non-linear models (for more details, see Casella and Berger, 2002). However, there is no warranty that this method will provide adequate estimates, and we do not recommend its use in the current context.

6.4 Interpretation for Log-Transformed Responses

In the case where a log transformation is appropriate, so that all assumptions of the analysis are met by the log-transformed response, we can interpret the back-transformed predictions in terms of a multiplicative model. This connection exists because the sum of logarithms of two (or more) values is equal to the logarithm of their product, expressed as

$$\log(a) + \log(b) = \log(a \times b) .$$

Now consider the CRD model (Equation 4.1) as applied to a $\log_e$-transformed response, $z_{jk} = \log_e(y_{jk})$, which takes the form

$$z_{jk} = \mu_j + e_{jk} .$$

We back-transform this model by applying the exponential function to both sides of the equation (see Table 6.4), to get

$$\exp(z_{jk}) = \exp(\mu_j + e_{jk}) . \tag{6.1}$$

From the original transformation, we know that $\exp(z_{jk}) = y_{jk}$. We can simplify the expression on the right-hand side further by using the mathematical property that

$$\exp(a + b) = \exp(a) \times \exp(b) ,$$

i.e. the exponential of a sum is equal to the product of the exponentials of the components of the sum. We can therefore rewrite Equation 6.1 as

$$y_{jk} = \exp(\mu_j) \times \exp(e_{jk}) .$$

This is now a multiplicative model on the natural scale: the components of the model are multiplied together rather than added together, with the log transformation providing this change in the form of relationship.

The predicted values from this model are the treatment means formed on the $\log_e$ scale. The predicted population mean for the jth treatment is calculated as

$$\hat{\mu}_j = \bar{z}_{j.} = \frac{1}{n} \sum_{k=1}^{n} \log_e(y_{jk}) = \log_e(\tilde{y}_{j.}) ,$$

which is the natural logarithm of the geometric mean for the jth treatment (here denoted $\tilde{y}_{j.}$, see Mathematical Aside 6.1). The back-transform of this prediction is therefore simply the geometric mean with respect to the original responses for the jth treatment group. One characteristic of the geometric mean is that it is always smaller than or equal to the corresponding arithmetic mean. The difference depends on the skewness of the sample: the more right-skewed the sample, the larger the discrepancy in these two measures of location. It follows that the back-transformed prediction for any treatment will always be smaller than the arithmetic sample mean for that treatment in the original data (as in Example 6.1B).

Differences between treatment predictions on the $\log_e$ scale can be interpreted in terms of ratios on the back-transformed scale, as a difference on the $\log_e$ scale between the jth and kth treatments, written as $\hat{\mu}_j - \hat{\mu}_k$, is back-transformed as

$$\exp(\hat{\mu}_j - \hat{\mu}_k) = \exp(\hat{\mu}_j)/\exp(\hat{\mu}_k) = \tilde{y}_{j.}/\tilde{y}_{k.} ,$$

using the general rule that $\exp(a - b) = \exp(a)/\exp(b)$. The back-transformation of a predicted difference between two treatments on the $\log_e$ scale is therefore equivalent to the ratio of the corresponding geometric means on the untransformed scale. We can also interpret the confidence interval (CI) for a difference in population means on the back-transformed scale. The limits of a $100(1 - \alpha_s)\%$ CI take the form

$$\left((\hat{\mu}_j - \hat{\mu}_k) - \mathrm{LSD}_{j,k}, \quad (\hat{\mu}_j - \hat{\mu}_k) + \mathrm{LSD}_{j,k} \right),$$

with back-transform

$$\left(\exp(\hat{\mu}_j - \hat{\mu}_k)/\exp(\mathrm{LSD}_{j,k}), \quad \exp(\hat{\mu}_j - \hat{\mu}_k) \times \exp(\mathrm{LSD}_{j,k}) \right).$$

The quantity $\exp(\mathrm{LSD}_{j,k})$ can therefore be interpreted as a multiplicative factor giving a range of plausible values on the original scale in terms of a percentage decrease or increase. This is illustrated in Example 6.1C.

If the underlying physical process is multiplicative, then the log transformation maps this onto an additive form, congruent with the form of the linear model. It is fairly common to find that an interaction that is statistically significant on the original scale becomes non-significant on the log scale. The logit transformation may play a similar role in transforming proportions or percentages onto a scale where an additive model is more appropriate.

Similar properties hold for log transformation to any base by substituting in the appropriate back-transformation function, for example, for the $\log_{10}$ transformation, substitute the function 10^z in place of $\exp(z)$.

EXAMPLE 6.1C: BEETLE MATING

The results from the transformed analysis in Example 6.1A indicate that there are differences between treatments – we now wish to interpret these differences with respect to a multiplicative model. On the $\log_{10}$ scale, the difference between the predicted population means (Table 6.3) for *P. vulg.* × intraspecies mating (treatment 4) and *P. vit.* × interspecies mating (treatment 1) is

$$\overline{Treatment_4} - \overline{Treatment_1} = 2.008 - 1.513 = 0.496 \,.$$

As expected, the back-transform of this difference ($10^{0.496} = 3.13$), is equivalent to the ratio of the geometric means on the original scale for each treatment (Table 6.1), with

$$\frac{\tilde{y}_{4\cdot}}{\tilde{y}_{1\cdot}} = \frac{102.0}{32.6} = 3.13 \,.$$

We therefore estimate that, on average, when both were mated with *P. vulgatissima* males, *P. vulgatissima* females laid 3.13 times as many eggs as *P. vitellinae* females.

The SED on the $\log_{10}$ scale is 0.0690 with 36 df, and $t_{36}^{[0.025]} = 2.028$ (Example 6.1B), so the 5% LSD is equal to $\mathrm{SED} \times t_{36}^{[0.025]} = 0.0690 \times 2.028 = 0.1399$. We can calculate a 95% CI for this difference on the transformed scale as

$$\left(\overline{Treatment_4} - \overline{Treatment_1} \right) \pm \mathrm{LSD} = 0.496 \pm 0.1399 = (0.356, 0.636) \,,$$

with back-transform $(10^{0.356}, 10^{0.636}) = (2.27, 4.32)$. The back-transform of the LSD is $10^{0.1399} = 1.38$, and it is straightforward to verify that the CI limits are equal to the back-transformed difference (3.13) divided or multiplied by 1.38 (an increase to 138% or decrease to $100/1.38 = 72\%$ of the predicted value). Our 95% confidence interval therefore states that *P. vulgatissima* females lay between 2.27 and 4.32 times as many eggs as *P. vitellinae* females when both are mated with *P. vulgatissima* males.

We shall return to this example in Chapter 8 where we look at the structure of the four treatments in more detail.

Mathematical Aside 6.1

The geometric mean for the *j*th treatment group with replication n is defined as

$$\tilde{y}_{j\cdot} = (y_{j1} \times y_{j2} \times \ldots \times y_{jn})^{1/n} = \sqrt[n]{\prod_{k=1}^{n} y_{jk}},$$

where the symbol 'Π' denotes the product of the values over the specified indices. It follows that the arithmetic mean of log-transformed values is equal to the logarithm of the geometric mean of the untransformed values, or

$$\frac{1}{n}\sum_{k=1}^{n}\log(y_{jk}) = \frac{1}{n}\log(y_{j1} \times y_{j2} \times \ldots \times y_{jn}) = \frac{1}{n}\log\left(\prod_{k=1}^{n} y_{jk}\right) = \log\left(\sqrt[n]{\prod_{k=1}^{n} y_{jk}}\right) = \log(\tilde{y}_{j\cdot}). \quad \blacksquare$$

6.5 Other Approaches

If a transformation is successful, so that all the assumptions of the analysis are satisfied, with good residual plots produced, then the results are likely to be reliable. However, in many circumstances this is not achievable. Some common situations where transformation is unlikely to be unsuccessful include

- Counts with many small or zero values
- Proportions calculated with respect to small samples (< 10)
- Proportions (or percentages) with many values at or close to the limits 0 or 1 (0 or 100)
- Proportions calculated with respect to samples of different sizes

In these cases, there is a better alternative to the use of transformations, which is the use of generalized linear models (GLMs) based on the assumption of a distribution other than Normal for the response. For example, insect counts might be assumed to follow a Poisson distribution which is defined for zero and positive integers, and is characterized by a variance equal to the expected response (so that as the mean increases so does the variance). When proportion responses have been calculated with respect to a sample of entities, the original counts might be assumed to follow a Binomial distribution. GLMs provide a flexible framework of models that use one of several statistical distributions for the response (including the Normal) and can take all features of the data into account. They also have the advantage that transformation of the model to an additive scale can be made independently of distributional assumptions (McCullagh and Nelder, 1989).

In other cases, there might be no simple alternative approach. In these cases, particularly when the sample size is small, so that it is difficult to deduce what distribution the response might follow, or when the responses are ranks or scores, it might be preferable not to make any distributional assumptions at all, in which case non-parametric methods are appropriate. The permutation tests described in Section 5.2.4 can also be useful in this context.

We postpone a discussion of GLMs until Chapter 18. For now, we simply state that all is not lost if your responses do not satisfy, and cannot be transformed to satisfy, the assumptions underlying the linear model and ANOVA.

EXERCISES

6.1   A study was conducted to estimate the abundance of rye-grass on three different sites. At each site, quadrats of size 0.1 m^2 were randomly placed (12 at sites 1 and 2, and 24 at site 3) and the number of rye-grass plants in each quadrat was recorded. The unit number (*DQuadrat*), site (factor Site) and rye-grass count (variate *Count*) for each quadrat are in file RYEGRASS.DAT. The objective of the study was to determine whether the abundance of rye-grass differed among the three sites. Plot the observed data and analyse them using a one-way ANOVA. Are there any indications that the data require transformation? Analyse the data on an appropriate alternative scale. Is there any evidence of site differences?[*]

6.2[*] A pilot study investigated the pattern of an insect pest (beetle) entering a susceptible field crop. It was suspected that the beetles entered the crop from the edge of the field and then progressed towards the centre. One field was surveyed periodically and, once the beetles were present in reasonable numbers, a transect was taken from the edge towards the centre of the field with samples taken at 2 m intervals. At each distance, beetle counts were made from four randomly selected plants, giving replicate measurements at each distance. The file TRANSECT.DAT contains the unit numbers (*DPlant*), distances (factor fDist) and beetle counts (variate *Count*). Analyse these data, using a transformation if necessary, to investigate whether there is any evidence that beetle numbers vary between sampling distances. What other hypotheses might you like to test?

6.3   A field experiment was carried out to investigate the effects of amount and timing of sulphur application on the level of scab disease in potatoes (Cochran and Cox, 1957, Table 4.1). Three doses of sulphur were used (300, 600 and 1200 lb/acre) and these were applied in either spring or autumn. Plots with no sulphur application were included as controls, giving seven treatments in total. The control treatment was replicated eight times and the six sulphur treatments each four times in a CRD with 32 plots in a four-row × eight-column layout. The average percentage surface area with scab for 100 potatoes per plot is the response to be analysed. The unit numbers (*Plot*), treatments applied (factor Treatment) and responses (variate *Scab*) can be found in file SCAB.DAT. Analyse these data on an appropriate scale using one-way ANOVA to compare the seven

---

[*] Data from S. Moss, Rothamsted Research.

treatments. Is there any evidence that the application of sulphur affects the incidence of scab? (We re-visit these data in Exercises 8.5 and 11.4.)

6.4   Re-analyse the data of Example 6.1 (Table 6.1 and file BEETLES.DAT) using one-way ANOVA with a square root rather than logarithmic transformation. Compare the two analyses. Which transformation is most appropriate?

6.5   An experiment investigated detection of IgG antibodies ingested by parasitoids with enzyme-linked immunosorbent assay (ELISA). At the start of the experiment, parasitoids were fed either honey spiked with antibodies or normal honey (negative control, labelled Control). Those fed spiked honey were either tested immediately afterwards (positive control to check that the antibodies had been ingested, labelled Day0) or after one, two or three days (labelled Day1, Day2, Day3, respectively), having been fed on normal honey in the interim. The five treatments were each allocated at random to 10 parasitoids as a CRD and the insect samples were placed into 50 wells of a standard 96-well microplate for testing. The resulting optical density readings (variate *OpticalDensity*) are in file PARASITOIDS.DAT with the unit number of each parasitoid (*DParasitoid*) and the treatment to which it was allocated (factor Treatment).

The main aim of the experiment was to assess for how long after ingestion the antibodies could be detected, i.e. comparisons between the negative control treatment and samples after one, two or three days. Analyse these data appropriately using one-way ANOVA and discuss whether this aim can be fully realized. What conclusions can you draw?[*]

6.6   The concentrations of several trace metals in a region of the Swiss Jura were quantified by a survey of soil samples at 366 sites (Atteia et al., 1994). The metals measured (in mg/kg) included cadmium (Cd), chromium (Cr), copper (Cu) and zinc (Zn). The full data set was published in Goovaerts (1997). Here, we consider a subset of 207 sample points on a square grid with approximately 250 m spacing. The land use at each sample point was classified into one of three categories (1 = forest, 2 = pasture, 3 = meadow). The unit number (*DSample*), spatial location (*x*- and *y*-coordinates in variates *X* and *Y*, respectively) and land-use category (factor LandUse) for each sample can be found in file METALS.DAT along with the concentrations of each metal at each location (variates *Cd*, *Cr*, *Cu* and *Zn*). Analyse the concentration of each metal on an appropriate scale to determine if there are differences among the land types. Are there any metals for which you cannot come to a reasonable conclusion? Plot the co-ordinates of the spatial locations, and consider how you might look for spatial dependence in the residuals. Can you implement your idea? Is there any evidence of spatial dependence?[†]

---

[*] Data from M. Torrance, Rothamsted Research.
[†] Data from R. Webster, Rothamsted Research & previously Ecole Polytechnique Fédérale de Lausanne.

7

Models with a Simple Blocking Structure

The design principle of blocking, used to control for known or expected heterogeneity (variability) among experimental units, was introduced in Section 3.1.3. The basic approach is to group, or block, together sets of experimental units expected to have similar responses in the absence of different treatments, and to separate those units expected to have different responses. Blocking is frequently used in designed experiments to account for heterogeneity due to the location or timing of measurements. For example, in a glasshouse experiment (in the northern hemisphere), we might expect plots closer to the south wall of a compartment to be warmer than plots closer to the north wall, and so, we group our experimental units based on their distance from the south wall. Similarly, if samples have to be processed, but only half can be done in the morning and the remainder done in the afternoon, then the morning and afternoon sessions might be used as two blocks to guard against systematic differences caused by any change in the background conditions, or a change of the experimenter. Blocking is also widely used in observational studies. For example, if an ecological study makes observations of the species present on pairs of fields (e.g. one growing wheat, another growing oilseed rape) on several farms, then the farms can be included as a blocking structure to account for the many expected differences (caused by a combination of location and management practices) between farms. Full specification of an experiment therefore requires knowledge of both the blocking and treatments present. In developing ideas for designs with blocking, we consider a single set of treatments, which may mean either imposed treatments in a designed experiment or groups in an observational study, as in the previous chapters.

The simplest layout that includes some form of blocking is the RCBD which was introduced in Section 3.3.2. In this design, the size of each block is equal to the number of treatments, with each treatment occurring exactly once in each block, and with treatments allocated at random to the units within each block (i.e. an independent randomization for each block). This chapter begins by describing the analysis of data from a RCBD. The first step in the analysis is to write down a model for the data (Section 7.1) and to obtain estimates of the model parameters (Section 7.2). A simple ANOVA is then used to obtain an estimate of the background variation and to test whether there are real differences between the treatments or groups (Section 7.3). These results can be combined to examine the treatment means together with appropriate estimates of error (Section 7.4). While in the analysis of the CRD (Chapter 4), there was only one factor to consider, in the analysis of the RCBD, there is a block factor in addition to the treatment factor. It is important to realize that within the model, the status of these two factors is different: the block factor is concerned with the structure (heterogeneity) of the units, and corresponds to the structural component of the model, while the treatment structure defines the different treatments (or treatment combinations) applied to the units, and corresponds to the explanatory component of the model (Section 1.3). Hence, the structural component allows us to assess different sources of natural variation among the experimental units and the explanatory component provides information about the differences in response caused by the different treatments, in particular allowing us to estimate the sizes of these differences. Recognition of the different roles played by these two components leads to the idea of a multi-stratum ANOVA which makes explicit the separation between them (Section 7.5). The translation of the simple ANOVA into a multi-stratum ANOVA can be easily demonstrated for the RCBD, with the idea then extended for more complex designs. The benefit of the multi-stratum ANOVA is that, given the correct specification of the structural and explanatory components for an experiment, the correct analysis follows. For example, this enables the automatic recognition of sub-sampling or pseudo-replication.

7.1 Defining the Model

In the model for a RCBD, there are now two factors to consider: one defining the block to which each experimental unit is allocated, and the other defining the treatment applied to each unit. We use a general notation that can be applied to any data set. Suppose there are t treatments, with each treatment equally replicated n times, such that there are n blocks, each consisting of t units with each treatment occurring once in each block. The simplest notation for the RCBD labels each observation by its block (index i) and treatment allocations (index j). Then y_{ij} represents the observation on the jth treatment in the ith block and the full set of observations can be denoted as y_{ij}, $i = 1 \ldots n$, $j = 1 \ldots t$ (see Section 2.1 for an overview of notation). The total number of observations is $N = n \times t$. Extending the notation presented for the CRD in Section 4.5, we can write the linear model for the data from a RCBD as

$$y_{ij} = \mu + b_i + \tau_j + e_{ij} \, , \tag{7.1}$$

where μ represents the overall population mean, b_i is the effect of the ith block (as a difference from the overall mean) and τ_j is the effect of the jth treatment (again as a difference from the overall mean), with deviations e_{ij} reflecting individual variation about the population values. In this model, the population mean for the jth treatment can be derived as $\mu + \tau_j$. The assumptions about properties of the deviations given in Section 4.1, including independence, homogeneity of variances and a Normal probability distribution, again apply to this model. As the treatment and block effects are expressed as differences from the overall population mean, it follows that they require the constraints $\Sigma_j \tau_j = 0$ and $\Sigma_i b_i = 0$. This is the sum-to-zero parameterization introduced in Section 4.5. Note that we use an italic Roman symbol (b) to denote block effects to emphasize that they are part of the structural component; we reserve Greek symbols (τ) to denote treatment effects in the explanatory component.

Labelling the units by their block and treatment allocation retains the simplicity of the notation as introduced for the CRD (Section 4.1), but again, information is lost with regard to the experimental layout, i.e. the randomized allocation of treatments to plots within blocks. We might alternatively write the model in terms of the block, plot and treatment relevant to each unit, but this extended notation is both more cumbersome and introduces an element of redundancy (given the design, we do not need to know both the plot number and the treatment applied to that plot). Therefore, we continue to use the simpler notation, but restate the importance of retaining all information in a data set, i.e. factors defining the blocks and units within blocks, as well as the treatment allocation, so that the full layout can be reconstructed when required.

Using our symbolic notation, we can write the model from Equation 7.1 as

Response variable: Y
Explanatory component: [1] + Treatment
Structural component: Block/Unit

where the variate Y holds the observed response, factor Treatment gives the allocation of observations to treatment groups, factor Block gives the allocation of observations to blocks and factor Unit labels the units within blocks. The term Block.Unit is associated with the deviations, e_{ij}. The term [1] was introduced in Section 4.5, and denotes a factor that takes value 1 everywhere, and is associated with the overall population mean, μ.

The model in Equation 7.1 is based on the assumption that the expected difference between any two treatments is the same in each block. In statistical parlance, we say that there is no interaction between blocks and treatments (the term 'interaction' is discussed in detail in Section 8.2), meaning that the treatment differences are independent of the block effects. In some cases, this assumption is not reasonable. For example, if blocks in a field experiment are assigned according to the soil's pH, and the treatments are expected to react differently according to the soil's pH, then strictly, this model – and hence this design – is inappropriate for the experiment (although see the remarks at the end of Section 7.3). An alternative design for this situation might have two replicates of each treatment within each block, allowing for estimation of treatment effects, effects of soil pH and the interaction between these factors, but a full consideration of this design approach is beyond the scope of this book (see, e.g. Mead et al., 2012, Chapter 7).

EXAMPLE 7.1A: POTATO YIELDS*

A field experiment designed as a RCBD to investigate the effects of four different types of fungicides (F1, F2, F3 and F4) on the yield of potatoes compared with untreated plots (negative control) was described in Example 3.5. The experiment was laid out as four blocks ($n = 4$) of five plots each ($t = 5$) with 20 units in total ($N = 20$). The plot yields are shown in field layout in Table 7.1 and can also be found in the file POTATO.DAT, which contains blocking factors Block (four levels) and Plot (five levels), treatment factor Fungicide (five levels, with labels 1 ... 5 corresponding to the control, F1, F2, F3 and F4 treatments, respectively) and response variate Yield.

The model for these data can be written in the mathematical form of Equation 7.1 as

$$Yield_{ij} = \mu + Block_i + Fungicide_j + e_{ij},$$

TABLE 7.1

Field Layout for the Potato Yields Trial with Potato Yield for Each Plot (Example 7.1A and File POTATO.DAT)

	Plot 1	Plot 2	Plot 3	Plot 4	Plot 5
Block 1	F3	Control	F2	F1	F4
	642	377	633	527	623
Block 2	F2	Control	F3	F4	F1
	600	408	708	550	604
Block 3	Control	F2	F3	F4	F1
	500	650	662	562	606
Block 4	F3	F2	F1	Control	F4
	504	567	533	333	667

with $Yield_{ij}$ representing the yield from the jth treatment in the ith block, $Fungicide_j$ ($j = 1 \ldots 5$) representing the effects of control and fungicides F1, F2, F3, F4, respectively, and $Block_i$ ($i = 1 \ldots 4$), representing the effects of the four blocks. As above, μ represents the overall population mean and e_{ij} is the deviation for the jth treatment in the ith block. In symbolic form, this is written as

Response variable: *Yield*
Explanatory component: [1] + Fungicide
Structural component: Block/Plot

7.2 Estimating the Model Parameters

The parameters associated with the RCBD model in Equation 7.1 are the overall population mean, μ, the treatment effects $\tau_j, j = 1 \ldots t$ and the block effects $b_i, i = 1 \ldots n$. The fitted value for the ijth observation, i.e. for the jth treatment in the ith block, denoted by $\hat{y}_{ij}$, consists of all components of the model except the deviations, with parameters replaced by their estimates, so that

$$\hat{y}_{ij} = \hat{\mu} + \hat{b}_i + \hat{\tau}_j \ .$$

We again estimate the parameters using the principle of least squares (see Section 4.2), by minimizing the residual sum of squares (ResSS)

$$\text{ResSS} \ = \ \sum_{i=1}^{n} \sum_{j=1}^{t} (y_{ij} - \hat{y}_{ij})^2 \ = \ \sum_{i=1}^{n} \sum_{j=1}^{t} (y_{ij} - \hat{\mu} - \hat{b}_i - \hat{\tau}_j)^2 \ ,$$

subject to the constraints on the parameters, $\Sigma_i \hat{b}_i = 0$ and $\Sigma_j \hat{\tau}_j = 0$. As a result of this procedure, the overall population mean, μ, for data from a RCBD is estimated by the sample grand mean,

$$\hat{\mu} \ = \ \bar{y} \ ,$$

and the effect of the jth treatment is estimated by the difference between the sample mean for that treatment and the sample grand mean,

$$\hat{\tau}_j \ = \ \bar{y}._j - \bar{y} \ ,$$

with the dot notation as introduced in Section 2.1. Similarly, the effect of the ith block is estimated by the difference between the sample mean for that block and the sample grand mean,

$$\hat{b}_i \ = \ \bar{y}_i. - \bar{y} \ .$$

The population mean for the jth treatment, denoted as μ_j, is then estimated as the sum of the estimates of the overall mean and the treatment effect, as

$$\hat{\mu}_j = \hat{\mu} + \hat{\tau}_j = \bar{y} + (\bar{y}_{\cdot j} - \bar{y}) = \bar{y}_{\cdot j} \, ,$$

i.e. the sample mean for the jth treatment. Similarly, the fitted value for each observation can be calculated as the sum of the estimates for all components of the model except the deviations,

$$\hat{y}_{ij} = \hat{\mu} + \hat{b}_i + \hat{\tau}_j = \bar{y} + (\bar{y}_{i\cdot} - \bar{y}) + (\bar{y}_{\cdot j} - \bar{y}) = \bar{y}_{i\cdot} + \bar{y}_{\cdot j} - \bar{y} \, .$$

Finally, the simple residuals are calculated as the discrepancy between the observations and the fitted values, namely

$$\hat{e}_{ij} = y_{ij} - \hat{y}_{ij} = y_{ij} - \bar{y}_{i\cdot} - \bar{y}_{\cdot j} + \bar{y} \, .$$

EXAMPLE 7.1B: POTATO YIELDS*

Table 7.2 lists the plot yields classified by blocks and treatments, with the block and treatment sample means and the sample grand mean. From these values and the formulae above, we can calculate the parameter estimates. In particular, the estimated population means for the treatments are equal to the treatment sample means given in Table 7.2. For example, the population mean for the control (treatment 1) is estimated as

$$\hat{\mu}_1 = \hat{\mu} + \widehat{Fungicide}_1 = 562.8 + (404.5 - 562.8) = 404.5 \, .$$

It appears that the mean yields for the four fungicide treatments are similar (between 567.5 and 629.0) and all are much greater than that for the untreated control (404.5). However, to draw sound conclusions about these differences, information on the background variation is required to calculate SEDs and hence LSDs for these comparisons.

TABLE 7.2

Plot Yields of Potatoes from a RCBD with Block and Treatment Means (Example 7.1B)

Block	Control	F1	F2	F3	F4	Block Mean $(\bar{y}_{i\cdot})$
1	377	527	633	642	623	560.4
2	408	604	600	708	550	574.0
3	500	606	650	662	562	596.0
4	333	533	567	504	667	520.8
Treatment mean $(\bar{y}_{\cdot j})$	404.5	567.5	612.5	629.0	600.5	$\bar{y} = 562.8$

7.3 Summarizing the Importance of Model Terms

As for the analysis of data from a CRD, the aim of ANOVA is to partition the total variation of the observations, quantified as sums of squares, into several components. For the RCBD, there are now two factors classifying the observations, corresponding to the blocks and treatments, and the analysis needs to account for the variation due to each of these sources. The total variation (TotSS) is therefore partitioned into the variation among blocks (BlkSS), the variation among treatments (TrtSS) and the residual variation (ResSS), with

$$\text{TotSS} = \text{BlkSS} + \text{TrtSS} + \text{ResSS} . \qquad (7.2)$$

Because of the difference in status between the block and treatment factors, we describe this analysis as **one-way ANOVA with blocks** (rather than two-way ANOVA, which we use to indicate the inclusion of two treatment factors; see Chapter 8). This clearly emphasizes the difference in the status of these two components in the model (see Section 7.5).

As for the CRD (see Equation 4.3), the total sum of squares (TotSS) is calculated as the sum, over all observations, of the squared differences between each observation and the sample grand mean,

$$\text{TotSS} = \sum_{i=1}^{n} \sum_{j=1}^{t} (y_{ij} - \bar{y})^2 .$$

The block sum of squares (BlkSS) measures the variation among blocks, and is calculated as the sum, over all observations, of the squared differences between the appropriate block mean and the sample grand mean,

$$\text{BlkSS} = \sum_{i=1}^{n} \sum_{j=1}^{t} (\bar{y}_{i\cdot} - \bar{y})^2 = t \sum_{i=1}^{n} (\bar{y}_{i\cdot} - \bar{y})^2 .$$

Similarly, the treatment sum of squares (TrtSS) is calculated as the sum, over all observations, of the squared differences between the appropriate treatment mean and the sample grand mean,

$$\text{TrtSS} = \sum_{i=1}^{n} \sum_{j=1}^{t} (\bar{y}_{\cdot j} - \bar{y})^2 = n \sum_{j=1}^{t} (\bar{y}_{\cdot j} - \bar{y})^2 .$$

The residual sum of squares (ResSS) is calculated as the sum, over all observations, of the squared differences between the observed and fitted values,

$$\text{ResSS} = \sum_{i=1}^{n} \sum_{j=1}^{t} (y_{ij} - \hat{y}_{ij})^2 = \sum_{i=1}^{n} \sum_{j=1}^{t} (y_{ij} - \bar{y}_{i\cdot} - \bar{y}_{\cdot j} + \bar{y})^2 ,$$

or, by rearranging Equation 7.2, as a simple subtraction of the block and treatment sums of squares from the total sum of squares,

$$\text{ResSS} = \text{TotSS} - \text{BlkSS} - \text{TrtSS} .$$

Alternatively, the sums of squares can be rewritten in terms of the sums of squared estimates

$$\text{BlkSS} = t \sum_{i=1}^{n} \hat{b}_i^2, \quad \text{TrtSS} = n \sum_{j=1}^{t} \hat{\tau}_j^2, \quad \text{ResSS} = \sum_{i=1}^{n} \sum_{j=1}^{t} \hat{e}_{ij}^2 .$$

Again, as for the CRD, there is a corresponding partition of the total degrees of freedom (TotDF) into those associated with variation among blocks (BlkDF), those associated with variation among treatments (TrtDF) and the remainder or residual (ResDF), as

$$\text{TotDF} = \text{BlkDF} + \text{TrtDF} + \text{ResDF} . \tag{7.3}$$

Using the recipe developed in Section 4.3, we can calculate the df associated with each sum of squares. The calculation of the TotSS uses the $N = n \times t$ observations (which can be described using a model with N parameters) with just the sample grand mean, an estimate of a single parameter, used for adjustment; hence

$$\text{TotDF} = N - 1 .$$

The calculation of the BlkSS uses block means (n values) adjusted by the sample grand mean; so

$$\text{BlkDF} = n - 1 .$$

By a similar argument, we have

$$\text{TrtDF} = t - 1 .$$

Rearranging Equation 7.3, the ResDF can be most easily obtained by subtraction as

$$\text{ResDF} = (N - 1) - (n - 1) - (t - 1) = N - n - t + 1 .$$

Using $N = nt$, we can also write this as

$$\text{ResDF} = nt - n - t + 1 = (n - 1)(t - 1) ,$$

i.e. one less than the number of blocks ($n - 1$) multiplied by one less than the number of treatments ($t - 1$).

As for the ANOVA for data from a CRD, we can put the sums of squares onto a common scale by dividing them by their degrees of freedom to produce mean squares. The residual mean square (ResMS) again provides an estimate of the background variation or noise, usually denoted as s^2. If there are no differences between treatments, then contributions to the treatment mean square arise from background variation alone and the treatment and residual mean squares should be of similar sizes, allowing for sampling variation. A similar argument follows for the comparison of the block mean square with the residual mean square if there are no differences between blocks. We can formalize these comparisons by considering the expected value of each of the respective mean squares, which are

$$E(\text{TrtMS}) = \sigma^2 + \frac{n}{t-1}\sum_{j=1}^{t}\tau_j^2 \, ,$$

$$E(\text{BlkMS}) = \sigma^2 + \frac{t}{n-1}\sum_{i=1}^{n}b_i^2 \, ,$$

$$E(\text{ResMS}) = \sigma^2 \, .$$

If the true population values of the treatment effects are zero, then the second term in the expression for E(TrtMS) is zero and the treatment mean square (TrtMS) has the same expected value as the ResMS. Similarly, if the true population values of the block effects are zero, then the block mean square (BlkMS) also has the same expected value as the ResMS. As for the CRD, this property of the TrtMS is the basis for a test of the null hypothesis of equal treatment population means, written for this parameterization as

$$H_0: \tau_1 = \tau_2 = \ldots = \tau_t = 0 \, ,$$

i.e. that all the treatment effects (the differences from the overall population mean) are zero. This is compared with the general alternative hypothesis of some non-zero treatment effects. The observed F-statistic is calculated as the variance ratio

$$F_{t-1,(n-1)(t-1)} = \text{TrtMS}/\text{ResMS} \, .$$

If the null hypothesis is true, then we expect the value of the variance ratio to be close to 1 (because the TrtMS and ResMS have the same expected value) and the test statistic follows an F-distribution with $t-1$ (TrtDF) and $(n-1)(t-1)$ (ResDF) df. If the observed statistic $F_{t-1,(n-1)(t-1)}$ is larger than the $100(1-\alpha_s)$th percentile of this F-distribution, then the null hypothesis is rejected at significance level α_s, and we have evidence of some variation among the treatment effects. Alternatively, the observed significance level can be calculated as

$$P = \text{Prob}(F_{t-1,(n-1)(t-1)} \ge F_{t-1,(n-1)(t-1)}) \, ,$$

where $F_{t-1,(n-1)(t-1)}$ is a random variable with an F-distribution on $t-1$ and $(n-1)(t-1)$ df. Using analogous reasoning, we can derive a test statistic for a null hypothesis of the block effects being equal to zero, using the ratio BlkMS/ResMS. This ratio also has an F-distribution under the null hypothesis, now with $n-1$ (BlkDF) and $(n-1)(t-1)$ (ResDF) df. However, we consider that blocks and treatments have a different status within the model, with blocks reflecting the experimental structure rather than treatments of interest; so, it follows that we might regard tests of hypotheses about block effects differently from tests of hypotheses about treatment effects. Treatments are imposed on the experiment

TABLE 7.3

Structure of the ANOVA Table for a RCBD with n Blocks (Factor **Block**) and t Treatments (Factor **Treatment**), and $N = n \times t$ Observations in Total (Observed Significance Levels Omitted)

Source of Variation	df	Sum of Squares	Mean Square	Variance Ratio
Block	$n - 1$	BlkSS	BlkMS = BlkSS/$(n - 1)$	BlkMS/ResMS
Treatment	$t - 1$	TrtSS	TrtMS = TrtSS/$(t - 1)$	TrtMS/ResMS
Residual	$(n - 1)(t - 1)$	ResSS	ResMS = ResSS/$(n - 1)(t - 1)$	
Total	$N - 1$	TotSS		

specifically to examine the differences between them; so, we always want to test the null hypothesis of no treatment differences. Blocks have been used to control inherent variation among the experimental units; so, we can use the F-statistic for blocks to evaluate the extent of this variation, and to help us design future similar experiments.

The information about degrees of freedom, sums of squares, mean squares and variance ratios is combined to form a simple ANOVA table for a RCBD as shown in Table 7.3.

EXAMPLE 7.1C: POTATO YIELDS*

Using the formulae presented earlier for sums of squares and for degrees of freedom, the ANOVA table for this trial takes the form shown in Table 7.4.

At this stage, we should validate the analysis using the residual plots described in Section 5.2. A composite set of residual plots using standardized residuals is shown in Figure 7.1. The Normal plot shows a reasonably straight line, and the histogram is not inconsistent with a Normal distribution (although note that with relatively few observations, the histogram provides a fairly poor diagnostic tool here). The fitted values plot has a slight suggestion of smaller variances for smaller fitted values, but this is difficult to judge because there are few observations in that region of the plot. Note that in the RCBD, the fitted values do not directly correspond to the treatment groups (as they do in the CRD) because both block and treatment effects contribute to the fitted values. We judge these plots to be acceptable and move on to interpret the analysis and draw conclusions.

The observed significance level (P) for the **Fungicide** variance ratio is obtained by comparison of the treatment variance ratio ($F_{4,12} = 9.576$) with the quantiles of the F-distribution with 4 (TrtDF) and 12 (ResDF) degrees of freedom. Here, $P = 0.001$, indicating that the observed treatment differences are very unlikely to have happened by chance if the null hypothesis is true; hence, we reject the null hypothesis at the 0.1% significance level and conclude that there are real differences among the set of treatment means.

TABLE 7.4

ANOVA Table for Potato Yields Trial Set Up as a RCBD with Four Blocks (Factor **Block**) and Five Treatments (Factor **Fungicide**) (Example 7.1C)

Source of Variation	df	Sum of Squares	Mean Square	Variance Ratio	P
Block	3	14,987.20	4995.73	1.434	0.283
Fungicide	4	133,419.20	33,354.80	9.576	0.001
Residual	12	41,796.80	3483.07		
Total	19	190,203.20			

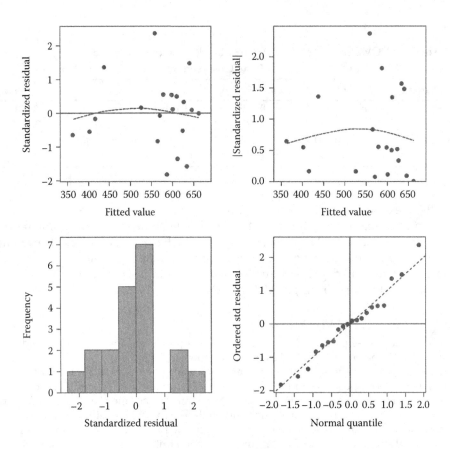

FIGURE 7.1
Composite set of residual plots using standardized residuals from the potato yields trial (Example 7.1C).

The observed significance level associated with the **Block** variance ratio ($F_{3,12} = 1.434$) is $P = 0.283$; so, the test statistic is consistent with the null hypothesis of no block differences. Even so, the BlkMS is larger than the ResMS; so, taking account of the block heterogeneity has reduced the ResMS, which in turn increases the precision of treatment comparisons. For this particular experiment, the advantage of using a RCBD instead of a CRD was small (based on a comparison of the relative sizes of the BlkMS and ResMS). However, field trials are notoriously heterogeneous; so, it would be unwise to use this result to abandon blocking for future similar experiments – sensible use of blocking still provides insurance against unit-to-unit heterogeneity. In other contexts, if prior knowledge or further experimentation indicated that there was generally little advantage in blocking for the type of experiment, then it would be sensible to weigh up the possible benefit of blocking against the cost in degrees of freedom – the reduction in ResDF could have a detrimental impact on the power of the experiment to detect treatment differences of interest. These issues are discussed further in Chapter 10.

One feature of the RCBD is that the two classifying factors – **Treatment** and **Block** – are **orthogonal** or independent. The mathematical definition of orthogonality is beyond the scope of this book (see, e.g. Bailey, 2008), but we can give some general intuitive insight into this property. If two factors are orthogonal, then the same ANOVA table is obtained regardless of the order in which the terms are fitted, and comparisons between the levels of one factor are unaffected by the levels of the other factor. This has the advantage that interpretation of the ANOVA table is unambiguous. Examples and consequences of non-orthogonality are discussed in Chapter 11.

Mathematical Aside 7.1

We can use the form of the model to gain further insight into this idea of orthogonality. For the RCBD, the treatment population means are estimated by the treatment sample means. We can derive an expression for the treatment sample means from the model for individual observations (Equation 7.1) by simply substituting the right-hand side of that equation for y_{ij} in the expression $\bar{y}_{\cdot j} = \frac{1}{n}\sum_{i=1}^{n} y_{ij}$, giving

$$\bar{y}_{\cdot j} = \frac{1}{n}\sum_{i=1}^{n}(\mu + b_i + \tau_j + e_{ij}).$$

Expanding the summation for each term separately simplifies the expression to give

$$\bar{y}_{\cdot j} = \mu + \frac{1}{n}\sum_{i=1}^{n} b_i + \tau_j + \frac{1}{n}\sum_{i=1}^{n} e_{ij},$$

and applying the constraint $\sum_{i=1}^{n} b_i = 0$, which we built into the model, causes the block effects to be removed from the expression, leaving

$$\bar{y}_{\cdot j} = \mu + \tau_j + \bar{e}_{\cdot j}.$$

So, because each treatment occurs once within each block, the treatment means do not depend on the block effects, and hence, the treatment means (and estimated treatment effects) are independent of (i.e. orthogonal to) the block effects. A similar derivation can be obtained for block sample means, and because each block contains one instance of each treatment, block means are independent of treatment effects. ∎

One assumption underlying the RCBD model (Section 7.1) is that there are no interactions between blocks and treatments, i.e. that the expected treatment differences are the same in all blocks. This assumption is required for an unambiguous analysis, as it is not possible to separate the block × treatment interaction from the model deviations in this design. However, it is technically possible to use the Treatment factor as a substitute for the Unit factor in the model specification, because the Block.Treatment combinations also uniquely label the full set of observations, and then the residual line in the ANOVA table may be labelled as Block.Treatment. We believe that the potential confusion caused by this approach makes it imperative to retain and use the full set of structural factors in each data set, and this is the reason why we use dummy structural factors where the true allocation is not available. In some situations, the presence of a block × treatment interaction cannot be discounted and may be expected to be much larger than the deviations from other sources (individual variation, measurement error, etc.). In this case, the residual error line in the ANOVA table might be legitimately labelled as Block.Treatment, and the Treatment variance ratio can then be considered as evaluating the consistency of the treatment effects across the set of conditions represented by the blocks.

7.4 Evaluating the Response to Treatments

The best estimate of the population mean for the jth treatment (denoted as μ_j) was identified in Section 7.2 as the treatment sample mean, i.e. $\hat{\mu}_j = \bar{y}._j$. Uncertainty associated with this estimate is measured by its estimated SE, the SEM,

$$\text{SEM} = \widehat{\text{SE}}(\hat{\mu}_j) = \sqrt{\frac{s^2}{n}},$$

with the estimate of background variation, $s^2 = \text{ResMS}$, in place of the unknown true value, σ^2. In this case, as all the treatments have equal replication, their SEs are also equal. As for the CRD, a $100(1 - \alpha_s)\%$ confidence interval for the population mean of the jth treatment can be calculated as

$$(\hat{\mu}_j - [t_{(n-1)(t-1)}^{[\alpha_s/2]} \times \text{SEM}], \quad \hat{\mu}_j + [t_{(n-1)(t-1)}^{[\alpha_s/2]} \times \text{SEM}]),$$

where $t_{(n-1)(t-1)}^{[\alpha_s/2]}$ is the $100(1 - \alpha_s/2)$th percentile of the t-distribution with $(n-1)(t-1)$ df.

The best estimate of a difference between two treatment population means is the difference between the two treatment sample means, for example, for the jth and kth treatments,

$$\hat{\mu}_j - \hat{\mu}_k = \bar{y}._j - \bar{y}._k.$$

The estimated standard error of this difference (denoted SED) takes the form

$$\text{SED} = \widehat{\text{SE}}(\hat{\mu}_j - \hat{\mu}_k) = \sqrt{\frac{2s^2}{n}},$$

and again is the same for any pair of treatments. Under the null hypothesis that the population means of the jth and kth treatments are equal, i.e. H_0: $\mu_j = \mu_k$, the statistic

$$t_{(n-1)(t-1)} = \frac{\hat{\mu}_j - \hat{\mu}_k}{\text{SED}} = \frac{\bar{y}._j - \bar{y}._k}{\text{SED}},$$

has a t-distribution with degrees of freedom equal to the ResDF, as denoted by its subscript. This statistic can be compared with the quantiles of this t-distribution to test the null hypothesis against a one- or two-sided alternative. A $100(1 - \alpha_s)\%$ confidence interval for the difference between the population means for these treatments can be computed as

$$((\hat{\mu}_j - \hat{\mu}_k) - \text{LSD}, (\hat{\mu}_j - \hat{\mu}_k) + \text{LSD}),$$

where $\text{LSD} = t_{(n-1)(t-1)}^{[\alpha_s/2]} \times \text{SED}$ is the least significant difference at significance level α_s.

EXAMPLE 7.1D: POTATO YIELDS*

From the ANOVA table obtained earlier (Table 7.4), the background variation is estimated as $s^2 = \text{ResMS} = 3483.07$. The treatment sample means were shown in Table 7.2. There are four blocks, $n = 4$; so, the standard errors of the means are equal to

$$\text{SEM} = \sqrt{\frac{s^2}{n}} = \sqrt{\frac{3483.07}{4}} = \sqrt{870.77} = 29.51 ,$$

with the SED for any pair of treatments equal to

$$\text{SED} = \sqrt{\frac{2s^2}{n}} = \sqrt{\frac{2 \times 3483.07}{4}} = \sqrt{1741.53} = 41.73 .$$

The residual df is 12; so, for $\alpha_s = 0.05$, $t_{12}^{[0.025]} = 2.179$ with LSD = 90.93. Confidence intervals can be derived from these values as shown in Examples 4.1D and E. It is clear that each of the fungicide treatments gives a statistically significant improvement in yield compared with the untreated control, but there appears to be little real difference among the four fungicide treatments.

7.5 Incorporating Strata: The Multi-Stratum Analysis of Variance

We use the term structural component to encompass all the structure within the set of observations. The structural component therefore includes both blocks imposed by the experimenter (e.g. units with similar time of processing, spatially grouped units within a field), and any other structure within the experiment (e.g. the nesting of fields within farms, or aphids within cages), including the sub-sampling of individual units (e.g. plants within a field plot) sometimes referred to as pseudo-replication.

The simple ANOVA table derived in Section 7.3 does not make any distinction between the explanatory and structural components of the model. The multi-stratum ANOVA table is an alternative, and more general, form that preserves the distinction between terms describing the underlying structure of the data (structural component) and those indicating the treatments applied (explanatory component). Again, the formal mathematical definition of **strata** is beyond the scope of this book (see, e.g. Bailey, 2008), but strata can be informally regarded as the different structural sources of variability among the experimental units (see also Section 3.2). Each term in the structural component generates a stratum; so, the RCBD has two strata: one corresponding to variation between blocks (generically the Block term), and one corresponding to variation between units within blocks (the Block.Unit term). If we sub-sample within units, then this adds another stratum, as in the following example.

EXAMPLE 7.2A: POTATO YIELDS USING ROW DATA*

Each plot of the RCBD potato fungicide trial (described in Example 7.1A) consisted of four rows, and the means of the measurements from these four rows were used as the yield observations in the previous analysis. The individual row yields are also available, giving a new data set with 80 observations. The field layout is shown in Table 7.5.

Guard rows were planted so that edge effects were absent, and row effects were expected to be local so that rows can be reasonably regarded as nested within plots.

TABLE 7.5

Field Layout for the Potato Yields Trial Showing Individual Row Yields (Example 7.2A and File POTATOROW.DAT)

		Plot 1	Plot 2	Plot 3	Plot 4	Plot 5
Block 1	Treatment	F3	Control	F2	F1	F4
	Row 1	720	348	652	635	642
	Row 2	528	405	658	512	639
	Row 3	678	364	569	536	642
	Row 4	642	391	653	425	569
Block 2	Treatment	F2	Control	F3	F4	F1
	Row 1	554	411	682	639	583
	Row 2	618	374	741	544	530
	Row 3	621	396	712	521	629
	Row 4	607	451	697	496	674
Block 3	Treatment	Control	F2	F3	F4	F1
	Row 1	561	555	638	505	598
	Row 2	491	633	712	597	620
	Row 3	429	715	633	607	596
	Row 4	519	697	665	539	610
Block 4	Treatment	F3	F2	F1	Control	F4
	Row 1	451	513	441	367	631
	Row 2	493	626	467	319	618
	Row 3	535	574	701	361	689
	Row 4	537	555	523	285	730

To fully describe the structure of this data set, a new factor **Row** is required to specify the allocation of observations to rows within plots, in addition to the factors **Block** and **Plot**. A model for these data can be written as

$$RowYield_{ijk} = \mu + Block_i + Fungicide_j + Block.Plot_{ij} + e_{ijk} ,$$

where $RowYield_{ijk}$ is the yield obtained from the kth row ($k = 1 \ldots 4$) in the plot with the jth treatment ($j = 1 \ldots 5$) in the ith block ($i = 1 \ldots 4$). The overall mean, μ, is now the population mean with respect to row yields, $Block_i$ is the effect of the ith block and $Block.Plot_{ij}$ is the effect of the plot with the jth treatment in the ith block. The deviations, e_{ijk}, now correspond to observations on rows nested within plots, which are in turn nested within blocks. This three-level nested structure is denoted in symbolic form as

Structural component: Block/Plot/Row

which can be expanded to individual model terms as

Structural component: Block + Block.Plot + Block.Plot.Row

In this case, there are three strata, which correspond to blocks (Block), plots within blocks (Block.Plot) and rows within plots (Block.Plot.Row), with the latter term corresponding to the model deviations. This illustrates a case where the blocking structure reflects both blocks imposed by the experimenter and other structure, in this case, the presence of rows within plots.

TABLE 7.6

Structure of the Multi-stratum ANOVA Table for a RCBD with n Blocks (Factor **Block**), t Units per Block (Factor **Unit**), t Treatments (Factor **Treatment**) and $N = n \times t$ Observations in Total

Source of Variation	df	Sum of Squares	Mean Square	Variance Ratio
Block stratum				
Residual	$n-1$	BlkSS	BlkMS = BlkSS$/(n-1)$	BlkMS/ResMS
Block.Unit stratum				
Treatment	$t-1$	TrtSS	TrtMS = TrtSS$/(t-1)$	TrtMS/ResMS
Residual	$(n-1)(t-1)$	ResSS	ResMS = ResSS$/(n-1)(t-1)$	
Total	$N-1$	TotSS		

The multi-stratum ANOVA approach results in an ANOVA table with separate components for each of the strata defined by the blocking structure. The variation within each stratum (i.e. at each level of the design) is then partitioned into sums of squares associated with the treatments that vary between units at that level of the design (if any) and a residual term. For example, the simple ANOVA table for the RCBD can be rewritten in the form shown in Table 7.6.

This ANOVA table has two strata, corresponding to the **Block** and **Block.Unit** terms in the structural component. The data in the **Block** stratum correspond to block totals calculated after subtraction of (i.e. adjustment for) the sample grand mean. This is sometimes referred to as **inter-block** information. Since every treatment has been applied once in each block, block differences cannot be attributed to treatment differences (see below for a more mathematical argument). So, variation in the **Block** stratum consists of only a residual term representing the background variation between blocks – which is exactly the same interpretation of the block sum of squares (BlkSS) as seen earlier. The data in the **Block. Unit** stratum correspond to the original observations adjusted for the relevant block means, sometimes referred to as **intra-block** information. In this stratum, every unit within a block has a treatment different from others, and so, variation between units includes variation due to treatments; hence, some variation within this stratum can be attributed to treatment differences. The **Block.Unit** stratum variation is thus partitioned as variation due to treatments (TrtSS) plus residual variation (ResSS).

Mathematical Aside 7.2

To establish in which strata different treatment effects are estimated, you should consider the form of data within each stratum of the design. In general, data at the top level of a structure correspond to unit totals within that stratum, calculated after subtraction of the sample grand mean; for example, for the RCBD, data in the **Block** stratum are the mean-adjusted block totals, calculated as

$$\sum_{j=1}^{t} (y_{ij} - \bar{y}) = y_{i\cdot} - t\bar{y} = y_{i\cdot} - \frac{1}{n} y_{\cdot\cdot}, \quad \text{for } i = 1 \ldots n .$$

At the next level down, data again correspond to unit totals within the stratum, calculated after subtraction of the means corresponding to units in higher strata; for example, for the RCBD, data in the Block.Unit stratum are the observations adjusted for the block means

$$y_{ij} - \bar{y}_{i \cdot} = y_{ij} - \frac{1}{t} y_{i \cdot}, \quad \text{for } i = 1 \ldots n, j = 1 \ldots t.$$

The data in each stratum can then be written algebraically in terms of model parameters by substitution of the model formula in place of the response values. When treatment effects are present in the algebraic expression for data within a particular stratum, it follows that there is information on these treatments within that stratum.

For the RCBD, the underlying model is given in Equation 7.1, and the expression for the mean-adjusted block totals can therefore be rewritten, after substitutions based on this model, as

$$y_{i \cdot} - \frac{1}{n} y_{\cdot \cdot} = \left(t\mu + tb_i + \sum_{j=1}^{t} \tau_j + \sum_{j=1}^{t} e_{ij} \right) - \frac{1}{n} \left(nt\mu + t \sum_{i=1}^{n} b_i + n \sum_{j=1}^{t} \tau_j + \sum_{i=1}^{n} \sum_{j=1}^{t} e_{ij} \right)$$

$$= tb_i + e_{i \cdot} - \frac{1}{n} e_{\cdot \cdot},$$

with the inbuilt model constraints $\sum_i b_i = 0$ and $\sum_j \tau_j = 0$ used to simplify the expression. This expression does not involve the treatment effects, and so, this stratum does not contain information about treatment differences. Variation at this level is related to block effects and the deviations. Similarly, the adjusted observations in the lower stratum (units within blocks) can be written as

$$y_{ij} - \frac{1}{t} y_{i \cdot} = (\mu + b_i + \tau_j + e_{ij}) - \frac{1}{t} \left(t\mu + tb_i + \sum_{j=1}^{t} \tau_j + \sum_{j=1}^{t} e_{ij} \right)$$

$$= \tau_j + e_{ij} - \frac{1}{t} e_{i \cdot},$$

again with the constraints used to simplify the expression. Units within a block clearly have different treatments applied, and this expression confirms that unit differences do hold information on treatment differences.   ∎

The multi-stratum ANOVA table for the RCBD rearranges the simpler form to reflect the structure of the experiment. However, the great advantage of the multi-stratum ANOVA is the recognition of the interplay between the blocking and treatment structures so that treatment effects are always allocated to the correct strata, and an appropriate measure of precision can be calculated for the comparison of treatment means, with the correct degrees of freedom. One example where this can be particularly important is where pseudo-replication is present in the structure (see Section 3.1.1), for example, where sub-sampling or technical replication (several measurements per experimental unit) have been used to reduce measurement error. An example of this situation is presented below.

EXAMPLE 7.2B: POTATO YIELDS USING ROW DATA*

Here, we analyse the individual row data (see Example 7.2A) as presented in Table 7.5. The full data set, including classifying factors Block, Plot and Row, can also be found in file POTATOROW.DAT. The multi-stratum ANOVA table for these data, corresponds to the model

Response variable:	*RowYield*
Explanatory component:	[1] + Fungicide
Structural component:	Block/Plot/Row

This ANOVA table is in Table 7.7 and has three strata corresponding to blocks (Block), plots within blocks (Block.Plot) and rows within plots (Block.Plot.Row). Treatments (factor Fungicide) are applied to plots; so, treatment differences are estimated from the differences between plots, and the TrtSS is a component of the variability within the Block.Plot stratum. The Block, Fungicide and Block.Plot residual sums of squares are equal to four times the BlkSS, TrtSS and ResSS from the analysis of plot means given in Example 7.1C. This multiplication by four for each term is due to the presence of four observations from each plot (i.e. from the four separate rows). As the degrees of freedom in the Block and Block.Plot strata are the same as in Table 7.4, the variance ratios for the Block residual and Fungicide terms are preserved. The conclusions with respect to the treatments are thus unchanged. Since treatments are applied to plots, the Block.Plot residual mean square (and not the Block.Plot.Row residual mean square) is the appropriate measure of background variation for estimates of Fungicide SEMs, SEDs and LSDs, with degrees of freedom equal to the residual df from the Block.Plot stratum.

As an illustration of the importance of specifying the correct blocking structure, suppose that the presence of sub-sampling was ignored, and that only the block and treatment factors (Block and Fungicide) were specified in the analysis. This would lead to the simple ANOVA presented in Table 7.8.

TABLE 7.7

Multi-stratum ANOVA Table for Potato Yields Trial Using Yields from Four Rows (Factor Row) per Plot (Factor Plot) (Example 7.2B)

Source of Variation	df	Sum of Squares	Mean Square	Variance Ratio	P
Block stratum					
Residual	3	59,948.80	19,982.93	1.434	0.283
Block.Plot stratum					
Fungicide	4	533,676.80	133,419.20	9.576	0.001
Residual	12	167,187.20	13,932.27	4.474	< 0.001
Block.Plot.Row stratum					
Residual	60	186,848.00	3114.13		
Total	79	947,660.80			

TABLE 7.8

Incorrect ANOVA Table (Ignoring Strata) for Potato Yields Trial Using Individual Row Yields (Example 7.2B)

Source of Variation	df	Sum of Squares	Mean Square	Variance Ratio	P
Block	3	59,948.80	19,982.93	4.064	0.010
Fungicide	4	533,676.80	133,419.20	27.133	< 0.001
Residual	72	354,035.20	4917.16		
Total	79	947,660.80			

Omission of information on the full structure (rows within plots within blocks) means that the two lower levels of background variation (due to plots within blocks, and rows within plots) cannot be separated and are combined in the analysis. This leads to an estimate of the background variation to be used for treatment comparisons that is much smaller than it should be (a residual mean square of 4917 rather than 13,932), with many more residual degrees of freedom (72 instead of 12). This is typical of a situation where pseudo-replication is ignored. Hence, compared with the correct analysis, the Block and Fungicide variance ratios are inflated, and treatment SEMs, SEDs and LSDs are greatly underestimated, leading to incorrect inferences.

In multi-stratum ANOVA tables, it is possible to test the null hypotheses associated with the structural terms using comparisons between nested strata. For example, in Example 7.2B, we can test whether there is any evidence of non-zero plot effects by comparing the residual mean square from the Block.Plot stratum with that from the Block.Plot.Row stratum. As stated previously, this is usually a side issue in the analysis, although the information may be useful when designing further experiments.

We take the opportunity to restate here that to preserve the distinction between the explanatory and structural components, it is necessary to store factors associated with both of these components within a data set. Although it is often possible to obtain the correct analysis without doing this, we believe that this loses information on the exact experimental layout and can lead to unnecessary confusion (see comments at the end of Section 7.3).

Unfortunately, the multi-stratum ANOVA table can be formed only when the explanatory and structural components obey certain conditions of balance, and the details are further discussed in Chapters 9, 11 and 16. The simplest case occurs when block and treatment factors are orthogonal as in the RCBD (see Section 7.3), so that each term can be estimated independently of the other.

EXERCISES

7.1* A controlled environment experiment to compare the effect of a diet on weight of three aphid species was conducted using a RCBD with three blocks.

   a. What are the null and alternative hypotheses for this experiment?

   b. Construct the ANOVA table given that BlkSS = 0.00317, TrtSS = 0.35106 and TotSS = 0.36195.

   c. What is the appropriate F-distribution for the treatment variance ratio under the null hypothesis? What is the 5% critical value from this distribution?

   d. Would we accept or reject the null hypothesis?

7.2* A field trial to test the response of a crop to five fertilizer treatments (0, 50, 100, 150 and 200 kg/ha of N) was designed as a RCBD with four blocks of five plots (factors Block and Plot, respectively). The yield at harvest was recorded for each plot. The file FERTILIZER.DAT contains the unit numbers (*ID*), structural factors (Block, Plot), applied rates of N (factor N) and the yields (variate *Yield*).

   a. Write down a mathematical model for the yields.

   b. Construct a multi-stratum ANOVA table by calculating the total, block, treatment and residual sums of squares and df and then deriving the other columns. Is there any evidence of differences in yield among the fertilizer treatments?

    c. Calculate the estimated mean for each treatment.

    d. Calculate the LSD at the 5% level for the difference between any two treatment means and use it to compare the yields obtained for 150 and 200 kg of N applied.

    e. Use residual plots to check whether the model assumptions are reasonable.

    f. Write a short summary of the results of the analysis.

7.3 A 2-year field experiment investigated the effects of soil cultivation on the activity of beneficial arthropods. Plots of winter oilseed rape were laid out as a RCBD with five blocks of three plots. Three soil cultivation treatments were to be compared: ploughing in both years, minimum tillage in both years and minimum tillage in year 1 followed by ploughing in year 2. We consider data from the first season, when the latter two treatments were equivalent resulting in two first-year treatments 'plough' ($n_1 = 5$, one plot per block) and 'minimum tillage' ($n_2 = 10$, two plots per block). The accumulated catch of three pitfall traps per plot during a 3-month period was recorded for various arthropod species; here, we analyse counts of spiders of the taxa *Oedothorax*. The plot-level unit numbers (*ID*), structural factors (Block, Plot), treatments applied (factor Treatment) and the total count data (variate *PlotCount*) can be found in the file OEDOPLOT.DAT.[*]

    a. Use multi-stratum ANOVA to determine whether these soil cultivation methods affect spider numbers. Obtain the standard errors for each treatment mean and the standard error of the difference between the two means (you will need to take into account the differing replication, as in Section 4.4). Produce and interpret a composite set of residual plots.

    b. The trap-level unit numbers (*ID*), structural factors (Block, Plot and Trap), treatments applied (factor Treatment) and individual counts from the three pitfall traps in each plot (variate *TrapCount*) can be found in file OEDOTRAP.DAT. Obtain the multi-stratum ANOVA table and residual plots for these data. Compare and contrast your results here with those obtained in part (a) and discuss any differences.

7.4 A controlled environment experiment investigated the impact of inoculation rate on leaf symptoms in oilseed rape. Four rates of inoculation were chosen (0.4, 4, 40 and 400) and each rate was tested on six oilseed rape cultivars. The experiment was carried out in three runs. In each run (or occasion), the 24 treatments were randomly allocated to 24 single plants in pots, and the average percentage area of leaf infected from two leaves per plant was recorded. The unit numbers (*ID*), structural factors (Occasion, Pot), explanatory factor (Treatment) and responses (variate *PInfected*) can be found in file INOCULA-TION.DAT. Analyse these data on an appropriate scale using ANOVA accounting for blocks and the set of 24 treatments. Is there any evidence that the area of leaf infected differs among 24 treatments? (We re-visit these data in Exercise 8.1.)[†]

---

[*] Data from A. Ferguson, Rothamsted Research.
[†] Data from N. Evans, Rothamsted Research.

8

Extracting Information about Treatments

In previous chapters, where the principles of designing experiments and several different designs were introduced, the focus was on understanding and specifying the structure of the experimental units, i.e. the structural component of the model. In all of these situations, the explanatory component of the model, consisting of two or more treatment groups, has been represented using a single factor. Testing the null hypothesis of ANOVA allows us to answer the broad question of whether the treatments differ from one another, but usually we are interested in more structured comparisons between treatments, and investigating these is the subject of this chapter.

The process of treatment selection begins with specification of the biological questions to be answered or hypotheses to be tested by the experiment, which in turn suggests a set of experimental treatments. Comparisons between these treatments can be turned into statistical hypotheses, so that it is clear which questions can be answered by statistical analyses. Often several different sets of treatments might be considered, each of which enables slightly different questions to be answered, and statistical considerations such as efficiency and precision can help to choose between the different sets. The role of statistical evaluation is therefore important even during the preliminary planning stages of an experiment. It is also helpful to realize that it is possible, and usually desirable, to address more than one hypothesis within a single experiment, and to appreciate that different aspects of a statistical analysis will be appropriate to address different types of question. For example, consider an experiment set up as a RCBD to compare the effects of three increasing doses of growth regulator with a control treatment (no regulator applied). The question of whether growth regulator affects yield can be directly answered by an F-test from an ANOVA table, but the more important question of how growth regulator affects yield is best addressed by examination of the pattern of response to dose.

In this chapter, we examine ways of translating questions about a set of treatments into a statistical analysis. Here, we do not emphasize the structure of the experiment, but remember throughout that specification of the correct structural model is required to obtain the correct analysis. The structure of this chapter is summarized in Table 8.1.

This chapter begins with an overview of several common types of question and the corresponding structure used for their statistical analysis (Section 8.1). We are often interested in comparisons relating to distinct factors underlying the set of treatments. The complete definition, analysis and interpretation of a crossed structure for two factors are described (Section 8.2) and then extended to three or more factors with the emphasis on interpretation (Section 8.3). In some circumstances, a nested structure is more appropriate (Section 8.4), and this structure can also be used to allow for the presence of control or standard treatments (Section 8.5). Often, specific treatment comparisons are required to answer scientific questions. These comparisons may be incorporated in the analysis via treatment contrasts (Section 8.6), which can also be used to model patterns of response for factors with quantitative values (Section 8.7). Following analysis, it can be useful to make comparisons between treatments based on tables of predicted means. We describe some methods for making specific types of comparisons and then discuss some issues associated with this approach (Section 8.8).

TABLE 8.1

Location of Topics Discussed in Chapter 8

Section	Topic
8.1	Overview
8.2	Crossed explanatory structure for two factors
8.3	Crossed explanatory structure with three or more factors
8.4	Nested explanatory structure
8.5	Nested structure to account for control or standard treatments
8.6	Use of contrasts to make specific comparisons
8.7	Use of polynomial contrasts to model response to quantitative factors
8.8	Making treatment comparisons from predicted values

8.1 From Scientific Questions to the Treatment Structure

The structure of the treatments in an experiment should relate to the scientific questions addressed. Many experiments are concerned with assessing how several different types of treatment affect the response of a biological system. For example, in an experiment to investigate storage conditions for potatoes, the interest might be in establishing the effect of different temperatures (3°C, 5°C or 7°C) and humidities (low, 92.5%, and normal, 95%) on subsequent frying quality. The temperatures and humidities can be considered as two different types of treatments that have been combined together. At this point, it is helpful to clarify some terminology. A **treatment factor** is a group of treatments of a common type, for example, in the above experiment, the temperatures correspond to one treatment factor, and the humidity to a second treatment factor. The **factor levels** correspond to the groups labelled by each treatment factor (e.g. the three individual temperatures), and an **experimental treatment** is a combination made by taking one level from each of the treatment factors used in the experiment (e.g. temperature 5°C with 95% humidity). A **factorial treatment structure** consists of all possible experimental treatments constructed by taking one level from each of the treatment factors. So in the example above, a factorial structure consists of six treatments: all three temperatures tested at both humidity levels. This structure is often called a 3×2 factorial, i.e. a factorial structure with two factors, one with three levels and the other with two levels. This is also sometimes referred to as a two-way structure, i.e. a structure with two factors, leading to a two-way ANOVA. The concept of a factorial structure can be extended to any number of factors (an r-way structure for r factors). For example, a $3 \times 3 \times 2$ factorial contains three factors, two with three levels and one with two levels, giving a three-way structure with 18 experimental treatments. For the moment, we consider two treatment factors only.

If an experiment has a factorial treatment structure then usually the scientific questions relate to both the overall effects of each treatment factor, and whether the different treatment factors act independently or interact. This requires the use of a crossed structure (see Section 3.2), which is expressed with the explanatory component of the model. The potato storage experiment described above has this structure: both the temperature and humidity treatments are of individual interest, as is the question of whether changing the temperature also changes, or interacts with, the effect of the humidity treatment. Using the obvious symbolic names, we can write this structure as

Explanatory component: [1] + Temperature*Humidity

                                = [1] + Temperature + Humidity + Temperature.Humidity

Recall that [1] represents a factor with only one group and that this term is associated with the overall mean, μ. In this context, the terms Temperature and Humidity are called main effects and the term Temperature.Humidity represents their interaction. This crossed treatment structure is explained in detail in Sections 8.2 and 8.3.

Occasionally, a nested treatment structure will be more appropriate (see Section 3.2). This happens when the levels of one treatment factor have no real meaning when considered alone, but have meaning when considered in conjunction with another treatment factor. In this case, the two factors are not independent, and we denote this hierarchical relationship in terms of a parent factor and a nested factor. For example, consider a small variety trial designed to test four families each with six lines. This can be considered as a 4×6 factorial structure for factors Family and Line, with lines numbered 1–6 within families. But there is no connection between the nested lines with label 1 (or 2 or ... 6) across families and so there is no interest in the overall effect of each line number. There is interest in the overall effect of each family, as these constitute separate groups, and also in whether there are differences between lines within families. Hence, the appropriate treatment structure takes the form

> Explanatory component: [1] + Family/Line
> = [1] + Family + Family.Line

Here, there is only one main effect, for the parent factor Family, and the term Family.Line represents the nested effects of lines within families. This nested treatment structure is explained further in Section 8.4. The nested structure can also be useful when control or standard treatments are included within the set of experimental treatments, but direct comparison with other treatments is not of major interest, or when a control or standard is added onto a factorial set. In this case, a nested structure can be used to partition the experimental treatments into sets, and comparisons of interest are then made across and within the sets. This structure is explained further in Section 8.5.

If the scientific questions do not relate directly to an underlying crossed or nested structure, then testing specific hypotheses about differences between treatment effects, known as contrasts, can often be an efficient approach. For example, consider a repellence screening experiment in which four compounds were tested: standard, A alone, A with B, A with C. The questions of interest are 'is the compound A as good as or better than the standard?' and 'are either of the combinations A + B or A + C better than A alone?' We can construct contrasts to incorporate these hypotheses into our analysis, and this matter is examined in more detail in Section 8.6. Contrasts can also be used to make specific comparisons within a crossed or nested structure. As an alternative to embedding contrasts within the analysis, we can apply contrasts to tables of predicted means after the analysis, and this is discussed in Section 8.8.

Finally, if one or more of the treatment factors have levels related to an underlying quantitative scale, for example, amount of fertilizer applied, sowing date, dose of chemical or temperature, then it may be desirable to model the response on that quantitative scale. Polynomial contrasts can be used to build simple empirical models for the response, and this can also be done within a crossed or nested treatment structure. For example, consider a field trial to investigate factors affecting wheat establishment that tests five sowing dates (each two weeks apart) for six different varieties. This is a 5×6 factorial experiment, with the appropriate explanatory model being a crossed treatment structure. Polynomial contrasts can be used to model the response as a linear or quadratic function of sowing date, and to test whether this function is consistent across varieties. This approach is illustrated in Section 8.7.

8.2 A Crossed Treatment Structure with Two Factors

A **crossed treatment structure** for two factors allows variation within the full set of treatments to be partitioned into the overall (or main) effects of the individual factors and the interaction between those factors. The **main effects** for a factor represent the common effect for each of its levels, when averaged over all levels of the other factor. For example, consider a controlled environment (CE) experiment to investigate the infectivity of four different strains of a pathogen species (labelled A–D), for artificially damaged (wounded) or unwounded leaves. This is a 4×2 factorial structure and both treatment factors are to be evaluated individually, as well as in combination, so a crossed treatment structure is appropriate here. The main effect for each pathogen strain represents an average effect across wounded and unwounded leaves. Conversely, the main effect of wounding represents an average effect across all strains.

Two factors are considered to act independently if the effect of applying them together is equivalent to adding together their main effects; any deviation from this pattern is known as an **interaction**. The concept of interaction is most easily represented in a graphical context, and we illustrate it using the CE experiment with different pathogen strains and wounding, where the response is a measure of pathogen growth within the leaf. Here, we consider the true, but unknown, response (rather than observed responses) in terms of population parameters, in the context of two different scenarios. The first scenario assumes independent action of the two factors and is represented in Figure 8.1a.

The main effect of each pathogen strain is based on the average of its growth across the wounded and unwounded treatments and it is clear that there is some difference in average virulence among the strains. The individual main effects are considered as deviations from the overall mean, so that the main effect for strain A is a small negative value, for B is a large negative value, and both C and D are positive, with C larger than D. Similarly, the main effect for unwounded plants is based on the average growth across strains for this condition, expressed as a difference with the overall mean (represented at the right-hand side of Figure 8.1a). The main effect for unwounded plants is negative and that for wounded plants is positive, i.e. pathogen growth is greater in wounded leaves, and the absolute values of these two effects are equal. If there is no interaction, then the growth for each treatment combination should arise solely from the main effects. This would imply that the difference in growth between wounded and unwounded plants should be the same for all strains and so the pattern of growth across strains should be similar for both wounded and unwounded plants. This is seen most clearly if we draw lines for growth across strains in the wounded and unwounded plants separately: if there is no interaction,

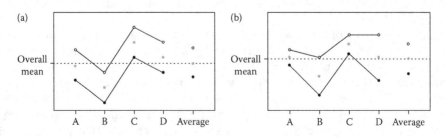

FIGURE 8.1

Pattern of response for four pathogen strains (A–D) with (o) or without (•) wounding in the case of (a) no interaction and (b) strong interaction present. • indicates mean for each pathogen strain.

then these lines should be parallel, as in Figure 8.1a. This model is variously known as the independence, main effects or additive model.

In the second scenario, we assume the presence of some interaction between the factors, which is represented in Figure 8.1b. The main effects are calculated and represented as in the previous scenario. Again, there are differences in average virulence between the strains (C largest, B smallest with A and D intermediate) and more pathogen growth in wounded than in unwounded leaves. However, here it is clear that the difference in growth between the wounded and unwounded leaves changes according to the strain. For example, the difference is small for strains A and C but much larger for strains B and D. In consequence, the pattern of growth across strains for wounded leaves is substantially different from the pattern for unwounded leaves, and the two lines are no longer parallel (Figure 8.1b). The interaction effect for each treatment combination is the difference between the growth expected under the independence model and the actual value.

In practice, of course, we do not know the true population response and have only a sample of observations for each experimental treatment, which introduces variation. We can plot the observed treatment means to get some insight into the presence of an interaction, but these means are subject to uncertainty. As in previous chapters, we can use ANOVA to obtain an estimate of background variation and use this to judge whether the observed interaction effects are real, or if they can be attributed to background variation.

EXAMPLE 8.1A: BEETLE MATING

Consider the beetle mating experiment described in Example 6.1 (data in file BEETLES.DAT). Females from two species of willow beetle (*P. vitellinae* and *P. vulgatissima*) were mated with males from either their own species (intraspecies mating) or the other species (interspecies mating). There were 10 replicates of each of the four treatment combinations, and the number of eggs laid by each female was recorded. Example 6.1 established that a log transformation is required to homogenize the variance (we used a $\log_{10}$ transformation), and considered the structure as a single set of four treatments. Here, we use a crossed treatment structure to address the question of interest, namely the viability of interspecies mating. We represent the four treatments as a factorial combination of two factors, the species of the female and the type of mating.

By using this structure, we partition the variation between treatments into that due to each of the two main effects and their interaction. The main effect of species represents the logged number of eggs produced by females of each species, averaged across mating types. Conversely, the main effect of mating type represents the logged number of eggs produced for each type of mating, averaged across species. The interaction examines whether there is any difference in logged numbers of eggs between types of mating across females of the two species. The observations are plotted with the four treatment means in Figure 8.2.

It is clear that females of species *P. vulgatissima* appear more fecund and that interspecies mating is generally less productive than intraspecies mating. The lines that join means for the same mating type across species are not parallel, suggesting that an interaction may be present, and the loss of productivity due to interspecies mating appears smaller for species *P. vulgatissima* than for species *P. vitellinae*. However, it is also clear that there is much variation in the observed logged numbers, and the apparent interaction must be evaluated in the context of this background variation.

8.2.1 Models for a Crossed Treatment Structure with Two Factors

In the previous section, a set of experimental treatments was decomposed into components associated with the underlying treatment factors. In this section, we write this decomposition in the form of a statistical model. For simplicity, here we assume that the experimental

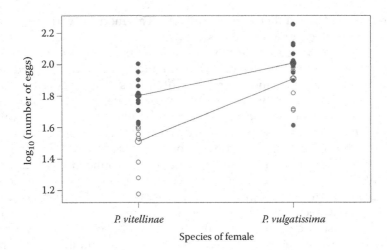

FIGURE 8.2
Observed productivity, measured as $\log_{10}$(number of eggs), for inter- (o) and intraspecies (•) mating for two species of willow beetle, with means for inter- (larger o) and intraspecies (larger •) mating joined across species (Example 8.1A).

units are unstructured, i.e. with no structural component, and that the treatments form a factorial set with equal replication. A model for the observations in terms of a single set of treatments can be written as in Equation 4.6 as

$$y_{jk} = \mu + \tau_j + e_{jk}, \tag{8.1}$$

where y_{jk} is the observed response for the kth replicate of the jth treatment group, for $j = 1 \ldots t$ and $k = 1 \ldots n$, with deviation e_{jk}, μ is the overall mean and τ_j is the effect of the jth treatment group, expressed as a deviation from the overall mean. The assumptions of Section 4.1 all still apply. Recall from Section 4.5 that this form of the model uses the sum-to-zero constraint $\Sigma_j\tau_j = 0$ to avoid over-parameterization. If we denote the responses as variate Y and the factor labelling the treatment groups as Treatment, then we can write this model in symbolic form as

Response variable: Y
Explanatory component: [1] + Treatment

The term [1] is associated with the overall mean, and the term Treatment is associated with the set of treatment effects, τ_j, $j = 1 \ldots t$.

To write the model in terms of the individual factors, we must relabel the observations in terms of those individual factors. In the case of a generic crossed structure constructed from two factors, we denote these factors as A and B. Factor A has t_A levels and factor B has t_B levels and their product gives the total number of treatments, i.e. $t = t_A \times t_B$. The subscripts r and s are used to indicate the level of factors A and B, respectively, present on each unit. The statistical model above can then be rewritten, with this new labelling, as

$$y_{rsk} = \mu + \tau_{rs} + e_{rsk}.$$

The jth treatment group is now acknowledged as arising from the combination of the rth level of treatment factor A with the sth level of treatment factor B but, apart from this cosmetic change, this is exactly the same linear model. This model gives exactly the same ANOVA table and estimates as in the previous form but now labelled by the underlying factors. Here, the estimate of each treatment effect is the deviation of the observed treatment mean from the sample grand mean, as obtained previously, and can be written as

$$\hat{\tau}_{rs} = \bar{y}_{rs\bullet} - \bar{y} .$$

The crossed structure described above is written in symbolic form as

Explanatory component: $[1] + A^*B$
$= [1] + A + B + A.B$

To implement this form, we decompose the unstructured treatment effects into main and interaction effects as

$$\tau_{rs} = \alpha_r + \beta_s + (\alpha\beta)_{rs} , \qquad (8.2)$$

where α_r is the main effect for the rth group in factor A, β_s is the main effect for the sth group in factor B and $(\alpha\beta)_{rs}$ is the interaction effect for the rth group in factor A with the sth group in factor B in term A.B. The composite symbol $(\alpha\beta)$ is used to show clearly which terms the interaction has arisen from. This expression can be rearranged as

$$(\alpha\beta)_{rs} = \tau_{rs} - (\alpha_r + \beta_s) ,$$

so the interaction can be considered as the difference between the original treatment effects and the additive model based on the assumption that factors act independently. Alternatively, the interaction can be considered as the treatment effects adjusted for all terms in the model that are marginal to it. A term is considered **marginal** to all terms of which it is a sub-term; for example, terms A and B are both marginal to A.B. By convention the overall mean, represented symbolically here as [1], is considered as marginal to all other terms. We use this important concept both for calculating parameter estimates and in identifying terms to be used for prediction.

The full model can then be written with a crossed treatment structure as

$$y_{rsk} = \mu + \alpha_r + \beta_s + (\alpha\beta)_{rs} + e_{rsk} .$$

In the sum-to-zero parameterization (introduced in Section 4.5), the main effects are written as deviations about the sample grand mean, with the resulting constraints $\Sigma_r \alpha_r = 0$ and $\Sigma_s \beta_s = 0$. The interaction effects are in turn written as deviations from the main effects, with the resulting constraints $\Sigma_r(\alpha\beta)_{rs} = 0$ for $s = 1 \ldots t_A$ and $\Sigma_s(\alpha\beta)_{rs} = 0$ for $r = 1 \ldots t_B$. Application of these constraints prevents the model from becoming over-parameterized. This and other parameterizations are discussed further in Section 8.2.6.

8.2.2 Estimating the Model Parameters

As usual, the model parameters are estimated by the method of least squares. Here, we consider the simplest case of a full factorial structure with equal replication, which gives an orthogonal structure; other cases are dealt with in Chapter 11. In this case, main effects and interactions can be expressed as functions of the unstructured treatment effects, τ_{rs}, and their estimates can be similarly derived from estimates of the unstructured treatment effects, $\hat{\tau}_{rs} = \bar{y}_{rs\bullet} - \bar{y}$.

From Equation 8.2 and the parameter constraints $\Sigma_s\beta_s = 0$ and $\Sigma_s(\alpha\beta)_{rs} = 0$, it follows that

$$\frac{1}{t_B} \sum_{s=1}^{t_B} \tau_{rs} = \alpha_r .$$

Hence, we can derive estimated main effects for factor A as

$$\hat{\alpha}_r = \frac{1}{t_B} \sum_{s=1}^{t_B} \hat{\tau}_{rs} = \frac{1}{t_B} \sum_{s=1}^{t_B} (\bar{y}_{rs\bullet} - \bar{y}) = \bar{y}_{r\bullet\bullet} - \bar{y} .$$

We can calculate this quantity by taking the marginal means of the unstructured treatment effects across levels of factor B. Similarly, we can estimate the main effects for factor B by taking the marginal means of the unstructured treatment effects across levels of factor A, giving

$$\hat{\beta}_s = \frac{1}{t_A} \sum_{r=1}^{t_A} \hat{\tau}_{rs} = \bar{y}_{\bullet s\bullet} - \bar{y} .$$

The estimated interaction effect is equal to the unstructured treatment effect adjusted for both main effects, as

$$(\widehat{\alpha\beta})_{rs} = \hat{\tau}_{rs} - (\hat{\alpha}_r + \hat{\beta}_s) = (\bar{y}_{rs\bullet} - \bar{y}) - [(\bar{y}_{r\bullet\bullet} - \bar{y}) + (\bar{y}_{\bullet s\bullet} - \bar{y})] = \bar{y}_{rs\bullet} - \bar{y}_{r\bullet\bullet} - \bar{y}_{\bullet s\bullet} + \bar{y} .$$

We can easily derive these estimates from a two-way table of treatment means, after adjusting for marginal terms. This is demonstrated in Example 8.1B.

EXAMPLE 8.1B: BEETLE MATING

The crossed model for the beetle mating experiment can be written in symbolic form with factors Species (two levels) and MateType (two levels) as

Response variable: *logEggs*
Explanatory component: [1] + Species*MateType
                                       = [1] + Species + MateType + Species.MateType

In mathematical form, this model becomes

$$logEggs_{rsk} = \mu + Species_r + MateType_s + (Species.MateType)_{rs} + e_{rsk} ,$$

where $logEggs_{rsk}$ is the $\log_{10}$-transformed number of eggs for the kth replicate measurement (for $k = 1 \ldots 10$) of a female of species r ($r = 1$ for *P. vitellinae*, 2 for *P. vulgatissima*) with mating of type s ($s = 1$ for interspecies, 2 for intraspecies). The main effect for species r is denoted *Species$_r$*, with *MateType$_s$* as the main effect for mating of type s, and (*Species.*

MateType)$_{rs}$ being the interaction between species r and mating type s. The overall mean, μ, and deviation, e_{rsk}, are as described above, and the sum-to-zero constraints take the form $\Sigma_r Species_r = 0$, $\Sigma_s MateType_s = 0$, $\Sigma_s(Species.MateType)_{rs} = 0$ for $r = 1, 2$, and $\Sigma_r(Species.MateType)_{rs} = 0$ for $s = 1, 2$.

The treatment means on the $\log_{10}$ scale for the four combinations of species and mating type were presented in Table 6.3. This table is reformatted as a two-way table with marginal means in Table 8.2a. As we have equal replication, the estimates take a simple form. The estimate of the overall mean is the overall mean of the table, i.e. $\hat{\mu} = 1.8085$. We can obtain estimates of the unstructured treatment effects by subtracting the overall mean from each of the other cells, as in Table 8.2b. The main effect estimates, which consist of marginal means adjusted for the overall mean, are now present in the margins. We then subtract the two marginal means from each of the internal cells, as in Table 8.2c, to obtain the interaction effects. The full set of estimates is shown in Table 8.2c. Notice that the values within each set of effects (main effects or interactions) have the same absolute value but differ in sign. This is a direct consequence of the sum-to-zero constraints, and always occurs for factors with two levels.

TABLE 8.2a

Treatment and Marginal Means for $\log_{10}$(Number of Eggs) in the Beetle Mating Experiment (Example 8.1B)

		Mating Type		
		Interspecies	Intraspecies	Average
Species	P. vit.	1.5129	1.8036	1.6582
of female	P. vulg.	1.9089	2.0085	1.9587
	Average	1.7109	1.9060	1.8085

TABLE 8.2b

Subtract Overall Mean from All Other Cells in Table 8.2a to Get Estimates of Main Effects in Margins

		Mating Type		
		Interspecies	Intraspecies	Average
Species of female	P. vit.	−0.2956	−0.0049	$\widehat{Species}_1 = -0.1503$
	P. vulg.	0.1005	0.2000	$\widehat{Species}_2 = 0.1503$
	Average	$\widehat{MateType}_1 = -0.0976$	$\widehat{MateType}_2 = 0.0976$	$\hat{\mu} = 1.8085$

TABLE 8.2c

Subtract Row and Column Marginal Means from Internal Cells in Table 8.2a to Get Estimated Interaction Effects. $(Sp.MT)_{rs}$ is an Abbreviation of $(Species.MateType)_{rs}$

		Mating Type		
		Interspecies	Intraspecies	Average
Species of female	P. vit.	$(\widehat{Sp.MT})_{11} = -0.0478$	$(\widehat{Sp.MT})_{12} = 0.0478$	$\widehat{Species}_1 = -0.1503$
	P. vulg.	$(\widehat{Sp.MT})_{21} = 0.0478$	$(\widehat{Sp.MT})_{22} = -0.0478$	$\widehat{Species}_2 = 0.1503$
	Average	$\widehat{MateType}_1 = -0.0976$	$\widehat{MateType}_2 = 0.0976$	$\hat{\mu} = 1.8085$

8.2.3 Assessing the Importance of Individual Model Terms

In Section 4.3, the one-way ANOVA table was obtained for an unstructured set of treatments that partitioned the total sum of squares into the treatment and residual sums of squares. The partitioning of the treatment effects into main effects and their interaction leads to a similar partitioning of the treatment sum of squares into components relating to the three model terms, as

$$\text{TrtSS} = \text{SS(A)} + \text{SS(B)} + \text{SS(A.B)} ,$$

where SS(A) and SS(B) are the sums of squares for the main effects of factors A and B, respectively, and SS(A.B) is the sum of squares for the interaction. In general, we use the form $\text{SS}(+A.B|A+B)$ for the interaction sum of squares, to denote that the interaction has been added into the model after both main effects. This emphasises the fact that the value of the interaction sum of squares depends on which other terms have been previously fitted in the model, and this is discussed further in Section 11.2.2. These sums of squares are calculated as

$$\text{SS(A)} = n \times t_B \sum_{r=1}^{t_A} (\bar{y}_{r..} - \bar{y})^2, \quad \text{SS(B)} = n \times t_A \sum_{s=1}^{t_B} (\bar{y}_{.s.} - \bar{y})^2 ,$$

$$\text{SS(A.B)} = n \sum_{r=1}^{t_A} \sum_{s=1}^{t_B} (\bar{y}_{rs.} - \bar{y}_{r..} - \bar{y}_{.s.} + \bar{y})^2 ,$$

or, equivalently, as the sums of squares of the parameter estimates

$$\text{SS(A)} = n \times t_B \sum_{r=1}^{t_A} \hat{\alpha}_r^2, \quad \text{SS(B)} = n \times t_A \sum_{s=1}^{t_B} \hat{\beta}_s^2, \quad \text{SS(A.B)} = n \sum_{r=1}^{t_A} \sum_{s=1}^{t_B} \widehat{(\alpha\beta)}_{rs}^2 .$$

The total treatment degrees of freedom (equal to $t - 1$) are partitioned in a similar manner. There are $\text{df(A)} = t_A - 1$ df associated with factor A, $\text{df(B)} = t_B - 1$ df associated with factor B and the interaction df is calculated by subtraction as

$$\text{df(A.B)} = (t - 1) - (t_A - 1) - (t_B - 1) = (t_A \times t_B) - t_A - t_B + 1 = (t_A - 1) \times (t_B - 1) .$$

This leads to the generic form of ANOVA table shown in Table 8.3.

TABLE 8.3

Generic Form of ANOVA Table for a Crossed Treatment Structure with Two Factors A and B, with t_A and t_B Levels, Respectively

Source of Variation	df	Sum of Squares	Mean Square	Variance Ratio
A	$t_A - 1$	SS(A)	MS(A)	$F^A = \text{MS(A)}/\text{ResMS}$
B	$t_B - 1$	SS(B)	MS(B)	$F^B = \text{MS(B)}/\text{ResMS}$
A.B	$(t_A - 1)(t_B - 1)$	SS(A.B)	MS(A.B)	$F^{A.B} = \text{MS(A.B)}/\text{ResMS}$
Residual	$N - t$	ResSS	ResMS	
Total	$N - 1$	TotSS		

Note: No structural component present, all $t (= t_A \times t_B)$ treatment combinations have n replicates giving $N = n \times t$ observations.

As previously, mean squares are calculated by division of the sums of squares by their degrees of freedom. There are now three variance ratios, formed by division of the mean squares for each of the two main effects and the interaction by the residual mean square (ResMS). Note that the ResMS and ResDF take the same values as when the treatments were considered as an unstructured set.

We have approached the factorial structure by building a crossed model for the treatment effects using the underlying factors. In terms of understanding the biological system, it is helpful to identify as simple a model as possible for prediction, subject to its being consistent with the observed data. We start with the most complex model, containing all of the terms, and try to identify terms that can be ignored for prediction. We consider the interaction term first, which holds information on dependencies in the response across the factors. If this term is not statistically significant then we can predict from the main effect terms only, asserting independent action of the two factors, and we might even be able to simplify the model further. If the interaction is statistically significant then prediction from a simpler model is not sensible.

The variance ratio for the interaction, denoted $F^{A.B}$ in Table 8.3, can be used to test the null hypothesis that all of the interaction effects equal zero, or $H_0: (\alpha\beta)_{rs} = 0$ for all $r = 1 \ldots t_A$, $s = 1 \ldots t_B$, against the general alternative hypothesis that the interaction effects are not all zero. Under the null hypothesis, the variance ratio $F^{A.B}$ has an F-distribution with $(t_A - 1) \times (t_B - 1)$ numerator and $N - t$ denominator df. Recall that we often specify these two df as a subscript, and we shall sometimes also abbreviate the factor names in the superscript for brevity. If $F^{A.B}$ exceeds the chosen critical value of this F-distribution, then we have statistical evidence for an interaction, which should not be ignored. If the interaction is not statistically significant then patterns in the response can be adequately represented by the main effects alone. Further simplification may be possible, however, and so we next examine each of the main effects in turn.

Because the structure is orthogonal when all treatment combinations are equally replicated, the order in which these terms are examined is not important, but we choose to work our way up the ANOVA table. The variance ratio for factor B, denoted F^B, can be used to test the null hypothesis that the main effects for this factor are all equal to zero, or $H_0: \beta_s = 0$ for all $s = 1 \ldots t_B$, against the general alternative that they are not all zero. Under the null hypothesis, the variance ratio F^B has an F-distribution with $t_B - 1$ numerator and $N - t$ denominator df. If F^B exceeds the chosen critical value of this distribution, this gives statistical evidence for the presence of main effects for factor B. A similar process is followed to test the main effects for factor A, using the mean square and df associated with that factor.

EXAMPLE 8.1C: BEETLE MATING

Table 8.4 is the ANOVA table for the crossed treatment structure in the beetle mating experiment. As expected, TotSS and ResSS and their degrees of freedom are the same as for the unstructured analysis shown in Table 6.2, and the sums of squares for the main effects and the interaction in Table 8.4 together add up to the TrtSS in Table 6.2, i.e. $1.37515 = 0.90307 + 0.38073 + 0.09135$. Similarly, the main effect and interaction df in Table 8.4 add up to the TrtDF in Table 6.2, i.e. $3 = 1 + 1 + 1$. This verifies numerically that our decomposition into main effects and the interaction is a partitioning of the total treatment information.

The variance ratio for the Species.MateType interaction is $F^{S.M}_{1,36} = 3.837$, with observed significance level $P = 0.058$. Taking a strict approach to hypothesis testing with a 5% significance level, then the null hypothesis would not be rejected, and we should conclude that there is no statistical evidence of an interaction. However, taking a more pragmatic approach, there is some suggestion that an interaction might be present and we might

TABLE 8.4

ANOVA Table for $\log_{10}$(Number of Eggs) from the Beetle Mating Experiment Using a Crossed Treatment Structure (Factors Species and MateType) (Example 8.1C)

Source of Variation	df	Sum of Squares	Mean Square	Variance Ratio	P
Species	1	0.9031	0.9031	$F^S = 37.932$	< 0.001
MateType	1	0.3807	0.3807	$F^M = 15.992$	< 0.001
Species.MateType	1	0.0913	0.0913	$F^{S.M} = 3.837$	0.058
Residual	36	0.8571	0.0238		
Total	39	2.2322			

consider the situation further. In this case, the value of the variance ratio for the interaction is much smaller than those for the main effects, suggesting that even if an interaction is present, its effect is relatively small. These issues are discussed further at the end of Section 8.3. For now, we decide that we shall not greatly misrepresent the data by omitting the interaction, and so the model of independent action is appropriate. As both main effects are highly statistically significant, we conclude that both the species and type of mating affect the expected $\log_{10}$-number of eggs, but the expected decrease due to interspecies mating is similar for both species.

8.2.4 Evaluating the Response to Treatments: Predictions from the Fitted Model

The ANOVA table is used to identify the subset of model terms that best describe the pattern of response across treatment groups. This subset can then be used to predict the expected response for any treatment combination or the expected difference in response between treatment combinations. If there is evidence of an interaction between the factors, then predictions must be based on the full model. If the interaction is not statistically significant, then it can be ignored for the purposes of prediction, together with any non-significant main effect(s). The remaining terms are used as the model for prediction, and there are several ways to approach this. One way is to refit the model containing the selected terms only, then obtain predictions from the revised model. This approach is always necessary for non-orthogonal structures, as discussed in Chapter 11, as the value of the predictions will depend on the model terms fitted. However, one unsatisfactory aspect of this procedure is that it involves re-estimation of the background variation by the pooling of true background variation (based on differences between replicates) with that from terms dropped from the model (based on differences between treatment combinations). In orthogonal structures, such as a factorial with equal replication (or certain forms of unequal replication, see Section 11.1) there is an alternative and more appropriate method which gives direct estimates of prediction standard errors based on the original estimate of background variation. This method uses the result that the multi-way table of observed treatment means, and its margins, give direct estimates of certain population treatment means, and that standard errors of these means, and their differences, are easy to derive. This is the approach that we outline below.

We extend our previous notation for population treatment means (Section 4.1) so that μ_{rs} denotes the population mean for the rth level of factor A and the sth level of factor B, which implies

$$\mu_{rs} = \mu + \alpha_r + \beta_s + (\alpha\beta)_{rs} .$$

Then $\mu_{r\cdot}$ is defined as the expected mean for the rth level of factor A, averaged over the levels of factor B present, and $\mu_{\cdot s}$ is defined similarly as the expected mean for the sth level of factor B, averaged over the levels of factor A present in the experiment.

If the interaction is significant, then the full model is used for prediction, with

$$\hat{\mu}_{rs} = \hat{\mu} + \hat{\alpha}_r + \hat{\beta}_s + \widehat{(\alpha\beta)}_{rs} = \bar{y}_{rs\cdot} \ .$$

In this case, the prediction for each treatment combination is equal to the observed treatment mean. The variance of this mean is equal to the background variation divided by its replication, and the set of such means are mutually independent. As in Section 4.3, we estimate the background variation using $s^2 = \text{ResMS}$, and so we can estimate the SE of a prediction (previously denoted SEM), or a difference between predictions for two treatment combinations (previously denoted SED), as

$$\widehat{\text{SE}}(\hat{\mu}_{rs}) = \sqrt{\frac{s^2}{n}}; \quad \widehat{\text{SE}}(\hat{\mu}_{rs} - \hat{\mu}_{ij}) = \sqrt{\frac{2s^2}{n}} \ .$$

The second expression only holds for treatment combinations ij such that $ij \neq rs$. If the interaction is not statistically significant, but both main effects are significant, then predictions for each treatment combination are made as

$$\hat{\mu}_{rs} = \hat{\mu} + \hat{\alpha}_r + \hat{\beta}_s = \bar{y}_{r\cdot\cdot} + \bar{y}_{\cdot s\cdot} - \bar{y} \ .$$

We then note that prediction for the rth level of factor A (averaged across all levels of factor B, and using $\Sigma_s \hat{\beta}_s = 0$) is the mean of these predictions across levels of factor B, giving

$$\hat{\mu}_{r\cdot} = \hat{\mu} + \hat{\alpha}_r = \bar{y}_{r\cdot\cdot} \ .$$

These quantities are the observed marginal means for factor A, taken across levels of factor B. Again, we have simple expressions for the estimated SEMs and SEDs as

$$\widehat{\text{SE}}(\hat{\mu}_{r\cdot}) = \sqrt{\frac{s^2}{n \times t_B}}; \quad \widehat{\text{SE}}(\hat{\mu}_{r\cdot} - \hat{\mu}_{i\cdot}) = \sqrt{\frac{2s^2}{n \times t_B}} \ .$$

Conversely, prediction for the sth level of factor B (averaged across all levels of factor A) is

$$\hat{\mu}_{\cdot s} = \hat{\mu} + \hat{\beta}_s = \bar{y}_{\cdot s\cdot} \ ,$$

with

$$\widehat{\text{SE}}(\hat{\mu}_{\cdot s}) = \sqrt{\frac{s^2}{n \times t_A}}; \quad \widehat{\text{SE}}(\hat{\mu}_{\cdot s} - \hat{\mu}_{\cdot j}) = \sqrt{\frac{2s^2}{n \times t_A}} \ .$$

In fact, these results for main effects hold whether or not the interaction is retained in the predictive model. The SEDs can be used to construct LSDs and confidence intervals as previously shown in Section 4.4. Again, the df of the t-statistic used to calculate the LSDs and confidence intervals is equal to the ResDF from the ANOVA table.

EXAMPLE 8.1D: BEETLE MATING

In Example 8.1C, we found that the interaction between species and mating type was not statistically significant and so we do not use this term for prediction. The model predictions take the form

$$\hat{\mu}_{rs} = \hat{\mu} + \widehat{Species}_r + \widehat{MateType}_s ,$$

and we can summarize our results by presenting the predicted means for each factor separately, i.e. averaging over the other factor. We are interested in the difference between the two levels in each case and so use SEDs as a measure of uncertainty. We form 95% CIs for the differences and back-transform these onto the original scale for interpretation. These results are presented in Table 8.5.

The expected logged number of eggs for females of species *P. vitellinae* (Species level 1) is 0.301 units smaller than for females of species *P. vulgatissima* (level 2), and the expected logged number of eggs is 0.195 units smaller for interspecies mating (MateType level 1) than for intraspecies mating (level 2). The estimated SEDs use $s^2 = \text{ResMS} = 0.0238$ and $n \times t_A = n \times t_B = 20$. The LSD calculated at a 5% significance level requires the 97.5th percentile of a t-distribution on ResDF = 36 df, i.e. $t_{36}^{[0.025]} = 2.028$, hence LSD = SED $\times t_{36}^{[0.025]} = 0.0990$. The 95% CI for each difference is equal to its estimate plus or minus the LSD, and the back-transformation for prediction $\hat{\mu}_{rs}$ on the $\log_{10}$ scale is $10^{\hat{\mu}_{rs}}$ (see Table 6.4).

As discussed in Section 6.4, a difference on any logarithmic scale back-transforms to a ratio on the original scale. From the back-transformed CIs, we can conclude that females of species *P. vulgatissima* lay on average 1.59–2.52 times as many eggs as females of species *P. vitellinae*. Similarly, intraspecies mating produces 1.25–1.97 times as many eggs as interspecies mating. These results are consistent with the back-transformed individual treatment means presented in Table 6.5, but are more readily interpreted in terms of the underlying factors.

8.2.5 The Advantages of Factorial Structure

The use of factorial treatment structures leads to experiments with clear conclusions, and they are to be recommended for this fact alone. However, they also enable us to design

TABLE 8.5

Marginal Means, $\log_{10}$(Number of Eggs), for Species and Mating Type with Estimated Differences, SEDs and Back-Transformed 95% Confidence Intervals (CI) (Example 8.1D)

	Species	MateType
Prediction level 1	$\hat{\mu}_{1\cdot} = 1.658$	$\hat{\mu}_{\cdot 1} = 1.711$
Prediction level 2	$\hat{\mu}_{2\cdot} = 1.959$	$\hat{\mu}_{\cdot 2} = 1.906$
Difference (level 2 − level 1)	$\hat{\mu}_{2\cdot} - \hat{\mu}_{1\cdot} = 0.301$	$\hat{\mu}_{\cdot 2} - \hat{\mu}_{\cdot 1} = 0.195$
SED	$\widehat{SE}(\hat{\mu}_{2\cdot} - \hat{\mu}_{1\cdot}) = 0.0488$	$\widehat{SE}(\hat{\mu}_{\cdot 2} - \hat{\mu}_{\cdot 1}) = 0.0488$
95% CI for difference	(0.202, 0.399)	(0.096, 0.294)
Back-transformed CI	(1.59, 2.52)	(1.25, 1.97)

efficient experiments whether interactions are present or not. For example, suppose that the beetle mating experiment had been done as two separate experiments: one to compare inter- versus intraspecies mating for *P. vitellinae* and the other to make the same comparison for *P. vulgatissima*, each using 10 replicates per treatment as in the original experiment. These two separate experiments use the same total number of experimental units as the factorial experiment. However, because these new experiments were run at different times, there is no way to assess the interaction, i.e. whether the decrease in the logged number of eggs laid after interspecies mating differs between the two species, as any difference may be due to some change in the background environment rather than the change in species. In contrast, the interaction can be tested directly in the factorial experiment. In addition, the replication for testing differences between main effects is 20 for both factors Species and MateType (as in Example 8.1D), rather than 10 for the individual experiments, which gives an increase in both precision and power. Finally, the main effects in a factorial experiment are tested over a range of conditions (corresponding to the other factor levels) and any main effects that emerge must be consistent over these conditions. Hence, a factorial experiment arguably provides a more broadly applicable estimate of main effects than an experiment that tests a single factor with other conditions held fixed.

8.2.6 Understanding Different Parameterizations

In Section 4.5, we described several different forms of parameterization for a model with an unstructured set of treatments. Similar forms of parameterization can be applied to structured models, such as the crossed models discussed in this section.

The sum-to-zero parameterization described in Section 8.2.1 has $1 + t_A + t_B + (t_A \times t_B)$ parameters, from the overall mean, two sets of main effects and the interaction, respectively. As we have only $t_A \times t_B$ treatment groups, not all of these parameters can be estimated uniquely and so constraints are imposed. The sum-to-zero parameterization imposes $1 + t_A + t_B$ constraints: $\Sigma_r \alpha_r = 0$ (1 constraint), $\Sigma_s \beta_s = 0$ (1 constraint), $\Sigma_r (\alpha\beta)_{rs} = 0$ (t_B constraints) and $\Sigma_s (\alpha\beta)_{rs} = 0$ (t_A constraints). This looks like $2 + t_A + t_B$ constraints, but in fact the latter two sets contain one dependency, so there are only $1 + t_A + t_B$ separate constraints in total. We therefore obtain a total of $t_A \times t_B$ unconstrained parameters, equal to the number of separate groups, as required.

As in the one-way unstructured model, we can also use the first-level-zero parameterization introduced in Section 4.5 for the crossed model with two factors. In this case, we write the model in the form

$$y_{rsk} = \mu_{11} + \eta_r + \zeta_s + (\eta\zeta)_{rs} + e_{rsk} ,$$

with constraints $\eta_1 = 0$, $\zeta_1 = 0$, $(\eta\zeta)_{r1} = 0$ and $(\eta\zeta)_{1s} = 0$ for $r = 1 \ldots t_A$, $s = 1 \ldots t_B$. Again there is one duplicate constraint within the latter two sets, so the total number of constraints is equal to $1 + t_A + t_B$ as required. We use different symbols here to emphasize the fact that interpretation of the effects differs from that in the sum-to-zero parameterization. Here, the parameter μ_{11} represents the population mean for a group with the first level of both factors. In terms of the sum-to-zero parameterization, we can write this as

$$\mu_{11} = \mu + \alpha_1 + \beta_1 + (\alpha\beta)_{11} .$$

Then η_r, $r = 1 \ldots t_A$, represents the difference between the rth and first levels of factor A at the first level of factor B, i.e.

$$\eta_r = \alpha_r - \alpha_1 + (\alpha\beta)_{r1} - (\alpha\beta)_{11} .$$

Similarly ζ_s, $s = 1 \ldots t_B$ represents the difference between the sth and first levels of factor B at the first level of factor A, i.e.

$$\zeta_s = \beta_s - \beta_1 + (\alpha\beta)_{1s} - (\alpha\beta)_{11} .$$

The effects associated with the interaction term A.B,

$$(\eta\zeta)_{rs} = (\alpha\beta)_{rs} - (\alpha\beta)_{r1} - (\alpha\beta)_{1s} + (\alpha\beta)_{11} ,$$

can be thought of as deviations relative to the first row and column in the two-way table of unstructured treatment effects. If we omit the interaction term, so that the model becomes

$$y_{rsk} = \mu_{11} + \eta_r + \zeta_s + e_{rsk} ,$$

then interpretation of some parameters changes: now $\eta_r = \alpha_r - \alpha_1$ is the expected difference between observations with the rth and first levels of factor A for any given level of factor B; similarly $\zeta_s = \beta_s - \beta_1$ is the expected difference between observations with the sth and first levels of factor B for any given level of factor A. This change in interpretation, dependent on whether an interaction is present in the model or not, is a major disadvantage of this parameterization. Despite this, first-level-zero parameterization is commonly used to fit linear models to non-orthogonal or unbalanced experiments. This is described in more detail in Section 11.2.

To some extent, the parameterization used is unimportant, as the sum-to-zero and first-level-zero parameterizations both result in the same predictions, fitted values and ANOVA table (although the sums of squares are no longer equal to sums of squared parameter estimates when using first-level-zero constraints). However, we might be somewhat confused when looking at individual parameter estimates if we do not understand the nature of the parameterization, and the default parameterization can vary between statistical packages, or even between different commands within the same package. This is a good reason for making inferences on population means rather than on individual parameters, as described in Section 8.2.4.

Other parameterizations are also used, the most common being last-level-zero constraints (see Section 4.5), which are closely related to the first-level-zero constraints but constrain parameters associated with the last level of each factor rather than the first level.

8.3 Crossed Treatment Structures with Three or More Factors

Multi-way crossed structures are formed from all combinations of levels of three or more factors. The same logic and modelling process followed for the two-way ANOVA also apply here, but interpretation of higher-order interactions, i.e. interactions among three or more factors, becomes more difficult. In this section, the analysis for a three-way crossed structure is demonstrated, followed by a discussion of some issues with multi-way factorial designs. Again, it is assumed that we have a full factorial structure with all experimental treatment combinations present with equal replication so that the structure is orthogonal (see Chapter 11 for details for non-orthogonal structures). As before, it is necessary to relabel the observations, here in terms of three individual factors: factor A with t_A levels, factor B with t_B levels and factor C with t_C levels so the total number of treatments is $t = t_A \times t_B \times t_C$. The subscripts r, s and u are used to indicate the level of factors A, B and C present on each unit, respectively, and the subscript k is used to distinguish units with the same treatment combination. The model in terms of an unstructured treatment set can then be written as

$$y_{rsuk} = \mu + \tau_{rsu} + e_{rsuk} .$$

As for the crossed model for two factors, we can now proceed to decompose the treatment effects in terms of a main effect for each factor and their interactions. We can do this by writing

$$\tau_{rsu} = \alpha_r + \beta_s + \gamma_u + (\alpha\beta)_{rs} + (\alpha\gamma)_{ru} + (\beta\gamma)_{su} + (\alpha\beta\gamma)_{rsu} ,$$

where α_r, β_s and γ_u are main effects for the rth group in factor A, sth group in factor B, and uth group in factor C, respectively. There are three two-factor interactions, the A.B interaction $(\alpha\beta)_{rs}$, the A.C interaction $(\alpha\gamma)_{ru}$, and the B.C interaction $(\beta\gamma)_{su}$, and a three-factor A.B.C interaction $(\alpha\beta\gamma)_{rsu}$. The three-factor interaction can be considered as the pattern remaining once the main effects and all two-factor interactions have been removed. The presence of a three-factor interaction implies a complex inter-dependency between levels of all three factors that can be hard to interpret. The full three-way crossed model can be written in symbolic form as

Explanatory component: [1] + A*B*C
$$= [1] + A + B + C + A.B + A.C + B.C + A.B.C$$

We again use sum-to-zero constraints, and parameter estimates can again be derived from the estimates of the unstructured treatment means. The TrtSS and TrtDF are now each partitioned into seven components as in Table 8.6.

TABLE 8.6

Generic Form of ANOVA Table for a Three-Way Crossed Treatment Structure with Factors A, B and C, with t_A, t_B and t_C Levels, Respectively

Source of Variation	df	Sum of Squares	Mean Square	Variance Ratio
A	$t_A - 1$	SS(A)	MS(A)	$F^A = MS(A)/ResMS$
B	$t_B - 1$	SS(B)	MS(B)	$F^B = MS(B)/ResMS$
C	$t_C - 1$	SS(C)	MS(C)	$F^C = MS(C)/ResMS$
A.B	$(t_A - 1)(t_B - 1)$	SS(A.B)	MS(A.B)	$F^{A.B} = MS(A.B)/ResMS$
A.C	$(t_A - 1)(t_C - 1)$	SS(A.C)	MS(A.C)	$F^{A.C} = MS(A.C)/ResMS$
B.C	$(t_B - 1)(t_C - 1)$	SS(B.C)	MS(B.C)	$F^{B.C} = MS(B.C)/ResMS$
A.B.C	$(t_A - 1)(t_B - 1)(t_C - 1)$	SS(A.B.C)	MS(A.B.C)	$F^{A.B.C} = MS(A.B.C)/ResMS$
Residual	$N - t$	ResSS	ResMS	
Total	$N - 1$	TotSS		

Note: No structural component present, all t $(= t_A \times t_B \times t_C)$ treatment combinations have n replicates giving $N = n \times t$ observations.

8.3.1 Assessing the Importance of Individual Model Terms

As in the two-factor case, we might hope that we can find a simple model for prediction by ignoring one or more interaction terms. We still examine the most complex terms in the ANOVA table first, working upwards through the table. The procedure runs as follows. We first examine the three-way interaction. If it is significant then we cannot ignore any terms, and predictions are made from the multi-way table of means for all three treatment factors ($A \times B \times C$). If the three-way interaction is not significant then we proceed to examine the two-way interactions. If these are all significant then the model cannot be simplified further, and predictions can be made from the three two-way tables of marginal means ($A \times B$, $A \times C$ and $B \times C$). If one of the two-way interactions is not significant, say $A \times B$, then predictions can be made from the other two-way tables of means ($A \times C$ and $B \times C$). If only one of the two-way interactions is significant, say $A \times C$, then predictions are made from that table of means and, if the B main effect is significant, the one-way table of means for factor B. If none of the two-way interactions is significant, then the main effects are examined, and predictions made from the marginal means for the treatment factors with significant main effects are reported.

This procedure may seem complicated, but it can be formalized with a diagram and the principle of marginality introduced in Section 8.2.1. Recall that all sub-terms of a model term are considered to be marginal to it; for example, terms A and B are marginal to term A.B. The **principle of marginality** requires that for each term in a model, all sub-terms should also be included. This is illustrated for a three-way factorial structure with factors A, B and C in Figure 8.3.

The testing procedure starts at the bottom of the structure, with the three-way interaction. We go through an iterative process: at each step, we examine only terms that have no arrows leading away from them, i.e. that are not marginal to other terms. We test these terms, and we erase any non-significant terms and the arrows that lead to them. This process is repeated until no further progress can be made, and predictions are made using the remaining terms. Figure 8.4 illustrates the case where only the two-way interaction A.C and the main effect B are significant. At step 1, only the three-way interaction is tested. At step 2, the three-way interaction has been found non-significant, so all two-way

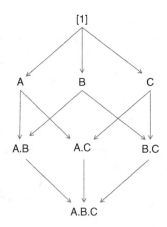

FIGURE 8.3

Marginality relationships for a three-way crossed structure with factors A, B and C. Arrows lead away from each term towards any other terms to which it is marginal.

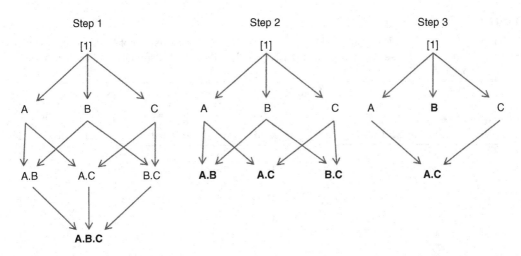

FIGURE 8.4
Eligibility of terms for testing in a three-way crossed structure as interactions are eliminated. Eligible terms at each step are highlighted in bold. At each step arrows lead away from each term towards any other remaining terms to which it is marginal.

interactions can be tested, and only A.C is found significant. At step 3, the main effect of B can now also be tested and is found significant and the process is finished. Predictions are made from all tables of means corresponding to terms without arrows leading away from them, here the A.C and B tables.

This approach can be extended to multi-way factorial tables and treatment structures of any kind. We reiterate here that in this orthogonal scenario we do not refit the model, we merely select terms to use for prediction.

EXAMPLE 8.2: LADYBIRD TRANSMISSION OF FUNGUS

An experiment was done to investigate the transmission of fungus by ladybirds onto aphids on two types of host plant (beans or birdsfoot trefoil; Ekesi et al., 2005). The experimental units were containers, each holding one plant with 20 aphids, and the space available was sufficient for 36 containers. A number of sporulating aphid cadavers (5, 10 or 20) were distributed on plants to provide different loads of infective material, with six plants of each host type at each load (36 containers). In three of the containers for each host × load combination, a ladybird was allowed to forage for four hours. The treatment allocations were made completely at random and the numbers of live and infected aphids per plant were counted after seven days. Because much variation was expected, this procedure was repeated, giving two runs each with a CRD structure, with six replicates of the 12 experimental treatments and 72 observations in total. The explanatory component for this experiment corresponds to a three-way crossed structure with treatment factors host type (factor Host with two levels), number of infective cadavers (factor Cadaver with three levels) and absence or presence of ladybirds (factor Ladybird with two levels). The two replicates introduce structure into the set of experimental units, and are labelled with factor Run (with two levels). As the actual randomization of plants within runs is not available we have arbitrarily labelled plants within each run using factor DPlant (with 36 levels). The data are shown in Table 8.7 and held in file LADYBIRD.DAT.

There was some predation by the ladybirds, so there were fewer than 20 live aphids (variate *Live*) in these containers (minimum 12), and the number of infected aphids (variate *Infected*) could not be directly compared across treatments. The percentage of

TABLE 8.7

Number of Infected and Live Aphids Used to Investigate the Transmission of Fungus (Cadaver Dose) by Ladybirds (Presence/Absence) on Different Host Plants (Birdsfoot Trefoil or Beans) (Example 8.2 and file LADYBIRD.DAT)

| | | Host: Birdsfoot Trefoil | | | | Host: Bean | | | |
| | | Run 1 | | Run 2 | | Run 1 | | Run 2 | |
Ladybird Presence	Cadaver Dose	Infected	Live	Infected	Live	Infected	Live	Infected	Live
+	5	1	15	2	18	5	18	5	15
+	5	1	13	1	13	3	20	3	17
+	5	2	16	1	15	7	17	2	18
+	10	2	12	2	17	10	17	8	19
+	10	2	16	1	18	3	15	7	15
+	10	3	15	3	16	2	14	6	16
+	20	7	14	9	18	9	19	11	16
+	20	8	17	6	19	6	18	9	17
+	20	7	16	7	15	12	19	11	14
−	5	1	20	0	20	1	20	1	20
−	5	1	20	0	20	6	20	2	20
−	5	2	20	1	20	7	20	2	20
−	10	2	20	0	20	3	20	7	20
−	10	1	20	2	20	4	20	5	20
−	10	2	20	2	20	5	20	5	20
−	20	2	20	3	20	4	20	9	20
−	20	3	20	2	20	8	20	8	20
−	20	3	20	2	20	5	20	5	20

Source: Data from Rothamsted Research (J. Pell).

infected aphids was therefore used as a measure of transmission. Preliminary analysis showed some variance heterogeneity, and so a logit transformation was applied to percentages of infection after adjustment for zero counts, i.e. *Logitp* = logit(*Percent*) where *Percent* = 100 × (*Infected* + 1)/(*Live* + 2). In symbolic form, the full model for this experiment can be written as

> Response variable: *Logitp*
> Explanatory component: [1] + Host*Cadaver*Ladybird
> Structural component: Run/DPlant

This model can be written in mathematical form as

$$Logitp_{rsukl} = \mu + Run_k + Host_r + Cadaver_s + Ladybird_u + (Host.Cadaver)_{rs}$$
$$+ (Host.Ladybird)_{ru} + (Cadaver.Ladybird)_{su} + (Host.Cadaver.Ladybird)_{rsu} + e_{rsukl} ,$$

where $Logitp_{rsukl}$ is the logit-transformed percentage of infection for the *l*th replicate measurement ($l = 1 \ldots 3$) in the *k*th experimental run ($k = 1, 2$) for the *r*th host ($r = 1$ for beans, 2 for birdsfoot trefoil) with the *s*th cadaver dose ($s = 1, 2, 3$ for 5, 10 or 20 cadavers) with ladybirds absent ($u = 1$) or present ($u = 2$). The structural component generates effects for each run (from term Run), denoted Run_k for $k = 1, 2$, and for each plant within each run (from term Run.DPlant), which are the model deviations (here equivalently labelled by treatments, runs and replicates). The main effect for the rth host is denoted $Host_r$ with $Cadaver_s$ as the main effect for the sth cadaver dose and $Ladybird_u$ as the main effect for ladybird absence or presence. The definition of the interaction terms and constraints follow as described above.

The multi-stratum ANOVA table generated by this model is in Table 8.8. There are two strata, representing variation between runs (Run stratum) and variation within runs (Run.DPlant stratum). Treatments are estimated from comparisons within runs and thus appear in the lower (Run.DPlant) stratum. The ResDF differ from those in Table 8.6 as they have been adjusted, i.e. reduced by one, to account for the presence of runs. A composite set of residual plots (Figure 8.5) gives no evidence of departures from the model assumptions.

Working upwards from the bottom of the ANOVA table, we see that the three-way interaction is not statistically significant ($F_{4,59}^{H.C.L} = 0.435$, $P = 0.649$) and so can be excluded for prediction. Of the three two-way interactions, only the Cadaver.Ladybird interaction is significant ($F_{2,59}^{C.L} = 3.774$, $P = 0.029$) so we can also test the Host main effect, which is highly significant ($F_{1,59}^H = 59.172$, $P < 0.001$). The model predictions therefore use all main effects and the two-way Cadaver.Ladybird interaction, in addition to the Run term from the structural component, giving the mathematical form

$$\hat{\mu}_{rsuk} = \hat{\mu} + \widehat{Run}_k + \widehat{Host}_r + \widehat{Cadaver}_s + \widehat{Ladybird}_u + \widehat{(Cadaver.Ladybird)}_{su} .$$

This gives a prediction for a specific run, but it is usually sensible to average over terms in the structural component which, using the sum-to-zero constraints, gives

$$\hat{\mu}_{rsu\bullet} = \hat{\mu} + \widehat{Host}_r + \widehat{Cadaver}_s + \widehat{Ladybird}_u + \widehat{(Cadaver.Ladybird)}_{su} .$$

We can therefore predict patterns of transmission by looking at the two-way Cadaver × Ladybird table of means (averaged across hosts) and the marginal table of means for Host (averaged across cadaver concentrations and presence or absence of ladybirds) shown in Table 8.9.

TABLE 8.8

Multi-Stratum ANOVA Table for the Logit-Transformed Percentage of Infected Aphids from the Ladybird Transmission Experiment Performed in Two Blocks (Factor Run) Each Using 36 Plants (Factor DPlant) with a Three-Way Crossed Treatment Structure (Factors Host, Cadaver and Ladybird) (Example 8.2)

Source of Variation	df	Sum of Squares	Mean Square	Variance Ratio	P
Run stratum					
Residual	1	0.0677	0.0677	0.294	0.589
Run.DPlant stratum					
Host	1	13.5992	13.5992	$F^H = 59.172$	< 0.001
Cadaver	2	17.0274	8.5137	$F^C = 37.044$	< 0.001
Ladybird	1	11.0907	11.0907	$F^L = 48.257$	< 0.001
Host.Cadaver	2	0.3078	0.1539	$F^{H.C} = 0.670$	0.516
Host.Ladybird	1	0.2279	0.2279	$F^{H.L} = 0.992$	0.323
Cadaver.Ladybird	2	1.7349	0.8675	$F^{C.L} = 3.774$	0.029
Host.Cadaver.Ladybird	2	0.1999	0.1000	$F^{H.C.L} = 0.435$	0.649
Residual	59	13.5596	0.2298		
Total	71	57.8151			

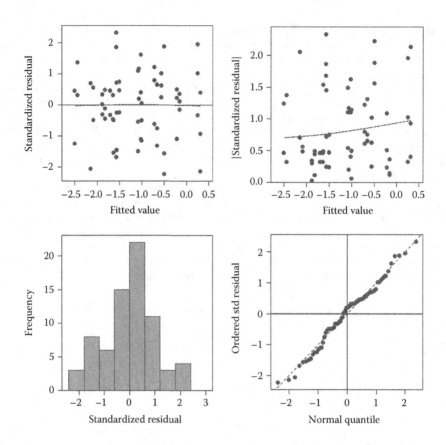

FIGURE 8.5

Composite set of residual plots for the ladybird transmission of fungus experiment (Example 8.2).

The mean logit percentage of infected aphids was greater when the host plants were beans rather than birdsfoot trefoil, when ladybirds were present and as the concentration of cadavers increased. The Cadaver.Ladybird interaction is caused by a larger increase in transmission (on the logit scale) due to ladybird presence at a concentration of 20 cadavers per plant than at smaller concentrations. These patterns are easier to see if we plot the predictions, as shown in Figure 8.6. This figure also shows predictions calculated from the full set of model terms, i.e. including the terms found to be not statistically significant, and it is clear that discrepancies between the two sets are small.

TABLE 8.9

Tables of Predicted logit(%Infection) (with Back-Transform as Percentage) for Cadaver × Ladybird Interaction and Main Effect of Factor Host (Example 8.2)

Ladybird Foraging	Cadaver Concentration			Host Plant	
	5	10	20		
Present	−1.454 (18.9)	−1.033 (26.3)	0.044 (51.1)	Trefoil	−1.641 (16.2)
Absent	−2.038 (11.5)	−1.580 (17.1)	−1.179 (23.5)	Bean	−0.772 (31.6)
		SED = 0.1957			SED = 0.1130

Note: SEDs apply to logit scale.

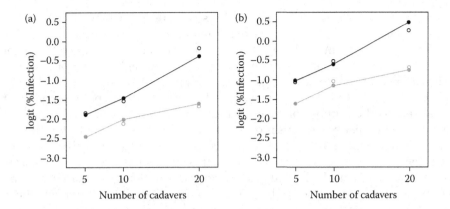

FIGURE 8.6
Logit(%Infection) predicted, for (a) birdsfoot trefoil and (b) beans, from the model with all main effects and the Cadaver × Ladybird interaction for presence (•) or absence (∘) of ladybird foraging, or from the full three-way crossed model (∘∘) (Example 8.2).

This confirms that the terms used for prediction give a simple but accurate summary of patterns of response. Finally, these predictions can be back-transformed to predict % infection for each treatment combination, as in Table 8.9.

8.3.2 Evaluating the Response to Treatments: Predictions from the Fitted Model

We have provided a recipe for the identification of model terms for use in prediction, but common sense is also required. In many practical situations, the size of the main effects are large compared with the size of the interactions, which are sometimes regarded as modifications to (or departures from) the main effects model. Using this argument, we can expect that the sizes of the effects decrease as the order of interactions increases. We might therefore find ourselves in a situation where a high-order interaction is statistically significant but its effects are so small compared with the main effects that ignoring them has little impact on the biological conclusions. It can therefore be helpful to discuss the relative impact of terms when you report results. In an extreme situation, particularly where the number of ResDF are very large, one might detect high-order interactions where the effects are so small as to have no biological relevance. (In such cases, equivalence testing, described in Section 10.5, can be useful to establish whether any biologically meaningful differences are present.) These considerations might suggest that it is more appropriate to work down the ANOVA table, starting with the main effects and respecting marginality, i.e. only testing terms for which all sub-terms are significant. However, in practice we have seen cases in which a two- (or higher) way interaction is significant but the corresponding main effects are negligible, and working down the table would then result in our drawing the wrong conclusions.

It is also possible that the presence of interactions depends on the scale of analysis: this usually becomes apparent when different transformations are applied to tackle problems with the assumptions seen in the residual plots (Chapter 6). This behaviour is expected: there may be some scales on which the assumption of additivity (no interaction) is valid, and a simpler model can be found. This is useful in practice only if the model assumptions are also met on the same scale. Validating the assumptions should always take precedence over simplifying the model.

In Section 8.2.5, the efficiency of a factorial structure for two treatment factors was discussed in comparison to two separate experiments testing each factor in isolation. Part of this efficiency arises from our being able to test for interactions between the factors and part arises from the additional (sometimes called hidden) replication available when no interaction is present. These advantages also apply to higher-order factorial structures, with three or more factors. However, the number of treatment combinations increases multiplicatively as the number of treatment factors increases, which can result in very many experimental treatments. This is often seen as a disadvantage of factorial structures, particularly if interactions between treatments are expected to be either small or non-existent. Several strategies can be used to tackle this problem. If the approach of a factorial structure is desirable, but the full replicated experiment would be too large to manage due to constraints of either space or time, then it may be possible to complete the experiment in several runs, each of which contains the full set of experimental treatments. This strategy was used in Example 8.2, in which six replicates of each treatment combination were required due to high variation but there was only enough growing space to manage three replicates at a time. The different time periods introduce structure into the experimental units that must be taken into account in the analysis, as was done in Example 8.2. This approach is feasible only if the experimental system is stable over time, so that treatment × time interactions are unlikely and the background variation will not change. If the experimental material cannot provide sets of homogeneous units large enough to contain the full set of experimental treatments, then designs are required in which each block contains only a subset of treatments. Designs with efficient blocking for factorial treatment structures are discussed in Section 11.3.2. Finally, if many of the possible interactions, especially higher-order interactions, are expected to be absent then it would be wasteful to replicate all treatment combinations. The class of fractional factorial designs was developed to deal with these cases and is discussed in Section 11.3.1.

In this and the previous section, we have assumed that the full factorial set of treatments is present and equally replicated. If this is the case then the structure is orthogonal, so that the estimates for each main effect are the same whether or not the other factor (and the interaction) is included in the model (see Section 11.1 for more details). This is the reason we can use a subset of terms for prediction. If treatment combinations are missing or unequally replicated, then the structure may become non-orthogonal and the statistical analysis becomes more complex, as described in Chapter 11. If one or more treatment combinations have been omitted, perhaps for sound practical reasons but without regard to the overall structure, then the individual factors are likely to become non-orthogonal. There are then several options. The treatments can always be analysed as an unstructured set, with contrasts used to explore specific comparisons (see Section 8.6), but this loses the advantages of the factorial structure. If the structure is still close to orthogonal, then a factorial analysis will often still be useful and the analysis for this situation is discussed further in Section 11.2. However, there are some special cases that deserve further attention here. Some schemes of unequal replication retain many of the properties of the full factorial structure. For example, if one level of a treatment factor has additional replication such that all treatment combinations involving that level have the same replication, then the structure remains orthogonal (see Section 11.1) and the analysis can proceed as described above. The only change is that estimated SEs for treatment means or differences involving that factor level will become smaller because of the additional replication. Similarly, sometimes subsets of experimental treatments can be omitted whilst preserving some of the factorial structure. This often applies when controls are present. For example, consider an experiment to compare the efficacy of different pesticides at several doses. The control, consisting of no dose, is the same across all pesticides so does not need to be replicated for each pesticide. If we regard this as a crossed structure of pesticide × dose plus an added control, then we retain the factorial structure in which we are interested, although a slightly more complex analysis is required, as described in Section 8.5.

8.4 Models for Nested Treatment Structures

In previous sections, we constructed models for treatment effects in terms of main effects and interactions for the treatment factors using a crossed structure. This structure is not always appropriate, and sometimes a nested structure is preferred. In these cases, the treatment factors usually fall into a natural hierarchy. For example, a forestry trial might assess a set of clones taken from a small set of mothers, and it is natural to think of clones as nested within mothers. Similarly, a laboratory trial to test the pathogenicity of a set of fungal isolates might use several isolates from several different races, with isolates considered to be nested within races. Or a study to examine aphid colonization of different hosts might use varieties from wheat, barley and oats as three different crop species, with varieties nested within species.

These types of trials require a somewhat different approach to partitioning the treatment information. Usually, there is interest in whether there is any significant variation between the higher-level grouping factor, for example, mothers, races or species in the examples above, in addition to variation at the lower level within groups. There are several ways to exploit this structure. We again start with a set of unstructured treatments corresponding to the model presented in Equation 8.1. To partition the treatment information, we relabel the treatments in terms of the groups, and then number the group members. Suppose there are t_G groups, labelled by index $r = 1 \ldots t_G$, and that the rth group contains t_r members, labelled by index $s = 1 \ldots t_r$. It is not necessary for all groups to have the same number of members. The relabelled treatment effects, τ_{rs}, are then partitioned into two terms, and written in mathematical form as

$$\tau_{rs} = \gamma_r + \delta(\gamma)_{rs} . \tag{8.3}$$

In this equation, we call γ_r the **parental effect**, which is associated with the factor at the top level of the hierarchy. This is the average effect of all treatments in the rth group, expressed as a deviation about the overall mean. Note that the term parental here does not imply any genetic relationship, but just denotes the top level of a hierarchical nested structure. We call $\delta(\gamma)_{rs}$ the **nested effect** of the sth member of the rth group, and this is expressed as a deviation about the group mean. This interpretation uses sum-to-zero constraints, which implies $\Sigma_r \gamma_r = 0$ and $\Sigma_s \delta(\gamma)_{rs} = 0$. The first model term allows for differences among group means and the second allows for differences among group members about their group mean. We use the notation $\delta(\gamma)_{rs}$ to make a distinction between this nested term and the interaction term in the crossed models of Section 8.2 to emphasize the difference in interpretation due to the marginal terms present in each model. The term interaction implies that both marginal terms are present in the model, and interaction effects are deviations from the additive model that includes both main effects; the term nested effect implies that only one marginal term is present, and nested effects are deviations from that term only.

We can again illustrate this decomposition in terms of the original unstructured treatment estimates $\hat{\tau}_{rs} = \bar{y}_{rs.} - \bar{y}$. Assuming equal replication for each of the members, we can again estimate the group parental effects as marginal means of the original treatment effects, as

$$\hat{\gamma}_r = \frac{1}{t_r} \sum_{s=1}^{t_r} \hat{\tau}_{rs} = \bar{y}_{r..} - \bar{y} \,,$$

with the nested effects calculated as the remainder after removing the parental effects, as

$$\widehat{\delta(\gamma)}_{rs} = \hat{\tau}_{rs} - \hat{\gamma}_r = \bar{y}_{rs.} - \bar{y}_{r..} \,.$$

Two factors are required to express this model in symbolic form. The first factor, denoted Group, labels the groups and the second, denoted Member, labels members nested within groups; hence, the explanatory component of the model can then be written as

Explanatory component: [1] + Group/Member
                         = [1] + Group + Group.Member

Unfortunately, for the case of unequal numbers of members within groups, this expression will not generate a direct translation of Equation 8.3 in most statistical software, because the number of effects generated by the nested term Group.Member is equal to $t_G \times \max(t_r)$, i.e. the number of groups multiplied by the maximum number of members in a group. As there are no data on the absent factor combinations, it is not possible to estimate their effects and they are effectively ignored, so that estimates for present combinations are calculated as above. In statistical software, nested effects for absent combinations may be represented as zero or as a missing value.

The TrtSS and TrtDF in the ANOVA table are partitioned according to these two terms. The ANOVA sums of squares can again be calculated as the sum of squares of the effects for each term (for present combinations only). We do not give further details here, as the ANOVA table, estimates and SEs can be obtained from statistical software once the model has been correctly specified. This is illustrated in the example below.

EXAMPLE 8.3A: SCREENING FOR PATHOGENICITY*

An experiment was done to screen a set of fungal isolates for pathogenicity on seedlings of oilseed rape. The isolates were collected from two different species of *Brassica*, labelled as A and B in factor Species, with several different isolates from each species being tested (nine in group A and four in group B), labelled by factor Isolate (with nine levels). The experiment was run in three replicates across time (factor Rep), with a tray of 22 (replicate 2) or 23 seedlings (replicates 1 and 3) being tested against each isolate in each run (factor Tray, with 13 levels). The number of seedlings tested was stored in variate *Seedlings*. The number of resistant seedlings, i.e. those showing no signs of infection (variate *Resistant*), was recorded five days after the isolates were applied. The percentage of resistant seedlings is the response to be analysed. The number responding in each tray is shown in Table 8.10 and the full data set is given in file BRASSICA.DAT. A preliminary analysis of these percentages showed heterogeneity of variance and so the percentage response, adjusted for zero counts as $P = 100 \times (Resistant + 1)/(Seedlings + 2)$, was logit-transformed to $Logitp = \log_e(P/(100 - P))$, which improved the residual plots (not shown).

TABLE 8.10

Number of Plants Showing Resistance to Isolates in the Pathogenicity
Screening Experiment (Example 8.3A and File BRASSICA.DAT)

	Replicate 1		Replicate 2		Replicate 3	
Tray	Isolate	Resistant	Isolate	Resistant	Isolate	Resistant
1	B3	3	A3	2	A3	3
2	A6	14	A5	2	A7	5
3	A4	5	A8	8	A2	3
4	B1	2	A9	1	A5	2
5	A7	6	A4	1	B1	1
6	B2	2	A1	3	A8	16
7	A1	3	B2	2	A6	15
8	A9	1	B4	0	A1	4
9	A5	2	A7	4	B3	4
10	A2	3	B1	1	A9	0
11	A8	15	A2	1	B4	4
12	A3	4	B3	2	A4	4
13	B4	2	A6	9	B2	1

Note: Isolates are here labelled using combinations of the levels of factors Species (A or
B) and Isolate (1–9 for species A, 1–4 for species B).

A model in symbolic form for these responses could be written as

Response variable: *Logitp*
Explanatory component: [1] + Species/Isolate
Structural component: Rep/Tray

The Species.Isolate term generates $2 \times 9 = 18$ effects, but those corresponding to species B with isolate numbers 5–9 are absent and so ignored. This model can be written in mathematical form as

$$Logitp_{rsk} = \mu + Rep_k + Species_r + Isolate(Species)_{rs} + e_{rsk} \, ,$$

where $Logitp_{rsk}$ is the logit-transformed percentage of resistant seedlings in the kth replicate ($k = 1 \ldots 3$) for the rth species ($r = 1, 2$ for species A and B) with the sth isolate (for $s = 1 \ldots t_r$ where $t_1 = 9$ and $t_2 = 4$). The structural component generates the replicate effects, denoted Rep_k for $k = 1 \ldots 3$ and the deviations e_{rsk} (equivalent to the Rep.Tray effects). The parental effect of the rth species is denoted $Species_r$, with $Isolate(Species)_{rs}$ being the nested effect of the sth isolate within the rth species. As usual, the overall mean is denoted by μ. The sum-to-zero constraints take the form $\Sigma_k Rep_k = 0$, $\Sigma_r Species_r = 0$ and $\Sigma_s Isolate(Species)_{rs} = 0$.

Table 8.11 shows the estimated parental (Species) and nested (Species.Isolate) effects derived from the unstructured set of treatment effects as described above. The multistratum ANOVA for the logit-transformed percentages is shown in Table 8.12. The Species sum of squares is equal to the sum of the squared parental effects (across all units), and the Species.Isolate sum of squares is equal to the sum of the squares of the estimated nested effects (across all units). The nested sum of squares here represents the accumulated within-species variation.

There is strong evidence of an overall difference in resistance to isolates from species A and B ($F_{1,24}^{S} = 29.841, P < 0.001$) and also of variation in resistance to the isolates

TABLE 8.11

Calculation of Species Parental Effects as the Mean of the Unstructured Treatment
Effects for Each Group, and of Species.Isolate Nested Effects as the Difference
between Unstructured Treatment Effects and Parental Effects (Example 8.3A)

Species	Isolate	Unstructured Treatment Effect	Species Parental Effect	Species.Isolate Nested Effect
A	1	0.011	0.221	−0.210
A	2	−0.342	0.221	−0.563
A	3	−0.101	0.221	−0.322
A	4	−0.083	0.221	−0.304
A	5	−0.415	0.221	−0.636
A	6	1.777	0.221	1.555
A	7	0.418	0.221	0.197
A	8	1.835	0.221	1.613
A	9	−1.110	0.221	−1.331
B	1	−0.715	−0.498	−0.218
B	2	−0.565	−0.498	−0.068
B	3	−0.101	−0.498	0.397
B	4	−0.609	−0.498	−0.112

within species ($F_{11,24}^{S.I} = 15.219$, $P < 0.001$). The treatment mean for isolates from species
A was −1.341 on the logit scale (back-transformed to 20.7%), and the mean for isolates
from species B was −2.060 (back-transformed to 11.3%), with SED = 0.1315, indicating
that fewer plants were resistant to isolates arising from species B.

One way to avoid the generation of effects for treatment groups that are absent is to use
a factor (called AllMembers, say) to label the full set of members across all groups (e.g. like
the combined levels of factors Isolate and Species given in Table 8.10). The explanatory
component can then be specified as

Explanatory component: [1] + Group + AllMembers

The first term identifies the parental effects, as before, and the second term identifies all
of the nested combinations present. This specification gives the same predictions and

TABLE 8.12

Multi-Stratum ANOVA Table for the Logit-Transformed Percentage of Resistant
Seedlings from the Pathogenicity Screening Experiment with Three Blocks (Factor
Rep) of 13 Trays (Factor Tray), Two Species (Factor Species) and Several Isolates
(Factor Isolate) per Species (Example 8.3A)

Source of Variation	df	Sum of Squares	Mean Square	Variance Ratio	P
Rep stratum					
Residual	2	1.8664	0.9332	6.492	0.006
Rep.Tray stratum					
Species	1	4.2896	4.2896	$F^S = 29.841$	< 0.001
Species.Isolate	11	24.0655	2.1878	$F^{S.I} = 15.219$	< 0.001
Residual	24	3.4500	0.1437		
Total	38	33.6715			

ANOVA table, and it avoids the generation of absent combinations, but is not completely satisfactory because the nested structure is no longer apparent in the form of the model. For this reason, we prefer the nested specification.

Using either specification, we encounter a slight complication if any group has only one member, as then the parental and nested effects for that individual refer to exactly the same subset of observations and are **aliased**, i.e. it is not possible to separate the two effects. In this case, the estimates can still be calculated as above and the first of the two terms fitted, i.e. the parental or Group effect, estimates the combined group and member effect. There is then no additional information left to contribute to the nested Group.Member effect, which is estimated as zero.

The presence of a statistically significant variance ratio for a nested term (as in Example 8.3A) is evidence that the nested effects are not all equal to zero (the null hypothesis). In this context, it is possible that the variation between members is present within some groups but not others, and it may be relevant to identify these groups. We can achieve this by splitting the Group.Member sum of squares into separate terms corresponding to the different groups. We can do this by defining a new set of factors, one for each group, here called Set1, Set2 and so on. The new factor for the rth group has levels 1 to t_r corresponding to the members of that group, and adds an extra level, for example, $t_r + 1$, for members of other groups. For example, with only two groups, the explanatory component of the model can then be written as

Explanatory component: [1] + Group/(Set1 + Set2)
                         = [1] + Group + Group.Set1 + Group.Set2

This specification introduces absent combinations into the model, for example, the factors are constructed so that there are no members of the second group with level 1 in factor Set1. We can ignore these combinations in calculating estimates although they will be generated (with value zero or missing) by some statistical software. And again we can reduce the number of missing combinations by writing the explanatory component as

Explanatory component: [1] + Group + Set1 + Set2

which gives an equivalent model but no longer emphasizes the nested structure. Both specifications have aliasing present between the individual group effects and the extra levels for each set. As the Group term is fitted first, there is no information left on the aliased levels in the Group.Set (or Set) terms, which are estimated as zero. The parental effects and each set of nested effects are estimated as outlined previously. This is illustrated in Example 8.3B.

EXAMPLE 8.3B: SCREENING FOR PATHOGENICITY*

Table 8.13 shows the definition of factors TypeA, which labels individual isolates 1–9 within species A (with level 10 for isolates from species B), and TypeB, which labels isolates 1–4 within species B (with level 5 for isolates from species A). These factors are also listed in file BRASSICA.DAT.

A within-group nested model for resistance scores could be written in symbolic form as

Response variable: *Logitp*
Explanatory component: [1] + Species/(TypeA + TypeB)
Structural component: Rep/Tray

The groups from the combination of factors Species and TypeA are the individual isolates within species A plus the whole set of isolates from species B. The Species.TypeA

TABLE 8.13

Calculation of Nested Effects for Each Type of Isolate within Each Species (Example 8.3B)

Species	TypeA	TypeB	Treatment Effects	Species Effects	Species.TypeA Nested Effects	Species.TypeB Nested Effects
A	1	5	0.011	0.221	−0.210	0
A	2	5	−0.342	0.221	−0.563	0
A	3	5	−0.101	0.221	−0.322	0
A	4	5	−0.083	0.221	−0.304	0
A	5	5	−0.415	0.221	−0.636	0
A	6	5	1.777	0.221	1.555	0
A	7	5	0.418	0.221	0.197	0
A	8	5	1.835	0.221	1.613	0
A	9	5	−1.110	0.221	−1.331	0
B	10	1	−0.715	−0.498	0	−0.218
B	10	2	−0.565	−0.498	0	−0.068
B	10	3	−0.101	−0.498	0	0.397
B	10	4	−0.609	−0.498	0	−0.112

effects for species A are therefore equal to the previous nested effects for this group (Table 8.10), and those for species B are zero (Table 8.13). The Species.TypeA sum of squares is then calculated from these estimates, and hence has zero contribution from species B, and so quantifies variation about the mean within species A only. The Species.TypeA mean square can be used to test the null hypothesis that all of the nested effects within species A are equal to zero. A similar argument follows for the Species.TypeB term. The ANOVA for this model based on the logit-transformed percentages is in Table 8.14. Again, the sums of squares correspond to sums of squared estimated effects (from Table 8.13) taken over all units.

The sum of squares for the Species main effect factor has not changed, as expected. The sum of squares and df for the term Species.Isolate from the previous analysis (Table 8.12) have both been partitioned into components for Species.TypeA and Species.TypeB. The variance ratio for Species.TypeA shows strong evidence of variation between isolates within species A ($F_{8,24}^{S.TA} = 20.349$, $P < 0.001$) but that for Species.TypeB gives no

TABLE 8.14

Multi-Stratum ANOVA Table for the Logit-Transformed Percentage of Resistant Seedlings from the Pathogenicity Screening Experiment Using a Within-Group Nested Structure (Factors TypeA and TypeB for the Two Species, Respectively) (Example 8.3B)

Source of Variation	df	Sum of Squares	Mean Square	Variance Ratio	P
Rep stratum					
Residual	2	1.8664	0.9332	6.492	0.006
Rep.Tray stratum					
Species	1	4.2896	4.2896	$F^S = 29.841$	< 0.001
Species.TypeA	8	23.4012	2.9251	$F^{S.TA} = 20.349$	< 0.001
Species.TypeB	3	0.6643	0.2214	$F^{S.TB} = 1.541$	0.230
Residual	24	3.4500	0.1437		
Total	38	33.6715			

evidence of variation between isolates within species B ($F_{3,24}^{S.TB}$ = 1.541, P = 0.230). We can therefore conclude that there was some variation in resistance to isolates from species A, but no significant variation in resistance to isolates from species B.

We have shown above how to identify and express a nested structure, including the attribution of variation between members to the individual parental groups. This principle can be applied to more complex explanatory structures. For example, consider an experiment which uses a treatment factor crossed with a nested structure, for example, Treatment*(Group/Member). In the context of Example 8.3, this structure would arise if each isolate had been tested with two treatments.

8.5 Adding Controls or Standards to a Set of Treatments

Many experiments include one or more control or standard treatments, and these can play several different roles. The concepts of negative and positive controls were briefly introduced in Section 3.1, and both are often intended as validation of the experimental process. A negative control is usually a null treatment that is included as a measure of baseline response, often used to demonstrate that other treatments have had a real effect. For example, consider a trial set up in glasshouse compartments to evaluate the effect of some new biocontrol agents on a glasshouse pest. In this case the negative control is a null treatment. If infestation in untreated compartments is small, then the experiment may be regarded as unsuccessful, as there is little scope to show any effect of the new agents. If, on the other hand, infestation is large in untreated compartments, then any effect of the new agents is more likely to be observed. A positive control is usually a treatment with a known effect that is included as a baseline for a good response. In our example, this might be an effective chemical control strategy. Finally, standard treatments may be defined for certain types of experiment and included as a means of comparing the response across several experiments and of providing a common reference point across experiments. This practice is common in variety trials, where some varieties are included in all trials across several years, with this standard set slowly evolving to reflect current elite varieties. It is also common in many laboratory procedures, where the standards are samples that are re-used either within or across experiments as quality controls, and might not have been part of the original experiment. The advantage of this approach is that behaviour of the standard is well-known, so any deviation from the expected response on these samples can give an immediate indication of problems in the experimental procedure. In this section, we use the term 'control' to also refer to standard treatments.

The correct approach to analysis when controls are present depends on both the aims of the experiment and the purpose of the control. If the main purpose of the experiment is the direct comparison of individual treatments with the controls, for example, when screening a set of chemicals or varieties as to whether they are comparable to one or more positive controls, or better than a negative control, then the controls can be regarded as an integral part of an unstructured set of treatments, and comparisons can be made as described in Section 8.8.4. If the main purpose of the experiment is comparisons between non-control treatments, then the approaches described in the remainder of this section might be helpful. This often requires the definition of a complex explanatory model containing both crossed and nested structures.

EXAMPLE 8.4A: POTATO YIELDS*

This experiment was introduced in Example 3.5 and the data were analysed according to a one-way treatment structure in Example 7.1. It consisted of a RCBD with four blocks to compare the yields of potatoes treated with four different fungicide sprays (F1, F2, F3, F4) with an unsprayed treatment (Control, negative control). The layout was shown in Table 7.1 (data in file POTATO.DAT). The analysis in Chapter 7 showed differences between the treatments, with the control giving smaller yields than the four fungicide sprays. It would be useful to refine this analysis specifically to evaluate whether there are any differences in yield between the fungicide sprays.

We recommend that controls always be included in the analysis of an experiment, except in the special case where the controls are uninformative. This might be the case if the control is not to be compared with any other treatment and the background variation within the control is quantitatively different from that of other treatments. For example, consider a glasshouse trial designed to test the resistance of several varieties to a fungal disease, where inoculum has been sprayed onto the leaves to provide a consistent infection. Two types of negative control have been included in the trial: a susceptible variety sprayed with inoculum to show that conditions are suitable for disease progression, and the same variety sprayed with clean water to show that there has been no additional infection or cross-infection during the trial. If the experiment is successful then all of the plants sprayed with water should show no sign of disease and have a consistent zero response. Because there is no variation within this group, including these plants in the statistical analysis will decrease the ResMS so that it underestimates the true extent of background variation. It is therefore legitimate to exclude these plants from the analysis. However, plants of the susceptible variety sprayed with inoculum should be retained in the analysis, because they provide real quantitative information on the biological system. An assessment as to whether controls are informative or not must be done on a case-by-case basis, and requires real understanding of both the experimental system and the statistical analysis.

Having decided to retain the controls within the statistical analysis, we need to decide which comparisons are of most interest. If the controls and treatments are analysed as a single unstructured set, i.e. labelled by a single factor, then the one-way ANOVA will provide only an overall test of variation within the full set. Treatment differences can then be extracted from the pairwise comparisons of predicted means, but there are dangers in this approach that are described in Section 8.8. Where the controls are expected to be substantially different from the treatments, or where comparisons within the set of treatments are the main purpose of the experiment, it can be helpful to partition the joint variation within the full set of treatments and controls into two components: one accounting for variation between the controls and the average treatment effect, and the other accounting for variation within the set of treatments. In the case of a single control, we start with the model in terms of the full set of control and treatment effects as presented in Equation 8.1, except that we allocate the first label, $j = 1$, to the control and allocate the remaining labels to the $t - 1$ non-control treatments. The treatment effects are then partitioned with a nested structure as

$$\tau_j = \gamma_r + \delta(\gamma)_{rj} \,,$$

where index r takes value 1 when $j = 1$ (control) and takes value 2 otherwise (treated). This requires the definition of a new factor, here denoted Type, that labels the control and treatment sets. The explanatory model for this structure can then be written as

Explanatory component: [1] + Type/Treatment
                                     = [1] + Type + Type.Treatment

where the factor Treatment denotes the full unstructured set of treatment groups. The process of estimating effects and sums of squares is exactly the same as for the nested structures in Section 8.5, and the same issues of missing combinations and aliasing arise.

EXAMPLE 8.4B: POTATO YIELDS*

The nested control structure can be represented by two factors (Type and Fungicide) as defined in Table 8.15. These factors are also given in file POTATO.DAT.

The model can then be written in symbolic form as

Response variable: *Yield*
Explanatory component: [1] + Type/Fungicide
Structural component: Block/Plot

This model can equivalently be written in mathematical form as

$$Yield_{irs} = \mu + Block_i + Type_r + Fungicide(Type)_{rs} + e_{irs} \, ,$$

where $Yield_{irs}$ is the yield in the ith block ($i = 1 \ldots 4$) for treatment of type r ($r = 1$ for control, $r = 2$ for fungicide treatments) with the sth fungicide ($s = 1 \ldots t_r$ with $t_1 = 1$ and $t_2 = 4$). Fungicide Control is of type 1 (Control), and fungicides F1, F2 ... F4 are of type 2 (Treated) as shown in Table 8.15, where we have omitted parameters corresponding to missing combinations. The structural component generates the block effects, denoted $Block_i$ for $i = 1 \ldots 4$, and the deviations e_{irs} (equivalent to term Block.Plot). The parental effect of the rth type (control or treated) is denoted $Type_r$, with $Fungicide(Type)_{rs}$ being the nested effect of the sth fungicide within the rth type.

The Type effects are estimated as the means of the control and treated groups. The Type.Fungicide groups with data present are the control treatment plus the individual fungicide treatments. The nested effect for the control is aliased with the control group parental effect, so the control nested effect is equal to zero. The nested effects for the fungicide treatments are differences from their group mean. The Type.Treatment sum of squares is then calculated from these estimates, and hence has zero contribution from the control and so quantifies variation about the mean within the fungicide treatments only, as required.

The resulting ANOVA table is Table 8.16. As expected, the variance ratio for factor Type ($F_{1,12}^{T} = 35.972, P < 0.001$) gives strong evidence of a difference between the control and average of the fungicide treatments. The variance ratio for the nested term Type.Fungicide ($F_{3,12}^{T.F} = 0.778, P = 0.529$) gives no evidence of any differences among the four fungicide treatments. This gives a quantitative confirmation of the tentative conclusions of Example 7.1D.

TABLE 8.15

Calculation of Type Parental Effects, and Nested Type.Fungicide Effects (Example 8.4B)

Fungicide	Type	Treatment Effects	Type Effects	Type.Fungicide Effects
Control	Control	−158.3	−158.3	0
F1	Treated	4.7	39.6	−34.9
F2	Treated	49.7	39.6	10.1
F3	Treated	66.2	39.6	26.6
F4	Treated	37.7	39.6	−1.9

TABLE 8.16

Multi-Stratum ANOVA Table for RCBD Potato Yield Trial with Treatment Effects (Factor Fungicide) Partitioned into 'Control vs Treated' (Factor Type) Plus Nested Variation among Fungicide Treatments (Type.Fungicide) (Example 8.4B)

Source of Variation	df	Sum of Squares	Mean Square	Variance Ratio	P
Block stratum					
Residual	3	14,987.20	4995.73	1.434	0.283
Block.Plot stratum					
Type	1	125,294.45	125,294.45	$F^T = 35.972$	< 0.001
Type.Fungicide	3	8124.75	2708.25	$F^{T.F} = 0.778$	0.529
Residual	12	41,796.80	3483.07		
Total	19	190,203.20			

This approach can be extended for more complex treatment structures with one or more controls, such as a factorial structure with added control. This type of structure may require a mixture of nested effects (to partition out the control) and crossed effects (to model the factorial structure) to extract information efficiently. If more than one control is present, then the structure can be extended in several different ways. If comparisons with these controls are unimportant, then it is sufficient to add one extra level to the Type factor for each type of control. The Type factor then evaluates differences among the individual controls and the average of the other treatments, and the Type.Treatment interaction evaluates variation among the non-control treatments.

8.6 Investigating Specific Treatment Comparisons

In previous sections, we have defined new factors to enable partitioning of a set of structured treatment effects into meaningful comparisons, often with the aim of finding the simplest possible description of patterns within the set. Contrasts provide an alternative way of partitioning a set of treatment effects. A **contrast** translates a specific hypothesis about treatment effects into mathematical form. There are two approaches to dealing with contrasts. The first approach involves building the contrast into the ANOVA and the second involves evaluating contrasts from tables of predicted means. In this section we use the former approach, and the latter is discussed in Section 8.8.

For example, consider an experiment set up as a RCBD with three blocks, investigating the resistance of six wheat varieties to virus transmission by aphids, measured in terms of virus concentration in the plant at the end of the experiment. A model for these data can be written in mathematical form as

$$y_{ij} = \mu + b_i + \tau_j + e_{ij}, \tag{8.4}$$

where b_i is the effect of the ith block, $i = 1, 2, 3$, τ_j is the effect of the jth variety, $j = 1 \ldots 6$, and all other terms are as defined previously. Throughout this section, we assume that we are using sum-to-zero constraints, so that τ_j represents the deviation from the overall mean due to the effect of the jth variety. If the first two varieties are related through a known resistant ancestor, it might be of particular interest to evaluate whether there is any difference in resistance between them. This question can be expressed as 'Is the treatment effect for variety 1 equal to that for variety 2?' so, in mathematical terms, we want to test the proposition H_0: $\tau_1 = \tau_2$. In practice, we rewrite this in a form such that, if the null hypothesis is true, then the value is equal to zero, which means reformulating the proposition as H_0: $\tau_1 - \tau_2 = 0$; this is now in the form of a linear contrast. If we build this contrast into our analysis, we can form a test for this hypothesis as part of our ANOVA table.

In general, and working in terms of a set of treatment effects $\tau_1 \ldots \tau_t$, a **linear contrast**, denoted ψ, is defined as a linear function of the treatment effects, i.e. of the form

$$\psi = l_1\tau_1 + l_2\tau_2 + \ldots + l_t\tau_t = \sum_{j=1}^{t} l_j\tau_j \, ,$$

such that the sum of the contrast coefficients, the set l_j for $j = 1 \ldots t$, is equal to zero, i.e. $\Sigma_j \, l_j = 0$. In our example above, $\psi = \tau_1 - \tau_2$ with $l_1 = 1$, $l_2 = -1$ and $l_3 = l_4 = \ldots = l_t = 0$.

EXAMPLE 8.4C: POTATO YIELDS*

In the potato yield trial described in Example 8.4A, four fungicide treatments were tested with a negative control (no fungicide treatment). However, fungicides F1 and F4 use one mode of action (mode A) and fungicides F2 and F3 use another (mode B), and it is of interest to evaluate whether there is any overall difference between the two modes of action. The linear model for this RCBD trial is equivalent to Equation 8.4 with four replicates and five treatment effects $\tau_1 \ldots \tau_5$ referring to the control and fungicides F1 ... F4, respectively. Equality of the two modes of action can be expressed in words as 'Is the average effect of mode A fungicides equal to the average effect of mode B fungicides?'. The average effect of mode A fungicides is the average of the effects associated with F1 and F4, or $\frac{1}{2}(\tau_2 + \tau_5)$. Similarly, the average effect of mode B fungicides (F2 and F3) is equal to $\frac{1}{2}(\tau_3 + \tau_4)$. Equality between the two quantities can then be written as

$$\frac{1}{2}(\tau_2 + \tau_5) = \frac{1}{2}(\tau_3 + \tau_4) \, .$$

We can rearrange this expression into a contrast by subtracting $\frac{1}{2}(\tau_3 + \tau_4)$ from both sides of the equation, to obtain

$$\psi = \frac{1}{2}(\tau_2 + \tau_5) - \frac{1}{2}(\tau_3 + \tau_4) = (0 \times \tau_1) + \left(\frac{1}{2} \times \tau_2\right) - \left(\frac{1}{2} \times \tau_3\right) - \left(\frac{1}{2} \times \tau_4\right) + \left(\frac{1}{2} \times \tau_5\right) \, .$$

In this case, the contrast coefficients are: $l_1 = 0$, $l_2 = l_5 = 0.5$, $l_3 = l_4 = -0.5$.

Because the true values of the treatment effects are unknown, so too is the true value of the contrast. The least-squares estimate is obtained by substitution of the estimated treatment effects in place of the unknown true values, so

$$\hat{\psi} = l_1\hat{\tau}_1 + l_2\hat{\tau}_2 + \ldots + l_t\hat{\tau}_t = \sum_{j=1}^{t} l_j\hat{\tau}_j \, .$$

For a RCBD or CRD with equal replication of all treatment groups, the variance of the contrast is equal to the sum of the squared coefficients multiplied by the background variance and divided by the replication, which is written as

$$\text{Var}(\hat{\psi}) \;=\; \frac{\sigma^2}{n} \sum_{j=1}^{t} l_j^2 \,.$$

As usual, we estimate the unknown background variation, σ^2, using s^2, the residual mean square from the ANOVA table. The estimated contrast standard error, $\widehat{\text{SE}}(\hat{\psi})$, is calculated as the square root of its estimated variance. Under the null hypothesis that the true value of the contrast is equal to zero, i.e. H_0: $\psi = 0$, the ratio of the contrast to its estimated standard error, i.e.

$$t = \frac{\hat{\psi}}{\widehat{\text{SE}}(\hat{\psi})} \,,$$

has a t-distribution with df equal to the ResDF from the ANOVA table. For a two-sided test, if the absolute value of the ratio exceeds the $100(1 - \alpha_s/2)$th percentile of this t-distribution, then there is statistical evidence (at significance level α_s) that the true value of the contrast is not equal to zero. The associated $100(1 - \alpha_s)\%$ confidence interval can be formed as

$$\hat{\psi} \pm (t_{\text{ResDF}}^{[\alpha_s/2]} \times \widehat{\text{SE}}(\hat{\psi})) \,.$$

We can construct an equivalent test by partitioning the TrtSS in the ANOVA table into a component corresponding to the contrast and a remainder. The contrast sum of squares can be written as

$$n \left(\sum_{j=1}^{t} l_j \hat{\tau}_j \right)^2 \Bigg/ \left(\sum_{j=1}^{t} l_j^2 \right) = \sigma^2 \left(\frac{\hat{\psi}}{\widehat{\text{SE}}(\hat{\psi})} \right)^2 \,.$$

As the contrast sum of squares has 1 df, it is equal to the contrast mean square. Under the null hypothesis, the variance ratio of the contrast mean square to the ResMS has an F-distribution with numerator df equal to 1 and denominator df equal to the ResDF. The portion of TrtSS left over is called the remainder sum of squares. Under the null hypothesis that the contrast has accounted for all of the treatment variation, the remainder mean square has an F-distribution on $t - 2$ and ResDF df. If there is no evidence of variation in the remainder, then the contrast alone can be used to describe treatment differences.

Both the ratio of the contrast to its SE and the contrast sum of squares are invariant to re-scaling, for example, if the coefficients for a contrast are all multiplied by 2, the ratio and contrast sum of squares are unchanged. To simplify computation, some software packages therefore automatically standardize contrasts by re-scaling so that the sum of the squared contrast coefficients is equal to 1, i.e. $\Sigma_j \, l_j^2 = 1$.

EXAMPLE 8.4D: POTATO YIELDS*

The estimated treatment effects for this trial were shown in the third column of Table 8.15. To compare the fungicide modes of action we calculate the contrast using $l_1 = 0$, $l_2 = l_5 = 0.5$, and $l_3 = l_4 = -0.5$. This contrast has $\sum_j l_j^2 = 1$ and is estimated as

$$\hat{\psi} = (0 \times -158.3) + (0.5 \times 4.7) + (-0.5 \times 49.7) + (-0.5 \times 66.2) + (0.5 \times 37.7)$$
$$= 0 + 2.35 - 24.85 - 33.10 + 18.85$$
$$= -36.75 ,$$

with estimated variance equal to $s^2/n = 3483/4 = 870.75$. The contrast SE is then equal to the square root of its variance at 29.51. The ratio of the contrast to its SE is $-36.75/29.51 = -1.245$. Compared to a t-distribution on 12 df, this gives $P = 0.237$ for a two-sided test. The contrast sum of squares is then $4 \times (-36.75)^2 = 5402.25$, and results in the same conclusion as the ANOVA shown in Table 8.17 ($F_{1,12}^{AvB} = 1.551, P = 0.237$). Hence, there is no evidence of any difference in yield between fungicides with different modes of action. The remainder mean square ($F_{3,12}^{Rem} = 12.251, P < 0.001$) indicates the presence of treatment variation not accounted for by this contrast.

Typically there are two or more comparisons of interest, generating a number of different contrasts. We label the ith contrast as ψ_i, with contrast coefficients $l_{i1} \ldots l_{it}$. In this situation, the concept of orthogonality becomes important, because it affects the interpretability of the contrasts. We construct the product of two contrasts, here denoted $\psi_i \times \psi_k$, by taking the pair of coefficients relating to each treatment effect, multiplying these together, and summing over all treatment effects, so

$$\psi_i \times \psi_k = \sum_{j=1}^{t} l_{ij} l_{kj} .$$

Two contrasts are said to be **orthogonal contrasts** if their product is zero. Orthogonal contrasts are also statistically independent with zero covariance.

In general, the sum of squares associated with a set of t treatment groups with $t - 1$ df can be partitioned into $t - 1$ orthogonal contrasts each with 1 df. The use of orthogonal con-

TABLE 8.17

Multi-Stratum ANOVA Table for Potato Yields with Treatment Effects (Factor Fungicide) Partitioned into a Contrast to Compare Fungicides of Modes A and B Plus a Remainder (Example 8.4D)

Source of Variation	df	Sum of Squares	Mean Square	Variance Ratio	P
Block stratum					
Residual	3	14,987.20	4995.73	1.434	0.283
Block.Plot stratum					
Fungicide	4	133,419.20	33,354.80	9.576	0.001
Contrast: mode A vs mode B	1	5402.25	5402.25	$F^{AvB} = 1.551$	0.237
Remainder	3	128,016.95	42,672.32	$F^{Rem} = 12.251$	< 0.001
Residual	12	41,796.80	3483.07		
Total	19	190,203.20			

trasts has the advantage that the ANOVA table is invariant to the order in which contrasts are added into the model, and different information is contributing to each contrast.

EXAMPLE 8.4E: POTATO YIELDS*

The analysis in Example 8.4D ignored our previous partitioning of the control as separate from the fungicide treatments. We can reintroduce this partition via a contrast that compares the control with the mean of the fungicide treatments, as $\psi_1 = \tau_1 - \frac{1}{4}(\tau_2 + \tau_3 + \tau_4 + \tau_5)$. We denote our previous contrast for comparison of modes as ψ_2. The coefficients for each of the contrasts ψ_1 and ψ_2 are shown in Table 8.18. Their product is calculated as

$$\psi_1 \times \psi_2 = (1 \times 0) + (-0.25 \times 0.5) + (-0.25 \times -0.5) + (-0.25 \times -0.5) + (-0.25 \times 0.5) = 0$$

and so these two contrasts are orthogonal. Furthermore, we can make comparisons between fungicides within each mode of action using contrast $\psi_3 = \tau_2 - \tau_5$ to compare F1 with F4 and contrast $\psi_4 = \tau_3 - \tau_4$, to compare F2 with F3. The coefficients for these contrasts are also in Table 8.18, and it is straightforward to verify that any pair of these four contrasts is orthogonal.

Table 8.19 is the ANOVA table with the TrtSS partitioned into single df terms for the contrasts fitted in order $\psi_1, \psi_2, \psi_3, \psi_4$, and it is straightforward to verify that the contrast sums of squares do not change if these contrasts are fitted in a different order. Each contrast is independently summarizing a different aspect of the treatment information.

TABLE 8.18

Coefficients for Four Orthogonal Treatment Contrasts for the Potato Yield Trial (Example 8.4E)

		Control τ_1	F1 τ_2	F2 τ_3	F3 τ_4	F4 τ_5
Contrast	ψ_1	1	−0.25	−0.25	−0.25	−0.25
	ψ_2	0	0.5	−0.5	−0.5	0.5
	ψ_3	0	1	0	0	−1
	ψ_4	0	0	1	−1	0

TABLE 8.19

Multi-Stratum ANOVA Table for Potato Yields with Treatment Effects (Factor Fungicide) Partitioned into Four Orthogonal Contrasts: $\psi_1 \ldots \psi_4$ (Example 8.4E)

Source of Variation	df	Sum of Squares	Mean Square	Variance Ratio	P
Block stratum					
Residual	3	14,987.20	4995.73	1.434	0.283
Block.Plot stratum					
Fungicide	4	133,419.20	33,354.80	9.576	0.001
Contrast ψ_1	1	125,294.45	125,294.45	$F^{\psi 1} = 35.972$	< 0.001
Contrast ψ_2	1	5402.25	5402.25	$F^{\psi 2} = 1.551$	0.237
Contrast ψ_3	1	2178.00	2178.00	$F^{\psi 3} = 0.625$	0.444
Contrast ψ_4	1	544.50	544.50	$F^{\psi 4} = 0.156$	0.699
Residual	12	41,796.80	3483.07		
Total	19	190,203.20			

The only contrast giving evidence against its null hypothesis is contrast ψ_1, which compares the negative control with the fungicide treatments. In fact, contrast ψ_1 here is equivalent to the use of factor Type in Example 8.4B, giving the same sum of squares and variance ratio (Table 8.16). Contrast ψ_1 is estimated as -177.0 (SE 29.51), indicating that the yield of the control treatment is 177.0 units less than the average yield of the fungicide treatments with 95% CI calculated as $-177.0 \pm (29.51 \times 2.179) = (-241.3, -112.7)$. The predictive model for this experiment can therefore be reduced to this contrast, agreeing with the conclusion of Example 8.4C.

Factors with two levels can always be represented by an interpretable single contrast constructed as the difference between the two levels, i.e. with contrast coefficients $l_1 = -1$ and $l_2 = 1$. Factors with t levels can be represented by $t - 1$ contrasts, but it is not always possible to construct orthogonal contrasts that ask sensible questions about the treatments. Sometimes it is reasonable to use contrasts to pick out a few comparisons of interest but not decompose the remainder. Within a factorial structure it may be sensible to partition one or more of the factors into one or more contrasts plus a remainder. This structure is then propagated into interactions involving those factors. Both of these approaches are illustrated in Example 8.5.

EXAMPLE 8.5: HERBICIDE EFFICACY

A factorial experiment was done to compare the general efficacy of three herbicides (factor Herbicide) against nine populations of black-grass (factor Population). Two of the herbicides (labelled A and C here) are from the same group (type 1 in factor Type), the third (labelled B) is from a different group (type 2). The design was arranged as a RCBD with five blocks (factor Rep), each containing 27 pots (dummy factor DPot). Six plants were grown in each pot and their combined fresh weight (*g*, variate Fwt) was recorded at the end of the study. The data are listed in Table 8.20 and held in file HERBICIDE.DAT. Preliminary analysis indicated the need for a transformation of the fresh weight and the square root transformation, calculated as *sqrtFwt* = sqrt(*Fwt*), gave reasonable residual plots. There is interest in whether there is any systematic difference in herbicide effect both between and within the herbicide groups, and in whether this changes across the populations.

Here, a crossed treatment structure is appropriate, as the main effects of both herbicide and population are of interest. The full model can be written in symbolic form as

Response variable:	*sqrtFwt*
Explanatory component:	[1] + Herbicide*Population
Structural component:	Rep/DPot

The mathematical model can be written as

$$y_{rsk} = \mu + Rep_k + Herbicide_r + Population_s + (Herbicide.Population)_{rs} + e_{rsk} ,$$

where y_{rsk} is the response in the *k*th block ($k = 1 \ldots 5$) for the *r*th herbicide ($r = 1, 2, 3$ for A, B, C) and the *s*th population ($s = 1 \ldots 9$) with associated deviation e_{rsk}, μ is the overall mean, Rep_k is the effect of the *k*th block, $Herbicide_r$ is the main effect of the *r*th herbicide, $Population_s$ is the main effect of the *s*th population and $(Herbicide.Population)_{rs}$ is the interaction between the *r*th herbicide and the *s*th population. Within the herbicide main effect, the two types of herbicide can be compared using a contrast of the form

$$\psi_1 = \frac{1}{2} (Herbicide_1 + Herbicide_3) - Herbicide_2$$
$$= 0.5 \times Herbicide_1 - Herbicide_2 + 0.5 \times Herbicide_3 .$$

TABLE 8.20

Fresh Weight (g) from a Pot Experiment Testing the Efficacy of Three Herbicides on Nine Populations of Black-Grass Using a RCBD with Five Blocks (Example 8.5 and File HERBICIDE.DAT)

| Population | Herbicide | Fresh Weight (g) | | | | |
		Block 1	Block 2	Block 3	Block 4	Block 5
P1	A	5.94	3.63	5.56	4.09	3.65
P2	A	3.88	2.17	0.63	2.82	1.73
P3	A	3.55	5.16	5.17	1.07	2.61
P4	A	6.45	5.56	1.99	5.21	2.51
P5	A	0.10	0.31	3.69	4.56	0.16
P6	A	4.94	5.21	2.51	3.76	1.90
P7	A	4.07	3.74	4.67	3.41	5.73
P8	A	2.13	6.46	5.02	2.36	2.88
P9	A	2.24	2.85	0.63	1.39	3.14
P1	B	1.25	1.01	0.92	0.98	0.26
P2	B	1.55	1.44	0.90	1.12	1.74
P3	B	1.53	4.21	3.39	4.13	1.85
P4	B	2.56	1.49	1.09	2.37	0.66
P5	B	4.96	5.11	4.84	4.64	4.96
P6	B	1.89	3.00	3.05	1.22	1.58
P7	B	0.67	0.39	0.25	0.44	0.40
P8	B	0.47	0.51	0.37	0.27	0.40
P9	B	0.53	0.66	1.70	0.70	0.17
P1	C	5.37	3.96	4.05	3.37	4.11
P2	C	3.60	1.81	1.82	5.21	2.13
P3	C	4.77	3.46	5.58	4.45	2.80
P4	C	4.10	4.48	7.92	5.46	3.97
P5	C	2.96	5.14	3.16	4.46	2.45
P6	C	2.59	5.33	5.38	5.13	2.61
P7	C	5.17	5.66	4.84	4.47	3.44
P8	C	2.63	5.93	4.74	4.71	4.63
P9	C	4.48	2.90	2.91	5.73	3.71

Source: Data from R. Hull, Rothamsted Research.

Similarly, herbicides A and C are compared via an orthogonal contrast of the form

$$\psi_2 = Herbicide_1 - Herbicide_3 \ .$$

Within the interaction, the contrasts are applied to each population as

$$\psi_{1s} = 0.5 \times (Herbicide.Population)_{1s} - (Herbicide.Population)_{2s}$$
$$+ \ 0.5 \times (Herbicide.Population)_{3s}$$
$$\psi_{2s} = (Herbicide.Population)_{1s} - (Herbicide.Population)_{3s}$$

for $s = 1 \ldots 9$, giving the estimates shown in Table 8.21. At this level, the interest is in whether the value of the contrast varies between populations. For example, consistency across populations in differences between types of herbicide corresponds to the null hypothesis H_0: $\psi_{1s} = 0$ for $s = 1 \ldots 9$.

TABLE 8.21

Estimated Effects and Contrasts for Main Effect (Herbicide and Population) and Interaction (Herbicide.Population) Terms (Example 8.5)

	Estimated Effects				Estimated Contrasts	
	Herbicide (Type)			Main		
Population	A (1)	B (2)	C (1)	Effect	Type 1 vs 2	A vs C
P1	0.340	−0.320	−0.020	0.039	0.480	0.360
P2	−0.069	0.188	−0.119	−0.231	−0.283	0.050
P3	−0.130	0.308	−0.178	0.199	−0.463	0.048
P4	0.097	−0.145	0.048	0.191	0.218	0.048
P5	−0.750	0.948	−0.198	0.069	−1.422	−0.552
P6	0.006	0.117	−0.123	0.129	−0.176	0.128
P7	0.350	−0.523	0.173	−0.026	0.785	0.176
P8	0.256	−0.457	0.202	−0.108	0.686	0.054
P9	−0.099	−0.116	0.214	−0.263	0.174	−0.313
Main effect	0.093	−0.457	0.363	1.654	0.685	−0.270

The ANOVA table is Table 8.22; it partitions the Herbicide main effect and Herbicide.Population interaction sums of squares into components associated with the two contrasts.

Variance ratios for all but one of the treatment mean squares are statistically significant. The Herbicide sum of squares is partitioned into the two contrasts, which are both highly significant ($F_{1,104}^{1v2} = 114.714$, $F_{1,104}^{AvC} = 13.392$, both $P < 0.001$) The type 1 versus 2 contrast is estimated as 0.685 (SE 0.0640), indicating that herbicides of type 1 (A and C) yielded on average 0.685 units more on the square root scale than those of type 2 (B). The herbicide A versus C contrast is estimated as −0.270 (SE 0.0739), indicating that herbicide A yielded 0.27 units (on the square root scale) less than C on average. These patterns can be seen in the full table of predicted means plotted in Figure 8.7.

TABLE 8.22

Multi-Stratum ANOVA Table for Black-Grass Fresh Weights (Square Root Scale) from the Herbicide Efficacy Experiment (Example 8.5)

Source of Variation	df	Sum of Squares	Mean Square	Variance Ratio	P
Rep stratum					
Residual	4	1.3111	0.3278	2.671	0.036
Rep.DPot stratum					
Herbicide	2	15.7207	7.8604	64.053	< 0.001
Type 1 vs 2	1	14.0774	14.0774	$F^{1v2} = 114.714$	< 0.001
Herbicide A vs C	1	1.6434	1.6434	$F^{AvC} = 13.392$	< 0.001
Population	8	3.5080	0.4385	$F^P = 3.573$	0.001
Herbicide.Population	16	13.9437	0.8715	7.102	< 0.001
(Type 1 vs 2).Population	8	12.4676	1.5585	$F^{1v2.P} = 12.700$	< 0.001
(Herbicide A vs C).Population	8	1.4761	0.1845	$F^{AvC.P} = 1.504$	0.165
Residual	104	12.7626	0.1227		
Total	134	47.2461			

Note: Treatment (factor Herbicide) sum of squares partitioned into comparisons between herbicides of types 1 (A and C) and 2 (B), and between herbicides A and C.

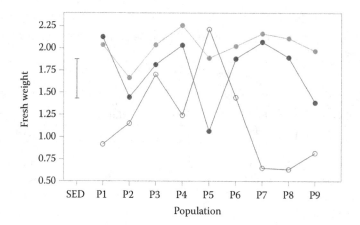

FIGURE 8.7
Predicted fresh weight (g, square root scale) with SED for nine black-grass populations (P1 ... P9) treated with herbicides A (•), B (o) and C (◦) (Example 8.5).

The main effect of population is highly significant ($F^{P}_{8,104} = 3.573, P = 0.001$), reflecting overall differences in fresh weight obtained from the different populations (averaged over herbicides). The Herbicide.Population interaction term is partitioned into the interactions of the two contrasts with the Population factor. The interaction of the type 1 versus 2 contrast is highly significant ($F^{1v2.P}_{8,104} = 12.700, P < 0.001$), indicating that the difference between the two types changes across populations, and indeed Figure 8.7 illustrates that this difference is strongly positive for populations P7 and P8, but negative for P5. The interaction of the A versus C contrast is not significant ($F^{AvC.P}_{8,104} = 1.504, P = 0.165$) indicating that the difference between these two herbicides is reasonably consistent across the populations. Again, this pattern can be observed in Figure 8.7. We can conclude that the relative effectiveness of the different herbicide types depends on the population considered but that within herbicides of type 1, herbicide A is generally more effective (lower fresh weight) than herbicide C.

We have seen in Example 8.5 that contrasts can be used to partition treatment information within a two-way crossed structure. This principle can be extended to nested or higher-level crossed structures and contrasts may be used to simplify the model terms required for prediction. If all of the significant treatment variation can be captured by a small set of contrasts, then a simplified model based on those contrasts can be used for prediction. This procedure is thus qualitatively different from the evaluation of treatment comparisons from the predictive model, which are discussed in Section 8.8.5. The approach is most useful when it is possible to write the treatment structure as a set of meaningful pairwise comparisons.

8.7 Modelling Patterns for Quantitative Treatments

When the groups associated with a treatment factor correspond to some real numeric (quantitative) scale, we might think of building a model to describe the trend in the response in terms of that numeric scale, for example, in Example 1.1, plant height increased linearly in relation to dose. If we can describe this linear trend, then we can use it to predict the response for any intermediate dose. Some responses are more complex, requiring a curve: plant yield tends to respond linearly to nitrogen application initially then tail off; fungal infection rates on plants tend to increase up to some optimal temperature and then decrease for higher temperatures. In this section, we examine the use of contrasts for fitting simple polynomial models to quantitative factors, i.e. factors where the groups correspond to positions on some underlying numeric scale.

Here, we consider the numeric levels of a quantitative factor on each unit as a variate, x. A **polynomial model** consists of several terms, each of which is a power of x multiplied by a coefficient. The order of the polynomial is equal to the highest power of x present, so a first-order polynomial describes a linear relationship. A second-order polynomial, or **quadratic model**, also includes the second power or square of the explanatory variate, and takes the generic form

$$f(x_i) = \alpha + \beta_1 x_i + \beta_2 x_i^2 .$$

This equation consists of three terms, and can be considered as three components: a constant term (α), a linear term ($\beta_1 x_i$) and a quadratic term ($\beta_2 x_i^2$). This model can be considered as an example of polynomial regression (as presented in Section 17.1.2), but here we use polynomial contrasts to fit models of this form.

As an example, consider an experiment set up as a RCBD with four blocks, looking at the response of hydroponic plant growth (measured as biomass) to eight relative concentrations of nutrient solution (0.25, 0.5, 0.75, 1, 1.25, 1.5, 1.75, 2). A model for these data can be written in mathematical form as

$$y_{ij} = \mu + b_i + \tau_j + e_{ij} ,$$

where b_i is the effect of the ith block, $i = 1 \ldots 4$, τ_j is the effect of the jth concentration, $j = 1 \ldots 8$, and all other terms are as defined previously. Again, we use sum-to-zero constraints. The second-order polynomial model is applied to the set of treatment effects, τ_j, $j = 1 \ldots 8$. We form one contrast for each polynomial term (here constant, linear and quadratic) using the appropriate power of x to give the contrast coefficients. The constant contrast corresponds to $\Sigma_j \tau_j$ but, because of the sum-to-zero constraints, this is equal to zero and so is omitted. The linear contrast takes the form

$$(0.25 \times \tau_1) + (0.50 \times \tau_2) + (0.75 \times \tau_3) + (1.00 \times \tau_4) + (1.25 \times \tau_5) + (1.50 \times \tau_6) + (1.75 \times \tau_7) + (2.00 \times \tau_8) ,$$

with the concentration values being used as the contrast coefficients. The quadratic contrast takes the form

$$(0.25^2 \times \tau_1) + (0.50^2 \times \tau_2) + (0.75^2 \times \tau_3) + (1.00^2 \times \tau_4) + (1.25^2 \times \tau_5) + (1.50^2 \times \tau_6) + (1.75^2 \times \tau_7) + (2.00^2 \times \tau_8)$$
$$= (0.0625 \times \tau_1) + (0.25 \times \tau_2) + (0.5625 \times \tau_3) + (1.00 \times \tau_4) + (1.5625 \times \tau_5) + (2.25 \times \tau_6)$$
$$+ (3.0625 \times \tau_7) + (4.00 \times \tau_8) ,$$

with the square of the concentration values now being used as the contrast coefficients.

Unfortunately, this approach results in contrasts that are non-orthogonal, and so the apparent importance of each term can depend on the order in which it is fitted, and the estimated contrast value depends on which other contrasts are fitted. This non-orthogonality can be seen in Figure 8.8a – all powers of x show an increasing pattern for $x > 0$ and

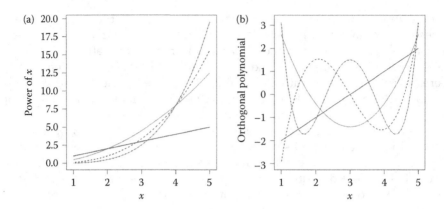

FIGURE 8.8
(a) Simple powers of explanatory variate: x (—), $x^2/2$ (···), $x^3/8$ (- -), $x^4/32$ (-··-); (b) orthogonal polynomials of explanatory variate x of order 1 (—), 2 (···), 3 (- -) or 4 (-··-).

so have strong positive correlations across this range. This problem can be avoided by the use of **orthogonal polynomials**, rather than simple powers. Orthogonal polynomials are constructed so that the qth function is of order q, and is orthogonal to all of the lower order functions. This means that the contrast for each component picks out the elements of the pattern that are unique to that power. Figure 8.8b shows a set of orthogonal polynomials: correlations within this set are all zero. The form of the orthogonal polynomials also illustrates the complexity allowed within these models: a second-order polynomial can accommodate one turning or inflexion point, a third-order model can have two turning points and so on.

Calculation of contrast coefficients for orthogonal polynomials is less straightforward than for simple powers and these coefficients depend on both the quantitative factor levels and their replication. In practice, statistical software will calculate the necessary contrast coefficients. Once the contrast coefficients have been calculated, inference follows as described in the previous section.

In theory, it is possible to fit $t - 1$ orthogonal polynomials for a quantitative factor with t levels, i.e. a polynomial of order $t - 1$. However, this polynomial model would give exactly the same fit as use of the factor itself, and interpolation between factor levels would be uninformative – this is illustrated in Section 17.1.2 in the context of polynomial regression. The usual aim is to find a low-order (i.e. parsimonious) polynomial to describe the general trend across factor levels. Variation due to the quantitative factor that is not accounted for by these lower-order contrasts is usually allocated as a remainder term that amalgamates variation associated with higher-order polynomial terms. This remainder can then be tested against the appropriate residual term to ensure that there is no statistically significant variation associated with the higher-order terms. This remainder is sometimes also called 'lack of fit' and is discussed further in Section 12.8 in the context of regression models.

EXAMPLE 8.6: VOLTAGE RESPONSE

An experiment was conducted to investigate the affinity of a sugar transporter protein for a substrate within plant cells. A range of voltages associated with different sugar concentrations was tested, and the response was measured in terms of electric current (variate *Km*). Nine different voltages were used, in increasing steps from −160 to 0 mV (factor **Voltage**). The experiment was set up as a RCBD, with blocks corresponding to two different occasions (factor **Rep**) with one replicate of each voltage measured during each occasion (factor called **DUnit**, as the actual randomization of plants within runs is not available). The data are listed in Table 8.23 and held in file VOLTAGE.DAT.

A natural logarithm transformation, *logKm* = log$_e$(*Km*), was used to stabilize the variances. The model for the data can be written in symbolic form as

Response variable:	*logKm*
Explanatory component:	[1] + Voltage
Structural component:	Rep/DUnit

The corresponding mathematical model is written as

$$logKm_{ij} = \mu + Rep_i + Voltage_j + e_{ij} \, ,$$

where *logKm$_{ij}$* is the log$_e$-transformed observed current in the *i*th replicate (*i* = 1, 2) for the *j*th level of voltage applied (*j* = 1 ... 9 for −160, −140 ... 0 mV, respectively). The structural component generates the replicate effects, denoted *Rep$_i$* for *i* = 1, 2, and the deviations *e$_{ij}$* (equivalent to term **Rep.DUnit**). The effect of the *j*th voltage is denoted *Voltage$_j$*. The sum-to-zero constraints take the form $\Sigma_i Rep_i = 0$ and $\Sigma_j Voltage_j = 0$.

The predicted treatment means for this model (presented with the data in Figure 8.9a) show a broadly linear pattern of increase in response as voltage increases with a suggestion of slight curvature. This pattern can be investigated further by use of linear and quadratic polynomial contrasts. Table 8.24 lists the estimated treatment effects and coefficients for orthogonal linear and quadratic polynomial contrasts across voltages.

We evaluate the contrasts by multiplying the coefficients by the estimated voltage effects, which give the linear contrast equal to 162.6 and the quadratic contrast equal to 1218.8, but these values must be re-scaled (to correspond to standardized contrasts) before they can be related to the polynomial model for the response, and this is done automatically by statistical software. Table 8.25 is the ANOVA table with the **Voltage** sum of squares partitioned into components corresponding to the linear contrast, the

TABLE 8.23

Electric Current (km) Observed in Plant Cells as a Response to Different Voltages Applied, Using a RCBD with Two Replicates (Example 8.6 and File VOLTAGE.DAT)

Voltage (mV)	km	
	Rep 1	**Rep 2**
−160	0.234	0.219
−140	0.320	0.227
−120	0.326	0.282
−100	0.327	0.277
−80	0.331	0.343
−60	0.489	0.386
−40	0.437	0.421
−20	0.786	0.476
0	0.842	0.611

Source: Data from Rothamsted Research (T. Miller).

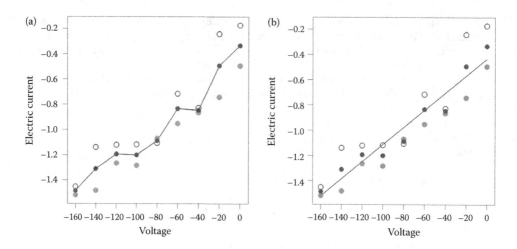

FIGURE 8.9

(a) Observed electric current (o rep 1, • rep 2; $\log_e$(Km)) and fitted treatment means (•) for different levels of voltage and with (b) fitted linear component of trend (solid line) (Example 8.6).

TABLE 8.24

Calculation of Coefficients for Orthogonal Linear and Quadratic Polynomial Contrasts for Electrical Response to Voltages (Example 8.6)

Voltage	Estimated Voltage Effects	Coefficients for Linear Trend	Coefficients for Quadratic Trend
−160	−0.5097	−80	3733
−140	−0.3353	−60	933
−120	−0.2175	−40	−1067
−100	−0.2249	−20	−2267
−80	−0.1120	0	−2667
−60	0.1422	20	−2267
−40	0.1294	40	−1067
−20	0.4843	60	933
0	0.6435	80	3733

TABLE 8.25

Multi-Stratum ANOVA Table for the Electrical Responses from the Voltage Experiment with Treatment (Factor **Voltage**) Sum of Squares Partitioned into Components for Linear and Quadratic Trend and a Remainder (Example 8.6)

Source of Variation	df	Sum of Squares	Mean Square	Variance Ratio	P
Rep stratum					
Residual	1	0.17622	0.17622	12.185	0.008
Rep.DUnit stratum					
Voltage	8	2.33653	0.29207	20.195	< 0.001
Linear contrast	1	2.20458	2.20458	$F^{Lin} = 152.435$	< 0.001
Quadratic contrast	1	0.06029	0.06029	$F^{Quad} = 4.169$	0.075
Remainder	6	0.07166	0.01194	$F^{Rem} = 0.826$	0.581
Residual	8	0.11570	0.01446		
Total	17	2.62845			

quadratic contrast and a remainder. The variance ratio for the remainder term is not significant ($F_{6,8}^{Rem} = 0.83$, $P = 0.581$), indicating no need for higher-order terms. The quadratic component is close to significant ($F_{1,8}^{Quad} = 4.17$, $P = 0.075$), indicating weak evidence for a quadratic component of trend. However, this is small compared with the linear component of trend ($F_{1,8}^{Lin} = 152.44$, $P < 0.001$), which clearly dominates the pattern. Given the small size of the quadratic component compared with the linear component of trend, we can ignore the quadratic component and allocate it to the remainder (which must then be recalculated). The fitted linear trend model is shown in Figure 8.9b and takes the form (see Exercise 15.8)

$$\hat{\mu}._j = -0.434 + 0.0068 \times Voltage_j .$$

In general, fitting quantitative trends by polynomial contrasts is a much less direct approach than regression (Chapters 12 to 15 and 17). However, it can be difficult to account adequately for structure within the regression context (Section 11.6). For quantitative factors in designed experiments, it is therefore usually advantageous to start with polynomial contrasts to investigate the presence and complexity of trend. If the observations have no structure, this process can be followed by regression analysis, allowing for other treatment factors present (as in Chapter 15). If there is structure present then linear mixed models (Chapter 16) can be used as a framework for regression modelling that includes a structural component.

8.8 Making Treatment Comparisons from Predicted Means

In this section, we consider issues that arise in making treatment comparisons from tables of predicted means. These methods should be used following analysis with an appropriate explanatory model that reflects the experimental aims, as described in the preceding sections. The most common type of comparison is a simple pairwise difference of two treatments but more complex functions, such as contrasts, may also be of interest. We first consider the case of simple pairwise comparisons and return to contrasts in Section 8.8.5.

In the simplest case, the aim is to test a null hypothesis of equality between a pair of treatment population means, for example, H_0: $\mu_i = \mu_j$ for treatments i and j or, equivalently, to form a CI for the difference $\mu_i - \mu_j$. Recall that $\hat{\mu}_j$ denotes the predicted mean for the jth treatment group for $j = 1 \ldots t$. As introduced in Section 4.4, these hypotheses can be investigated with statistics of the form

$$t = \frac{\hat{\mu}_i - \hat{\mu}_j}{\widehat{SE}(\hat{\mu}_i - \hat{\mu}_j)} ,$$

where the numerator is a difference between the predicted treatment means and the denominator is their SED. This statistic has a t-distribution with df equal to the ResDF from the ANOVA table. The test is evaluated against a two-sided alternative hypothesis, H_1: $\mu_i \neq \mu_j$, at a specified significance level α_s, typically $\alpha_s = 0.05$, known as the **comparison-wise** significance level. As indicated in Section 2.3.2, α_s is the Type I error, the probability of rejecting the null hypothesis when in fact it is true. The Type I error can therefore also be interpreted as the probability of obtaining a single false-positive result, i.e. declaring a difference significant when in fact it is zero. It is important to realize that the Type I error rate applies to *each* individual hypothesis test done as part of a statistical analysis, and if we perform several tests then the probability of a false-positive result increases with the number of tests; this is sometimes referred to as the problem of **multiple testing**. If we make m independent tests at significance level α_s, then we can regard the number of false-positive results as having a Binomial distribution (Section 2.2.1) with m trials and success probability α_s. It follows that

$$\text{Prob(at least one false positive)} = \alpha_f = 1 - (1 - \alpha_s)^m \,,$$

where α_f is known as the **experiment-wise Type I error**. For example, if we do 15 independent tests with $\alpha_s = 0.05$, then $\alpha_f = 1 - (0.95)^{15} = 1 - 0.463 = 0.537$, i.e. a 53.7% chance of one or more false-positive results. However, in our context of treatment comparisons from a single experiment, the tests are not independent because their denominators, the SED for each comparison, are based on the same ResMS. When hypothesis tests are not independent, there is less certainty about how the Type I error rate accumulates, as this depends on the degree of dependence between the tests: the greater the degree of dependence, the smaller the rate of increase in experiment-wise error rate, with

$$\alpha_s \leq \alpha_f \leq m \times \alpha_s \,.$$

The lower limit holds only in the case when the tests are perfectly correlated.

Here, we first consider two general approaches for dealing with multiple tests: the Bonferroni correction (Section 8.8.1) and the false discovery rate (Section 8.8.2). We then go on to discuss some more specific approaches for some common scenarios: pairwise comparison of all means within a table (often called multiple comparisons, Section 8.8.3); comparison of a set of treatments against a control or standard (Section 8.8.4); and evaluation of a pre-planned set of comparisons or contrasts (Section 8.8.5).

8.8.1 The Bonferroni Correction

The Bonferroni correction adjusts the Type I error rate for each comparison, α_s, downwards. The adjustment is based on the number of comparisons, m, to be evaluated, and aims to achieve the desired experiment-wise error, α_f. The Bonferroni inequality was used above to put an upper limit on the experiment-wise error rate for m comparisons each made at significance level α_s, namely $\alpha_f \leq m \times \alpha_s$.

The Bonferroni correction uses significance level $\alpha_s^* = \alpha_f / m$ for each individual comparison, so that the experiment-wise error rate becomes bounded above by α_f. For example, if we make 15 comparisons with a comparison-wise significance level of $\alpha_s^* = 0.003333$, then the experiment-wise error rate is $\leq 15 \times \alpha_s^* = 0.05$. Use of α_s^* in place of α_s means that the critical value of the test statistic required to obtain a significant result for any individual comparison increases. For example, if our case of 15 comparisons has 18 ResDF, then the critical value of the t-distribution moves from 2.10 to 3.38, i.e. absolute treatment differences need to be 1.6 times larger to be significant; however, we shall control the number of false positives.

The main disadvantage of this approach is that, where many comparisons are made, absolute treatment differences often have to become very large to exceed the Bonferroni-corrected critical value, and so power (the probability of detecting a difference if one is present, Section 10.3) is likely to fall considerably.

8.8.2 The False Discovery Rate

The false discovery rate, FDR, introduced by Benjamini and Hochberg (1995), is a different type of approach that does not attempt to control for the experiment-wise error rate, but instead seeks to quantify the expected proportion of Type I errors within the set of rejected hypotheses. So an FDR of 0.05 means that 5% of the differences that have been found statistically significant are expected to be false-positive results.

There are two ways to calculate the FDR for a given set of comparisons. One method fixes the significance level α_s for individual tests and then calculates the observed FDR. The other calculates the required significance level, α_s^*, for individual tests required to achieve a pre-specified value of FDR. Both methods are applied after the test results have been obtained, and we outline them both below.

We start by calculating the observed FDR for m comparisons made with comparison-wise significance level α_s. The observed FDR is calculated as the ratio of the expected number of significant results under the null hypothesis to the observed number of statistically significant results, s, or

$$\text{FDR} = \frac{m \times \alpha_s}{s} .$$

For example, suppose that we make 200 comparisons at $\alpha_s = 0.05$, of which 24 gave a significant result. Then FDR = $200 \times 0.05/24 = 0.417$, i.e. it is expected that 41.7% of the 24 significant comparisons, i.e. approximately 10 of them, will correspond to false-positive results.

The procedure to set the comparison-wise significance level to obtain a given level of FDR is a little more complicated. First, we rank the observed significance levels from the individual comparisons in ascending order as

$$P_{(1)} \leq P_{(2)} \leq \ldots \leq P_{(m)} ,$$

where subscript (i) indicates the ith most significant test (i.e. the ith smallest observed significance level). We then calculate the values $m \times P_{(k)}/k$ for $k = 1 \ldots m$. For control of the false discovery rate at level FDR, we find the largest value k such that

$$m \times P_{(k)}/k \leq \text{FDR} ,$$

and reject all null hypotheses with $P_{(j)} \leq P_{(k)}$. If there is no k that satisfies that condition, then none of the hypotheses are rejected.

This procedure can be followed for any set of m independent tests, and for dependent tests that meet certain conditions (see Benjamini and Yekutieli, 2001 for more details). This includes most situations of pairwise comparisons (Section 8.8.3) and comparisons of treatments with control (Section 8.8.4). For other sets of dependent tests the expressions above are modified by replacing the total number of comparisons, m, by m^* which is calculated as

$$m^* = m \sum_{j=1}^{m} \frac{1}{j} .$$

The FDR approach seems a good compromise between two extremes: either ignoring the problem of multiple testing (as when we use unadjusted LSDs, Section 8.8.3.1), which may lead to many (unrecognized) false positives; or specification of the experiment-wise error, which may give a loss of power (as when we use the Bonferroni correction).

8.8.3 All Pairwise Comparisons

In this section, we consider several different methods used for making all pairwise comparisons (often called **multiple comparisons**) within a table of predicted means. This is most commonly used for an unstructured set of treatments, but it may also be used to investigate a table of means from a structured set of treatments. There are $t \times (t-1)/2$ pairwise comparisons for a set of t means, and the number of comparisons thus increases proportionally to the square of the number of means. For example, for four treatment groups there are six possible pairwise comparisons, but for 10 treatment groups there are 45 possible pairwise comparisons. The set of tests associated with these comparisons are not independent. Here, we consider the use of the LSD, multiple range tests and Tukey's simultaneous confidence intervals for pairwise treatment comparisons. We assume that all of the treatment comparisons are estimated with equal precision, i.e. that a single common SED applies to the table of predicted means, with associated residual df denoted ResDF.

In the context of multiple comparisons, we often rank the t predicted treatment means as

$$\hat{\mu}_{(1)} \geq \hat{\mu}_{(2)} \geq \ldots \geq \hat{\mu}_{(t)} \, ,$$

where the subscript (i) denotes the ith largest mean, and differences within this ordered set are then examined. The statistical properties of this ordered set differ from those of a random sample and inference requires the distribution of the range of an ordered set under the null hypothesis that the population effects are all equal; this is known as the Studentized range distribution. Quantiles for this distribution are available in most statistical software. We denote the $100(1 - \alpha_s)$th percentile of the Studentized range distribution for t groups with ResDF residual df as $q_{t,\text{ResDF}}^{[\alpha_s]}$.

8.8.3.1 The LSD and Fisher's Protected LSD

The least significant difference, LSD, was introduced in Section 4.4. For two treatments labelled as i and j, the LSD was defined as the smallest absolute difference that would result in rejection of the null hypothesis H_0: $\mu_i = \mu_j$ at significance level α_s, and was calculated as

$$\text{LSD} = t_{\text{ResDF}}^{[\alpha_s/2]} \times \text{SED} \, ,$$

where $t_{\text{ResDF}}^{[\alpha_s/2]}$ is the $100(1 - \alpha_s/2)$th percentile of the t-distribution with ResDF df.

The unprotected LSD approach to multiple comparisons rejects the null hypothesis for any pair of treatments whose absolute difference exceeds the LSD, i.e. where $|\hat{\mu}_i - \hat{\mu}_j| \geq \text{LSD}$. This approach provides no control of the experiment-wise error rate, which therefore increases with the total number of pairwise comparisons as described above, although the Type I error rate for each individual comparison is maintained at level α_s.

The protected LSD procedure differs only in its requirement that the overall F-test for the null hypothesis H_0: $\mu_1 = \ldots = \mu_t$ must be rejected before any individual comparisons are evaluated. However, this procedure gives no additional control of experiment-wise error. It is possible (although uncommon) to obtain a significant F-statistic without any of the pairwise treatment differences exceeding the LSD. Conversely, it is also possible to obtain a non-significant F-statistic when there is one or more significant pairwise differences within the set of treatment comparisons. The protection afforded by the F-test may therefore be illusory.

Because these procedures provide no control of the experiment-wise error rate, they should be used only when this control is regarded as unimportant. A simple way of introducing this experiment-wise control suggested by Hsu (1996, Section 4.1.8) is to use an adjusted version, the aLSD, calculated as

$$\mathrm{aLSD} = \frac{q^{[\alpha_s]}_{t-1,\mathrm{ResDF}}}{\sqrt{2}} \times \mathrm{SED}.$$

8.8.3.2 Multiple Range Tests

Multiple range tests work on the ranked set of predicted means and are used to identify groups of treatments with a similar response. The main difference between the most common procedures is in the significance level used at each stage. Here, we consider the Newman–Keuls and Duncan's multiple range tests, as being among those most commonly used in practice. These procedures define a comparison-wise significance level α_s^* (defined below) then run as follows.

Step 1: Compare the t-statistic for the largest and smallest means, i.e. $t_{(1)(t)} = (\hat{\mu}_{(1)} - \hat{\mu}_{(t)})/\mathrm{SED}$, with $q^{[\alpha_s^*]}_{t,\mathrm{ResDF}}$, the $100(1 - \alpha_s^*)$th percentile of the Studentized range distribution for t groups with ResDF df. If $t_{(1)(t)} \leq q^{[\alpha_s^*]}_{t,\mathrm{ResDF}}$, then we conclude that there are no differences within this set of means and stop. Otherwise, we conclude that some differences are present, and move onto step 2.

Step 2: Repeat the procedure on the test statistics $t_{(1)(t-1)} = (\hat{\mu}_{(1)} - \hat{\mu}_{(t-1)})/\mathrm{SED}$ and $t_{(2)(t)} = (\hat{\mu}_{(2)} - \hat{\mu}_{(t)})/\mathrm{SED}$, adjusting the number of groups to $t-1$ for the Studentized range distribution and testing whether $t_{(1)(t-1)} \leq q^{[\alpha_s^*]}_{t-1,\mathrm{ResDF}}$ or $t_{(2)(t)} \leq q^{[\alpha_s^*]}_{t-1,\mathrm{ResDF}}$. If differences are present within a set, then we proceed to test subsets of $t-2$ adjacent means within that set.

The procedures continue in this manner, working with progressively smaller subsets until all differences are less than the required value, giving groups of means that can be considered as not significantly different. For each subset so identified, a common letter is allocated to all members. The only exception is that a new letter is not allocated to any subset of a group already found to contain no differences. Any mean not allocated to a group at the end of the procedure is assigned its own letter. This process is illustrated in Table 8.26. In this example the groups are distinct, but in many cases they will overlap.

The Newman–Keuls method uses $\alpha_s^* = \alpha_s$ at each step, where α_s is the comparison-wise error rate; this results in an experiment-wise error rate greater than α_s. Duncan's multiple range test uses $\alpha_s^* = 1 - (1 - \alpha_s)^{u-1}$, where u is the size of the subset being tested at each stage. These values are much larger than α_s when u is large, so this procedure is more lax than Newman–Keuls at the initial stages. This approach is intended to preserve the comparison-wise error rate at α_s. The experiment-wise error rate for Duncan's multiple range test can therefore be considerably larger than α_s.

In both cases, the actual experiment-wise error is difficult to determine (other than that it is greater than α_s) and may depend on the unknown configuration of the true population means. This ambiguity in the experiment-wise error rate is a major drawback to these methods and we therefore do not recommend their use.

TABLE 8.26

Schematic Representation of a Multiple Range Test with Five Treatments

Step	Difference Tested	$\hat{\mu}_{(1)}$ $\hat{\mu}_{(2)}$ $\hat{\mu}_{(3)}$ $\hat{\mu}_{(4)}$ $\hat{\mu}_{(5)}$	Set	Result	Action
1	$\hat{\mu}_{(1)} - \hat{\mu}_{(5)}$		0	$t_{(1),(5)} > q_{5,\mathrm{ResDF}}^{[\alpha^*]}$	Significant, test subsets.
2	$\hat{\mu}_{(1)} - \hat{\mu}_{(4)}$		0.1	$t_{(1),(4)} > q_{4,\mathrm{ResDF}}^{[\alpha^*]}$	Significant, test subsets.
	$\hat{\mu}_{(2)} - \hat{\mu}_{(5)}$		0.2	$t_{(2),(5)} > q_{4,\mathrm{ResDF}}^{[\alpha^*]}$	Not significant, stop.
3	$\hat{\mu}_{(1)} - \hat{\mu}_{(3)}$		0.1.1	$t_{(1),(3)} > q_{3,\mathrm{ResDF}}^{[\alpha^*]}$	Significant, test subsets.
	$\hat{\mu}_{(2)} - \hat{\mu}_{(4)}$		0.1.2	$t_{(2),(4)} > q_{3,\mathrm{ResDF}}^{[\alpha^*]}$	Not significant, stop.
4	$\hat{\mu}_{(1)} - \hat{\mu}_{(2)}$		0.1.1.1	$t_{(1),(2)} > q_{2,\mathrm{ResDF}}^{[\alpha^*]}$	Significant, no subsets, stop.
	$\hat{\mu}_{(2)} - \hat{\mu}_{(3)}$		0.1.1.2	$t_{(2),(3)} > q_{2,\mathrm{ResDF}}^{[\alpha^*]}$	Not significant, stop.
Final		a b b b b			Assign letters.

Note: Steps, and tests within steps, are executed in sequential order. Set numbers relate to those from the preceding step, for example, 0.1.1 and 0.1.2 arise as subsets from set 0.1. Rectangles represent the ordered predicted means $\hat{\mu}_{(1)}$ to $\hat{\mu}_{(5)}$. Means compared in each step are shown as filled bars; means found to differ are coloured in black; those that do not are coloured grey. At final step, groups of treatments found not to differ are assigned a common letter.

8.8.3.3 Tukey's Simultaneous Confidence Intervals

Finally in this section, we describe the use of Tukey's simultaneous confidence intervals, where the coverage probability applies to the full set of intervals. For treatments i and j, the $100(1 - \alpha_s)\%$ confidence interval for the comparison $\mu_i - \mu_j$ is

$$(\hat{\mu}_i - \hat{\mu}_j) \pm \frac{q_{t,\mathrm{ResDF}}^{[\alpha_s]}}{\sqrt{2}} \times \mathrm{SED} .$$

This approach is perhaps more useful than simple testing, as it provides a range of plausible values for each comparison. Both the position and length of these confidence intervals may give useful information on treatment differences.

All of the formulae given in this section assume that treatment groups have equal replication and hence equal precision. Calculations become more complex when groups have unequal replication, and hence comparisons have unequal precision. In particular, this may give some inconsistencies in the groups formed at different steps within the multiple range tests.

8.8.4 Comparison of Treatments against a Control

The comparison of treatments against a control or standard treatment is a common requirement in screening trials, where a set of new treatments is evaluated against standard practice. This set of comparisons is a subset of all pairwise comparisons, but we can achieve more power by recognizing the structure of the subset. Dunnett's method is simple and constructs a set of confidence limits for the comparison of each new treatment population mean (μ_j, $j = 2 \ldots t$) with the control (μ_1).

If the aim is to detect treatments that give a larger value than the control (a one-sided test) then the method generates lower limits for the difference of treatments with the control as

$$(\hat{\mu}_j - \hat{\mu}_1) - d^{[\alpha_s]}_{t-1,\text{ResDF}} \times \text{SED} ,$$

where $d^{[\alpha_s]}_{t-1,\text{ResDF}}$ is the $100(1 - \alpha_s)$th percentile of Dunnett's distribution for $t - 1$ treatment groups (excluding the control) and ResDF df for the SED. Quantiles of Dunnett's distribution are available in most statistical software. Any treatment with a lower limit greater than zero can then be considered as larger than the control at significance level α_s. If the aim is the detection of treatments that give a smaller value than the control (another one-sided test) then the method generates upper limits as

$$(\hat{\mu}_j - \hat{\mu}_1) + d^{[\alpha_s]}_{t-1,\text{ResDF}} \times \text{SED} .$$

Any treatment with an upper limit less than zero can be considered as smaller than the control. For a two-sided test, you should calculate both limits after adjusting the critical value, using

$$(\hat{\mu}_j - \hat{\mu}_1) \pm d^{[\alpha_s/2]}_{t-1,\text{ResDF}} \times \text{SED} ,$$

and consider any treatment with either a lower limit greater than zero or an upper limit less than zero as different from the control.

8.8.5 Evaluation of a Set of Pre-Planned Comparisons

In some situations, there may be a pre-planned subset of treatment comparisons that are of particular interest. To qualify as pre-planned, the comparisons must be determined before any results are obtained; this matter is discussed further in Section 8.8.6. In this more general situation, it is difficult to obtain an optimal strategy, and so we deal with the problem of multiple testing by using the methods of Sections 8.8.1 and 8.8.2. If controlling Type I error is the main concern, then a Bonferroni correction to the significance level would be appropriate. If we wish to retain power, but with some insight into the false-positive rate, then use of the FDR may be more appropriate.

In this context, comparisons that are more complex than simple differences, such as treatment contrasts, may be of interest. Here, a contrast is defined as a linear function of the population means, i.e. of the form

$$l_1\mu_1 + l_2\mu_2 + \ldots + l_t\mu_t = \sum_{j=1}^{t} l_j\mu_j \; ,$$

which is estimated by substitution of the predicted means in place of the unknown true population means. Estimation of the SE for this contrast is more difficult than with effects (Section 8.6) because correlations between the predicted means must be accounted for, but this can be done with statistical software.

Before our final summing up, we consider an example in which a set of pre-planned comparisons are of interest, and use all of the methods discussed in this section to demonstrate some of the differences between them.

EXAMPLE 8.7: LUPIN VARIETY TRIAL

A field trial was set up to evaluate the overall performance of a set of lupin breeding lines. The experiment was laid out as a RCBD with three blocks of 14 plots (factors Block and Plot). Fourteen different lines were tested (factor Line), comprising 12 dwarf lines (DTN lines) and two non-dwarf lines (CH-304 lines). Performance across a range of characteristics, including the average number of plants per square metre (variate *NPlant*) and oil yield (t/ha, variate *OilYield*), was to be compared with the candidate variety for release, line DTN20. Here, we analyse oil yields. The data are held in file LUPINTRIAL.DAT and listed in Table 8.27.

TABLE 8.27

Average Number of Plants (NPlant) and Oil Yield (t/ha,Yield) from a RCBD with Three Blocks and 14 Lupin Breeding Lines (Example 8.7 and File LUPINTRIAL.DAT)

	Block 1			Block 2			Block 3		
Plot	Line	NPlant	Yield	Line	NPlant	Yield	Line	NPlant	Yield
1	DTN84	16.68	0.36	DTN84	31.13	0.34	DTN31	26.68	0.21
2	DTN108	24.46	0.58	DTN12	31.13	0.36	DTN78	28.90	0.36
3	DTN78	37.80	0.39	DTN04	24.46	0.33	DTN10	28.90	0.32
4	DTN19B	37.80	0.38	DTN11	55.58	0.38	CH304-70	26.68	0.24
5	CH304-73	22.23	0.37	DTN19B	37.80	0.33	DTN84	26.68	0.58
6	DTN10	37.80	0.30	DTN10	28.90	0.34	DTN108	6.67	0.56
7	DTN11	24.46	0.29	DTN108	26.68	0.54	DTN20	24.46	0.41
8	DTN19A	26.68	0.17	DTN20	24.46	0.35	DTN19B	40.02	0.32
9	DTN04	35.57	0.26	DTN31	31.13	0.18	DTN19A	8.89	0.30
10	DTN31	24.46	0.19	DTN01	37.80	0.37	DTN01	31.13	0.38
11	CH304-70	26.68	0.22	CH304-73	35.57	0.24	DTN04	28.90	0.23
12	DTN20	31.13	0.32	DTN19A	24.46	0.23	DTN11	46.69	0.28
13	DTN12	20.01	0.31	DTN78	35.57	0.59	DTN12	31.13	0.35
14	DTN01	42.24	0.32	CH304-70	17.79	0.24	CH304-73	26.68	0.29

Source: Data from I. Shield, Rothamsted Research.

The full model can be written in symbolic form as

Response variable: *OilYield*
Explanatory component: [1] + Line
Structural component: Block/Plot

The mathematical model for these observations is written as

$$OilYield_{ij} = \mu + Block_i + Line_j + e_{ij} \, ,$$

where $OilYield_{ij}$ is response in the ith block ($i = 1 \ldots 3$) for the jth line ($j = 1 \ldots 14$). The structural component generates the block effects, denoted $Block_i$ for $i = 1 \ldots 3$, and the deviations e_{ij} (equivalent to term Block.Plot). The effect of the jth line is denoted $Line_j$. The sum-to-zero constraints take the form $\Sigma_i \, Block_i = 0$ and $\Sigma_j \, Line_j = 0$.

The oil yields did not require transformation, and the predicted treatment means are calculated as

$$\hat{\mu}_j = \hat{\mu} + \widehat{Line}_j \, .$$

These predicted means are listed in Table 8.28. The ResMS obtained from ANOVA was $s^2 = 0.0039$ on 26 df, leading to SEDs for treatment comparisons calculated as $\sqrt{(2 \times 0.0039/3)} = 0.0509$.

The aim of the analysis is comparison of other lines with line DTN20 (with predicted mean 0.360), which we can consider as comparisons with a control, as described in Section 8.8.4, and so we start by using Dunnett's method. For a two-sided test, Dunnett's method for significance level $\alpha_s = 0.05$ requires the 97.5% critical value of Dunnett's distribution for 13 treatments and 26 df, i.e. $d_{13,26}^{[0.025]} = 3.004$. By rearranging the formula in Section 8.8.4, we find that treatments different to DTN20 must satisfy one of the following conditions:

$$\hat{\mu}_j > \hat{\mu}_1 + d_{13,16}^{[0.025]} \times SED = 0.360 + 3.004 \times 0.0509 = 0.513$$
$$\hat{\mu}_j < \hat{\mu}_1 - d_{13,16}^{[0.025]} \times SED = 0.360 - 3.004 \times 0.0509 = 0.207$$

Any line with predicted oil yield > 0.513 can be considered to yield more than DTN20, and any line with predicted oil yield < 0.207 can be considered to have a lower yield

TABLE 8.28

Predicted Means for 14 Breeding Lines (SED = 0.0509 on 26 df) in the Lupin Variety Trial (Example 8.7)

Line	Predicted Mean	Line	Predicted Mean
CH304-70	0.233	DTN12	0.340
CH304-73	0.300	DTN19A	0.233
DTN01	0.357	DTN19B	0.343
DTN04	0.273	DTN20	0.360
DTN10	0.320	DTN31	0.193
DTN108	0.560	DTN78	0.447
DTN11	0.317	DTN84	0.427

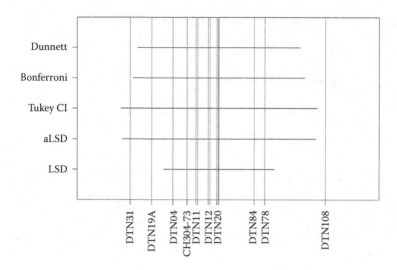

FIGURE 8.10
Range of predicted oil yields of a set of lupin breeding lines considered not different from line DTN20 under different tests (Example 8.7). Lines not labelled: CH304-70 (= DTN19A), DTN10 (> DTN11), DTN19B (> DTN12) and DTN01 (< DTN20).

than DTN20. This range is shown in Figure 8.10 (labelled Dunnett), and only lines DTN31 and DTN108 are outside of this range.

Instead of using Dunnett's method, we might have considered this as an arbitrary set of 13 pre-planned comparisons, and used a Bonferroni correction to the critical value. Instead of using $\alpha_s = 0.05$ for each individual comparison, we would then use $\alpha_s^* = 0.05/13 = 0.00385$. For a two-sided test, we use $\alpha_s^*/2$ then $t_{26}^{[0.00198]} = 3.174$ and lines with predicted means outside the range (0.199, 0.521) can be considered different from DTN20. This range is very close to that obtained from Dunnett's method, and is also shown in Figure 8.10 (labelled Bonferroni), leading to the same conclusions.

Instead of using the structure of the method, we might think of treating this as a problem of multiple comparisons, and extract conclusions for the tests we are interested in. The range of values considered not different from DTN20 using Tukey's simultaneous confidence intervals are shown in Figure 8.10 (labelled Tukey CI). The range for this method is greater, because it allows for $14 \times 13/2 = 91$ tests to have taken place, whereas we are interested in only 13 of them (each line vs DTN20). This test identifies only DTN108 as having a yield different to DTN20. If instead we use the LSD (Figure 8.10, labelled LSD) then there is no allowance for multiple testing, hence no adjustment to the significance level and lines DTN31, DTN19A, CH304-70 and DTN108 are identified as different to DTN20. However, if we use the adjusted LSD by substituting the Studentized range distribution in place of the t-distribution then the results are similar to Tukey's simultaneous confidence intervals (Figure 8.10, labelled aLSD). In this context, none of these procedures takes account of the number of tests of interest, and may either over- or under-estimate the number of differences. This mismatch illustrates the benefit of considering the structure of the problem rather than automatic use of multiple comparison procedures.

As another possibility, we might use the unadjusted LSD in combination with the FDR to give insight into the expected number of false-positive results. Testing at $\alpha_s = 0.05$, we identify four significant results out of 13, so the expected false-discovery rate is $100 \times 13 \times 0.05/4 = 16.25\%$. If we wish to restrict our false-positive rate to 5%, then the first step is to rank the 13 tests in order of the observed significance levels associated with the t-tests, as shown in Table 8.29. Calculating $13 \times P_{(k)}/k$ for each rank, i.e. for

TABLE 8.29

Calculation of Significance Level Required to Obtain FDR of 5% (Example 8.7)

Rank (k)	Line	Predicted Mean	Difference from DTN20	t	P	$13 \times P/k$
1	DTN108	0.560	0.200	3.933	0.001	0.007
2	DTN31	0.193	−0.167	−3.277	0.003	0.019
3	CH304-70	0.233	−0.127	−2.491	0.019	0.084
4	DTN19A	0.233	−0.127	−2.491	0.019	0.063
5	DTN04	0.273	−0.087	−1.704	0.100	0.261
6	DTN78	0.447	0.087	1.704	0.100	0.217
7	DTN84	0.427	0.067	1.311	0.201	0.374
8	CH304-73	0.300	−0.060	−1.180	0.249	0.404
9	DTN11	0.317	−0.043	−0.852	0.402	0.581
10	DTN10	0.320	−0.040	−0.787	0.439	0.570
11	DTN12	0.340	−0.020	−0.393	0.697	0.824
12	DTN19B	0.343	−0.017	−0.328	0.746	0.808
13	DTN01	0.357	−0.003	−0.066	0.948	0.948
–	DTN20	0.360	0	–	–	–

$k = 1 \ldots 13$, gives the last column in Table 8.29. Only the first two values are ≤ 0.05, so to obtain an FDR of 5% we reject only the two null hypotheses corresponding to the two smallest observed significance levels, which, in this case, matches the conclusions from Dunnett's test.

8.8.6 Summary of Issues

We have presented several approaches to multiple testing and considered specific methods for multiple comparisons and comparisons against a control. The statistical literature contains many more methods and it can be difficult to decide which procedure is the most appropriate. Miller (1981) and Hsu (1996) provide a more detailed account of the subject. To finish, we discuss some controversial issues.

There is a school of thought that states that multiple comparisons should never be performed for experiments where there is some structure within the set of treatments (see e.g. Bondari, 1999; Cousens, 1988; Gates, 1991; Gilligan, 1986; Madden, 1982; Pearce, 1993; Perry, 1986; Webster, 2007). The main concern of these authors is that multiple procedures ignore this structure, and we agree that this is a grave error. However, once the treatment structure has been used to obtain a set of predictive terms, it can be helpful to evaluate comparisons within these terms, and the issue of multiple testing should then be considered. For example, in a crossed three-way treatment structure with a significant three-way interaction, it may be useful to use multiple comparisons to help disentangle patterns within the three-way table. If only main effects were significant, then comparisons within predictive tables for those main effects should be evaluated, not comparisons on the whole three-way table.

Somewhat different considerations arise when the response is examined after the experiment is done to decide which comparisons to test, i.e. *a posteriori*. In any set of treatments, even if there are no differences between the true population means, some groups will have smaller responses and some larger responses, purely because of random sampling variation. The eye is drawn to comparisons comprising the larger differences, and this bias must be taken into account. This is done by adjustment of the experiment-wise error to control for all possible tests of pairs of treatments within the set, and this adjustment is necessary even if all of the tests are not made.

EXERCISES

8.1   Re-analyse the data of Exercise 7.4 taking into account the crossed treatment struc-
      ture (additional factors Cultivar and Rate can be found in file INOCULATION.DAT).
      Write down the model for this analysis in both mathematical and symbolic form.
      Is there any evidence for differences among cultivars or inoculation rates, and do
      these factors act independently? Compare this analysis with that from Exercise 7.4:
      has the crossed structure clarified your results?

8.2   An experiment designed as a RCBD with three blocks and a $2 \times 2 \times 2$ factorial
      structure investigated the effect of three factors and their interactions on the rate
      of callus growth on wheat seeds. Wheat seeds were placed in separate isolation
      containers with sets of eight containers, one for each of the eight treatments, kept
      together in holding trays (factor Tray). The treatment factors were age of the seed
      ('old' or 'young', factor Age), concentration of growth media (2.5 or 5 mg, factor
      Conc) and type of growth promoter (Cutlass or Rapier, factor Type). Seeds were
      weighed (variate *Weight*) after they had been in the media for 15 days. Analyse
      the seed weights; the data set is in file CALLUS.DAT. Remember to check the model
      assumptions. What conclusions can you draw from this experiment?[*]

8.3   A field experiment investigated the effect of two seed rates (40 or 80 seeds/
      m^2, factor Rate) and two row spacings (12 or 36 cm, factor Spacing) on the
      performance of four lupin genotypes: two determinate genotypes, A and
      B, and two new dwarf-determinate genotypes, C and D (factor Genotype).
      The experiment was designed as a three-block RCBD with 16 plots per block
      and a $4 \times 2 \times 2$ factorial treatment structure. At harvest the number of lupin
      plants per m^2 was recorded (variate *NoPlants*). The data set can be found in
      file LUPINDENSITY.DAT. Analyse the densities at harvest and identify and inter-
      pret a suitable predictive model. Do the determinate and dwarf-determinate
      genotypes behave differently?[†]

8.4   A field experiment to investigate the effect of weed competitors on yield of win-
      ter wheat was set up as a RCBD with three blocks of 18 plots. Three weed species
      were used: chickweed (CW), black-grass (BG) and cleavers (CL). Target weed
      densities were 0, 40, 80, 160, 320 and 640 plants per m^2 for CW and BG, and 0,
      3, 6, 12, 24 and 48 plants per m^2 for CL. However, the weed densities achieved
      were lower and differed among species. The unit numbers (*ID*), structural fac-
      tors (Block, Plot), species sown (factor Weed), density achieved (variate *Density*)
      and the final yields at harvest (variate *Yield*, tonnes/hectare at 85% dry matter)
      are given in file WEEDCOMPETITION.DAT. Consider whether it is appropriate to
      consider density as crossed with or nested within weed species, and construct a
      suitable factor for the density treatment. Analyse the data and interpret the tests
      generated from your ANOVA table. What conclusions can you draw from this
      trial?[‡]

---

[*] Data from M. Wilkinson, Rothamsted Research.
[†] Data from I. Shield, Rothamsted Research.
[‡] Data from P. Lutman, Rothamsted Research.

8.5   Re-analyse the data of Exercise 6.3 using new factors (and information supplied in file SCAB.DAT) to answer the following questions: Does the addition of sulphur affect the level of scab? Does either of the rate or timing of application affect the level of scab? Do these two factors act independently? (We re-visit these data in Exercise 11.4.)

8.6   An experiment assessed the effect of two lectins, Con-A and GNA, on nematode motility. Nematodes were incubated overnight with one of the two lectins or a buffer solution (PBS) as a control (factor Treatment). Nematodes were placed in the centre of Petri dishes, with four dishes allocated to each treatment completely at random. Here, we analyse the total distance moved by the nematodes in each dish after 40 min. File NEMATODES.DAT contains the unit numbers (*DDish*), explanatory factor (Treatment) and distances moved (variate *Distance*). Analyse these data and construct contrasts to assess whether (a) addition of lectins affects nematode movement and (b) the two lectins have similar effects on movement.[*]

8.7   An experiment at Rothamsted Research in 1996 investigated the yield response of forage maize to nitrogen fertilizer. The experiment was designed as a RCBD with three blocks of four plots, with nitrogen fertilizer rates of 0, 70, 140 and 210 kg N. The whole crop forage yields from each plot (at 100% dry matter in tonnes/hectare) are shown in Table 15.11. File FORAGE.DAT contains unit numbers (*ID*), structural factors (Block, Plot), explanatory factor N and the final yields (variate *Yield*). Analyse these data using ANOVA and incorporate a first-order polynomial (linear trend) in nitrogen fertilizer rate. State your conclusions from this analysis.[†]

8.8* Consider the data from the calcium pot trial of Example 4.1 (Table 4.1 and file CALCIUM.DAT). In this trial, the treatments A, B, C and D were concentrations of calcium in the soil, measured as relative concentrations of 1, 5, 10 and 20, respectively. Re-analyse these data using polynomial contrasts. Which low-order polynomial provides the best fit to these data?

8.9   In Example 8.7, a three-block RCBD lupin breeding line experiment was described and the resulting oil yields analysed. Now analyse the average number of plants per square metre (variate *NPlant*) in a similar way. The data can be found in Table 8.27 and file LUPINTRIAL.DAT. Compare the plant density of other lines with the line DTN20.[‡]

---

[*] Data from R. Curtis, Rothamsted Research/Bionemax.
[†] Data from P. Poulton, Rothamsted Research.
[‡] Data from I. Shield, Rothamsted Research.

9

Models with More Complex Blocking Structure

In this chapter we present the analysis of some common designs, introduced in Chapter 3, that have blocking structures somewhat more complex than that of the randomized complete block design (RCBD). Recall that the structure of an experiment describes sources of heterogeneity among the experimental units (Section 3.1.3). Blocking is included within the structural component of the model, which also encompasses other aspects such as the presence of technical replicates. The blocking structure may be nested, crossed or contain both nested and crossed components (Section 3.2). To recap, nested structures comprise multiple units at lower levels of the experimental structure associated with a single unit at a higher level, with no relationship between lower level units contained in different higher level units. For example, in an experiment laid out as a RCBD (Chapter 7), plots are regarded as lower level units, and blocks as higher level units; it is assumed that there is no association between plots in different blocks and so plots are considered to be nested within blocks. On the other hand, crossed blocking structures occur when lower level units are simultaneously included within two independent higher level units associated with different factors. For example, consider a rectangular layout of pots in a glasshouse experiment, with both rows and columns of the layout regarded as blocking factors. Each pot is in both a row and a column; each row contains pots from each of the columns, and vice versa. Rows and columns are therefore considered as crossed blocking factors in this context.

Several commonly used designs incorporate specific forms of blocking structure, and here we consider in detail three designs already introduced in Chapter 3: the Latin square (Section 3.3.3), split-plot (Section 3.3.4) and balanced incomplete block (Section 3.3.5) designs. For each we give a general description, state the underlying model, present the analysis of variance, describe the comparison of treatment means, and briefly discuss some common extensions to or variations on the basic design. We omit the mathematical expressions for some of the parameter estimates and sums of squares in this chapter (the calculations follow from the principles introduced in Chapters 4, 7 and 8 for simpler designs), and instead emphasize the interpretation of results produced by statistical software. The Latin square design, which uses two crossed blocking factors, is described first in Section 9.1, followed by the split-plot design which, in its standard form, uses three nested blocking factors with two crossed treatment factors applied to different levels of experimental unit (Section 9.2). Finally, details are given for the balanced incomplete block design, a useful variant of the RCBD when the block size is smaller than the number of treatments (Section 9.3).

9.1 The Latin Square Design

The Latin square (LS) design was introduced in Section 3.3.3 and is used where heterogeneity is associated with two crossed blocking factors, both with the same numbers of levels. This design was originally used for field experiments with plots laid out on a square grid, with heterogeneity expected across both rows and columns of the grid. The blocking factors are therefore often referred to as rows and columns and we use these generic terms in this section. However, the blocking factors may correspond to any two crossed sources of heterogeneity, such as time of day or observer. Some common situations where a crossed blocking structure may be appropriate, and where a LS design could be used, include the following:

- Field experiments laid out with plots on a square grid, with both rows and columns of the grid expected to contribute to heterogeneity between plots. Factors influencing the heterogeneity could include soil characteristics (e.g. fertility or position on a slope), management practices, or the (potential) direction of influx of pests and diseases.

- Experiments in a glasshouse, controlled environment (CE) room or growth cabinet where the positioning of benches, shelves and so forth, with respect to walls, doors or light sources may introduce systematic variability (e.g. related to temperature, humidity or light) in different directions, for example, from left to right and from back to front (or, possibly, top to bottom).

- Laboratory experiments where there are two potential sources of variability, for example, scientists and machines (see Example 3.3), and we are concerned about the impacts of variation from the two sources.

The LS design is the simplest crossed blocking design suitable for such situations, and is a special case of a more general class known as **row–column designs** (see Section 9.1.5). For a LS design, the number of rows and columns (i.e. the number of levels of each blocking factor) must equal the number of treatments and also the number of replicates of each treatment; we denote this number by t. The treatment allocation is such that each treatment appears exactly once in each row and once in each column, with each row and each column containing the complete set of treatments. Estimates of treatment effects are then independent of differences between either rows or columns, and the row, column and treatment factors are mutually orthogonal (see Section 11.1). Overall, there are a total of $N = t \times t = t^2$ experimental units, and each treatment is replicated exactly t times. The treatment structure associated with a LS design may comprise a single factor (with t levels), or any structure with a total of t treatment combinations (see Chapter 8).

EXAMPLE 9.1A: LUPIN TRIAL

In Example 3.6, we introduced an experiment devised to investigate the effects of soil type and water availability on the growth of individual lupin plants in pots. Because of potential systematic trends due to temperature and light, the rows and columns of the square array of pots were considered as crossed blocking factors using a LS design. The treatments corresponded to a 2×2 factorial structure with two soil types (factor Soil; clay, C, or sand, S) combined with two levels of water supply (factor Water; low, L, or high, H). Initially, we consider a single set of four treatment combinations, coded in factor Treatment (with labels 1 = CH, 2 = CL, 3 = SH, 4 = SL). Plant heights (cm) were measured for each pot at the end of the experiment. The experimental layout and plant heights are shown in Table 9.1, with data held in file LUPIN.DAT.

For this experiment the number of treatments is $t = 4$ with a total of $t \times t = 16$ experimental units (pots). It is easy to verify from Table 9.1 that each treatment combination is present once in each row and once in each column, such that each row (or column) contains all four treatment combinations.

TABLE 9.1

Experimental Plan and Observed Response (Plant Heights, cm) for a LS Design with Two Treatment Factors: Soil Type (C = Clay, S = Sand) and Water Availability (L = Low, H = High) (Example 9.1A and File LUPIN.DAT)

	Column 1	Column 2	Column 3	Column 4
Row 1	CH	SL	CL	SH
	19.6	23.5	21.7	19.0
Row 2	CL	SH	CH	SL
	15.5	22.4	23.2	19.3
Row 3	SH	CH	SL	CL
	18.5	23.5	26.4	19.0
Row 4	SL	CL	SH	CH
	19.8	19.8	23.9	20.8

Source: Data from I. Shield, Rothamsted Research.

9.1.1 Defining the Model

A model for observations from a LS design with a single treatment factor with t levels takes the form

$$y_{ijk} = \mu + r_i + c_j + \tau_k + e_{ijk} \,, \tag{9.1}$$

where y_{ijk} is the observed response for the kth treatment in the ith row and jth column, μ the overall population mean, r_i the effect of the ith row, c_j the effect of the jth column, τ_k the effect of the kth treatment and e_{ijk} the deviation associated with that observation. Note that the treatment allocated to unit ij (in the ith row and jth column) is actually determined by the randomization of treatments to units, so in theory we do not need another subscript to indicate the treatment applied. However, this would make the notation more complex and so for simplicity we use the extra subscript. All of the subscripts i, j and k run from 1 to t but because we have only t^2 units, not all of the combinations are present. For example, for each ij combination, only one value of k, corresponding to the treatment applied to that unit, will be valid. We use sum-to-zero constraints such that $\Sigma_i r_i = 0$, $\Sigma_j c_j = 0$ and $\Sigma_k \tau_k = 0$. This model can be written in our symbolic notation as

Explanatory component: [1] + Treatment
Structural component: Row*Column
    = Row + Column + Row.Column

where factor Treatment labels the treatments, factor Row labels the level of the first (row) blocking factor and factor Column labels the level of the second (column) blocking factor present on each unit. The Row.Column term labels the individual observations and corresponds to the model deviations. In practice, the single treatment term will often be partitioned to investigate crossed or nested structures in terms of underlying factors as described in Chapter 8.

9.1.2 Estimating the Model Parameters

The parameters associated with the model in Equation 9.1 are the overall population mean, μ, the treatment effects τ_k, $k = 1 \ldots t$, and the row and column effects, r_i and c_j for $i, j = 1 \ldots t$.

The effects are estimated with the principle of least squares (see Section 4.2), and they take the same general form, i.e. sample means for treatments adjusted by the sample grand mean, as in the CRD and RCBD. The orthogonality of the design means that the effects of treatment, row and column are independent of one another. As previously, the overall population mean, μ, is estimated by the sample grand mean,

$$\hat{\mu} = \bar{y} .$$

The effect of the kth treatment is then estimated by the difference between the sample mean for that treatment and the sample grand mean,

$$\hat{\tau}_k = \bar{y}_{..k} - \bar{y} ,$$

although here we take means only over the combinations of i and j present for each treatment. Similarly, the effect of the ith row (jth column) is estimated by the difference between the sample mean for that row (column) and the sample grand mean,

$$\hat{r}_i = \bar{y}_{i..} - \bar{y} ; \quad \hat{c}_j = \bar{y}_{.j.} - \bar{y} .$$

Again, in these equations, we take means only across combinations of subscripts that are present in the design.

The quantity $\mu_k = \mu + \tau_k$ represents the population mean for the kth treatment. The best estimate of this population mean is then

$$\hat{\mu}_k = \hat{\mu} + \hat{\tau}_k = \bar{y} + (\bar{y}_{..k} - \bar{y}) = \bar{y}_{..k} ,$$

i.e. the sample mean for the kth treatment.

EXAMPLE 9.1B: LUPIN TRIAL

The model of Equation 9.1 applies to this experiment and can be written with a single set of treatments in symbolic form as

Response variable:	*Height*
Explanatory component:	[1] + Treatment
Structural component:	Row*Column

The sample grand mean for the lupin trial is 20.99. Parameter estimates derived from the sample means are listed in Table 9.2.

9.1.3 Assessing the Importance of Individual Model Terms

Like the RCBD (Section 4.3.1), the LS is an orthogonal design, so it is possible to uniquely partition the total sum of squares of the observations (TotSS) into components due to the different sources of variation: here, rows (RowSS), columns (ColSS), treatments (TrtSS) and background variation (ResSS), so that

$$\text{TotSS} = \text{RowSS} + \text{ColSS} + \text{TrtSS} + \text{ResSS} .$$

TABLE 9.2

Parameter Estimates (Row, Column and Treatment
Effects) for the Lupin Trial (Example 9.1B)

Rows		Columns		Treatments	
$\hat{r}_1$	−0.04	$\hat{c}_1$	−2.64	$\hat{\tau}_1$	0.78
$\hat{r}_2$	−0.89	$\hat{c}_2$	1.31	$\hat{\tau}_2$	−1.99
$\hat{r}_3$	0.86	$\hat{c}_3$	2.81	$\hat{\tau}_3$	−0.04
$\hat{r}_4$	0.08	$\hat{c}_4$	−1.47	$\hat{\tau}_4$	1.26

These sums of squares can be written in terms of the parameter estimates, as

$$\text{RowSS} = t\sum_{i=1}^{t} \hat{r}_i^2; \quad \text{ColSS} = t\sum_{j=1}^{t} \hat{c}_j^2; \quad \text{TrtSS} = t\sum_{k=1}^{t} \hat{\tau}_k^2; \quad \text{ResSS} = \sum_{i,j,k} \hat{e}_{ijk}^2,$$

where the summation for ResSS is made over combinations of i, j and k that are present in the design; this can be achieved by summation over any two of the indices. There is a corresponding partition of the total degrees of freedom as

$$\text{TotDF} = \text{RowDF} + \text{ColDF} + \text{TrtDF} + \text{ResDF} .$$

The total number of df are computed as

$$\text{TotDF} = N - 1 = t^2 - 1 ,$$

with the same number of df for each of the row, column and treatment terms, i.e.

$$\text{RowDF} = \text{ColDF} = \text{TrtDF} = t - 1 .$$

The residual df (ResDF) are most easily obtained by subtraction as

$$\text{ResDF} = (t^2 - 1) - (t - 1) - (t - 1) - (t - 1) = t^2 - 3t + 2 = (t - 1) \times (t - 2) .$$

As usual, we calculate the mean square for each term by division of its sum of squares by its degrees of freedom (Section 4.3). If any of the sets of row, column or treatment effects are uniformly zero, then the corresponding mean squares are attributable solely to background variation. The variance ratios required to test null hypotheses of zero effects are therefore calculated as the appropriate mean square divided by the residual mean square, and these variance ratios are compared with the percentiles of an F-distribution with $t - 1$ numerator and $(t - 1) \times (t - 2)$ denominator df.

To construct a multi-stratum ANOVA table, we need to recognize the strata in this design (see Section 7.5 for an introduction to this concept). The LS design has three strata, corresponding to rows (factor Row), columns (factor Column) and the individual units (Row.Column). The multi-stratum ANOVA table shown in Table 9.3 is partitioned according to this structure. Because the complete set of treatments appears in each row and in each column of the design, there is no information on treatment effects in comparisons

TABLE 9.3

Structure of the Multi-Stratum ANOVA Table for a LS Design with t Rows (Factor Row),
Columns (Factor Column) and Treatments (Factor Treatment), and a Total of $N = t^2$ Units

Source of Variation	df	Sum of Squares	Mean Square	Variance Ratio
Row stratum				
Residual	$t-1$	RowSS	RowMS = RowSS/$(t-1)$	RowMS/ResMS
Column stratum				
Residual	$t-1$	ColSS	ColMS = ColSS/$(t-1)$	ColMS/ResMS
Row.Column stratum				
Treatment	$t-1$	TrtSS	TrtMS = TrtSS/$(t-1)$	TrtMS/ResMS
Residual	$(t-1)(t-2)$	ResSS	ResMS = ResSS/ $[(t-1)(t-2)]$	
Total	$N-1$	TotSS		

between either rows or columns, and hence within either the Row or Column strata. As treatments are applied to the units defined by the combinations of rows and columns, variation in the Row.Column stratum is partitioned into the variation associated with the Treatment factor and residual variation.

Recall that, in addition to testing hypotheses about treatment effects, we can also use the multi-stratum ANOVA table to test hypotheses about terms in the structural component. Units within a higher-level stratum can always be constructed from units at some lower level (e.g. rows consist of a set of plots). If there is no heterogeneity at the higher level, then the ratio of the residual variances for these strata, as estimated by their mean squares, should be close to unity. In the LS design, we can compare variation between rows with background variation using the variance ratio RowMS/ResMS (Table 9.3). Under the null hypothesis that the row effects are all zero, this variance ratio is distributed as an F-distribution with $t-1$ numerator and $(t-1) \times (t-2)$ denominator df. An analogous test, based on ColMS/ResMS, can be made for column effects. Recall that these tests are made to give information on the major sources of variation present in the structure, and can give information useful in designing future experiments; they are not used to refine the predictive model.

EXAMPLE 9.1C: LUPIN TRIAL

The multi-stratum ANOVA table for this model is shown in Table 9.4.

The residual plots (not shown) indicate no obvious violations of the assumptions (Section 5.2). The variance ratio for the treatment mean square gives strong evidence of differences between treatments ($F^T_{3,6} = 12.667$, $P = 0.005$).

There is no evidence of differences between rows (corresponding to an expected light gradient, $F^R_{3,6} = 3.165$, $P = 0.107$), but there is strong evidence of differences between columns (corresponding to an expected temperature gradient, $F^C_{3,6} = 38.478$, $P < 0.001$). This suggests that columns are likely to be an important source of structural variation for any future experiments in this environment. Although row variation is not significant for these data, it should still be allowed for in future experiments if previous experience suggests that it is sometimes substantial.

This analysis ignores the underlying treatment structure, and a more appropriate analysis uses a two-way crossed explanatory structure (Section 8.2) in terms of the underlying factors, Soil and Water, written as

Explanatory component: [1] + Soil*Water
= [1] + Soil + Water + Soil.Water

TABLE 9.4

Multi-Stratum ANOVA Table for the Lupin Trial with Four Rows, Columns and Treatments (Factors Row, Column and Treatment, Respectively) (Example 9.1C)

Source of Variation	df	Sum of Squares	Mean Square	Variance Ratio	P
Row stratum					
Residual	3	6.162	2.054	$F^R = 3.165$	0.107
Column stratum					
Residual	3	74.912	24.971	$F^C = 38.478$	< 0.001
Row.Column stratum					
Treatment	3	24.662	8.221	$F^T = 12.667$	0.005
Residual	6	3.894	0.649		
Total	15	109.629			

The treatment sum of squares can then be partitioned into components for the two main effects and the interaction. The resulting ANOVA table is shown in Table 9.5. The sum of the main effect and interaction sums of squares is equal to the combined TrtSS in Table 9.4.

Starting at the bottom of the ANOVA table, there is strong evidence of a treatment interaction ($F_{1,6}^{S.W} = 25.588$, $P = 0.002$), indicating that the effect of soil type depends on the amount of water supplied. The presence of this interaction means that predictions should be based on all model terms, and that main effects might not be easily interpreted. Nevertheless, growth appears to differ between soil types ($F_{1,6}^S = 9.062$, $P = 0.024$), with no overall effect of water supply ($F_{1,6}^W = 3.352$, $P = 0.117$).

9.1.4 Evaluating the Response to Treatments: Predictions from the Fitted Model

For an unstructured set of treatments, the best estimate of the population mean for the kth treatment is the treatment sample mean, i.e. $\hat{\mu}_k = \bar{y}_{..k}$ (Section 9.1.2). Uncertainty associated with this estimate is measured by its estimated SE,

$$\widehat{SE}(\hat{\mu}_k) = \sqrt{\frac{s^2}{t}},$$

TABLE 9.5

Multi-Stratum ANOVA Table for the Lupin Trial Using the Two-Way Crossed Explanatory Structure Soil*Water (Example 9.1C)

Source of Variation	df	Sum of Squares	Mean Square	Variance Ratio	P
Row stratum					
Residual	3	6.162	2.054	$F^R = 3.165$	0.107
Column stratum					
Residual	3	74.912	24.971	$F^C = 38.478$	< 0.001
Row.Column stratum					
Soil	1	5.881	5.881	$F^S = 9.062$	0.024
Water	1	2.176	2.176	$F^W = 3.352$	0.117
Soil.Water	1	16.606	16.606	$F^{S.W} = 25.588$	0.002
Residual	6	3.894	0.649		
Total	15	109.629			

using the estimate of background variation, $s^2 = \text{ResMS}$, in place of the unknown true value, σ^2. This can be used to form a $100(1 - \alpha_s)\%$ CI for the treatment population mean as

$$\left(\hat{\mu}_k - \left[t^{[\alpha_s/2]}_{(t-1)(t-2)} \times \widehat{\text{SE}}(\hat{\mu}_k) \right], \quad \hat{\mu}_k + \left[t^{[\alpha_s/2]}_{(t-1)(t-2)} \times \widehat{\text{SE}}(\hat{\mu}_k) \right] \right),$$

where $t^{[\alpha_s/2]}_{(t-1)(t-2)}$ is the $100(1 - \alpha_s/2)$th percentile of the Student's t-distribution with $(t - 1) \times (t - 2)$ df.

The best estimate of the difference between population means for the kth and sth treatments is provided by the difference between their respective sample means, i.e.

$$\hat{\mu}_k - \hat{\mu}_s = \bar{y}_{\cdot\cdot k} - \bar{y}_{\cdot\cdot s},$$

and the estimate of the standard error of this difference is

$$\text{SED} = \widehat{\text{SE}}(\hat{\mu}_k - \hat{\mu}_s) = \sqrt{\frac{2s^2}{t}},$$

with corresponding $\text{LSD} = \text{SED} \times t^{[\alpha_s/2]}_{(t-1)(t-2)}$. As usual, the statistic

$$t = \frac{\hat{\mu}_k - \hat{\mu}_s}{\text{SED}}$$

has a t-distribution with degrees of freedom equal to the ResDF, here $(t - 1) \times (t - 2)$, and can be used to evaluate the null hypothesis of equality of the two treatment population means against a two-sided alternative hypothesis. The corresponding $100(1 - \alpha_s)\%$ confidence interval for this treatment difference can be computed as

$$\left((\hat{\mu}_k - \hat{\mu}_s) - \text{LSD}, \quad (\hat{\mu}_k - \hat{\mu}_s) + \text{LSD} \right).$$

Predictions for crossed or nested structures within the set of treatments can be derived from the individual predictions as described in Chapter 8.

EXAMPLE 9.1D: LUPIN TRIAL

Table 9.6 shows predicted population means for each treatment combination. As the interaction term is statistically significant, this table is the most appropriate summary of this experiment (see Section 8.2.4), and the predictions and their SE are the same as would be obtained from use of the combined factor **Treatment**. The SE for the individual predictions is equal to 0.403, calculated using the ResMS = s^2 = 0.649 from Table 9.5 as

$$\widehat{\text{SE}}(\hat{\mu}_k) = \sqrt{\frac{s^2}{t}} = \sqrt{\frac{0.649}{4}} = 0.403.$$

The ResDF is $(t - 1) \times (t - 2) = 3 \times 2 = 6$ df. The SED for comparisons between pairs of individual treatments is $\sqrt{(2 \times 0.649/4)} = 0.570$, with a 5% (two-sided) LSD calculated as

TABLE 9.6

Predicted Population Means (SE = 0.403, SED = 0.570 on 6 df) for the Lupin Trial (Example 9.1D)

Soil Type	Water Availability	
	H	L
C	21.77	19.00
S	20.95	22.25

$$\text{LSD} = t_6^{[0.025]} \times \text{SED} = 2.447 \times 0.570 = 1.394 \, .$$

In combination with the LSD, the predictions in Table 9.6 indicate that there is no real difference in growth in sandy soil (Soil = S) across the two levels of water availability, but that growth is significantly reduced in clay soil (Soil = C) when less water is available (Water = L).

9.1.5 Constraints and Extensions of the Latin Square Design

The LS design has several serious disadvantages that make its use impractical in many circumstances. The numbers of rows, columns and treatments must be equal, but in practice it might not be possible to construct realistic blocks of the required size in both dimensions (rows and columns). In addition, when the number of treatments is small, the ResDF are also small. For example, if the number of treatments is three, four or five we have two, six and 12 ResDF, respectively. This means that the estimate of background variability is likely to be poor and the power to detect real treatment differences will be reduced (see Chapter 10 for more discussion about statistical power). However, there are various extensions of LS designs that ease these restrictions, and we discuss some briefly here.

When the number of treatments is small, one way to increase the ResDF is to use **multiple squares** of the same size, with each square having a different randomization. The squares may either be considered as independent, with separate rows and columns, or as linked, with rows or columns shared across the squares to form a **Latin rectangle design**. The advantage of linked squares is that common effects can be used for the shared rows or columns, which further increases ResDF, as demonstrated in the following examples.

EXAMPLE 9.2: INDEPENDENT LATIN SQUARES

An experiment was set up to investigate the effect of petal colour on the influx of pollen beetles into a crop of oilseed rape. Five different shades of petal colour were considered, and a LS design was used to account for the unknown direction of migration into the crop. Previous studies had found much spatial variation in beetle counts, and so two replicates of the LS design were used to increase the precision of treatment comparisons. The two squares had the same orientation in adjacent fields, but common row or column effects could not reasonably be expected, and so the squares were regarded as independent. The experimental plan is shown in Table 9.7a. The structural component of the model takes the form

Structural component: Field/(Row*Column)
                   = Field + Field.Row + Field.Column + Field.Row.Column

TABLE 9.7

Designs using (a) Two Independent LSs (Example 9.2) with Separate Rows and Columns in Different Fields; (b) Two Linked LSs (Example 9.3) with Position within Stacks Considered as Common across Replicates

(a)

Field 1   Column

		1	2	3	4	5
	1	E	C	A	D	B
	2	A	D	B	E	C
Row	3	B	E	C	A	D
	4	C	A	D	B	E
	5	D	B	E	C	A

(b)

Rep 1   Position within stack

		1	2	3	4	5
	1	E	D	A	B	C
	2	D	C	E	A	B
Stack	3	A	E	B	C	D
	4	B	A	C	D	E
	5	C	B	D	E	A

Field 2   Column

		1	2	3	4	5
	1	E	C	D	B	A
	2	A	D	E	C	B
Row	3	D	B	C	A	E
	4	C	A	B	E	D
	5	B	E	A	D	C

Rep 2   Position within stack

		1	2	3	4	5
	1	A	C	E	D	B
	2	D	A	C	B	E
Stack	3	E	B	D	C	A
	4	C	E	B	A	D
	5	B	D	A	E	C

A dummy multi-stratum ANOVA table, showing the sources of variation with their df, is presented in Table 9.8a. There are four strata, corresponding to the two fields (Field), rows and columns within fields (Field.Row and Field.Column) and the individual plots within fields (deviations, indexed by Field.Row.Column combinations). The ResDF is now 28 compared with only 12 for a single 5×5 square.

EXAMPLE 9.3: LINKED LATIN SQUARES

An experiment was required to investigate the growth of different strains of fungus on a new substrate. Five strains of the fungus were available, each applied to 10 dishes. The

TABLE 9.8

Dummy Multi-Stratum ANOVA Tables for Two Replicates of a 5×5 LS with (a) Independent Squares (Example 9.2) in Different Fields and (b) Linked Squares Using Common Position Effects within Stacks (Example 9.3)

(a) Source of Variation	df	(b) Source of Variation	df
Field stratum		Rep stratum	
Residual	1	Residual	1
Field.Row stratum		Position stratum	
Residual	8	Residual	4
Field.Column stratum		Rep.Stack stratum	
Residual	8	Residual	8
Field.Row.Column stratum		Rep.Stack.Position stratum	
Treatment	4	Treatment	4
Residual	28	Residual	32
Total	49	Total	49

dishes were to be held in vertical stacks of five dishes within a CE cabinet. The investigator expected that position within each stack would affect growth rates, as well as the location of each stack on the shelf. Stack and Position were two independent sources of heterogeneity, and so the experiment was designed as two replicates of a 5×5 LS, with both replicates placed on the same shelf. The experimental plan is set out in Table 9.7b. Since the effect of position within stack was expected to be the same across all stacks, the two squares were considered linked, with position effects held in common. The structural component takes the form

Structural component: Rep + Position + Rep.Stack + Rep.Stack.Position

A dummy multi-stratum ANOVA table is Table 9.8b. This table also has four strata, corresponding to the two replicates (Rep), stacks within replicates (Rep.Stack), position within stack (Position), and the individual dishes (indexed by Rep.Stack.Position combinations). As position effects are held in common across squares (replicates), fewer effects are fitted than would be the case if these effects were expected to differ across squares (replicates), and the additional df are passed into the ResDF in the lowest stratum; now the ResDF equals 32 compared with 28 for the independent squares of Example 9.2.

There are some situations in which we can extend the constraints provided by a LS design to provide real benefits in controlling the impacts of adjacent treatments. For example, in insect pheromone trials, neighbouring treatments can interfere because of movement of the pheromone plumes by wind. **Neighbour-balanced LS designs**, also known as complete or quasi-complete LS designs, are useful in such situations where rows and columns reflect the physical layout of the experiment. By balancing the occurrence of neighbouring pairs of treatments (so that each treatment occurs adjacent to each other treatment the same number of times within both rows and columns), these designs ensure that no individual treatment has an unfair advantage (or disadvantage) over others due to a lucky (or unlucky) allocation of neighbours.

EXAMPLE 9.4: NEIGHBOUR-BALANCED LATIN SQUARE

A design was required to investigate strategies for pest control on a crop of field beans. The treatments were two semio-chemicals with repellent qualities, three field margin mixtures as a trap crop, and an untreated control. If one uses small plots, there is a danger of interaction (or contamination) between neighbouring treatments. For example, if the semio-chemicals are very effective, they might also repel pests from neighbouring plots. Conversely, a large pest population on the untreated control plots might start moving into neighbouring plots. A neighbour-balanced LS was used to even out any such interference. The experimental plan is set out in Table 9.9. It is straightforward to verify that each pair of treatments occurs as neighbours twice within rows and twice within columns.

Another design extending the constraints of the LS is the **Graeco–Latin** (or **Euler**) **square design**, which is an orthogonal combination of two LS designs. This design allows the independent assessment of the effects of two t-level treatment factors, with each of the t^2 treatment combinations occurring once within a single Graeco–Latin square. The main effects of the two t-level treatment factors can be estimated but, as there is no replication of the individual treatment combinations, it is not possible to test for the presence of a treatment interaction. This design should therefore be used only when prior knowledge suggests that no interaction will occur. A common use of these designs is for perennial

TABLE 9.9

Neighbour-Balanced LS Design for Six Treatments: Control (C), Three Margin Mixtures (M1, M2, M3) and Two Repellent Semio-Chemicals (S1, S2) (Example 9.4)

	Column					
Row	1	2	3	4	5	6
1	S2	S1	M3	M1	M2	C
2	S1	M1	S2	C	M3	M2
3	M3	S2	M2	S1	C	M1
4	M1	C	S1	M2	S2	M3
5	M2	M3	C	S2	M1	S1
6	C	M2	M1	M3	S1	S2

crops, where different treatments may be applied in consecutive years, with the possibility of carry-over effects from treatments applied in previous years, but where we expect no interaction between previous and current treatments. These designs can also be adapted to situations where there are three (crossed) blocking factors and one treatment factor, all with t levels. Graeco–Latin square designs are even more restrictive than LS designs, and have even fewer residual degrees of freedom – for example, a single 4×4 Graeco–Latin square design has only 3 ResDF. Note that for some values of t (e.g. 6 and 10) it is not possible to construct Graeco–Latin square designs.

The LS design is a particular example of a row–column design, a class that includes more general designs with two crossed blocking factors. General row–column designs need not have equal numbers of treatments, rows and columns, and usually do not have an orthogonal structure. The simplest modification to a LS design might involve the removal of just a single row (or column) to produce an **incomplete LS design**, whilst the addition of a single row or column produces an **extended LS design**. Whilst these designs with one row or column deleted (or added) are no longer orthogonal, they are still balanced for treatment comparisons, and so can be analysed by standard multi-stratum ANOVA algorithms (see Section 11.6). As the discrepancy between the number of treatments and the number of replicates grows, however, it can be more challenging to find balanced row–column designs. Further details on the construction and analysis of general row–column designs can be found in Mead et al. (2012, Chapter 8).

9.2 The Split-Plot Design

In some experimental situations, the natural scale of experimental unit varies between different treatment factors – examples include the following:

- Field experiments in which machinery constraints apply, for example, irrigation treatments often have to be set up for a large area, but varieties can be sown on much smaller plots.

- CE experiments where different regimes of temperature or lighting must be applied to whole rooms (or cabinets) whilst levels of other factors, such as plant variety or watering, can be applied within rooms (e.g. to plants in pots).

One approach in these situations would be to apply all treatment factors at the coarser scale, but this quickly leads to experiments requiring substantially more resources than are usually available. A better alternative is the split-plot (SP) design, introduced in Section 3.3.4, which is appropriate for factorial experiments where levels of one or more factors must be applied to larger experimental units while the levels of the other factor(s) can be applied to smaller units. It is worth noting, however, that a SP design should be used only when real constraints on the scale of treatment application are present, as the design is less efficient than the corresponding RCBD based on the same number of experimental units. SP designs occasionally arise as a means of modifying an existing experiment to enable the addition of a new treatment factor.

Similarly to the LS, the SP design had its origins in field experimentation, but it is much more widely applicable. We first consider one form of SP design, which we call the standard form, and discuss variations later. The standard SP design uses two treatment factors, A and B, with a factorial structure and a three-level nested structure for the experimental units. We assume that factor A can be applied only to large units but that factor B can be applied to smaller units, and that a crossed model (Section 8.2) is appropriate for these factors. The highest level of the structure corresponds to complete replicates of the set of treatments, and we denote this level as blocks. Each block is then divided into several whole plots (sometimes called main plots), with levels of treatment factor A randomized to the whole plots separately within each block (equivalent to the randomization of a single treatment factor in a RCBD). Finally, each whole plot is divided into several subplots, and the levels of factor B are randomized onto subplots within each whole plot (just as if we were considering the whole plots as blocks in a RCBD for factor B). Because the two treatment factors, A and B, are applied within different strata, the main effects of factors A and B are assessed against different levels of background variation (between whole plots and between subplots, respectively, see Section 9.2.2) and hence estimated with different precision.

In general notation, in the standard SP design there are t treatments, formed from all factorial combinations of two treatment factors, A and B, where factor A has t_A levels and factor B has t_B levels, and $t = t_A \times t_B$. The number of blocks is denoted m, and the number of whole plots in each block must be equal to t_A (one for each level of factor A) giving a total of $m \times t_A$ whole plots. The number of subplots per whole plot must then be equal to t_B (one for each level of factor B) giving a total of $N = m \times t_A \times t_B$ subplots. Each level of factor A will be present on one whole plot in each of the blocks, and each level of factor B will be present on one subplot within each whole plot. The replication for each level of factor A is m main plots whilst the replication for each level of factor B is $m \times t_A$ subplots. Finally, the replication for each of the t individual treatment combinations is m subplots.

EXAMPLE 9.5A: WEED COMPETITION EXPERIMENT

A field experiment using a SP design to investigate the competitive effects of weeds, with and without irrigation, on the yield of winter wheat was introduced in Example 3.7. The experiment used two irrigation regimes (non-irrigated or irrigated) in combination with three different weed species: *Alopecurus myosuroides* (black-grass), *Galium aparine* (cleavers) and *Stellaria media* (chickweed), abbreviated to Am, Ga and Sm, respectively, and a negative control (no weeds). The SP design was used because the irrigation regimes could be applied only to larger areas of land. The experiment had four blocks ($m = 4$), with irrigation regimes applied to whole plots within each block (i.e. two whole plots per block, $t_A = 2$), and different weed species were sown in subplots within the whole plots (i.e. four subplots per whole plot, $t_B = 4$). The layout and data for this experiment are shown in Table 9.10.

TABLE 9.10

SP Layout of the Weed Competition Experiment (Example 9.5A and File COMPETITION.DAT)

	Block 1		Block 2		Block 3		Block 4	
Whole plot 1	– 7.92	Am 3.62	– 8.02	Ga 5.72	Sm 4.91	Ga 2.20	Am 3.71	– 7.16
	Sm 5.70	Ga 4.49	Sm 6.32	Am 3.19	– 5.54	Am 1.97	Ga 6.51	Sm 6.65
Whole plot 2	– 9.11	Sm 6.77	Am 2.52	Ga 4.70	Am 2.92	Sm 6.64	Ga 4.91	Sm 5.78
	Ga 7.59	Am 4.12	– 7.05	Sm 5.91	Ga 6.90	– 8.18	Am 2.73	– 8.22

Source: Data from P. Lutman, Rothamsted Research.

Note: Whole plots are shaded grey (irrigated) or white (non-irrigated) and each whole plot contains four subplots to which weed species (Am, Sm, Ga or no weeds, –) are applied.

For this experiment, the replication for each level of the irrigation factor is four (one whole plot in each block), and the replication for each of the four weed treatments is eight (one subplot in each whole plot in each block). Each individual treatment (combinations of irrigation regime and weed species) is replicated four times.

9.2.1 Defining the Model

For this standard SP design we have a nested structure with three strata (blocks, whole plots within blocks, and subplots within whole plots within blocks) and a crossed treatment structure with two factors. Ideally, we should label the whole plots and subplots for each observation according to the experimental plan, so as to maintain the distinction between the treatment and blocking structures. However, for simplicity of notation here we label the whole plots within blocks such that the jth whole plot has the jth level of treatment factor A applied, and label subplots within whole plots such that the kth subplot has the kth level of treatment factor B applied. The linear model for this design can then be written as

$$y_{ijk} = \mu + b_i + \alpha_j + w_{ij} + \beta_k + (\alpha\beta)_{jk} + e_{ijk} \ ,$$

where y_{ijk} is the observed response on the kth subplot within the jth whole plot within the ith block. Parameter μ represents the overall population mean, b_i is the effect of the ith block, α_j the effect of the jth level of treatment factor A, w_{ij} the effect associated with the jth whole plot located in the ith block, β_k the effect of the kth level of treatment factor B, $(\alpha\beta)_{jk}$ the interaction effect for the jth and kth levels from treatment factors A and B, respectively, and e_{ijk} the model deviation. The subscripts range over $i = 1 \ldots m$, $j = 1 \ldots t_A$ and $k = 1 \ldots t_B$. Sum-to-zero constraints are applied as $\Sigma_j \alpha_j = \Sigma_k \beta_k = \Sigma_j (\alpha\beta)_{jk} = \Sigma_k (\alpha\beta)_{jk} = 0$, and $\Sigma_i b_i = \Sigma_j w_{ij} = 0$. Parameter estimation for a SP design again follows the principles of least-squares estimation. If we use the symbolic names Y for the response, Block to label the blocks, WholePlot to label the whole plots within blocks, and Subplot to label the subplots within whole plots within blocks, then this design can be represented in our symbolic notation as

Response variable:	Y
Explanatory component:	[1] + A*B
	= [1] + A + B + A.B
Structural component:	Block/WholePlot/Subplot
	= Block + Block.WholePlot + Block.WholePlot.Subplot

The term Block.WholePlot.Subplot defines the individual units and so corresponds to the model deviations, e_{ijk}.

9.2.2 Assessing the Importance of Individual Model Terms

We omit formulae for the sums of squares for the SP design, but instead concentrate on the general form and interpretation of the multi-stratum ANOVA table that can be obtained from statistical software. The total sum of squares of the observations (TotSS) is partitioned into six components of variation: blocks (BlkSS), the main effect of treatment factor A (SS(A)), whole plots (WPtSS), the main effect of treatment factor B (SS(B)), the A.B interaction (SS(A.B)) and subplot or background variation (ResSS), giving the following relationship:

$$TotSS = BlkSS + SS(A) + WPtSS + SS(B) + SS(A.B) + ResSS .$$

As usual, there is a corresponding partition of the total degrees of freedom. An important aspect of the analysis of a SP design is the partitioning of treatment variation among strata; variation among main effects for treatment factor A, quantified by SS(A), must be compared with the background variation at the whole-plot level, represented by WPtSS. However, variation among main effects for treatment factor B and variation among the interaction effects, quantified by SS(B) and SS(A.B), respectively, must be compared with background variation at the subplot level, represented by ResSS. Within the ANOVA table, SS(A) therefore appears within the Block.WholePlot stratum, and SS(B) and SS(A.B) appear within the Block.WholePlot.Subplot stratum. The multi-stratum ANOVA table for this standard SP design is Table 9.11. As usual, each mean square is calculated by division of the corresponding sum of squares by its degrees of freedom.

TABLE 9.11

Structure of the Multi-Stratum ANOVA Table for a Standard SP Design with m Blocks (Factor Block), t_A Whole Plots per Block (Factor WholePlot) and t_B Subplots (Factor Subplot) per Whole Plot and a Total of $N = m \times t_A \times t_B$ Units

Source of Variation	df	Sum of Squares	Mean Square	Variance Ratio
Block stratum				
Residual	$m - 1$	BlkSS	BlkMS	BlkMS/WPtMS
Block.WholePlot stratum				
A	$t_A - 1$	SS(A)	MS(A)	MS(A)/WPtMS
Residual	$(t_A - 1) \times (m - 1)$	WPtSS	WPtMS	WPtMS/ResMS
Block.WholePlot.Subplot stratum				
B	$t_B - 1$	SS(B)	MS(B)	MS(B)/ResMS
A.B	$(t_A - 1) \times (t_B - 1)$	SS(A.B)	MS(A.B)	MS(A.B)/ResMS
Residual	$t_A \times (t_B - 1) \times (m - 1)$	ResSS	ResMS	
Total	$N - 1$	TotSS		

Note: Treatment factor A (t_A levels) is applied to whole plots within blocks, treatment factor B (t_B levels) is applied to subplots within whole plots.

In addition to evaluating the variance ratios for treatment terms, we can consider the variation due to the experimental structure. The block mean square is compared with the whole-plot residual mean square (WPtMS), because if the block effects are all zero then variation between blocks arises from whole-plot variation alone. Similarly, the whole-plot residual mean square is compared with the subplot residual mean square. Note that the ResDF for the Block.WholePlot stratum will always be smaller than that for the Block.WholePlot.Subplot stratum, so that the background variation in the lowest stratum is always estimated with greater precision. If the whole plots form natural blocks, so that units within the same whole plot are more similar than units in different whole plots, then we expect the Block.WholePlot stratum residual mean square to be larger than the Block.WholePlot.Subplot stratum residual mean square. These two facts together mean that the effects of treatment factor B and the A.B interaction are usually estimated with more precision than the effects of treatment factor A.

EXAMPLE 9.5B: WEED COMPETITION EXPERIMENT

To analyse the data from this experiment, we require factors to define the blocking structure corresponding to the physical layout, here called Block, WholePlot and Subplot and the two treatment factors, here called Irrigation and Species. Note the distinction between the levels of WholePlot and Irrigation (and similarly Subplot and Species) as the blocking factors represent the full field plan labelled systematically (shown in Table 9.10) to which the treatment factors have been randomized. These factors can be found in file COMPETITION.DAT, along with a variate called *Grain* containing the response (weight of grain at 85% dry matter in tonnes/hectare). The full model written in symbolic notation is

Response variable:	*Grain*
Explanatory component:	[1] + Irrigation*Species
Structural component:	Block/WholePlot/Subplot

The multi-stratum ANOVA for these data is shown in Table 9.12. As expected, there are three strata in the ANOVA table, the top stratum (Block) contains no treatment information, the middle stratum (Block.WholePlot) comprises variation due to the main effect of Irrigation and the whole-plot residual, and the lowest stratum (Block.WholePlot.Subplot) contains variation due to the Species main effect, the variation due to the Irrigation.Species interaction and the subplot residual. The associated residual plots (not shown) indicate no major violations of the model assumptions, so we can evaluate the treatment terms, as usual working upwards from the bottom of the ANOVA table. The interaction is highly significant ($F_{3,18}^{I.S} = 5.582$, $P = 0.007$) indicating that the effect of irrigation on competition varies among weed species. The presence of this interaction indicates that we cannot simplify our model for prediction, which needs to use all the explanatory terms, but out of interest we still examine the main effects to assess the relative importance of the different model terms. There is very strong evidence for overall differences in the competitive effects of the weed species ($F_{3,18}^S = 109.726$, $P < 0.001$). The size of this F-statistic suggests that the interaction may be relatively small compared to these main effects. Despite the low ResDF in the whole-plot stratum, there is some evidence for overall differences between irrigation regimes ($F_{1,3}^I = 9.480$, $P = 0.054$). Patterns of response in the predictions are explored further in Example 9.5C.

Examination of variation associated with the experimental structure indicates large variation between different whole plots ($F_{3,18}^{B.W} = 5.751$, $P = 0.006$) but little additional variation among blocks ($F_{3,3}^B = 1.476$, $P = 0.378$). These results suggest that spatial variation within the field occurs at reasonably fine (i.e. whole-plot) scales.

TABLE 9.12

Multi-Stratum ANOVA Table for Grain Weight from the Weed Competition Experiment (Example 9.5B)

Source of Variation	df	Sum of Squares	Mean Square	Variance Ratio	P
Block stratum					
Residual	3	6.6473	2.2158	$F^B = 1.476$	0.378
Block.WholePlot stratum					
Irrigation	1	14.2311	14.2311	$F^I = 9.480$	0.054
Residual	3	4.5035	1.5012	$F^{B.W} = 5.751$	0.006
Block.WholePlot.Subplot stratum					
Species	3	85.9257	28.6419	$F^S = 109.726$	< 0.001
Irrigation.Species	3	4.3714	1.4571	$F^{I.S} = 5.582$	0.007
Residual	18	4.6986	0.2610		
Total	31	120.3776			

Note: Two irrigation regimes (factor Irrigation) were applied to whole plots (factor WholePlot) within four blocks (factor Block), three weed species and a negative control (no weeds) (factor Species) were applied to subplots (factor Subplot) within whole plots.

9.2.3 Evaluating the Response to Treatments: Predictions from the Fitted Model

In Section 8.2.4, we stated our policy for making predictions from a crossed structure with two factors, namely, that predictions are made from all model terms if the interaction term is significant, and that predictions are made from any significant main effects when the interaction is not significant. Predictions from the standard SP design take the same form (i.e. treatment sample means) as described in Section 8.2.4, but the SEs and SEDs for estimates of treatment effects made within the whole-plot stratum take a slightly different form. The predictions are summarized in Table 9.13, together with their standard errors and associated df.

The estimated SE for predictions for factor B (averaged across all levels of factor A, denoted $\hat{\mu}_{\cdot k}$) are calculated from the ResMS, denoted s^2 as usual, and the associated df are the ResDF from the subplot (Block.WholePlot.Subplot) stratum (ResDF). The estimated SE for predictions for factor A (averaged across all levels of factor B, denoted $\hat{\mu}_{j\cdot}$) take the same

TABLE 9.13

Form of Predicted Population Means for Combinations of Treatment Factors Applied to Whole Plots (Factor A) and Subplots (Factor B) in the Standard SP Design with Estimated SEs and Associated Df

Description	Population Mean	Prediction	$\widehat{SE}$ of Prediction	df for $\widehat{SE}$
jth level of A	$\mu_{j\cdot}$	$\hat{\mu}_{j\cdot} = \bar{y}_{\cdot j\cdot}$	$\sqrt{s_w^2 / (t_B \times m)}$	$(t_A - 1) \times (m - 1)$
kth level of B	$\mu_{\cdot k}$	$\hat{\mu}_{\cdot k} = \bar{y}_{\cdot\cdot k}$	$\sqrt{s^2 / (t_A \times m)}$	$t_A \times (t_B - 1) \times (m - 1)$
jth level of A with kth level of B	μ_{jk}	$\hat{\mu}_{jk} = \bar{y}_{\cdot jk}$	$\sqrt{(s_w^2 + (t_B - 1) \times s^2) / (t_B \times m)}$	Equation 9.2

form but are calculated from the residual mean square of the whole-plot (Block.WholePlot) stratum, estimated as s_w^2 = WPtMS, with replication expressed in terms of the number of subplots present with each level of factor A (as $m \times t_B$). The df associated with these SE are the ResDF from the whole-plot stratum. The estimated SE for prediction of an individual treatment combination, denoted $\hat{\mu}_{jk}$, takes a more complex form. These predictions are calculated as a mean of responses from subplots in different whole plots, and both the estimated whole-plot and subplot stratum variances, i.e. s_w^2 and s^2, contribute to the variance of this mean. The stratum variances are combined, taking account of the number of subplots per whole plot, as $(s_w^2 + (t_B - 1) \times s^2)/t_B$. This must then be divided by the replication of each treatment combination, m, before taking the square root to obtain the SE given in Table 9.13. The df for these SE take account of the contributions of the two stratum variances, using **Satterthwaite's formula** (Satterthwaite, 1946), as

$$\text{df} = \frac{(m-1)[s_w^2 + (t_B - 1)s^2]^2}{s_w^4/(t_A - 1) + (t_B - 1)s^4/t_A} . \tag{9.2}$$

This quantity lies between the smaller of the ResDF for the two strata, here $(m - 1) \times (t_A - 1)$, and the sum of the ResDF for the two strata, here $(m - 1) \times (t_A t_B - 1)$. It approaches its minimum value when the whole-plot residual mean square is very much larger than the subplot residual mean square, and takes the maximum value when $s_w^2 = (t_A - 1)s^2/t_A$, i.e. when the whole-plot residual mean square, WPtMS, is a specific proportion of the ResMS. These df will usually be non-integer: statistical software can calculate critical values for non-integer df, but if statistical tables are to be used, then the df should be rounded down to the nearest integer. Confidence intervals for predictions can be obtained from the SEs and their associated df in the usual manner.

The difference between two population treatment means is as usual estimated by the difference in the two respective sample treatment means. Again, calculation of the estimated SEs for treatment comparisons (SEDs) is more complex where the comparison involves contributions from more than one stratum. Comparisons across individual treatment combinations with the same level of treatment factor A, for example, $\mu_{jk} - \mu_{js}$ with $k \neq s$, are made entirely within the subplot stratum with the estimated SE calculated as

$$\widehat{SE}(\hat{\mu}_{jk} - \hat{\mu}_{js}) = \sqrt{2s^2/m}$$

and associated df equal to the subplot ResDF, $t_A \times (t_B - 1) \times (m - 1)$. Comparisons across different levels of treatment factor A, for example, $\mu_{jk} - \mu_{rs}$ with $j \neq r$, involve contributions from different whole plots and have their estimated SE calculated as

$$\widehat{SE}(\hat{\mu}_{jk} - \hat{\mu}_{rs}) = \sqrt{2[s_w^2 + (t_B - 1) \times s^2]/(t_B \times m)} ,$$

with associated df given by the Satterthwaite formula in Equation 9.2. This SE is valid whether the comparison is for the same level of treatment factor B ($k = s$) or for different levels ($k \neq s$). Comparisons of predictions for different levels of factor A averaged over all levels of factor B are made entirely within the whole-plot stratum, with estimated SE

$$\widehat{SE}(\hat{\mu}_{j\cdot} - \hat{\mu}_{r\cdot}) = \sqrt{2s_w^2/(t_B \times m)}$$

and associated df equal to the whole-plot ResDF, $(t_A - 1) \times (m - 1)$. Finally, comparisons of predictions for different levels of factor B averaged over all levels of factor A are made entirely within the subplot stratum, with estimated SE

$$\widehat{SE}(\hat{\mu}_{\cdot k} - \hat{\mu}_{\cdot s}) = \sqrt{2s^2/(t_A \times m)}$$

and associated df equal to the ResDF, $t_A \times (t_B - 1) \times (m - 1)$. As usual a t-statistic, LSD or $100(1 - \alpha_s)\%$ CI associated with the null hypothesis of no difference between the population means can be calculated from the appropriate SED and df.

EXAMPLE 9.5C: WEED COMPETITION EXPERIMENT

A statistically significant interaction between irrigation regime and weed species was found in the ANOVA table (Table 9.12) so the predictive model must use all of the explanatory terms, i.e. both main effects and the interaction. The predictions are listed in Table 9.14 and shown in Figure 9.1.

For this experiment we have $m = 4$, $t_A = 2$ and $t_B = 4$, with $s_w^2 = 1.5012$ and $s^2 = 0.2610$ (see Table 9.12). The estimated SEs for these predictions can therefore be calculated as

$$\widehat{SE}(\hat{\mu}_{jk}) = \sqrt{\frac{s_w^2 + (t_B - 1)s^2}{t_B \times m}} = \sqrt{\frac{1.5012 + (3 \times 0.2610)}{4 \times 4}} = \sqrt{\frac{2.2843}{16}} = 0.3778,$$

with associated df calculated from Equation 9.2 as

$$\frac{(m-1)[s_w^2 + (t_B - 1)s^2]^2}{s_w^4/(t_A - 1) + (t_B - 1)s^4/t_A} = \frac{3 \times [1.5012 + 3 \times 0.2610]^2}{1.5012^2 + \frac{3}{2} \times 0.2610^2} = \frac{3 \times 2.2843^2}{2.3558} = 6.64 \text{ df}.$$

SEDs between predictions for different species (labelled k and s) within the jth irrigation regime are estimated as

$$\widehat{SE}(\hat{\mu}_{jk} - \hat{\mu}_{js}) = \sqrt{2s^2/m} = \sqrt{2 \times 0.2612/4} = 0.3613,$$

TABLE 9.14

Predicted Grain Weight for All Combinations of Irrigation Regime and Weed Species, with Comparisons across Irrigation Regimes within Species (Example 9.5C)

| | Irrigation | | Difference |
Species	No	Yes	No − Yes
–	8.117	7.182	0.935
Am	3.485	2.710	0.775
Ga	6.680	4.075	2.605
Sm	6.595	5.575	1.020

Note: Prediction SE = 0.3778 with 6.64 df. SED for comparisons within irrigation regime = 0.3613 on 18 df, SED for comparisons across irrigation regimes = 0.5344 on 6.64 df.

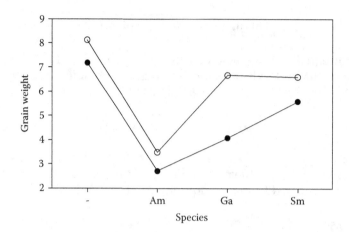

FIGURE 9.1
Predicted grain weight (t/ha) for each combination of species (*x*-axis; – = no weeds) and irrigation regime (o = no irrigation, ● = irrigation) in the weed competition experiment (Example 9.5C). SED for comparisons within each irrigation regime = 0.361 on 18 df, SED for comparisons across irrigation regimes = 0.534 on 6.64 df.

with 18 df, and SEDs for different irrigation regimes (labelled *j* and *r*) within the *k*th species are estimated as

$$\widehat{\text{SE}}(\hat{\mu}_{jk} - \hat{\mu}_{rk}) = \sqrt{\frac{2(s_w^2 + (t_B - 1)s^2)}{t_B \times m}} = \sqrt{\frac{2(1.5012 + 3 \times 0.2610)}{4 \times 4}} = \sqrt{\frac{2 \times 2.284}{16}} = 0.5344 \,,$$

with 6.64 df (from the calculation above). In this case, SEs for comparisons across irrigation levels are substantially larger and considerably less precise than comparisons within each irrigation regime, because of relatively large variation between whole plots and the small ResDF at that level.

To interpret the patterns of yield response, we are interested in comparing the effect of irrigation on the competitive effects of each species (four comparisons, shown in Table 9.14), and in comparing grain yields with each species present against the control in the absence of irrigation (another three comparisons). To allow for the number of tests (seven), we use a Bonferroni correction to the significance level (Section 8.8.1) giving adjusted significance level $\alpha_s^* = 0.05 / 7 = 0.007$. The critical value of the t-distribution at significance level $\alpha_s^*/2$ for 6.64 df is 3.830 and for 18 df is 3.034. The LSD for differences across species for no irrigation is then 1.0960 (= 3.034 × 0.3613) and the LSD for comparisons within species across irrigation regimes is 2.0457 (= 3.830 × 0.5344). It is clear that all of the weed species reduce grain yield in the absence of irrigation, with the reduction (competitive effect) being greatest for Am (black-grass). The application of irrigation further increases the competitive effects for Ga (cleavers) but has no real impact on yield in other cases.

9.2.4 Drawbacks and Variations of the Split-Plot Design

The most important criterion that determines whether a SP design should be used is the existence of practical or operational limitations on experimental units to which the proposed treatments can be applied. If the same experimental unit can reasonably be used for all factors and suitable (homogeneous) blocks are available, then the RCBD is almost always a better design because all treatment comparisons then have the same precision.

The main drawback of the SP design is in the small number of ResDF available within the whole-plot stratum, which limits the precision of any comparisons made across different levels of treatments applied to whole plots.

We have described one standard form of the SP design in which complete replicates of the design are arranged in separate blocks. There are several common variations on this design. For example, a SP structure may also be implemented without partitioning replicates into separate blocks. In this case, the design corresponds to a CRD at the whole-plot level, rather than a RCBD, with the whole plots split into subplots as in the standard design. This type of design may be useful if there is no evidence or expectation of heterogeneity between whole plots assigned to different replicate blocks, as omitting the blocks increases the ResDF in the whole-plot stratum by $m - 1$ df. The model for this design is similar to that in Section 9.2.1 but excludes the block effect (b_j), and the ANOVA table hence also omits the block stratum. We illustrate this situation in the following example.

EXAMPLE 9.6: TREE SEEDLING GROWTH*

The effects of temperature and soil substrate on growth of tree seedlings are of interest. A glasshouse containing six temperature-controlled beds, each of which can be set at only one temperature at a time, will be used for this experiment. Each bed can accommodate two trays of plants, and substrates can be applied to individual trays. Previous experiments have shown no evidence of coarse-scale spatial heterogeneity within the glasshouse, hence a SP design without blocks is appropriate, with beds as whole plots and trays within beds as subplots. A randomized allocation of three replicates each of two temperatures (15°C and 20°C) is made onto the six beds. One tray in each bed contains plants growing in a vermiculite-based substrate and the other contains plants growing in a chipped-wood-based substrate, using a randomized allocation within beds. A schematic layout for this experiment is set out in Table 9.15.

With an obvious nomenclature for factors, the form of the symbolic model is

Explanatory component: [1] + Substrate*Temperature
Structural component: Bed/Tray

A dummy ANOVA table for this design is shown in Table 9.16. The **Bed** stratum has four ResDF here, which is very low, but a blocked form of this design would have only two ResDF at this level. This therefore seems a more sensible design if there is no previous evidence of heterogeneity across beds within the glasshouse.

TABLE 9.15

Layout of Tree Seedling Growth Trial (Example 9.6)

	Tray 1	Tray 2
Bed 1	20°C/Chips	20°C/Vermiculite
Bed 2	20°C/Vermiculite	20°C/Chips
Bed 3	15°C/Chips	15°C/Vermiculite
Bed 4	15°C/Vermiculite	15°C/Chips
Bed 5	15°C/Chips	15°C/Vermiculite
Bed 6	20°C/Chips	20°C/Vermiculite

Note: Two temperatures (15°C or 20°C) each randomly allocated to three beds and two substrates (chips or vermiculite) randomly allocated to trays in each bed.

TABLE 9.16

Dummy Multi-Stratum ANOVA Table for Tree
Seedling Growth Trial (Example 9.6)

Source of Variation	df
Bed stratum	
Temperature	1
Residual	4
Bed.Tray stratum	
Substrate	1
Substrate.Temperature	1
Residual	4
Total	11

Note: Two temperatures (factor **Temperature**) were allocated randomly to beds and two substrates (factor **Substrate**) randomly to trays within beds.

Another common variant of the SP design occurs with CE cabinets (or rooms) or glass-house compartments where the whole-plot treatment is an environmental condition (e.g. temperature, humidity or CO_2 level). In this situation the cabinet is considered as the whole plot, but often the number of cabinets available is limited so that different replicate blocks comprise separate runs in different time periods using the same set of cabinets. To even out any bias associated with individual cabinets, a row–column design can be used for the allocation of treatments to cabinets across the runs (considering runs as the row blocking factor and cabinet as the column blocking factor). Ideally, the application of whole-plot treatments is balanced across the cabinets as well as across runs. If the design is orthogonal, for example, if a LS design is used at the whole-plot level, then variation associated with individual cabinets and individual runs can be separated from the remaining variation between whole plots, potentially leading to a more precise evaluation of the whole-plot treatment effects (if sufficient ResDF are present at that level). Example 9.7 illustrates this situation and shows the structure of the multi-stratum ANOVA table.

EXAMPLE 9.7: ENRICHED CO_2 TRIAL*

An experiment was conducted to investigate the variation in growth rates of varieties of spring wheat under ambient CO_2 compared with two richer CO_2 conditions in a set of three growth cabinets. Within each cabinet, pots of six separate varieties were grown. As differences were expected to be small, each CO_2 treatment was repeated six times, and the limited number of growth cabinets meant that this replication could be implemented only by repeating the experiment over time. The allocation of CO_2 levels to cabinets was assigned as two linked replicates of a 3×3 LS design (a Latin rectangle). A schematic layout for this experiment is shown in Table 9.17.

The structure of this experiment can be written as

Explanatory component:	[1] + CO$_2$*Variety
Structural component:	Square + Square.Run + Cabinet
	+ Square.Run.Cabinet + Square.Run.Cabinet.Pot

where factor **Square** (two levels) labels the two repeats of the LS part of the design and runs (factor **Run**, three levels) are labelled sequentially within each **Square**. The same cabinets are used in both squares and are expected to have a consistent effect.

TABLE 9.17

Design for Enriched CO_2 Trial, with CO_2 Treatments (Ambient or Two Levels of Enrichment) Allocated to Cabinets as Two Replicates of a 3×3 LS (Example 9.7)

		Cabinet 1	**Cabinet 2**	**Cabinet 3**
	Run 1	Enriched level 1	Ambient CO_2	Enriched level 2
Square 1	Run 2	Enriched level 2	Enriched level 1	Ambient CO_2
	Run 3	Ambient CO_2	Enriched level 2	Enriched level 1
	Run 1	Enriched level 2	Enriched level 1	Ambient CO_2
Square 2	Run 2	Enriched level 1	Ambient CO_2	Enriched level 2
	Run 3	Ambient CO_2	Enriched level 2	Enriched level 1

Note: Six varieties allocated to positions within each cabinet at random (not shown).

In the terminology of a SP design, the cabinets within each run (Square.Run.Cabinet term) correspond to the whole plots, and pots within cabinets correspond to subplots (term Square.Run.Cabinet.Pot). A dummy ANOVA table for this structure is shown in Table 9.18.

The CO_2 treatments were applied to the individual cabinets within each run of the experiment, and so the CO_2 sum of squares appears in the Square.Run.Cabinet stratum. Due to the repetition over time, this experiment has achieved eight ResDF in this stratum for testing the CO_2 main effects. The different varieties were applied to pots within cabinets; hence, the Variety and Variety.CO_2 sums of squares appear in the bottom stratum, which corresponds to the Square.Run.Cabinet.Pot combinations.

TABLE 9.18

Dummy Multi-Stratum ANOVA Table for Enriched CO_2 Trial with Three CO_2 Treatments (Factor CO_2, Applied to Cabinets) on Growth of Six Plant Varieties (Factor Variety, Applied to Pots) (Example 9.7)

Source of Variation	df
Square stratum	
Residual	1
Square.Run stratum	
Residual	4
Cabinet stratum	
Residual	2
Square.Run.Cabinet stratum	
CO_2	2
Residual	8
Square.Run.Cabinet.Pot stratum	
Variety	5
Variety.CO_2	10
Residual	75
Total	107

Note: The experiment used two replicate LSs (factor Square) each with three rows (factor Run) and columns (factor Cabinet) and with six pots in each cabinet in each run (factor Pot).

A third common extension to the standard SP design involves subdivision of the sub-plots into even smaller units (i.e. sub-subplots). This design, known as a **split-split-plot design**, corresponds to a blocking structure with four nested strata. In the simplest extension, a third treatment factor is applied to the sub-subplots, and the explanatory component becomes a three-way crossed structure. The analysis of data from such designs is a straightforward extension of the multi-stratum ANOVA table presented for SP designs in Section 9.2.2. In theory, any number of divisions and corresponding extra treatment factors may be included. However, the drawbacks of the standard SP design are amplified as new levels of subdivision are added, and so this type of design should be used only if the blocking structure matches real constraints on the experimental procedure. In addition, further complications arise from the calculations of SEs and SEDs because of the additional strata. Mead et al. (2012, Chapter 18) discuss variations on SP designs, including related designs such as the strip-plot or criss-cross design, in more detail.

Finally, multiple treatment factors can be included in one or more of the strata of a SP design. The analysis then involves partitioning the treatment variation between main effects and interactions corresponding to the underlying factors, following principles introduced in Chapter 8.

9.3 The Balanced Incomplete Block Design

In Section 3.3.5, the balanced incomplete block design (BIBD) was presented as a useful alternative to the RCBD for a situation in which the size of each homogeneous block is smaller than the number of treatments. As for the RCBD, the blocking structure associated with a BIBD consists of two nested strata, with the structural component written as

Structural component: Block/Unit
                        = Block + Block.Unit

using the symbolic names Block to label the blocks, and Unit to label the experimental units within blocks.

In general, an incomplete block design (IBD) is likely to be useful when either the number of treatments is very large, or the block size has to be very small. Some typical examples include

- *Variety trials.* Often many (> 100) varieties are grown within the same field trial. It is usually not possible to locate homogeneous blocks large enough to contain all varieties.

- *Two-colour microarray experiments.* In this experimental framework, two treatments – labelled with different dyes – can be applied to a microarray slide simultaneously. Where more than two treatments are investigated, the combinations of treatments applied to each slide should be carefully chosen because direct comparisons within the same slide are generally more reliable than indirect comparisons across different slides.

There are many classes of incomplete block design, but here we concentrate on the class of BIBDs. Some new notation is required to describe this type of design, and this is summarized in Table 9.19.

TABLE 9.19

Summary of Notation for Balanced Incomplete Block Designs (BIBD)

Symbol	Description
N	Total number of experiment units ($N = t \times n = m \times u$)
t	Number of treatments
n	Number of replicates of each treatment
m	Number of blocks
u	Number of units per block
λ	Number of times each treatment pair occurs together within a block

We still use t and n to denote the number of treatments and the number of replicates per treatment, respectively. The total number of experimental units is therefore equal to $N = n \times t$. These units will be arranged in a total of m incomplete blocks each consisting of u experimental units, where $u < t$, so the block size is smaller than the number of treatments. The total number of experimental units can then alternatively be written as $N = m \times u$. In a BIBD, each pair of treatments must occur together within blocks *exactly* the same number of times, denoted λ. This condition ensures that all treatment comparisons are evaluated with the same precision, as required by the definition of balanced designs.

EXAMPLE 9.8: GRAIN PROTEIN CONTENT*

In Example 3.8, we described an experiment to evaluate the grain protein content for six different varieties ($t = 6$), A to F, each with five replicates ($n = 5$). Protein content was measured during six sessions (blocks, $m = 6$), with five samples processed in each session ($u = 5$). Each treatment was omitted from just one of the sessions as shown in Table 4.6. In this design, each pair of treatments appears in only four of the six blocks. For example, both of treatments C and E appear in the first four blocks, and both of treatments B and D in the last four.

BIBDs cannot be constructed for every combination of treatment number, block size, and level of replication because of the requirement that all pairs of treatments must occur together within blocks the same number of times across the design. The following two relationships between the five design parameters must hold in order for a BIBD to exist (but do not guarantee that such a design does exist),

$$t \times n = m \times u,$$
$$n \times (u - 1) = (t - 1) \times \lambda.$$

The first relationship was introduced above. The second relationship concerns the number of within-block comparisons for each treatment. The left-hand side calculates the number of comparisons for a given treatment in terms of the number of blocks in which it appears (the number of replicates, n) multiplied by the number of comparisons in each block (one less than the number of units per block, $u - 1$). The right-hand side calculates this quantity using the number of times each treatment pair occur together within blocks (λ) multiplied by the number of other treatments ($t - 1$). Given the level of replication (n), number of treatments (t) and block size (u), the first relationship can be used to identify the number of blocks as $m = (n \times t)/u$. The second relationship can be rearranged as

$$\lambda = \frac{n(u-1)}{(t-1)} \, ,$$

to indicate the number of times each treatment pair occurs in a block together. As for all the other design parameters, λ must be a positive integer – any non-integer value indicates that a BIBD does not exist. In addition, Bailey (2008) quotes Fisher's inequality which states that, in a BIBD, the number of blocks must be greater than or equal to the number of treatments ($m \geq t$). Finally, a BIBD is called **resolvable** if the blocks of the design can be grouped into sets such that each set contains one replicate of each treatment. Obviously this can occur only when the number of treatments is exactly divisible by the number of units per block (i.e. t/u is an integer number).

Tables of BIBDs are given in some text books (e.g. Cochran and Cox, 1957; Fisher and Yates, 1963; Box et al., 1978), and some of these designs can be generated by statistical software.

EXAMPLE 9.9A: DESIGNING A BIBD EXPERIMENT FOR SEVEN TREATMENTS*

A scientist is interested in evaluating seven different treatments ($t = 7$), using blocks of size three or four ($u = 3$ or 4), with up to four replicates of each treatment ($n \leq 4$), giving a maximum of 28 units in total ($N \leq 28$). Because seven is a prime number, there are no resolvable BIBDs for this scenario. There are two possible BIBDs that fit these constraints, and both use seven blocks (the minimum possible number). The first design uses seven blocks of size three, 21 units in total, with three replicates of each treatment and each pair of treatments occurring together in just one of the blocks ($m = t = 7$, $u = n = 3$, $\lambda = 1$, see Table 9.20a). The second design uses seven blocks of size four, 28 units in total, with four replicates of each treatment and each pair of treatments occurring together in exactly two of the blocks ($m = t = 7$, $u = n = 4$, $\lambda = 2$, see Table 9.20b).

There is an obvious connection between these two designs: for each block in the first design, there is a corresponding block in the second design such that the pair of blocks contains the full set of treatments (e.g. block 1 in Table 9.20a and block 3 in Table 9.20b). Further information is required to make an informed decision on which design is more

TABLE 9.20

BIBDs for Seven Treatments in Seven Blocks with (a) Three Units per Block or (b) Four Units per Block (Example 9.9A)

	(a)			(b)			
	Unit 1	Unit 2	Unit 3	Unit 1	Unit 2	Unit 3	Unit 4
Block 1	1	5	7	4	1	2	7
Block 2	6	4	7	7	6	5	2
Block 3	2	5	4	3	6	2	4
Block 4	2	7	3	3	4	5	7
Block 5	4	3	1	6	5	1	4
Block 6	5	3	6	6	3	7	1
Block 7	2	6	1	3	5	1	2

appropriate for the experiment (see Example 9.9B), though the greater replication will generally provide more power, provided that the larger block size still contains homogenous units.

9.3.1 Defining the Model

The linear model associated with a BIBD takes the same form as for the RCBD (see Section 7.1). For a single treatment factor, the model can be written as

$$y_{ij} = \mu + b_i + \tau_j + e_{ij} ,$$

where y_{ij} represents the observation on the jth treatment in the ith block, μ the overall mean, b_i the effect of the ith block, τ_j the effect of the jth treatment and e_{ij} the deviation for this observation. Again, ideally we should label units within plots according to the experimental layout to maintain the distinction between the blocking and treatment structures, but for simplicity we omit this distinction. The subscript i runs from 1 to m, and the subscript j runs from 1 to t, note however that, because we have only $N = t \times n = m \times u$ units, not all combinations are present. In our usual symbolic notation, and with obvious definitions of the factors, the full description of the model is

Response variable:	Y
Explanatory component:	[1] + Treatment
Structural component:	Block/Unit

In this case the Block.Unit combinations label the full set of units and correspond to the model deviations.

Complexities arise in analysis of the BIBD because the block and treatment factors are not orthogonal, as only a subset of the full treatment set is present in each block. The simplest analysis, based on comparisons between treatments within blocks, and hence called the **within-block** or **intra-block analysis**, estimates treatment effects after adjustment for (elimination of) block effects. These treatment estimates take the form

$$\hat{\tau}_{j(w)} = \frac{1}{\text{EF}} \left(\bar{y}_{\cdot j} - \bar{B}_j \right) ,$$

where $\bar{y}_{\cdot j}$ is the sample mean for the jth treatment, and $\bar{B}_j$ the mean of all units in blocks that contain the jth treatment. For example, in the BIBD in Table 9.20a, treatment 3 occurs only in blocks 4, 5 and 6, so $\bar{B}_3$ would correspond to the average of all observations in those three blocks. Finally, EF is the **efficiency factor**, calculated as

$$\text{EF} = \frac{\lambda \times t}{n \times u} .$$

The EF is the proportion of the information on treatment differences available from the within-block analysis, with $0 \le \text{EF} \le 1$. The remainder of the information is available from comparisons across blocks, called the **between-block** or **inter-block analysis**. When the EF is less than 1, the inter-block estimates of treatment effects take the form

$$\hat{\tau}_{j(b)} = \frac{1}{(1 - \text{EF})}\left(\bar{B}_j - \bar{y}\right).$$

This estimate involves the jth treatment via $\bar{B}_j$, described above. The use of these intra- and inter-block estimates is discussed in Section 9.3.2.

9.3.2 Assessing the Importance of Individual Model Terms

The ANOVA table for the BIBD contains information on treatments, and therefore a sum of squares for treatments, in both the Block stratum (the between-block treatment sum of squares, BTrSS, corresponding to the inter-block estimates) and in the Block.Unit stratum (the within-block treatment sum of squares, WTrSS, corresponding to the intra-block estimates), see Table 9.21.

Within each stratum, $t - 1$ df are allocated for estimation of treatment effects and the ResDF are calculated by subtraction from the TotDF (which equal $m - 1$ df for the Block stratum and $N - m$ df for the Block.Unit stratum). As usual, the mean squares are obtained by division of each sum of squares by its df. The treatment mean squares are compared with the residual mean squares from the strata in which they occur, and the block mean square can be compared with the residual mean square from the Block.Unit stratum.

The immediate question on construction of this ANOVA table is: How do we reconcile the two separate variance ratios for treatments? If the EF is large (close to 1), then most of the treatment information lies within blocks. Since variation within blocks tends to be smaller (often much smaller) than variation between blocks, in these cases it makes sense to base inference on the within-block estimates of treatment effects. The population treatment means, in this case, are then estimated as

$$\hat{\mu}_{j(w)} = \bar{y} + \hat{\tau}_{j(w)} \quad \text{with} \quad \widehat{\text{SE}}(\hat{\mu}_{j(w)}) = \sqrt{(2 \times s^2)/(n \times \text{EF})} \,,$$

where, as usual, $s^2 = \text{ResMS}$ is the estimate of background variability at the lowest stratum. If the EF is small, then we might lose substantial information by ignoring treatment comparisons between blocks, and both sources of treatment information should be used.

TABLE 9.21

Structure of the Multi-Stratum ANOVA Table for a BIBD with m Blocks (Factor Block), u Units per Block (Factor Unit), t Treatments (Factor Treatment) and $N = m \times u$ Units in Total

Source of Variation	df	Sum of Squares	Mean Square	Variance Ratio
Block stratum				
Treatment	$t - 1$	BTrSS	BTrMS	BTrMS/BlkMS
Residual	$m - t$	BlkSS	BlkMS	BlkMS/ResMS
Block.Unit stratum				
Treatment	$t - 1$	WTrSS	WTrMS	WTrMS/ResMS
Residual	$N - m - t + 1$	ResSS	ResMS	
Total	$N - 1$	TotSS		

However, the two variance ratios may give contradictory indications of the importance of treatment effects, and in cases when $t = m$ (as in Example 9.9) the ResDF for the Block stratum will be zero, so that the variance ratio for treatments cannot be formed within that stratum. These problems can be solved by combination of information from the two strata to give a single estimate of the treatment effects, a single test statistic for the treatment terms and a revised estimate of the stratum variances. This **combined estimate** is a weighted mean of the two components, with the weighting determined by the (revised) stratum variances and their estimated df, and is most easily obtained from linear mixed models (see Chapter 16), although some implementations of the multi-stratum ANOVA (e.g. GenStat) can also provide these estimates.

EXAMPLE 9.9B: DESIGNING A BIBD EXPERIMENT FOR SEVEN TREATMENTS*

We now have more information to assess the proposed designs of Example 9.9A. The first design (Table 9.20a) has seven blocks of three units for seven treatments each with three replicates ($N = 21$, $t = m = 7$, $n = u = 3$), with each pair of treatments appearing together only in one of the blocks ($\lambda = 1$). Its EF is therefore $(\lambda \times t)/(n \times u) = 7/9 = 0.778$, so almost 78% of the treatment information is available within blocks. The SE for treatment comparisons based on the within-block estimates takes the form

$$\widehat{SE}(\hat{\mu}_{j(w)} - \hat{\mu}_{k(w)}) = \sqrt{(2 \times s^2)/(n \times EF)} = \sqrt{0.857 \times s^2} = 0.93s,$$

with $N - m - t + 1 = 21 - 7 - 7 + 1 = 8$ ResDF.

The second design (Table 9.20b) has seven blocks of size four for seven treatments each with four replicates ($N = 28$, $t = m = 7$, $n = u = 4$), with each pair of treatments appearing together in two of the blocks ($\lambda = 2$). Its EF is therefore $(\lambda \times t)/(n \times u) = 7/8 = 0.875$, so almost 88% of the treatment information is available within blocks. The SED for treatment comparisons takes the form

$$\widehat{SE}(\hat{\mu}_{j(w)} - \hat{\mu}_{k(w)}) = \sqrt{(2 \times s^2)/(n \times EF)} = \sqrt{0.571 \times s^2} = 0.76s$$

with $N - m - t + 1 = 28 - 7 - 7 + 1 = 15$ ResDF.

So, for a 33% increase in the number of units (from 21 to 28) we get a 13% increase in the proportion of information available within blocks (from 0.78 to 0.88) and an 18% reduction in the SE for treatment comparisons (if we assume that s would be similar across the two experiments). In addition, the ResDF has almost doubled from a value that is barely adequate (ResDF = 8) to a value (ResDF = 15) that is likely to give a reasonable estimate of the background variation. Given that the original experimental outline allowed for 28 pots, this gives several good reasons to choose the larger experiment (Table 9.20b).

9.3.3 Drawbacks and Variations of the Balanced Incomplete Block Design

The BIBD is usually a good design when the number of treatments is only a little larger than the number of units within each block. The main drawback of these designs is that a BIBD might not exist for any given combination of treatment number, replication level and block size. Even when a BIBD design does exist, it might require many more replicates of each treatment than is practicable (recall that the number of blocks must be at least equal to the number of treatments). The explanatory structure can be extended to accommodate factorial and other structures (see Chapter 8), although requirements of balance across individual model terms then impose an even greater restriction on the available designs. For these reasons, several different classes of **partially balanced incomplete block designs** have been developed, dividing the treatment pairs into two or more groups. Treatment pairs within each group then have a different value of λ, so that different groups of treatment comparisons are estimated with different levels of precision. These designs can relax some of the practical constraints imposed by BIBDs whilst retaining some advantages of a balanced design. Some classes of partially balanced incomplete block designs can be analysed with algorithms for multi-stratum ANOVA, but most can be analysed only with more general algorithms, such as those associated with linear mixed models (see Section 11.6 and Chapter 16). Issues of balance and orthogonality associated with these and other forms of design are discussed further in Chapter 11. Partially balanced incomplete block designs and unbalanced designs are discussed by Mead et al. (2012, Chapters 7 and 9).

EXERCISES

9.1 A 5×5 LS design was used to investigate the effect of sulphur fertilizer on the yield (tonnes/ha) of spring barley grown on a light soil. Five levels of fertilizer were applied (0, 10, 20, 30 and 40 kg S). File SULPHUR.DAT contains the plot numbers (*Plot*), structural factors (Row, Col), the treatment factor (Sulphur) and the grain yield (variate *Grain*). Write down the full model for the yields in both mathematical and symbolic form. Analyse the data and state your conclusions. What other hypotheses might you like to test? (We re-visit these data in Exercise 17.4.)[*]

9.2 An experiment used three incubators to compare growth of fungal colonies of *Metarhizium anisopliae* at three temperatures (23°C, 30°C and 35°C; Wright, 2013). Replication of temperatures was achieved by repeating the experiment on three occasions. Temperatures were allocated to incubators according to a 3×3 LS, so each incubator ran once at each temperature. Small fungal plugs were placed in Petri dishes and three dishes were placed in each incubator on each occasion. The sizes of the fungal colonies were recorded after four days. The dish numbers (*ID*), structural factors (Incubator, Occasion and Dish), explanatory factor (Temperature) and size measurements (variate *Size*) are given in file SIZE.DAT. Write down the structural component of the model for the colony sizes. Analyse the data and state your conclusions. What can you say about the effect of temperature on the growth of these fungal colonies?[†]

9.3 A three-year field trial was set up to investigate the susceptibility of six varieties of lily to the lily beetle, *L. lilii* (Salisbury et al., 2010). The trial was laid out as two independent 6×6 LSs. Regular counts of beetle adults, eggs and larvae were made between May and early August each year. The file LILY.DAT contains the unit numbers (*ID*), structural factors (Square, Row, Column), explanatory factor (Variety) and the total count of larvae observed during 2006 (variate *Larvae*). Analyse these data on an appropriate scale. Are these lily varieties equally susceptible?[‡]

---

[*] Data from S. McGrath, Rothamsted Research.
[†] Data from E. Wright, Rothamsted Research.
[‡] Data from A. Salisbury, Royal Horticultural Society/Rothamsted Research/Imperial College London.

9.4 A series of field experiments tested various 'push-pull' strategies to control insect pests in oilseed rape. In one experiment the use of turnip rape (TR) as an earlier flowering trap crop (the 'pull') was tested alongside use of an anti-feedant applied to oilseed rape in spring (S; the 'push'). Untreated oilseed rape (U) was included as a control. The experiment was set up as a 6 × 6 LS with two replicates of each of the three treatments per row and column. An assessment of adult pollen beetle numbers was made on 10 plants per plot in early April, one day post-spray of the anti-feedant. The unit numbers (*ID*), structural factors (Row, Column), treatment factor (Treatment) and mean pollen beetle count per plot (variate *Count*) are given in file POLLENBEETLES.DAT. Is there any evidence that either of the pull or push strategies works?[*]

9.5*A field experiment investigated the effect of four herbicides (A, B, C, D) on the yield of three varieties of onions (V1, V2, V3). The herbicides could only be applied to relatively large areas of land due to the width of the spray boom, so the experiment was set up as a SP design with three blocks of four main plots to which the herbicides were applied. Each main plot comprised three subplots to which the varieties were allocated. The final yield of onions per subplot was recorded at harvest. The file ONIONS.DAT contains the unit numbers (*ID*), structural factors (Block, MainPlot, Subplot), explanatory factors (Herbicide, Variety) and final yields (variate *Yield*). Write down the structural and explanatory components of the model for the onion yields. Analyse these data and summarize your conclusions.

9.6 A field experiment studied forms and rates of nutrient application and the effect on the yield of spring barley in the presence or absence of foliar diseases. Nitrogen fertilizer was applied either in a liquid form, alone (L) or with a nitrification inhibitor added (LI), or in a solid form, to the seedbed (SS) as a top-dressing (ST) or split (half to the seedbed and half as top-dressing, SST). Each form was applied at two rates (70 and 110 kg N/ha), giving 10 nutrient treatments in total. The occurrence of foliar diseases was intended to be manipulated by a 2 × 2 factorial in the presence or absence of a mildew fungicide (None, Tridemorph) and a rust fungicide, but no rust developed and so the latter fungicide was not applied. The trial used a SP design with two blocks. The 10 nutrient treatments were applied to main plots, each of which was split into four subplots, and the mildew fungicide was applied to two subplots in each main plot. The plot numbers (*ID*), structural factors (Block, MainPlot, Subplot), explanatory factors (NForm, NRate, MildewF) and yield at harvest (variate *Yield*, tonnes/hectare at 85% dry matter) are in file SPRINGBARLEY.DAT. Identify a suitable predictive model and comment on the comparison between liquid and solid forms of fertilizer.[†]

9.7 A field experiment compared the effects of three strains of barley yellow dwarf virus (BYDV, a virus transmitted by aphids) on yield of two varieties of winter barley, one (Vixen) with genetic resistance to BYDV and the other susceptible (Igri). The experiment aimed to test the efficacy of any resistance, its consistency across the strains, and the effectiveness of insecticide sprays at different times of the year (Cypermethrin in October or December, or Pirimicarb in March). A SP design was used with five blocks of six main plots each split into four

---

[*] Data from L. Smart, Rothamsted Research.
[†] Data from J. Jenkyn, Rothamsted Research.

subplots. The six combinations of variety and spray timing were applied to main plots within blocks. The virus strains (MAV, PAV or RPV) were applied to subplots by releasing infected aphids in the centre of the subplot; one subplot was left uninoculated in each main plot. The plot numbers (*ID*), structural factors (Block, MainPlot, Subplot), explanatory factors (Variety, Spray, Strain) and yield at harvest (variate *Yield*, tonnes/hectare at 85% dry matter) are in file BYDV.DAT. Analyse these data and relate your conclusions to the experimental aims stated above.[*]

9.8 An experiment assessed the response of two aphid clones to a foliar insecticide applied to cabbage plants. The experiment used two simulators, each containing six plants in individual pots. All plants in one simulator were sprayed with the insecticide and all plants in the other were sprayed with water only (control). Two weeks after spraying adult aphids were placed onto the plants using clip cages. Two clip cages were attached to each plant, one containing three aphids of a clone susceptible to the insecticide, the other containing three aphids of a moderately resistant clone. The number of nymphs produced by the adults in each clip cage was recorded after two days. The experiment was then repeated using two new simulators. File SIMULATOR.DAT contains the unit numbers (*ID*), structural factors (Expt, DSimulator, Plant, DCage), explanatory factors (Treatment, Clone) and the nymph counts (variate *Nymphs*). Determine the structural and explanatory components for this experiment, write down the full model in symbolic form and state the experimental units for the insecticide and clone treatments. Analyse the data and verify that the explanatory terms are tested in the correct strata. Identify and interpret the predictive model. (We re-visit these data in Exercise 16.1.)[†]

9.9 An experiment to compare yields of 13 varieties of corn was set up as a BIBD with 13 blocks, each containing four plots (Cochran and Cox, 1957, Table 11.2). File CORN.DAT contains the unit numbers (*ID*), structural factors (Block, Plot), explanatory factor (Variety) and plot yields (variate *Yield*, pounds per plot). Calculate λ and the efficiency factor for this design. Is the design resolvable? Is there any evidence of differences in yield among the varieties?

---

[*] Data from R. Plumb, Rothamsted Research.
[†] Data from S. Foster, Rothamsted Research.

10

Replication and Power

In Chapter 3, we examined the principles of replication, randomization and blocking that are central to the construction of efficient designs. However, in doing so we did not say how to choose the number of replicates to be used for each treatment. The question 'How many replicates do I need?' is probably the most common question posed to consultant statisticians, but the answer is rarely obvious! As the number of replicates increases, smaller differences among a set of treatments can be detected because more information becomes available. Conversely, with few replicates or large background variation, or both, we might not detect differences between treatment population means as statistically significant even if some of those differences are large and biologically meaningful. In general, the replication required in a study depends on numerous (possibly competing) features, such as

- The available resources (money, experimental material, space and time)
- Treatment structure
- Size of treatment difference(s) to be detected
- Relative importance of different treatment comparisons
- Risks associated with wrong decisions (false-positive or false-negative results)
- Variability associated with the experimental units and measurement process

The first five items in this list are usually either a matter of choice or restricted by practical considerations. The risks of making wrong decisions can be related to the ideas of hypothesis testing introduced in Section 2.3.2. However, the last item, namely the variation in the data, which we have called background variation and denoted σ^2, is not under the control of the experimenter, is inherent to the process under study and is often unknown.

In this chapter, we discuss how to determine the number of replicates to be included in an experiment. To assess the required replication we must specify the minimum size of a true treatment difference (i.e. the difference between population means for two treatments) that should be detected as statistically significant (for a given significance level α_s). First, we describe some simple approximate methods to determine the number of replicates required for an experiment, based on the required size of treatment difference and the estimated LSD (Section 10.1). These methods illustrate the importance of obtaining a good estimate of the background variation both before the experiment and within the analysis (Section 10.2). The important concept of the power of a design, which gives the probability of detecting a treatment difference of a given size, is then introduced (Section 10.3). An example is used to illustrate these ideas for a particular scenario (Section 10.4). Finally, the usual null hypothesis (of no treatment differences) is not useful when the purpose of an experiment is to illustrate the equivalence of, rather than the difference between, treatments. In this case an alternative strategy of two one-sided t-tests (TOST) is often used to give a more powerful test (Section 10.5).

10.1 Simple Methods for Determining Replication

In this section, we suppose that we wish to detect an observed difference of a given size between two treatments. We denote this observed difference as d to distinguish it from the true (but unknown) difference between the treatment population means, which we denote as δ. In the designs considered thus far, the least significant difference (LSD) (Section 4.4) has been defined as the smallest observed difference between two treatments that will be detected as statistically significant at a specified significance level. In this section, we determine replication in terms of the LSD, directly in Section 10.1.1 and indirectly, via the coefficient of variation, in Section 10.1.2. Both of these cases are illustrated for the CRD with equal replication, with extensions to other designs given in Section 10.1.3.

10.1.1 Calculations Based on the LSD

Initially, we focus on the CRD and consider an experiment with equal replication (n) for each of t treatments using a total of $N = n \times t$ experimental units. In Section 4.4, the LSD between two treatments in a CRD with equal replication was derived as

$$\text{LSD} = t_{N-t}^{[\alpha_s/2]} \times \text{SED} = t_{N-t}^{[\alpha_s/2]} \times \sqrt{\frac{2s^2}{n}}, \tag{10.1}$$

where $t_{N-t}^{[\alpha_s/2]}$ is the $100(1 - \alpha_s/2)$th percentile for the t-distribution with $N - t$ df (the residual df, ResDF, for the CRD). However, in making calculations prior to experimentation, the estimate s^2 is not available. Unfortunately, obtaining a realistic pre-experiment estimate of s^2 is often difficult, and we discuss strategies for overcoming this problem in Section 10.2. For now, we assume that an appropriate value is available.

The LSD indicates the size of estimated (or observed) treatment differences that should be detected as statistically significant (at significance level α_s) by ANOVA. If we wish to detect an observed difference d between two treatments as significant, then it follows that we want LSD $\leq d$ or, from Equation 10.1,

$$t_{\text{ResDF}}^{[\alpha_s/2]} \times \sqrt{\frac{2s^2}{n}} \leq d. \tag{10.2}$$

We can systematically evaluate the left-hand side of this inequality for increasing values of n: the required replication is the smallest value of n for which this inequality is satisfied.

> **EXAMPLE 10.1A: SAMPLE SIZE CALCULATIONS FOR A NEW CALCIUM POT TRIAL***
>
> A scientist is planning a follow-up experiment to the calcium pot trial presented in Example 4.1 to confirm these results. This new experiment will again be a CRD with $t = 4$ treatments and, based on the previous experiment, s^2 is expected to be approximately 75 ($s = 8.66$). Observed treatment differences of $d = 10$ cm are required to be detected as statistically significant with $\alpha_s = 0.05$. By systematic evaluation of the LSD for different values of n, as shown in Table 10.1, we find that a replication of 7 is the smallest value that gives a value of the LSD less than 10.

TABLE 10.1

Calculation of SED and LSD for a CRD with $t = 4$ Treatments, Varying Replication (n) and Estimated Residual Variance $s^2 = 75$ (Example 10.1A)

Replication (n)	Units ($N = n \times t$)	Residual df ($N - t$)	$t_{N-t}^{[0.025]}$	SED	LSD
2	8	4	2.776	8.66	24.04
3	12	8	2.306	7.07	16.31
4	16	12	2.179	6.12	13.34
5	20	16	2.120	5.48	11.61
6	24	20	2.086	5.00	10.43
7	28	24	2.064	4.63	9.55

Since the true value of the background variation is unknown, it is sensible to verify the impact of a range of values of s^2. For example, here we might realistically expect s^2 to lie between 50 and 100. As shown in Table 10.1, a CRD experiment with $n = 7$ replicates of each of $t = 4$ treatments has a 5% critical t-value of $t_{N-t}^{[\alpha_s/2]} = t_{24}^{[0.025]} = 2.064$. The possible range of the SED is then calculated as

$$\text{minimum(SED)} = \sqrt{(2 \times 50/7)} = 3.78 \, ,$$

$$\text{maximum(SED)} = \sqrt{(2 \times 100/7)} = 5.35 \, ,$$

and the corresponding LSD values are

$$\text{minimum(LSD)} = t_{N-t}^{[\alpha_s/2]} \times \text{minimum(SED)} = 2.064 \times 3.78 = 7.79 \, ,$$

$$\text{maximum(LSD)} = t_{N-t}^{[\alpha_s/2]} \times \text{maximum(SED)} = 2.064 \times 5.35 = 11.02 \, .$$

So, with seven replicates for each treatment, we might detect observed treatment differences in the range 7.8–11.0 cm. If this worst-case scenario is unacceptable then we might consider further increasing the replication, or take additional measures to reduce background variation (if this is possible).

If the replication required exceeds the resources available then some compromise must be found. For example, some treatments might be eliminated to enable increased replication of the remaining treatments, or reduced precision might be accepted. This is discussed further at the end of Section 10.3.

10.1.2 Calculations Based on the Coefficient of Variation

The coefficient of variation (%CV) for a sample is defined as

$$\%CV = 100 \times s/\bar{y} \, ,$$

where s is the unbiased sample standard deviation and $\bar{y}$ the sample mean (Section 2.1). The %CV can be a useful measure for evaluating the quality of experiments where the background variation increases with the mean, as the %CV is often quite stable for successful experiments. An increase in %CV then indicates an unexpectedly large value of background variation and hence some problem with the trial.

The LSD can be rewritten, in terms of the %CV and sample mean, by multiplying both the numerator and denominator of Equation 10.1 by 100, and dividing both by the sample mean to obtain

$$\text{LSD} = t_{N-t}^{[\alpha_s/2]} \times \sqrt{\frac{2}{n}} \times s = t_{N-t}^{[\alpha_s/2]} \times \sqrt{\frac{2}{n}} \times \frac{100s/\bar{y}}{100/\bar{y}} = \left(t_{N-t}^{[\alpha_s/2]} \times \sqrt{\frac{2}{n}} \times \frac{\%\text{CV}}{100} \right) \times \bar{y} \, .$$

The LSD can thus be evaluated as a proportion of the mean for different levels of replication. In this form, a suitable estimate of %CV rather than s^2 is required, so if acceptable ranges of %CV are well established, this may provide a more useable approach to the calculation of an appropriate replication, as illustrated in Example 10.1B.

EXAMPLE 10.1B: SAMPLE SIZE CALCULATIONS FOR A NEW CALCIUM POT TRIAL*

The %CV for the calcium pot trial of Example 4.1 was 14% and experience of similar experiments suggests that the %CV should be at worst 20%. Example 10.1A suggested that a follow-up experiment should have replication $n = 7$. The LSD between two treatment means each with seven replicates is estimated for %CV = 14 by

$$\text{LSD} = \left(t_{24}^{[0.025]} \times \sqrt{\frac{2}{n}} \times \frac{\%\text{CV}}{100} \right) \times \bar{y} = \left(2.064 \times \sqrt{\frac{2}{7}} \times \frac{14}{100} \right) \times \bar{y} = 0.15\bar{y} \, ,$$

and for %CV = 20 by

$$\text{LSD} = \left(t_{24}^{[0.025]} \times \sqrt{\frac{2}{n}} \times \frac{\%\text{CV}}{100} \right) \times \bar{y} = \left(2.064 \times \sqrt{\frac{2}{7}} \times \frac{20}{100} \right) \times \bar{y} = 0.22\bar{y} \, .$$

Hence, in the worst case, we expect to detect any observed difference between two treatments that is larger than 22% of the overall mean response as statistically significant (at the 5% level). If the new experiment is as precise as the previous experiment, with %CV = 14, this decreases to 15% of the overall mean response.

10.1.3 Unequal Replication and Models with Blocking

In the calculations above we assumed the simplest experimental design of a CRD with equal replication. In general, a more complex design might be used, perhaps with unequal replication. Calculations for the case of unequal replication follow directly from the formula for the SED between two treatments with replication n_i and n_j, respectively, i.e.

$$\text{SED} = \sqrt{s^2 \left(\frac{1}{n_i} + \frac{1}{n_j} \right)} \, .$$

The extension to more complex designs is similarly straightforward, requiring only that the appropriate form of the SED is substituted into Equations 10.1 and 10.2, and that the appropriate ResDF are applied to obtain the critical value of the t-distribution, which can be expressed in more general form as $t_{\text{ResDF}}^{[\alpha_s/2]}$. For example, for a RCBD with n blocks and t treatments, the residual df must be adjusted to ResDF = $(n-1) \times (t-1)$ (see Section 7.3), whilst for a LS design with t treatments the residual df must be adjusted to ResDF $= (t - 1) \times (t - 2)$ (see Section 9.1).

10.2 Estimating the Background Variation

The methods presented in Section 10.1 require a plausible value (or range of values) for the estimated background variation, s^2, to be available before the experiment is done. In some cases, the %CV can be a useful alternative, but this is often not available, and so a strategy to obtain a 'reasonable' estimate of s^2 (or the %CV) is required.

The simplest option is to obtain estimates of variation from previous studies that used similar experimental units under similar conditions. Ideally, experiments from the same institution (or laboratory) should be used, as long as the study conditions and protocols are analogous. Another, albeit more expensive, alternative is to do a preliminary (or pilot) study, using a subset of the proposed experimental treatments to establish the size and sources of variability (see e.g. Case Study 19.1). Such preliminary studies are often used in laboratory work to calibrate new experimental techniques. Published reports or papers describing similar experiments are another possible source of information, but these may provide less reliable estimates if insufficient detail is given or if the experimental conditions are different.

If none of these options is available then a mixture of common sense and good guesswork is required. If the expected range of values (for a single treatment) is known, and these observations are expected to follow a Normal distribution, then the properties of this distribution can be used. It is well known that 95% of the observations from a Normal distribution are found within approximately two standard deviations of the population mean (Figure 2.4). Therefore, if the likely minimum and maximum values for experimental units receiving the same treatment can be predicted then the population standard deviation σ can be *approximated* as

$$\sigma \approx \frac{\text{maximum} - \text{minimum}}{4},$$

and this can be substituted for the estimate s. If there is much uncertainty about the likely variation then consideration of a range of possible values may be helpful (as in Example 10.1A).

When we do an experiment, we obtain a new estimate of the background variation and the precision of that estimate increases as the residual df increases, reflecting the amount of information available. For this same reason, the critical value $t_{\text{ResDF}}^{[\alpha_s/2]}$ decreases as the residual df increases. As a rule of thumb, to ensure a reasonable estimate of the background variation the replication should be sufficiently large to give at least 10 residual df. As the gain in precision decreases as the residual df increases further, there is usually little advantage in having more than 20 residual df (see Chapter 19).

10.3 Assessing the Power of a Design

The calculations in Section 10.1 used the estimated LSD to assess whether an experiment would detect an observed treatment difference of a given size (denoted d). In practice, we are more interested in the true treatment difference (δ), which cannot be observed directly and is estimated with error. Because of the stochastic nature of this error, we can use probability calculations to evaluate whether a true treatment difference of size δ is likely to be detected. The probability that a true treatment difference of size δ will be detected as statistically significant is called the **power** of the test, and hence of the design. The power is a function of the size of the treatment difference δ.

For any statistical test, the **significance level** and power are related to errors of inference that may occur when a given hypothesis is tested. The terminology associated with these inferential errors and the related probabilities is summarized in Table 10.2.

A Type I error occurs when H_0 is rejected when it is true, i.e. a false-positive conclusion, for example, that the population means differ when in fact they are equal. The probability of a Type I error occurring is denoted α_s, i.e. **Prob(Type I error)** = α_s. As mentioned previously (Section 2.3.2), α_s is the pre-determined significance level (or size) of a test, and is often chosen to be 0.05.

A **Type II error** occurs when H_0 is not rejected when it is false, i.e. a false-negative conclusion, for example, that the population means are equal when in fact they differ. The probability of a Type II error occurring is denoted β_s, i.e. **Prob(Type II error)** = β_s. The **power of a test** is directly associated with the Type II error rate and is defined as the probability of making the correct decision to reject H_0 when H_0 is false, so power = $1 - \beta_s$. Tests with large power (and therefore small β_s) are preferred, as they give a larger chance of detecting treatment differences for a given design. However, there is a relationship between the Type I and Type II error rates (α_s and β_s) that usually makes some compromise on either significance or power inevitable.

To demonstrate these concepts we consider a test concerning population means, μ_1 and μ_2, for two equally replicated treatments assessed in an experiment with a CRD. The null hypothesis of no difference, H_0: $\mu_1 = \mu_2$ or H_0: $\delta = \mu_1 - \mu_2 = 0$, is to be tested against the one-sided alternative hypothesis that the population mean of the first group is larger, H_1: $\mu_1 > \mu_2$ or H_1: $\delta = \mu_1 - \mu_2 > 0$. We use a one-sided test here for simplicity, but the same concepts extend to the two-sided case. We make the usual assumptions about the deviations (see Section 4.1). Figure 10.1 illustrates the relationship between the significance level and the power of the t-test for assessing these hypotheses as the value of the true treatment difference, $\delta = \mu_1 - \mu_2$, varies. The curves represent the sampling distributions of the observed test statistic, i.e. the random variable t = $(\hat{\mu}_1 - \hat{\mu}_2)/\text{SED}$, in different situations.

In both graphs, the left-hand curve represents the situation when the null hypothesis is true ($\delta = 0$). This is a t-distribution with mean zero and df equal to the ResDF used to estimate the SED (in this figure, ResDF = 24). The right-hand curve shows the distribution of the observed t-statistic under the alternative hypothesis. In Figure 10.1a, the true difference is

TABLE 10.2

Terminology for Inferential Errors and Probabilities Associated with a Hypothesis Test

		Decision (Probability)	
		Accept H_0	Reject H_0
Null hypothesis (H_0)	True	Correct decision $(1 - \alpha_s)$	Incorrect decision (Type I error, α_s)
	False	Incorrect decision (Type II error, β_s)	Correct decision (Power, $1 - \beta_s$)

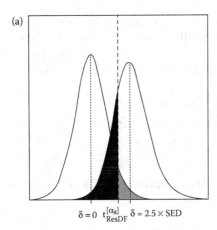
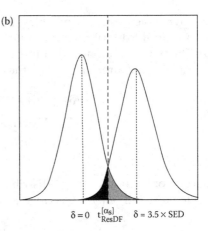

$$\delta = 0 \quad t^{[\alpha_s]}_{ResDF} \quad \delta = 2.5 \times SED \qquad\qquad \delta = 0 \quad t^{[\alpha_s]}_{ResDF} \quad \delta = 3.5 \times SED$$

FIGURE 10.1

Definition of Type I (grey area, α_s) and Type II (black area, β_s) error probabilities for the t-test of the null hypothesis of no treatment differences (H_0: $\delta = 0$) against a one-sided alternative hypothesis (H_1: $\delta > 0$) for the difference δ between two treatment population means: (a) $\delta = 2.5 \times SED$, (b) $\delta = 3.5 \times SED$. $t^{[\alpha_s]}_{ResDF}$ denotes the $100(1 - \alpha_s)$th percentile of a t-distribution with ResDF df.

$\delta = 2.5 \times SED$, and in Figure 10.1b, the difference is slightly larger, with $\delta = 3.5 \times SED$. These distributions are non-central t-distributions with non-centrality parameter δ/SED and ResDF df (see Section 2.2.4). The dotted vertical lines mark the median of each distribution. The dashed vertical line shows the critical value for the t-test, equal to the $100(1 - \alpha_s)$th percentile of the t-distribution with ResDF df, i.e. $t^{[\alpha_s]}_{ResDF}$. Any observed difference between the treatment means that is greater than this critical value (to the right of the dashed line) is declared as significantly different from zero, and any observed difference smaller than this critical value (to the left of the dotted line) is declared as not significantly different from zero. The grey-shaded area corresponds to the rejection region of size α_s, and the black-shaded area corresponds to the Type II error of size β_s in each case. The power is equal to $1 - \beta_s$ (i.e. the non-shaded area of the right-hand distribution). The grey area stays the same size whatever the true value of the difference, δ, whereas the size of the black area changes as δ changes: as δ increases, the black area (β_s) decreases, and the power increases; as δ decreases, the black area (β_s) increases and the power decreases. The **power function of a test** expresses the power as a function of δ. In practice, it is often easiest to state the size of difference δ that the test is required to detect and to calculate the corresponding power.

Figure 10.1 also indicates how characteristics of the test influence power. For example, increasing the significance level (i.e. decreasing α_s, e.g. $\alpha_s = 0.01$ instead of $\alpha_s = 0.05$), shifts the critical value (dashed line) to the right and increases the black area (β_s), thus reducing the power. A common compromise is to aim for a test with $\alpha_s = 0.05$ and $\beta_s = 0.20$ (power = 0.80) for a given treatment difference. A decrease in the SED, through a decrease in background variation, increase in replication or increase in the ResDF, makes the two distributions narrower so that their overlap decreases and hence the power increases.

For our example of a treatment comparison in a CRD, the calculations are straightforward. We calculate power in terms of the distribution of the test statistic under the alternative hypothesis, which is a non-central t-distribution with non-centrality parameter δ/SED and df equal to the ResDF (defined in Section 2.2.4). Probability functions for these distributions are present in most statistical software, although not commonly available in statistical tables. We denote the cumulative distribution function for the non-central t-distribution with non-centrality parameter c on D df at value x as $F_t(x, c, D)$. The power of a one-sided test with H_1: $\delta > 0$ is the probability of rejecting the null hypothesis if the alternative hypothesis is true, and can be calculated as

$$Power(\delta) = Prob(H_0 \text{ rejected} \mid \mu_1 - \mu_2 = \delta)$$
$$= Prob(t > t_{ResDF}^{[\alpha_s]} \mid \mu_1 - \mu_2 = \delta)$$
$$= 1 - F_t(t_{ResDF}^{[\alpha_s]}, \delta/SED, ResDF),$$

i.e. the portion of the non-central t-distribution that exceeds the critical value. The power of a one-sided test with H_1: $\delta < 0$ is calculated similarly as

$$Power(\delta) = F_t(-t_{ResDF}^{[\alpha_s]}, \delta/SED, ResDF),$$

i.e. the portion of the non-central t-distribution that is less than the critical value. This calculation uses the symmetry of the t-distribution to derive $t_{ResDF}^{[1-\alpha_s]} = -t_{ResDF}^{[\alpha_s]}$. Not surprisingly, the power of a two-sided test, with H_1: $\delta \neq 0$, combines these two expressions, having adjusted the critical value, and is calculated as

$$Power(\delta) = 1 - F_t(t_{ResDF}^{[\alpha_s/2]}, \delta/SED, ResDF) + F_t(-t_{ResDF}^{[\alpha_s/2]}, \delta/SED, ResDF),$$

i.e. the portion of the non-central t-distribution that lies outside of the two critical values.

EXAMPLE 10.1C: SAMPLE SIZE CALCULATIONS FOR A NEW CALCIUM POT TRIAL*

In Example 10.1A, we considered a follow-up experiment to the calcium pot trial originally introduced in Example 4.1. This had four treatments ($t = 4$), background variation of $s^2 = 75$ ($s = 8.66$), with a requirement to detect observed treatment differences of 10 cm at a significance level of $\alpha_s = 0.05$. The simple approach of Section 10.1 required replication of $n = 7$ for each treatment. Now we also want to consider the power associated with this design. For a two-sided test with ResDF = 24, the critical values are $t_{24}^{[0.975]}$ and $t_{24}^{[0.025]}$, equal to ±2.064. From Table 10.1, for seven replicates the SED is equal to 4.63. The non-centrality parameter is then $\delta/SED = 10/4.63 = 2.16$. The CDF of the non-central t-distribution satisfies

$$F_t(-2.064, 2.16, 24) < 0.0001, \quad F_t(2.064, 2.16, 24) = 0.455,$$

and hence the power for a two-sided test with $\delta = 10$ is $1 - 0.455 + 0.000 = 0.545$. This means that with seven replicates, we have only a 55% chance of detecting a true treatment difference of size 10 cm, given that our assumptions about the background variation are true. Table 10.3 shows the power for greater replication, and replication of $n = 13$ pots per treatment (with a total of $N = 52$ pots) is necessary to get power greater than 0.80 for a difference of 10 cm.

In principle, power can be calculated for any statistical test, but the calculations are often quite complex. In the context of ANOVA, we are often interested in the null hypothesis that a set of treatment population means are all equal against a general alternative hypothesis of some difference between population means, evaluated by using an F-test. Power

TABLE 10.3

Calculation of SED and Power for a Difference of Size $\delta = 10$ Units in a CRD with $t = 4$ Treatments, Varying Replication (n) and Estimated Residual Variance $s^2 = 75$ (Example 10.1C)

Replication (n)	Units ($N = n \times t$)	Residual df ($N - t$)	SED	δ/SED	$t_{N-t}^{[0.025]}$	Power
7	28	24	4.63	2.16	2.064	0.545
8	32	28	4.33	2.31	2.048	0.606
9	36	32	4.08	2.45	2.037	0.661
10	40	36	3.87	2.58	2.028	0.710
11	44	40	3.69	2.71	2.021	0.753
12	48	44	3.54	2.83	2.015	0.790
13	52	48	3.40	2.94	2.011	0.822

calculations for this test are more complicated and details can be found in Montgomery (1997); however, the concepts are similar and this problem corresponds to an extension of the situation illustrated above. Most statistical software contains facilities to determine the power of standard designs, such as those described in previous chapters.

It is always useful to calculate the power of potential designs for an experiment, preferably using a range of plausible values for background variation. If it is possible to use sufficient replication to give power > 0.8, then this should usually be done. However, huge replication is not always desirable: the treatment differences that can be detected might be too small to be biologically meaningful, which implies that the experiment is over-precise and potentially represents a waste of resources. Because of large background variation or limited experimental resources, or both, it is more common in much biological research to find that the intended design has weak power (< 0.5). In this case, there are several options open to the experimenter: the number of treatments tested might be reduced to allow the replication of the remaining treatments to be increased, or the experiment might be repeated at a later date. If neither of these options is available then the investigator must decide if it is worthwhile using resources to pursue an experiment that is unlikely to detect treatment differences of a given size even if they are present.

10.4 Constructing a Design for a Particular Experiment

In previous sections, we have considered how to calculate the power of a t-test within a given design. In practice, constructing the design for an experiment usually involves a compromise between several constraints (previously discussed in Chapter 3), of which power is only one. The first step in designing any experiment is to identify the experimental units that are to be used. Once these are identified, we need to determine any practical or physical constraints on the available resources, such as the maximum number of experimental units available (or affordable), and any physical or practical structures associating groups of these experimental units. These structures may arise from the intrinsic nature of the units, for example shelves within a CE cabinet, or from the way in which these units are used in the experiment, such as subsets of samples processed on different days. A parallel step is to consider the set of treatments to be tested, and to recognize any structure within this set. Recall from Chapter 8 that use of a factorial structure is generally more efficient if several treatment factors are to be included. Finally, these components can be combined to form one or more candidate designs, and these designs can be compared in terms of power.

EXAMPLE 10.2: COMPARING DESIGNS FOR AN IRRIGATION EXPERIMENT*

An experiment is required to screen a set of candidate willow varieties for susceptibility to drought. The experiment is to be on a site with good drainage and low rainfall, where drought stress would be expected to occur naturally in most years. The field plots are to be set up as four rows of six trees, with the eight trees in the centre of each plot being used for measurements. At most 72 field plots are available for the trial. Three irrigation treatments are to be applied: no irrigation, occasional (low) irrigation and frequent (high) irrigation. The irrigation treatments can be applied only to large blocks of land. A core set of four varieties must be included in the trial, but the scientists would like to include some of an additional set of six varieties if possible. The requirement to use larger blocks of land for irrigation suggests use of a split-plot design (Section 9.2), with the irrigation treatments applied to whole plots, and varieties applied to subplots within whole plots.

Drought stress is expected to reduce growth, and the primary aim of the experiment is to detect varieties badly affected by drought across a range of characteristics. A secondary aim is to quantify the typical response to the differences in water stress. Several variables are to be measured after three years, including the number of shoots, where all of the varieties are expected to have 15–20 shoots per tree in the absence of water stress. This variable is usually evaluated as the mean number of shoots per tree from the central eight trees in each plot, and analysed with a square root transformation. The design is required to be able to detect a 33% decrease in number of shoots per tree in both the irrigation main effect (secondary aim) and in the comparisons across irrigation regimes within variety (primary aim). Both tests are to use a significance level of 5%. For the square-root-transformed mean number of shoots per tree, previous trials have shown the estimated subplot variation, s^2, is usually close to 0.25, and that the whole-plot stratum variance increases with the size of the whole plots (number of subplots). Using these data from previous trials, it is estimated that the whole-plot stratum variance takes the approximate form $s_w^2 = [(0.1 \times t_B) + 1]s^2$, where t_B is the number of subplots in each whole plot (equal to the number of varieties).

The statistical task is to find the most powerful design that fits within the constraints. The first problem is to fit the scientist's question into the framework of the analysis. Most of the information is in a format that we can translate directly, except for the requirement to detect a 33% reduction in shoot numbers. This would translate into an additive difference if we were analysing data on the log-transformed scale, but we expect to use a square root transformation and on this scale there is no direct translation. However, a 33% decrease from the expected mean value of about 17.5 shoots per tree is 11.7 shoots, or on the square root scale a decrease from 4.2 to 3.4, or 0.8 units and so we shall look for decreases of this order, i.e. set $\delta = -0.8$.

There are two comparisons of interest, comparisons across irrigation regime within variety and overall comparisons of irrigation regime; we shall consider each in turn. We use the notation for split-plot designs introduced in Section 9.2. The population means for the different treatment combinations are labelled as μ_{jk}, where the first index $j = 1 \ldots t_A$ labels the irrigation treatments and the second index $k = 1 \ldots t_B$ labels the varieties. The number of irrigation treatments is fixed at $t_A = 3$ and the number of varieties is to be decided. The number of replicate blocks, denoted m, is also to be determined.

For the primary aim, we are interested in comparisons between irrigation treatments within a variety, i.e. comparisons of the form $\mu_{jk} - \mu_{rk}$ with $j \neq r$. From Section 9.2.3, we know that the SED of the estimated comparisons takes the form

$$\widehat{SE}(\hat{\mu}_{jk} - \hat{\mu}_{rk}) = \sqrt{2[s_w^2 + (t_B - 1) \times s^2)]/(t_B \times m)} \ .$$

We can use our estimate of $s^2 = 0.25$, with $s_w^2 = [(0.1 \times t_B) + 1]s^2$ to calculate

$$s_w^2 + (t_B - 1)s^2 = [(0.1 \times t_B) + 1]s^2 + (t_B - 1)s^2 = 1.1 \times t_B \times s^2 = 0.275 \times t_B \ ,$$

and simplify the SED as

$$\widehat{SE}(\hat{\mu}_{jk} - \hat{\mu}_{rk}) = \sqrt{2 \times 0.275 \times t_B/(t_B \times m)} = \sqrt{0.55/m} \ .$$

The associated degrees of freedom (Equation 9.2) can be slightly simplified (using $t_A = 3$ and omitting redundant '×' symbols) as

$$df = \frac{(m-1)[s_w^2 + (t_B-1)s^2]^2}{s_w^4/(t_A-1) + (t_B-1)s^4/t_A} = \frac{(m-1)[1.1t_Bs^2]^2}{\frac{1}{2}[(0.1t_B + 1)^2s^4] + \frac{1}{3}(t_B-1)s^4} = \frac{(m-1)[1.1t_B]^2}{\frac{1}{2}(0.1t_B + 1)^2 + \frac{1}{3}(t_B-1)} \ .$$

If we chose a design with $m = 5$ blocks, each containing $t_B = 4$ varieties, then the SED for variety comparisons is SED $= \sqrt{(0.55/5)} = \sqrt{0.11} = 0.332$ with

$$df = \frac{(m-1)[1.1t_B]^2}{\frac{1}{2}(0.1t_B + 1)^2 + \frac{1}{3}(t_B-1)} = \frac{4 \times [4.4]^2}{\frac{1}{2}(1.4)^2 + \frac{1}{3}(3)} = \frac{77.44}{1.98} = 39.11 \ .$$

We can use these values to calculate power as described in Section 10.3 and demonstrated in Example 10.1C. The critical value of the two-sided t-test under the null hypothesis is $t_{39.11}^{[0.025]} = 2.023$. Under the alternative hypothesis $\delta = -0.8$, the non-centrality parameter is then $\delta/\text{SED} = -0.8/0.332 = -2.41$. The CDF of the non-central t-distribution satisfies

$$F_t(-2.023, -2.41, 39.11) = 0.653, \quad F_t(2.023, -2.41, 39.11) = 1.000 \ ,$$

and hence the power for a two-sided test with $\delta = -0.8$ is equal to $1 - 1.000 + 0.653 = 0.653$. Similar calculations can be made for other numbers of blocks and varieties.

For the secondary aim, a similar process can be followed. Here, we are interested in overall comparisons between irrigation treatments, i.e. comparisons of the form $\mu_{j.} - \mu_{r.}$ with $j \neq r$. Again from Section 9.2.3, the SED of the estimated comparisons takes the form

$$\widehat{SE}(\hat{\mu}_{j.} - \hat{\mu}_{r.}) = \sqrt{2s_w^2/(t_B \times m)} = \sqrt{2s^2(0.1t_B + 1)/(t_B \times m)} = \sqrt{0.5 \times (0.1t_B + 1)/(t_B \times m)} \ ,$$

with $(t_A - 1) \times (m - 1) = 2 \times (m - 1)$ df. For the design with $m = 5$ blocks, each containing $t_B = 4$ varieties, then the SED for irrigation comparisons is

$$\widehat{SE}(\hat{\mu}_{j.} - \hat{\mu}_{r.}) = \sqrt{0.5 \times (0.1t_B + 1)/(t_B \times m)} = \sqrt{0.5 \times 1.4/20} = \sqrt{0.035} = 0.187 \ ,$$

with 8 df. The critical value of the two-sided t-test under the null hypothesis is $t_8^{[0.025]} = 2.306$. Under the alternative hypothesis $\delta = -0.8$, the non-centrality parameter is then $\delta/SED = -0.8/0.187 = -4.276$, and the CDF of the non-central t-distribution satisfies

$$F_t(-2.306, -4.276, 8) = 0.961, \quad F_t(2.306, -4.276, 8) = 1.000 .$$

Hence, the power for this two-sided test with $\delta = -0.8$ is equal to $1 - 1.000 + 0.961 = 0.961$.

Table 10.4 presents the results of similar power calculations for several designs that fit the experimental constraints. The upper limit of 72 field plots means that as the number of replicate blocks increases, the number of varieties that can be tested decreases. For both comparisons, the power is heavily influenced by the number of replicate blocks: as the number of blocks decreases, so does the power. The power for the main effects is good (> 0.75) for all of the designs with three or more blocks. The power for the comparison within varieties is much less, and exceeds 0.5 only for designs with four or more blocks, which allows a maximum of six varieties. The design with four blocks and six varieties appears promising: it uses all of the available plots, has high power for the irrigation main effect (0.95) and reasonable power for the interaction (0.56), and tests two additional varieties. If this power is insufficient, then the design with six blocks and four varieties might be preferred, as this has power of 0.99 for the main effect and 0.74 for the variety comparisons.

TABLE 10.4

Split-Plot Design with m Blocks, Three Whole Plots and t_B Subplots: SED, df and Power for Comparing Irrigation (Whole-Plot Treatment) within Varieties (Subplot Treatment) and the Main Effect of Irrigation (Example 10.2)

	Number of		Irrigation Comparison within Varieties			Irrigation Main Effect		
Varieties (t_B)	Blocks (m)	Units ($3 \times t_B \times m$)	SED	df	Power	SED	df	Power
4	6	72	0.30	48.89	0.736	0.17	10	0.987
4	5	60	0.33	39.11	0.653	0.19	8	0.961
4	4	48	0.37	29.33	0.550	0.21	6	0.887
4	3	36	0.43	19.56	0.427	0.24	4	0.701
4	2	24	0.52	9.78	0.281	0.30	2	0.335
5	4	60	0.37	36.92	0.556	0.19	6	0.927
5	3	45	0.43	24.61	0.435	0.22	4	0.762
5	2	30	0.52	12.31	0.291	0.27	2	0.372
6	4	72	0.37	44.35	0.560	0.18	6	0.951
6	3	54	0.43	29.57	0.440	0.21	4	0.807
6	2	36	0.52	14.78	0.297	0.26	2	0.405
7	3	63	0.43	34.42	0.440	0.20	4	0.839
7	2	42	0.52	17.21	0.302	0.25	2	0.432
8	3	72	0.43	39.18	0.445	0.19	4	0.864
8	2	48	0.52	19.59	0.306	0.24	2	0.454
9	2	54	0.52	21.92	0.308	0.23	2	0.474
10	2	60	0.52	24.20	0.311	0.22	2	0.491

10.5 A Different Hypothesis: Testing for Equivalence

In the context of hypothesis testing, it is important to remember that although one can obtain evidence against the null hypothesis, one cannot evaluate evidence in favour of the null hypothesis. If the null hypothesis is not rejected, there are two possible explanations: either the null hypothesis is true, or it is false but the background variation is sufficiently large to mask treatment differences (i.e. the experiment has insufficient power to detect the true treatment difference). The question of interest is therefore usually posed as the alternative hypothesis. However, in some cases, the question of interest is whether there is equality (or equivalence) of treatment population means, which corresponds to the usual null hypothesis. This situation often occurs when a new (sometimes faster or cheaper) treatment is compared with a standard; the aim is to show that the new treatment is equivalent to the standard so that it can be adopted. This scenario is known as **equivalence testing** and is widespread in pharmaceutical studies, although less well-established in plant science research. To test the question of equivalence, one must specify a region of equivalence and switch the roles of the two hypotheses. To illustrate these concepts we use a simple example with two treatment groups, with population means μ_1 and μ_2, respectively, and difference $\delta = \mu_1 - \mu_2$.

We first define a region of equivalence by specifying a quantity (c) such that a difference of c units between two population means is not considered biologically meaningful, so that two population means are considered equivalent if $|\mu_1 - \mu_2| = |\delta| \leq c$. In this context, the null hypothesis to be tested is that the two population means are different, or H_0: $|\delta| > c$, against the alternative hypothesis H_1: $|\delta| \leq c$. This is an **interval hypothesis**, i.e. the null hypothesis corresponds to a range of values. The equivalence testing procedure splits this null hypothesis into two one-sided components:

$$H_{0a}: \delta < -c ,$$

$$H_{0b}: \delta > c .$$

Each of these null hypotheses then has a corresponding alternative hypothesis, i.e. H_{1a}: $\delta \geq -c$ and H_{1b}: $\delta \leq c$. Each hypothesis can then be tested by a one-sided t-test, with test statistics t_a and t_b defined as

$$t_a = \frac{d + c}{\text{SED}}; \quad t_b = \frac{d - c}{\text{SED}} .$$

Here d is the observed treatment difference, $d = \hat{\mu}_1 - \hat{\mu}_2$, and SED is the estimated SE for the treatment comparison with associated df equal to ResDF. For a test with significance level α_s, we then reject null hypothesis H_{0a} if $t_a \geq t_{\text{ResDF}}^{[\alpha_s]}$. Similarly, we reject null hypothesis H_{0b} if $t_b \leq -t_{\text{ResDF}}^{[\alpha_s]}$. The overall null hypothesis, H_0: $|\delta| > c$, is rejected if *both* H_{0a} and H_{0b} are rejected, giving evidence in favour of the alternative hypothesis of equivalence between the treatment means. This procedure is generally referred to as **two one-sided t-tests** (TOST). As usual in hypothesis testing, there is a correspondence between the hypothesis test and a related confidence interval (CI). In this case, if the $100(1 - 2\alpha_s)\%$ confidence interval $d \pm (\text{SED} \times t_{\text{ResDF}}^{[\alpha_s]})$ is completely contained within the limits ($-c$, c), then the null hypothesis of inequivalence is rejected at significance level α_s and we have positive evidence of equivalence.

This procedure can be particularly useful as a secondary test where differences detected by ANOVA are considered to be biologically unimportant. Once an equivalence range has been defined, an equivalence test can establish whether there is evidence to reject a null hypothesis of inequivalence. For a very precise experiment, it is possible for small differences between treatments to be detected as significant but to then obtain evidence of equivalence. However, questions of power still arise, and a non-significant equivalence test does not prove inequivalence.

It is possible to switch the null and alternative interval hypotheses to obtain a direct **test for inequivalence**. In this case, the null hypothesis is H_0: $|\delta| \leq c$ which is tested against the alternative hypothesis H_1:$|\delta| > c$. The test statistics are the same as above, with the null hypothesis (equivalence) rejected if either $t_a < -t_{ResDF}^{[\alpha_s]}$ or $t_b > t_{ResDF}^{[\alpha_s]}$. This test rejects equivalence if the CI calculated as $d \pm (SED \times t_{ResDF}^{[\alpha_s]})$ has either its upper limit below $-c$ or its lower limit above c. McBride (1999) demonstrates the use of this test (and equivalence tests) in the context of environmental monitoring. Note that this interval-based null hypothesis is not the same as our usual point null hypothesis H_0: $\delta = 0$, and so the results of the two tests might not match.

EXAMPLE 10.3: MEASURING SOIL MICROBIAL BIOMASS

An experiment was done to investigate the effects of changing the procedure for processing samples to obtain measurements of carbon in soil microbial biomass (as mg C per kg soil). The protocol under examination used 200 g soil samples passed over a 2.5 mm sieve and shaken for 60 min. The experiment tested the effects of a larger sieve, two smaller sample weights and a reduced shaking time, giving a $2 \times 3 \times 2$ factorial structure. Each of the 12 treatment combinations was replicated four times in a CRD. The aim of analysis is to quantify the effects of the individual modifications, whether they interact, and to evaluate whether any of the modified procedures obtain results within 10% of the standard protocol. The data are listed in Table 10.5 and held in file BIOMASSC.DAT. The mean for the standard protocol is 1095.5 mg C/kg, so we consider differences smaller than 110 mg C/kg as unimportant.

Factors Size (sieve size), Weight (sample weight) and Time (shaking time) define the treatment combinations, with response variate C (microbial carbon biomass). There is no structural component of the linear model, which can be written as

Response variable: C

Explanatory component: [1] + Size*Weight*Time

The explanatory component is a three-way crossed structure, and the ANOVA table for this model is Table 10.6. There is no evidence of interactions between the different modifications, but strong evidence that increasing the sieve size and decreasing the shaking time both decrease the quantity of biomass C measured. However, these results do not establish equivalence (or inequivalence) of any of the 11 test combinations in relation to the standard protocol and to evaluate this we examine 95% CIs based on the interval hypothesis, H_0: $|\delta| < 110$, shown in Table 10.7.

The confidence limits are calculated using the 90th percentile of the t-distribution on 36 df ($t_{36}^{[0.10]} = 1.688$) with SED = 48.5. There is only one case (small sieve, 50 g weight, 60 min shaking) where there is evidence of equivalence; in this case, the 95% CI (−61.4, 102.4) is entirely contained within the limits (−110, 110). On the other hand, there is no evidence of inequivalence (no lower limits > 110 and no upper limits < −110). The absence of positive results here reflects the large amount of background variation and hence

TABLE 10.5

Biomass Carbon (C) Measurements on 48 Samples from a CRD for Different Combinations of Sieve Size (S = 2.5 mm, L = 12 mm), Sample Weight (g) and Shaking Time (min), Listed in Treatment Order (Example 10.3 and file BIOMASSC.DAT)

Size	Weight	Time	C	Size	Weight	Time	C	Size	Weight	Time	C
L	20	30	971	L	200	30	951	S	50	30	995
L	20	30	858	L	200	30	878	S	50	30	1177
L	20	30	984	L	200	30	882	S	50	30	951
L	20	30	900	L	200	30	918	S	50	30	1118
L	20	60	1062	L	200	60	974	S	50	60	1050
L	20	60	1028	L	200	60	1097	S	50	60	1196
L	20	60	1020	L	200	60	996	S	50	60	1116
L	20	60	1106	L	200	60	1048	S	50	60	1102
L	50	30	956	S	20	30	965	S	200	30	904
L	50	30	1083	S	20	30	1068	S	200	30	983
L	50	30	764	S	20	30	922	S	200	30	959
L	50	30	836	S	20	30	968	S	200	30	926
L	50	60	1030	S	20	60	1115	S	200	60	1050
L	50	60	1014	S	20	60	1123	S	200	60	1016
L	50	60	981	S	20	60	1167	S	200	60	1144
L	50	60	1065	S	20	60	1181	S	200	60	1172

Source: Data from Rothamsted Research (P. Brookes).

uncertainty in this experiment; the power to detect a difference of 110 mg between two treatment combinations is only 60%. Moreover, we have not adjusted for the number of tests (11) made, and so our overall rate of Type I error will be larger than the nominal value of 0.05 (see Section 8.8). Following McBride (1999), we could adjust the significance level using a Bonferroni correction, making the confidence limits even wider.

We can conclude from this experiment that shaking time and sieve size affect the quantity of biomass measured, but we require additional data to establish whether the different procedures give measurements within the 10% range specified.

TABLE 10.6

ANOVA Table for Soil Microbial Carbon Biomass Measured Using Two Sieve Sizes (Factor Size), Three Sample Weights (Factor Weight) and Two Shaking Times (Factor Time) (Example 10.3)

Source of Variation	df	Sum of Squares	Mean Square	Variance Ratio	P
Size	1	80,524.08	80,524.08	17.114	< 0.001
Weight	2	12,060.67	6030.33	1.282	0.290
Time	1	179,585.33	179,585.33	38.167	< 0.001
Size.Weight	2	10,543.17	5271.58	1.120	0.337
Size.Time	1	65.33	65.33	0.014	0.907
Weight.Time	2	8855.17	4427.58	0.941	0.400
Size.Weight.Time	2	5744.67	2872.33	0.610	0.549
Residual	36	169,385.50	4705.15		
Total	47	466,763.92			

TABLE 10.7

Treatment Means and Differences from Standard Protocol (Small Sieve, 200 g Sample Weight, 60 min Shaking Time) with 95% CI for the Differences Based on Interval Hypothesis H_0: $|\delta| < 110$ (Example 10.3)

Size	Weight	Time	Mean	Difference	Lower Limit	Upper Limit	Equivalent to Standard?
Large	20	30	928.25	−167.25	−249.1	−85.4	No
Large	20	60	1054.00	−41.50	−123.4	40.4	No
Large	50	30	909.75	−185.75	−267.6	−103.9	No
Large	50	60	1022.50	−73.00	−154.9	8.9	No
Large	200	30	907.25	−188.25	−270.1	−106.4	No
Large	200	60	1028.75	−66.75	−148.6	15.1	No
Small	20	30	980.75	−114.75	−196.6	−32.9	No
Small	20	60	1146.50	51.00	−30.9	132.9	No
Small	50	30	1060.25	−35.25	−117.1	46.6	No
Small	50	60	1116.00	20.50	−61.4	102.4	Yes
Small	200	30	943.00	−152.50	−234.4	−70.6	No
Small	200	60	1095.50	0	–	–	–

EXERCISE

10.1 You need to design an experiment in which you have to first make extracts from different cultivars and then process those extracts through a machine to compare the cultivars. You have four cultivars that you must test, and another four that you are quite interested in. It requires 10 plants (grown in the same pot) to make one extract to run through the machine and only one extract can be run at a time. You have the resources to make and process up to a total of 30 extracts. However, the machine needs resetting at least every eight runs, and the level of its readings may vary slightly each time it is reset. A batch of four to eight runs between resetting the machine can therefore be considered as a block. A pilot study has shown that the background variation across a set of four to eight runs is about 1 unit^2, and you wish to detect treatment differences of 2 units. Consider and compare possible designs for both stages of this experiment.

11

Dealing with Non-Orthogonality

This chapter explores the concept of orthogonality, its role in designs and the consequences of non-orthogonality, either between two (or more) treatment factors or between blocking and treatment factors. Non-orthogonality between explanatory variates, as may occur in regression models (Chapter 12), is usually termed collinearity and this concept is discussed in more detail in Chapter 14. A sufficient condition for two factors (or terms) to be orthogonal is given in Section 11.1. The procedure for analysis of a crossed model for two non-orthogonal treatment factors is then described in detail (Section 11.2). If two factors are non-orthogonal then parameter estimates may change according to the terms present in the model (Section 11.2.1) and a unique ANOVA table for the experiment no longer exists. Some consideration must be given to the order in which the factors are fitted and to the interpretation of the treatment sums of squares, giving rise to several possible sequential ANOVA tables (Section 11.2.2). This also results in different types of sums of squares (Section 11.2.3) and procedures for model selection (Section 11.2.4). The manner in which predictions are formed for individual treatments is also more complex, and affects their interpretation (Section 11.2.5).

Non-orthogonality between block and treatment factors can be planned, structured and exploited to obtain an efficient design (Section 11.3). Several classes of design for a factorial treatment structure exploit non-orthogonality to reduce the resources required for an experiment. Fractional factorial designs (FFDs) (Section 11.3.1) reduce the replication and may even omit certain treatment combinations to minimize the number of experimental units, although some knowledge of the system is required to obtain a meaningful analysis. Factorial designs with confounding enable the efficient allocation of treatment combinations to small blocks (Section 11.3.2). Often non-orthogonality is unplanned, because either there are missing values in the data (Section 11.4), treatment factors are accidentally misallocated, or unplanned events lead to additional (extraneous) factors in the model (Section 11.5).

Most statistical packages contain several algorithms that can be used to analyse a linear model, depending on features of the design, and most such algorithms can deal with an explanatory component alone. Options become more limited when both explanatory and structural components are present. Multi-stratum ANOVA algorithms can deal with separate explanatory and structural components, but require a balanced orthogonal structure. Linear mixed models provide a more complex alternative for non-orthogonal structures (Chapter 16). However, in some cases, one can combine the explanatory and structural components into a single model component and still obtain a valid analysis. This approach is often called an 'intra-block analysis' (Section 11.6).

11.1 The Benefits of Orthogonality

Two explanatory variables (or terms) in a linear model are said to be orthogonal if the estimated parameter effects and sum of squares for each term are the same regardless of whether the other term is included or not in the model. A more rigorous mathematical definition of orthogonality is beyond the scope of this book (details can be found in Bailey, 2008), but this definition will suffice here. For example, consider the case of two factors, A and B, in a design with equal replication of all factorial combinations and no experimental structure. Here, we consider the additive model, [1] + A + B, consisting of the overall constant and main effects of the two factors. In this case, estimates of the A and B main effects are the same regardless of whether the other main effect is fitted. Likewise, the sums of squares for each main effect term are the same whether the model is specified as [1] + A + B or as [1] + B + A, as illustrated in Example 11.1A. The main effect terms, corresponding to factors A and B, are thus orthogonal in this design.

EXAMPLE 11.1A: BEETLE MATING

Consider the 2×2 factorial beetle mating experiment described in Example 8.1 (data in file BEETLES.DAT) where females from two species of willow beetle (factor Species) mated with males from either their own species (intraspecies mating) or the other species (interspecies mating, factor MateType). The response analysed was the $\log_{10}$-transformed number of eggs laid by each female. The parameter estimates for the main effects were derived in Example 8.1B (see Table 8.2) from the margins of a two-way table of observed treatment means. This derivation does not depend on the order in which the terms are fitted, or on which terms are fitted in the model. Table 11.1 shows the ANOVA tables for the explanatory component specified either as [1] + MateType + Species or with factors in the other order as [1] + Species + MateType. The sums of squares for each factor are the same for both orders, confirming that these factors are orthogonal.

For two treatment factors, the easiest way to assess the orthogonality of the design is to obtain a two-way table containing counts of replicates for each combination of the two factors. The simplest case of an orthogonal design is where observations are present in all cells with equal replication. If some cells are empty, or if replication is unequal, then the factors will usually (but not always) be non-orthogonal. As a rule of thumb, if all marginal means for one factor in the two-way table involve equal representation from levels of the other factor, then the design will be orthogonal.

For two factors, we can write down a mathematical condition sufficient for orthogonality (Mead et al., 2012, Chapter 7). If n_{rs} is the replication for the rth level of the first factor and the sth level of the second factor, then the two factors are orthogonal if, for all combinations of r and s,

$$n_{rs} = \frac{n_{r\cdot} \times n_{\cdot s}}{N} , \qquad (11.1)$$

TABLE 11.1

ANOVA Tables for Main Effects Model Fitted in Two Different Orders for $\log_{10}$(Number of Eggs) in the Beetle Mating Experiment (Example 11.1A)

Sequence 1			Sequence 2		
Source of Variation	df	Sum of Squares	Source of Variation	df	Sum of Squares
MateType	1	0.3807	Species	1	0.9031
Species	1	0.9031	MateType	1	0.3807
Residual	36	0.9484	Residual	36	0.9484
Total	39	2.2322	Total	39	2.2322

where $n_{r.} = \Sigma_s\, n_{rs}$ is the total number of observations for the rth level of the first factor, $n_{.s} = \Sigma_r\, n_{rs}$ is the total number of observations for the sth level of the second factor and, as usual, N is the total number of observations.

EXAMPLE 11.1B: BEETLE MATING

Each of the four treatment combinations in the beetle mating experiment is replicated 10 times, so $n_{rs} = 10$ for $r = 1, 2$ and $s = 1, 2$. The total count for each level of the individual factors is 20, giving $n_{r.} = 20$ and $n_{.s} = 20$, with $N = 40$. Hence, the condition for orthogonality given in Equation 11.1 is satisfied as

$$\frac{n_{r.} \times n_{.s}}{N} = \frac{20 \times 20}{40} = 10 \quad \text{for any } r = 1, 2, s = 1, 2 .$$

11.2 Fitting Models with Non-Orthogonal Terms

In this section we demonstrate the process of fitting models and making statistical inferences for two non-orthogonal factors with a crossed treatment structure, paying particular attention to steps where the procedure or inference differs from that described in Chapter 8 for orthogonal factors. We illustrate these differences by comparing the analysis of Example 11.1, which has an orthogonal structure, with that of Example 11.2, which has a non-orthogonal structure. For simplicity, we have chosen examples with no structural component, but the same principles apply to investigation of the explanatory component when structure is present, within the context of a multi-stratum ANOVA.

EXAMPLE 11.2A: GENETICS OF ROOT GROWTH*

An experiment was conducted to investigate the genetic component of root growth in manipulated lines. Two male parents (factor Male, levels M1 and M2) were crossed with five female parents (factor Female, levels F1–F5) and eight seeds were to be grown from each cross in a CRD. Root growth (maximum length) was measured (mm) after three weeks (variate Root). Unfortunately, many of the seeds were not viable because of genetic incompatibilities, leading to reduced replication of some treatments with only 30 observations in total. The data are provided in file cross.dat and displayed in Table 11.2.

For this two-way factorial the pattern of replication is without structure, as shown in Table 11.3 and the Male and Female factors are non-orthogonal. This can be verified using the condition presented in Equation 11.1. For example, consider the replication for offspring of male parent M1 with female parent F1, with $n_{11} = 6$. The marginal replication for male parent M1 is $n_{1.} = 19$ and the marginal replication for female parent F1 is $n_{.1} = 8$, and there are 30 observations ($N = 30$). Orthogonality then requires replication of $19 \times 8/30 = 5.07$, but this is not an integer value and so cannot equal the actual replication, here $n_{11} = 6$, confirming that the structure is non-orthogonal.

11.2.1 Parameterizing Models for Two Non-Orthogonal Factors

Although sum-to-zero constraints are often used for balanced designs, this parameterization becomes much less convenient for non-orthogonal structures, and so it is more common to use first-level-zero (or last-level-zero) constraints in this context. First-level-zero

TABLE 11.2

Observed Root Growth (mm) from Offspring of Crosses between Five Female and Two Male Parents (Example 11.2A and File CROSS.DAT)

Female Parent	Male Parent	Root Growth	Female Parent	Male Parent	Root Growth	Female Parent	Male Parent	Root Growth
F5	M1	76	F3	M2	68	F1	M1	83
F1	M1	83	F4	M1	81	F4	M1	84
F1	M1	85	F3	M2	69	F2	M1	82
F3	M1	75	F5	M2	77	F2	M2	75
F1	M1	88	F2	M2	78	F5	M1	77
F1	M2	80	F2	M1	79	F2	M2	77
F1	M2	79	F4	M2	80	F1	M1	84
F4	M1	83	F1	M1	89	F3	M2	70
F3	M1	80	F4	M1	85	F4	M1	86
F2	M1	81	F5	M1	76	F3	M2	70

TABLE 11.3

Replication of Parental Combinations for Germinated Seed in the Root Growth Experiment (Example 11.2A)

		Female Parent					
		F1	F2	F3	F4	F5	Total
Male	M1	6	3	2	5	3	19
Parent	M2	2	3	4	1	1	11
	Total	8	6	6	6	4	30

parameterization was introduced for crossed models with two factors in Section 8.2.6, and is used throughout this chapter. Parameter estimates are again obtained by the method of least squares.

In this section, we consider a two-factor crossed treatment structure for factor A with t_A levels and factor B with t_B levels. As a preliminary step, we examine the two-factor additive model which excludes the interaction, i.e. [1] + A + B, to explain the parameterization and to demonstrate the difference between an orthogonal and a non-orthogonal structure. The additive model takes the general form

$$y_{rsk} = \mu_{11} + \eta_r + \zeta_s + e_{rsk} .$$

First-level-zero constraints are imposed as $\eta_1 = 0$, $\zeta_1 = 0$, and this model corresponds to

Explanatory component: [1] + A + B

In this parameterization, μ_{11} is the overall constant associated with the term [1]. Because of the constraints, this constant represents the population mean under this additive model for a unit with the first level of both factors. The parameters η_r ($r = 1 \ldots t_A$) are associated with factor A and can be thought of as the expected difference between observations with the rth and first levels of factor A for any given level of factor B. Similarly, parameters ζ_s ($s = 1 \ldots t_B$), associated with factor B, represent the expected difference between observations with the sth and first levels of factor B for any given level of factor A. (If the response to the level of factor A depends on the level of factor B, or vice versa, then we should need to include the interaction term, as described below.) If we simplify the model further, by dropping out factor B, then this becomes

$$y_{rsk} = \mu_1 + \eta_r + e_{rsk} ,$$

corresponding to the explanatory structure [1] + A, with constraint $\eta_1 = 0$. Here, we have relabelled the overall constant, associated with term [1], as μ_1 because it now represents the population mean for the first level of factor A. The parameters η_r $(r = 1 \ldots t_A)$ are now the expected difference between observations with the rth and first levels of factor A (regardless of factor B). An analogous model can be constructed for factor B with factor A omitted, i.e. explanatory structure [1] + B, as

$$y_{rsk} = \mu_1 + \zeta_s + e_{rsk} ,$$

with constraint $\zeta_1 = 0$. The overall constant, associated with term [1], now represents the population mean for the first level of factor B, and parameters ζ_s $(s = 1 \ldots t_B)$ are now the expected difference between observations with the sth and first levels of factor B. Note that the interpretation (and hence value) of μ_1 changes according to the terms present in the model.

If the factors A and B are orthogonal, then the estimated parameters associated with each factor term do not change when the other factor is added to or dropped from the model, as illustrated in Example 11.1C. The same does not hold for the constant term, which is marginal to both terms A and B: because the interpretation of the constant changes as terms are added or dropped, so does its estimated value. (Note that this was not the case for the sum-to-zero parameterization, in which the interpretation of the constant term as the overall mean was consistent across different models.)

EXAMPLE 11.1C: BEETLE MATING

Here, we obtain parameter estimates for first-level-zero parameterization (using the generic notation introduced above) for main effects models with one or both factors. The estimates for both single factor models and the two-factor additive model are listed in Table 11.4. Effects labelled as η are associated with the MateType factor, and those labelled ζ are associated with the Species factor.

TABLE 11.4

Estimated Parameters for Several Models for the $\log_{10}$(Number of Eggs) in the Beetle Mating Experiment, Using First-Level-Zero Parameterization, with $\eta_1 = 0$, $\zeta_1 = 0$, $(\eta\zeta)_{rs} = 0$ for $r = 1$ or $s = 1$ (Example 11.1C)

		Model			
Term	Parameter	[1] + M + S	[1] + M	[1] + S	[1] + M*S
[1]	μ_{11} or μ_1	1.561	1.711	1.658	1.513
MateType Intra	η_2	0.195	0.195	–	0.291
Species *P. vulg.*	ζ_2	0.301	–	0.301	0.396
MateType Intra. Species *P. vulg.*	$(\eta\zeta)_{22}$	–	–	–	–0.191

Note: In models M = MateType, S = Species.

As each factor has only two levels, the estimated parameters represent the expected difference between the second and first levels of each factor. It is straightforward to verify that these figures are consistent with main effect estimates given under the sum-to-zero parameterization in Table 8.2c. For example, under first-level-zero parameterization the estimate of the effect ζ_2, associated with Species *P. vulgatissima* and equal to 0.301, is the difference between the main effect estimates for Species under the sum-to-zero parameterization:

$$\widehat{Species_2} - \widehat{Species_1} = 0.1503 - (-0.1503) = 0.301 .$$

Since the MateType and Species factors are orthogonal, estimates associated with the individual factors are the same for both one-way models (i.e. models containing only one of these factors) and the additive model [1] + MateType + Species. These estimates are also unchanged if the order of the factors in the model is swapped to give model [1] + Species + MateType. As expected from its interpretation, the value of the estimated constant (labelled μ_{11} or μ_1) differs between models.

EXAMPLE 11.2B: GENETICS OF ROOT GROWTH*

We now repeat the analyses of Example 11.1C for this non-orthogonal data set, with the estimates obtained from both single factor models and the two-factor additive model listed in Table 11.5. Here, effects labelled as η are associated with the Female factor, and those labelled ζ are associated with the Male factor.

In these models the parameter ζ_2, associated with the second male parent (M2), represents the expected difference in root growth with respect to offspring of the first male parent (M1) for a given female parent. The effect of the rth female parent (η_r) represents the expected difference in root growth with respect to offspring of the first female parent (F1) for a given male parent. The two factors Male and Female are non-orthogonal, and so the estimates associated with each factor change in value when the other factor is added to or dropped from the model. For example, the estimated effect of the second male parent (M2) equals −4.8 mm in the additive model containing both factors, indicating 4.8 mm less root growth for offspring of the second male parent when compared with the first male parent. But this estimate becomes −7.1 mm in a model containing the

TABLE 11.5

Estimated Parameters for Several Models for Root Growth, Using First-Level-Zero Parameterization, with $\eta_1 = 0$, $\zeta_1 = 0$, $(\eta\zeta)_{rs} = 0$ for $r = 1$ or $s = 1$ (Example 11.2B)

Term	Parameter	Model [1] + F + M	[1] + F	[1] + M	[1] + M*F
[1]	μ_{11} or μ_1	85.1	83.9	81.9	85.3
Female F2	η_2	−4.0	−5.2	–	−4.7
Female F3	η_3	−9.9	−11.9	–	−7.8
Female F4	η_4	−1.1	−0.7	–	−1.5
Female F5	η_5	−7.4	−7.4	–	−9.0
Male M2	ζ_2	−4.8	–	−7.1	−5.8
Female F2. Male M2	$(\eta\zeta)_{22}$	–	–	–	1.8
Female F3. Male M2	$(\eta\zeta)_{32}$	–	–	–	−2.4
Female F4. Male M2	$(\eta\zeta)_{42}$	–	–	–	2.0
Female F5. Male M2	$(\eta\zeta)_{52}$	–	–	–	6.5

Note: In models F = Female, M = Male.

Male factor only, indicating a considerably larger difference. One must therefore establish a suitable model before reliable inferences can be made.

The full model for a crossed treatment structure with two factors includes an interaction term and is written as

$$y_{rsk} = \mu_{11} + \eta_r + \zeta_s + (\eta\zeta)_{rs} + e_{rsk} .$$

First-level-zero constraints are imposed as $\eta_1 = 0$, $\zeta_1 = 0$ and $(\eta\zeta)_{rs} = 0$ when $r = 1$ or $s = 1$. This model corresponds to the crossed explanatory structure

Explanatory component: [1] + A*B
$$= [1] + A + B + A.B$$

In this parameterization, μ_{11} is the overall constant associated with the term [1], which now represents the population mean under the crossed model for the first level of both factors. Interpretation of the other parameters also differs somewhat from that in the additive models described above. The parameters η_r $(r = 1 \dots t_A)$, associated with factor A, represent the difference between the rth and first levels of factor A at the first level of factor B. Similarly, parameters ζ_s $(s = 1 \dots t_B)$, associated with factor B, represent the difference between the sth and first levels of factor B at the first level of factor A. Because these parameters now represent different quantities, they also now take different values. The parameters $(\eta\zeta)_{rs}$ $(r = 1 \dots t_A, s = 1 \dots t_B)$ associated with the interaction term A.B can be thought of as deviations relative to the first row and column in the two-way table of unstructured treatment effects (see Section 8.2.6), and allow the response to a level of factor A to depend on the level of factor B, and vice versa.

This change in the interpretation of model parameters according to which terms are present can make the first-level-zero parameterization confusing. It is important to realize that the parameterization used is only a tool to facilitate estimation, and that any valid parameterization for a given model results in the same set of fitted values or predictions. The predicted value for a given treatment combination (rth level of factor A, sth level of factor B) is obtained by addition of the relevant parameter estimates, i.e. for the full crossed model

$$\hat{\mu}_{rs} = \hat{\mu}_{11} + \hat{\eta}_r + \hat{\zeta}_s + \widehat{(\eta\zeta)}_{rs} .$$

These predictions for the full crossed model will always be equal to the observed treatment means. Predictions from the simple additive models are obtained in a similar manner by adding together the estimated parameters for the terms present in that model.

EXAMPLE 11.1D: BEETLE MATING

Column 6 of Table 11.4 lists estimates obtained from first-level-zero parameterization for the full crossed explanatory component, [1] + MateType*Species. The estimated constant changes when the interaction is added into the model, and is equal to the observed mean for the first level of both factors (MateType inter, Species *P. vitellinae*). Although the MateType and Species factors are orthogonal, estimates of the main effects also change when the interaction is added into the model as the interpretation of these parameters

TABLE 11.6

Predicted Population Means from First-Level-Zero Parameterization for the Full Crossed Model for the Beetle Mating Experiment (Example 11.1D)

		Mating Type	
		Interspecies	Intraspecies
Species of Female	*P. vitellinae*	$\hat{\mu}_{11} + \hat{\eta}_1 + \hat{\zeta}_1 + (\widehat{\eta\zeta})_{11} = 1.513$	$\hat{\mu}_{11} + \hat{\eta}_1 + \hat{\zeta}_2 + (\widehat{\eta\zeta})_{12} = 1.804$
	P. vulgatissima	$\hat{\mu}_{11} + \hat{\eta}_2 + \hat{\zeta}_1 + (\widehat{\eta\zeta})_{21} = 1.909$	$\hat{\mu}_{11} + \hat{\eta}_2 + \hat{\zeta}_2 + (\widehat{\eta\zeta})_{22} = 2.008$

changes. The effect associated with intraspecies mating (MateType Intra, estimate 0.291) is equal to the difference between observed means for intra- and interspecies mating for the first level of Species, i.e. *P. vitellinae* (calculated as 1.804 − 1.513). Table 11.6 demonstrates that predictions from the crossed model are equal to the observed treatment means, previously shown in Table 8.1a.

EXAMPLE 11.2C: GENETICS OF ROOT GROWTH*

Column 6 of Table 11.5 lists estimates from first-level-zero parameterization for the full crossed explanatory component, [1] + Male*Female. These estimates can be used to obtain the predictions shown in Table 11.7, which are equal to the observed treatment means.

The estimated constant equals the observed mean for crosses derived from the first male parent (M1) and the first female parent (F1) (see Table 11.5). Other parameters are interpreted as described above. For example, the estimated effect associated with female parent F3 (−7.8) equals the difference between the observed means for crosses derived from male parent M1 with female parents F3 or F1 (77.5 − 85.3 = −7.8).

Interpretation of parameters under first-level-zero constraints becomes increasingly more complex as higher-order interactions are added into the model. As a general rule, individual parameter estimates are not of particular interest, except as components of predictions. We use ANOVA to determine which model terms are required to give a good description of a data set, and this also gives an estimate of background variation that can be used to estimate standard errors of parameter estimates and predictions.

TABLE 11.7

Predicted Population Means from First-Level-Zero Parameterization for the Full Crossed Model for the Root Growth Experiment (Example 11.2C)

		Male Parent	
		M1	M2
Female Parent	F1	$\hat{\mu}_{11} + \hat{\eta}_1 + \hat{\zeta}_1 + (\widehat{\eta\zeta})_{11} = 85.3$	$\hat{\mu}_{11} + \hat{\eta}_1 + \hat{\zeta}_2 + (\widehat{\eta\zeta})_{12} = 79.5$
	F2	$\hat{\mu}_{11} + \hat{\eta}_2 + \hat{\zeta}_1 + (\widehat{\eta\zeta})_{21} = 80.7$	$\hat{\mu}_{11} + \hat{\eta}_2 + \hat{\zeta}_2 + (\widehat{\eta\zeta})_{22} = 76.7$
	F3	$\hat{\mu}_{11} + \hat{\eta}_3 + \hat{\zeta}_1 + (\widehat{\eta\zeta})_{31} = 77.5$	$\hat{\mu}_{11} + \hat{\eta}_3 + \hat{\zeta}_2 + (\widehat{\eta\zeta})_{32} = 69.3$
	F4	$\hat{\mu}_{11} + \hat{\eta}_4 + \hat{\zeta}_1 + (\widehat{\eta\zeta})_{41} = 83.8$	$\hat{\mu}_{11} + \hat{\eta}_4 + \hat{\zeta}_2 + (\widehat{\eta\zeta})_{42} = 80.0$
	F5	$\hat{\mu}_{11} + \hat{\eta}_5 + \hat{\zeta}_1 + (\widehat{\eta\zeta})_{51} = 76.3$	$\hat{\mu}_{11} + \hat{\eta}_5 + \hat{\zeta}_2 + (\widehat{\eta\zeta})_{52} = 77.0$

11.2.2 Assessing the Importance of Non-Orthogonal Terms: The Sequential ANOVA Table

As in previous chapters, ANOVA is used to partition the total variation into components associated with individual model terms and background variation. We do not give details of how to calculate the ANOVA here; instead we obtain the tables directly from statistical software. For non-orthogonal designs, the sums of squares in the ANOVA table may depend on the order in which terms are added into the model. It is therefore necessary to consider several sequences of sub-models that add terms into the full model in different orders. When constructing sequences of sub-models we respect the principle of marginality (see Section 8.2.1) and add a term only if all possible sub-terms are already in the model.

Consider a two-way crossed treatment structure with factors A and B, as defined above (Section 11.2.1). We consider the two sequences of sub-models shown in Figure 11.1, both of which start with the baseline model containing the overall constant. In Sequence 1, we first add factor A, then factor B, and then the A.B interaction term to get the full explanatory component as [1] + A + B + A.B. In the second sequence, the roles of factors A and B are reversed to obtain [1] + B + A + A.B. The principle of marginality means that we cannot add the interaction A.B before either of the main effects A or B, and we must include the overall constant first, so only these two sequences are valid.

Before proceeding, we need to introduce some new terminology. For any sub-model, we define its model sum of squares (ModSS) as the sum of squares accounted for by that sub-model. We identify a model sum of squares by explicitly specifying the sub-model it refers to within parentheses, so ModSS([1] + A + B) is the sum of squares associated with sub-model [1] + A + B. Similarly, the model df (ModDF) is defined as the total df associated

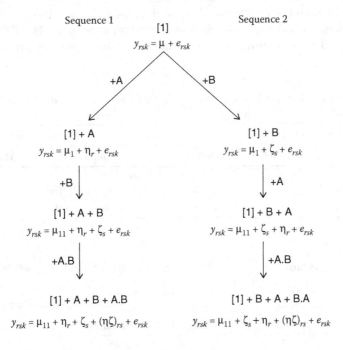

FIGURE 11.1
Symbolic and algebraic forms of models for a two-way crossed structure obtained by sequentially adding one term at a time, respecting marginality.

with the sub-model. Note that, because the overall mean is eliminated from the total sum of squares and the df, ModSS([1]) = 0 and ModDF([1]) = 0.

EXAMPLE 11.1E: BEETLE MATING

The model sums of squares and df are listed in Table 11.8. In this orthogonal design, we can calculate the sum of squares for the additive model [1] + Species + MateType (ModSS = 1.2838) by adding together the sums of squares associated with the two sub-models: [1] + Species (ModSS = 0.9031) and [1] + MateType (ModSS = 0.3807).

EXAMPLE 11.2D: GENETICS OF ROOT GROWTH*

The model sums of squares and df are listed in Table 11.9. In this non-orthogonal case, the sums of squares for the additive model [1] + Female + Male (ModSS = 747.11) is less than that obtained by addition of the sums of squares associated with the two sub-models [1] + Female (ModSS = 610.62) and [1] + Male (ModSS = 354.08).

The ModSS, ModDF and parameter estimates for two (sub-)models containing the same terms are always identical, regardless of the order of fitting. Hence, for a two-way crossed treatment structure, the sum of squares and df for the two versions of the full model, [1] + A*B and [1] + B*A, are equal. Similarly, the model SS, df and estimates for the additive model [1] + A + B are the same as for model [1] + B + A.

An ANOVA table for each sequence of sub-models, called a **sequential ANOVA table**, can be derived from the model sums of squares and df. When a new term is added into the model, changes in the ModSS and ModDF are attributed to that new term. These changes are called the **incremental** or **sequential sum of squares** and **df**, respectively. To avoid ambiguity, the incremental sum of squares and degrees of freedom (denoted as SS and df) are labelled by both the new term added and terms already in the model. For example, moving from model [1] + A to [1] + A + B we describe the change as +B | ([1] + A), to be read as 'adding factor B given that the overall mean and factor A are already in the model', or

TABLE 11.8

Model df (ModDF) and Sums of Squares (ModSS) for a Crossed Model and Its Sub-Models for the Beetle Mating Experiment (Example 11.1E)

Sequence 1			Sequence 2		
Model	ModDF	ModSS	Model	ModDF	ModSS
[1] + MateType	1	0.3807	[1] + Species	1	0.9031
[1] + MateType + Species	2	1.2838	[1] + Species + MateType	2	1.2838
[1] + MateType*Species	3	1.3751	[1] + Species*MateType	3	1.3751

TABLE 11.9

Model df (ModDF) and Sums of Squares (ModSS) for a Crossed Model and Its Sub-Models for the Root Growth Experiment (Example 11.2D)

Sequence 1			Sequence 2		
Model	ModDF	ModSS	Model	ModDF	ModSS
[1] + Female	4	610.62	[1] + Male	1	354.08
[1] + Female + Male	5	747.11	[1] + Male + Female	5	747.11
[1] + Female*Male	9	788.78	[1] + Male*Female	9	788.78

TABLE 11.10

Incremental Sums of Squares for a Two-Way Crossed Structure Fitted as [1] + A + B + A.B

Model	Change	Incremental Sum of Squares
[1] + A	+A | [1]	SS(+A | [1]) = ModSS([1] + A) − ModSS([1])
[1] + A + B	+B | ([1] + A)	SS(+B | [1] + A) = ModSS([1] + A + B) − ModSS([1] + A)
[1] + A*B	+A.B | ([1] + A + B)	SS(+A.B | [1] + A + B) = ModSS([1] + A*B) − ModSS([1] + A + B)

TABLE 11.11

Structure of the Sequential ANOVA Table for a Two-Way Crossed Structure Fitted as [1] + A + B + A.B

Term Added	Incremental Sum of Squares	Incremental df	Mean Square	Variance Ratio
+ A	SS(+A)	DF(+A)	MS(+A) = SS(+A)/DF(+A)	MS(+A)/ResMS
+ B	SS(+B)	DF(+B)	MS(+B) = SS(+B)/DF(+B)	MS(+B)/ResMS
+ A.B	SS(+A.B)	DF(+A.B)	MS(+A.B) = SS(+A.B)/DF(+A.B)	MS(+A.B)/ResMS
Residual	ResSS	ResDF	ResMS = ResSS/ResDF	
Total	TotSS	TotDF		

equivalently, 'adding factor B after eliminating the overall mean and factor A'. The forms of the incremental sums of squares for model sequence [1] + A + B + A.B are shown in Table 11.10. The incremental df are derived similarly from the ModDF.

In the context of a sequential ANOVA table, we can use some abbreviations by considering the table as a whole. For example, instead of listing the change *and* the terms already present in the model, we can deduce the terms already in the model from previous lines in the ANOVA table and just indicate the change by using ' + ' with the name of the term added into the model. For example, in Table 11.10, SS(+B) and DF(+B) can be used as shorthand to indicate SS(+B | [1] + A) and DF(+B | [1] + A), respectively.

We can now derive a full sequential ANOVA table. The residual sum of squares (ResSS) and df (ResDF) are those associated with full model, so here ResSS = TotSS − ModSS([1] + A*B), and ResDF = $N − 1 −$ ModDF([1] + A*B). Mean squares are calculated by division of the incremental sums of squares by their incremental df. The designs considered here have no structure, so the variance ratios for each term are all calculated with respect to the ResMS. The structure of one sequential ANOVA table for a two-way crossed structure is shown in Table 11.11. The first row gives the change on addition of term A to the baseline model containing only the overall constant term [1]. The second row corresponds to the change when term B is added to a model that already contains [1] + A, and is often described as 'the sum of squares for B after eliminating A'. The third row corresponds to the change when the interaction is added to a model containing both main effects ([1] + A + B). An analogous table can be derived for the other sequence. For each line of the ANOVA table, the variance ratio can be used as a test statistic for the null hypothesis that addition of the term gives no improvement to the current model (explains no additional variation). The df for the associated F-test are, as usual, given by the df associated with the numerator and denominator mean squares of the variance ratio.

EXAMPLE 11.1F: BEETLE MATING

The two sequential ANOVA tables for the full crossed model are in Table 11.12. In this orthogonal design, the incremental sums of squares are the same in both sequences.

TABLE 11.12

Sequential ANOVA Tables for a Crossed Model for the Beetle Mating Experiment (Example 11.1F)

	Sequence 1					Sequence 2			
Source	df	SS	MS	VR	Source	df	SS	MS	VR
+ M	1	0.3807	0.3807	15.992	+ S	1	0.9031	0.9031	37.932
+ S	1	0.9031	0.9031	37.932	+ M	1	0.3807	0.3807	15.992
+ M.S	1	0.0914	0.0914	3.837	+ S.M	1	0.0914	0.0914	3.837
Residual	36	0.8571	0.0238		Residual	36	0.8571	0.0238	
Total	39	2.2322			Total	39	2.2322		

Note: M denotes factor MateType and S denotes factor Species. df = incremental df, SS = incremental sum of squares, MS = mean square, VR = variance ratio.

For example, the sum of squares for Species eliminating MateType (i.e. +Species in Sequence 1) is the same as that for Species ignoring MateType (+Species in Sequence 2), and the conclusions are exactly the same for both model sequences.

EXAMPLE 11.2E: GENETICS OF ROOT GROWTH*

The two sequential ANOVA tables are in Table 11.13. In this non-orthogonal design, the incremental sum of squares on addition of the Male main effect (denoted M in Table 11.13) to the model differs according to whether the Female main effect (denoted F in Table 11.13) has already been added to the model (eliminated) or not (ignored). A similar pattern is present for the Female main effects but, as expected, the incremental sum of squares for the interaction term (Female.Male) is the same in both sequences. The conclusions are the same from both sequences: there is some evidence of an interaction ($F_{4,20}^{F.M} = F_{4,20}^{M.F} = 2.821$, $P = 0.053$), but there is very strong evidence from both sequences that both main effects are required in the model.

In some cases, interpretation of the ANOVA table is less straightforward. For example, suppose that observed significance levels from the ANOVA table for a two-way crossed model, [1] + A*B, took the values listed in Table 11.14. In this case, Sequence 1 gives strong evidence that factor A accounts for variation in the response but that, once this term has been taken into account, addition of factor B and the interaction term into the model accounts for no further variation. In contrast, Sequence 2 indicates that when factor A is ignored, there is evidence that factor B accounts for variation in the response, although the interaction is still not significant. We need to put this information together in a way that

TABLE 11.13

Sequential ANOVA Tables for a Crossed Model for the Root Growth Experiment (Example 11.2E)

	Sequence 1					Sequence 2			
Source	df	SS	MS	VR	Source	df	SS	MS	VR
+ F	4	610.62	152.66	41.324	+ M	1	354.08	354.08	95.849
+ M	1	136.48	136.48	36.945	+ F	4	393.02	98.26	26.597
+ F.M	4	41.68	10.42	$F^{F.M} = 2.821$	+ M.F	4	41.68	10.42	$F^{M.F} = 2.821$
Residual	20	73.88	3.69		Residual	20	73.88	3.69	
Total	28	862.67			Total	28	862.67		

Note: M denotes factor Male and F denotes factor Female. df = incremental df, SS = incremental sum of squares, MS = mean square, VR = variance ratio.

TABLE 11.14

Observed Significance Levels (*P*) from Sequential ANOVA
Tables for a Two-Way Crossed Model with Factors A and B

Sequence 1		Sequence 2	
Term Added	*P*	Term Added	*P*
+ A	0.01	+ B	0.04
+ B	0.08	+ A	0.02
+ A.B	0.32	+ B.A	0.32

makes sense. The fact that B is significant only when A is not in the model suggests that there is some association, or confounding, between levels of A and B.

In general, we prefer the simplest model that describes variation in the response – a **parsimonious** model. Here, this would be the model that contains factor A only, as Sequence 1 tells us that we do not need factor B or the interaction once we have factor A in the model. In selecting a model we also obey the principle of marginality; this implies that we work upwards from the bottom of the ANOVA table(s), as described in Section 8.3.

This process of model selection is reasonably straightforward for a model with two treatment factors, but becomes more complex when more factors and their interactions are present. The principle of marginality requires that we should fit the main effects before two-factor interactions, two-factor interactions before three-factor interactions and so on, but this may still result in a large number of valid sequences of sub-models to be compared (see Section 8.3). For this reason, different types of sum of squares have been developed to aid in model identification, and these are described in the next subsection.

11.2.3 Calculating the Impact of Model Terms

The incremental sums of squares described above, i.e. the change in the model sum of squares on addition of a new term into the model, are sometimes called the **Type I SS**. These sums of squares are widely used but, as shown above, they have the disadvantage that the value of the Type I SS for a given term changes according to the order in which the model terms are specified.

The **Type II SS** for a term is usually defined as the incremental sum of squares obtained when that term is added to a model that contains all terms marginal to itself. For example, consider a three-way crossed model containing all main effects and interactions for factors A, B and C. The Type II SS for the term A.B is then the incremental sum of squares obtained when term A.B is added to the model [1] + A + B. Sometimes the Type II SS is alternatively defined as the incremental sum of squares obtained when the term is added to a model that contains *all* other terms of lower or equal order. Using this second definition, the Type II SS for term A.B in our three-factor example is the incremental sum of squares obtained from adding the term A.B to the model [1] + A + B + C + A.C + B.C. Under both definitions, the Type II SS for term A.B.C would be the incremental sum of squares obtained when the three-way interaction is added to a model containing all main effects and two-factor interactions, i.e. [1] + A + B + C + A.B + A.C + B.C. These sums of squares can be useful in helping to establish a sensible model without having to refit terms in different orders, but they are not available in some statistical software.

The **Type III SS** (sometimes also called marginal or drop-one-out SS) are more complex, but broadly correspond to the change in the model sum of squares obtained when a term is dropped from the full model. Type IV SS are similar to Type III SS in principle, but use a slightly different calculation when there is no data for some treatment combinations. Further details of Type III and IV sums of squares can be found in Milliken and Johnson (2001). The Type I and Type III SS are always equal for the last term added into the model. We endorse the use of Type III SS whilst respecting the principle of marginality, as Type III SS may be inappropriate for any term that is a sub-term of one or more other terms in the model. We therefore should not use Type III SS to test terms that are marginal to other terms present in the model. For example, we should not calculate Type III SS for factor A whilst term A.B is present in the model.

11.2.4 Selecting the Best Model

The process of model selection applies to the explanatory component only; the structural component is used to obtain the correct strata and tests and so structural terms should never be dropped. Model selection for the explanatory component can proceed based on Type III SS, respecting marginality, as these allow us to start with the full model and to drop terms progressively. This process was described in Section 8.3 for an orthogonal three-way crossed structure. In that orthogonal case, the Type I and Type III SS are equivalent, and all tests could be obtained from a single ANOVA table. For non-orthogonal structures, the procedure is somewhat more complex. At each step, we use Type III SS to test all model terms that are not marginal to other terms still present in the model, i.e. only terms that are not contained within another term. For a fully crossed model, this means that, as a first step, we can test only the highest-order interaction. This process was illustrated in Figure 8.4. At each step, if any of the tested terms is not significant, then the least significant (largest observed significance level) can be dropped from the model. As each term is dropped, the model is refitted, and the process continues until no further terms can be dropped.

Because this process becomes more complex as the number of factors increases, sometimes automatic model selection procedures, such as stepwise selection, are advocated. These methods are discussed in detail in Section 14.9.1 in the context of regression models with quantitative explanatory variables and in Section 15.6 for models with factors. Here, we merely note that these methods must respect marginality to be valid, and that the procedure described above is equivalent to backward elimination. Forward selection is implemented by addition of terms based on the incremental sums of squares, again respecting marginality, and this method may be inadvisable if interactions are present in the absence of main effects, as the procedure will then stop too soon. In practice, the stepwise selection algorithms implemented in statistical software do not usually respect marginality, and so some intervention will often be required.

11.2.5 Evaluating the Response to Treatments: Predictions from the Fitted Model

Recall that prediction is the use of the fitted model to estimate functions of the explanatory variable(s). For example, we might want to predict the population mean for a given experimental treatment. In Section 8.2.4, for orthogonal structures, we used observed means to predict the effect of one factor whilst averaging across levels of other factors in the model. For a non-orthogonal structure this approach is not usually efficient, and is not possible when some treatment combinations are missing. A more general approach is therefore required, which also gives us the opportunity to consider other types of model predictions.

All predictions are based on the final fitted model. As we have already seen in Section 11.2.1, once a model has been selected, the predicted value for any treatment combination is equal to its fitted value based on this model. The full set of model predictions can be presented in a multi-way table classified by the treatment factors. To produce predictions for individual factors, or combinations of factors, we can take marginal means of this multi-way table. As in Chapter 8, we take account of the form of the final model: where significant interactions are present, we usually make predictions for combinations of factors rather than individual factors.

EXAMPLE 11.2F: GENETICS OF ROOT GROWTH*

In Example 11.2E, the interaction term was not quite significant for $\alpha_s = 0.05$ ($P = 0.053$), and the interaction mean square was much smaller than those for main effects, and so we decide to omit the interaction from the model. The explanatory structure of the selected model thus takes the form: [1] + Female + Male. Table 11.15 contains the predictions from this model for offspring of all possible crosses, formed as

$$\hat{\mu}_{rs} = \hat{\mu}_{11} + \hat{\eta}_r + \hat{\zeta}_s \; ,$$

using the estimates shown in column 3 of Table 11.5. Because the interaction term has been omitted from the final fitted model, the predictions are not equal to the observed treatment means listed in Table 11.7, although the differences are small since the interaction effects were also small. The precision of the predictions, represented by their standard errors, reflects the amount of information available for the individual treatment combinations.

Figure 11.2 plots these predictions, joining points by lines to make the pattern clear. In this additive model the lines are parallel, demonstrating that the expected difference in root growth between the two male parents is the same for each female parent. We can therefore simplify our summary of the model by taking marginal means of the predictions with respect to each factor separately; these predictions are also listed in Table 11.15. For example, the marginal predictions for females F1 and F3 show that offspring of female F3 tend to have on average 9.9 mm less root growth than offspring of female F1.

As well as using predictions to summarize the fitted model, we can also use them to estimate the expected outcome for specific scenarios. In this case, we might take marginal means even when interactions are present, and we might also use weights when taking marginal means. For example, consider an experiment done to evaluate the yield potential of several varieties (factor Variety) in several regions of a country (factor Region), with the aim of predicting potential yields under various scenarios of variety allocation. The experimental results can be summarized by use of the explanatory component [1] + Variety*Region. From the fitted model, we can produce a two-way table of predictions

TABLE 11.15

Predictions with Standard Errors (SE) and Replication (n_{rs}) from Explanatory Component [1] + Male + Female for the Root Growth Experiment (Example 11.2F)

Female	Male M1			Male M2			Margin		
	Prediction	SE	n_{1s}	Prediction	SE	n_{2s}	Prediction	SE	$n_{\cdot s}$
F1	85.1	0.81	6	80.3	1.03	2	82.7	0.81	8
F2	81.1	1.00	3	76.3	1.00	3	78.7	0.90	6
F3	75.2	1.08	2	70.4	0.94	4	72.8	0.91	6
F4	84.0	0.91	5	79.2	1.17	1	81.6	0.94	6
F5	77.7	1.12	3	72.9	1.29	1	75.3	1.12	4
Margin	80.6	0.53	19	75.8	0.70	11	78.2	0.43	30

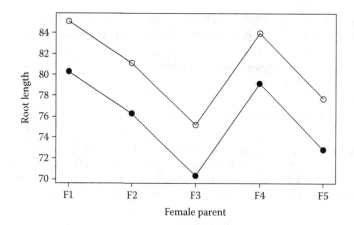

FIGURE 11.2
Predicted root length (mm) for offspring of each cross of two male (o M1, ● M2) and five female (F1–F5) parents from explanatory component [1] + Male + Female (Example 11.2F).

for the yield of each variety in each region. If we wish to compare expected regional yields, then we need somehow to form averages of the predicted means across the varieties within each region. If we simply take marginal means of the two-way table, then we are taking averages by using equal weights for each variety in each region. This may be unrealistic if the area suitable for each variety varies across regions. Therefore, we could use weights based on the expected area of each variety grown within each region, allowing this to vary across regions. Many different weighting schemes are possible, but the weights used should be appropriate to the type of prediction desired.

The process is further complicated if some of the predictions for individual treatment combinations cannot be estimated. This can happen when interactions are significant and so retained in the model but some of the corresponding treatment combinations are absent, for example if some varieties are not grown in some of the regions. In this case, it is not possible to obtain a reliable prediction for the missing variety × region combinations and so the multi-way table of predictions contains missing values. One can obtain marginal means by assigning zero weight to these missing combinations, but this action obviously affects both the composition and interpretation of the resulting predictions. Some thought is required to form appropriate predictions in this situation and some of these issues are discussed by Lane and Nelder (1982).

Any of the predictions described above can be written as a linear combination of the model parameters. Some algorithms produce predictions that are formulated as marginal means of the multi-way table of predictions classified by all factors in the model. Other algorithms require direct specification of the coefficients in the required linear combination; many statistical packages allow specification in either form.

11.3 Designs with Planned Non-Orthogonality

The previous sections have considered non-orthogonality between treatment factors and the problems that this can cause. Non-orthogonality can also occur between blocking and treatment factors and this again complicates the statistical analysis. Many of the designs we have considered so far avoid these complications by ensuring that block factors are orthogonal to treatment factors. For example, in the RCBD, LS and SP designs, the number of units per block is equal to the number of treatments, so that each treatment occurs once in each block and therefore, using the criteria of Section 11.1, the block and treatment factors are orthogonal. We have analysed these designs using multi-stratum ANOVA. When blocks and treatments are non-orthogonal, the multi-stratum ANOVA can be sensibly constructed only if the design is balanced. Recall that a design is called balanced if all treatment comparisons (differences) are estimated with the same precision (Section 3.2). Orthogonal designs with equal replication are balanced, because all treatments occur once in each block and thus all comparisons have the same precision, i.e. equal SEDs. The BIBD (see Section 9.3) is an example of a design with a non-orthogonal structure between blocking and treatment factors, as it uses blocks with fewer experimental units than the number of treatments. This design creates balance by ensuring that, when considered across the whole experiment, each pair of treatments occurs together equally often within the same blocks. Moreover, a certain proportion of the information on treatment differences can be obtained from intra-block comparisons, with the remainder being obtained from inter-block comparisons, and these proportions are the same for any treatment difference. This decomposition of the information on treatments can be summarized in a multi-stratum ANOVA table, as described in Section 9.3. The same principle can be extended to analyse partially balanced incomplete block designs (Section 9.3.3) within the framework of multi-stratum ANOVA. However, for a given block size, obtaining balanced (or partially balanced) incomplete block designs becomes more challenging as the number of experimental treatments increases. This is particularly relevant with factorial structures, because of the large number of treatment combinations that may be generated, and there are two classes of design that exploit planned non-orthogonality to reduce the resources required. **Fractional factorial designs** (Section 11.3.1) use a carefully chosen subset of the full set of treatment combinations to provide information about main effects and low-order interactions within a reduced number of experimental units, often with low levels of replication. **Factorial designs with confounding** enable the efficient allocation of factorial treatment combinations to small blocks (Section 11.3.2).

11.3.1 Fractional Factorial Designs

Fractional Factorial Designs (FFDs) are useful where the number of factorial treatment combinations is considered too large for full replication of all combinations, possibly because the number of treatment combinations is substantially larger than the number of experimental units in the natural blocking structures, and where interest is focussed on main effects and low-order interactions. The construction of a FFD involves the *a priori* assumption that some high-order interaction effects will be negligible, so that these can be aliased with the main effects and low-order interactions that are of interest. This usually requires some prior detailed knowledge and experience of the system under study. Given the natural block size, the challenge is then to identify an aliasing structure (i.e. the sets of high-order interaction effects that will be aliased with each main effect and low-order interaction effect) that allows all of the main effects and selected low-order interactions to be estimated. The aliasing structure defines the subset of treatment combinations to be used in the design. Often replication will still be possible, and the same subset of treatments will usually be repeated in each replicate block. Replication provides the main basis for the estimation of the background variation, although any estimates of high-order interactions can also be assigned to the estimation of the background variation, given that we have assumed that these effects are negligible. Further exploration of FFDs is beyond the scope of this book, but details can be found in Mead et al. (2012, Chapter 14).

11.3.2 Factorial Designs with Confounding

Whether the available resources allow replication of the full set of factorial combinations, or just some fractional subset (as discussed in Section 11.3.1), the natural block size may still be too small to include the full set, or subset, of factorial combinations. In this case, the approach of factorial designs with confounding can be used, which divides the full set, or subset, of factorial combinations into groups to be assigned to different blocks. As with the FFD discussed in Section 11.3.1, we have to make the *a priori* assumption that some high-order interaction effects are negligible, and we can then confound these effects with the differences between blocks. This means that it is not possible to identify whether block differences are due to structural variability or the confounded high-order interaction effect (assumed to be negligible). In this scenario, it is often necessary to assign some high-order interactions (assumed negligible) to provide our estimate of the background variation, though replication may also provide some information. The strategy for choosing the terms to be confounded with blocks depends on the relationship between the number of factorial treatment combinations included and the natural block size, and on the comparisons of most interest. Further exploration of this design approach is beyond the scope of this book, but, again, further details can be found in Mead et al. (2012, Chapter 14).

11.4 The Consequences of Missing Data

Missing responses occur frequently in scientific studies, sometimes because of an error in the experimental procedure so that a planned observation either has not been obtained or is identified as unreliable. Occasionally most, or even all, of the observations associated with one or more treatment combinations are missing. This may happen because of experimental error, but is also likely to occur where the treatment is partly or wholly unviable. Diagnostic checks (see Section 5.2) may also lead to observations being identified as outliers and omitted from analysis.

If the missing responses can be considered to occur at random independently of the treatment group or the (unseen) response, the pattern is known as **missing completely at random** (MCAR, for example, see Carpenter and Kenward, 2013). For example, if machine breakdown causes several field plot yields to be lost, this is likely to be unrelated to the treatment or the yields. If the missing responses can be considered to occur at random within each treatment group but independently of the (unseen) response, the pattern is known as **missing at random** (MAR). For example, suppose a study on seedling vigour uses 50 seeds from several varieties and measures biomass after seven days. Germination rates vary between varieties but are not thought to be related to seedling vigour. The values of biomass observed (conditional on germination) can then be considered as missing at random. In either of these cases (MCAR or MAR), it is valid to analyse the set of observed responses only and ignore the missing observations. If the pattern of missing responses is directly related to the (unseen) value of the response, then the pattern is known as **missing not at random** (MNAR, or sometimes NMAR). Here, missing responses cannot be ignored without the introduction of potential bias, and so it is not valid to analyse only the observed responses. For example, some machines have a lower limit of detection (LOD), below which responses may be recorded as missing. Because the allocation as missing is related directly to the response (i.e. small responses are more likely to be missing) these data are MNAR. Ignoring these low values leads to an over-estimate of the response, and this bias is larger for treatment groups with larger proportions of observations below the LOD. In this specific case, the missingness is related to censoring, and techniques to adjust for censoring are available (see Taylor, 1973). Another example occurs with the detection of outliers, which usually depend on both the treatment group and the observed response. Observations omitted because they have been identified as an outlier should therefore be considered as MNAR. This interpretation highlights the potential dangers associated with removal of outliers, and the limitations of any analysis excluding such outliers. In general, it is necessary to use more advanced modelling techniques to deal with observations MNAR (Carpenter and Kenward, 2013).

As stated above, where observations are MCAR or MAR, the subset of non-missing observations can be analysed. But, except in the case of an unstructured data set (e.g. CRD) with a single treatment factor, the presence of missing values usually leads to non-orthogonality, even if the original design was orthogonal. Several algorithms have been developed to preserve orthogonality for the case of a few missing values in orthogonal designs. One example is the algorithm of Healy and Westmacott (1956), where missing observations are estimated (by an iterative procedure) at the value of the treatment group mean, which results in a zero residual for the observation. With this algorithm, the TrtSS is inflated but the ResSS is not. The ANOVA table is then approximate, and the residual df must be adjusted (i.e. reduced) to account for the missing observations. Estimates of treatment effects and means are correct, but their standard errors do not take proper account of the missing data and so will be under-estimated. Use of this type of algorithm will often be satisfactory when the proportion of missing values is small, but not when a larger proportion of the observations are missing. Beware that some statistical software applies this type of algorithm automatically when missing values are present in the data.

EXAMPLE 11.3: ELISA CALIBRATION

A calibration experiment was done to establish a suitable protocol for an experimental procedure. Three methods of preparation (factor Prep) were tested in combination with four different initial concentrations (factor Conc), with two replicates of each combination. The solutions were applied in randomized order to an ELISA plate and processed. The measured absorbances (variate *Absorbance*) are listed in Table 11.16 and stored in file CALIBRATE.DAT.

This is a CRD so there is no structural component. The appropriate explanatory component is crossed, i.e. [1] + Prep*Conc, with specific interest in the interaction term: if this is large, then the preparation method has a differential effect on the response that depends on the concentration. The readings were transformed to logarithms before analysis. One reading (unit 9) was deemed invalid because of suspected contamination and set missing. ANOVA tables obtained with either the Healy–Westmacott algorithm or with missing responses ignored are in Table 11.17 and, in this case, the differences between the two analyses are small.

The Healy–Westmacott algorithm preserves the orthogonal structure, and so the ANOVA table is invariant to the order of the terms. This is not the case when missing values are omitted, but the degree of non-orthogonality for only one missing value is small, and hence the alternative sequential ANOVA table obtained by adding the factors in the other order is very similar to that shown.

TABLE 11.16

Absorbances (Abs) from the ELISA Calibration Study with Four Concentrations of Substrate (Conc) and Three Preparation Methods (Prep) (Example 11.3 and File CALIBRATE.DAT)

Unit	Prep	Conc	Abs	Unit	Prep	Conc	Abs	Unit	Prep	Conc	Abs
1	1	3	0.482	9	2	3	–	17	3	4	0.056
2	1	1	0.783	10	1	3	0.431	18	2	4	0.073
3	2	1	1.014	11	3	1	1.001	19	3	2	0.808
4	2	1	1.038	12	3	4	0.048	20	3	2	0.888
5	1	4	0.092	13	1	4	0.130	21	1	2	0.780
6	3	1	0.784	14	2	2	0.707	22	1	2	0.759
7	3	3	0.327	15	1	1	0.766	23	3	3	0.364
8	2	2	0.745	16	2	3	0.412	24	2	4	0.070

Source: Data from Rothamsted Research.

TABLE 11.17

Sequential ANOVA Table for the ELISA Calibration Study with One Missing Observation Using Either the Healy–Westmacott Algorithm or Ignoring the Missing Responses (Example 11.3)

		Healy–Westmacott				Ignoring Missing Responses			
Source	df	SS	MS	VR	P	SS	MS	VR	P
+ Prep	2	0.161	0.080	7.30	0.010	0.159	0.079	7.21	0.010
+ Conc	3	23.512	7.837	711.90	< 0.001	23.506	7.835	711.74	< 0.001
+ Prep.Conc	6	0.583	0.097	8.83	0.001	0.583	0.097	8.82	0.001
Residual	11	0.121	0.011			0.121	0.011		
Total	22	24.369				24.369			

Note: df = incremental df, SS = incremental sum of squares, MS = mean square, VR = variance ratio, P = observed significance level.

Later, another five observations (units 4, 11, 15, 17 and 20, all from different treatments) were identified as having been subject to contamination and so the analysis was rerun with these observations also set missing. Two sequential ANOVA tables for the full model [1] + Prep*Conc are in Table 11.18.

Both ANOVA tables have the same (correct) estimate of background variation derived from the ResMS. But the inflated interaction sum of squares in the approximate (Healy–Westmacott) ANOVA table suggest that there is strong evidence of an interaction,

TABLE 11.18

Sequential ANOVA Table for the ELISA Calibration Study with Six Missing Observations Using Either the Healy–Westmacott Algorithm or Ignoring the Missing Responses (Example 11.3)

		Healy–Westmacott				Ignoring Missing Responses			
Source	df	SS	MS	VR	P	SS	MS	VR	P
+ Prep	2	0.290	0.145	11.69	0.009	0.112	0.056	4.52	0.064
+ Conc	3	23.465	7.822	630.83	< 0.001	16.818	5.606	452.13	< 0.001
+ Prep.Conc	6	0.568	0.095	7.64	0.013	0.367	0.061	4.94	0.037
Residual	6	0.074	0.012			0.074	0.012		
Total	17	17.371				17.371			

Note: df = incremental df, SS = incremental sum of squares, MS = mean square, VR = variance ratio, P = observed significance level.

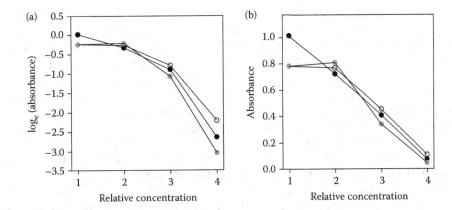

FIGURE 11.3
Predicted absorbance (a) on log scale and (b) back-transformed for four relative concentrations and three methods of preparation (o 1, ● 2, ● 3) (Example 11.3).

whereas the exact ANOVA table (ignoring the missing responses) has a larger observed significance level, although still statistically significant at the 5% level. This demonstrates the possible inflation of significance under the Healy–Westmacott algorithm as the number of missing values increases. Again, as the Healy–Westmacott analysis preserves the orthogonal structure, its ANOVA table is invariant to the order of fitting. This is not the case when the missing values are omitted, and the sequential ANOVA table obtained by fitting factor Conc first is now quite different (not shown) although the same sum of squares and observed significance level is obtained for the interaction term.

The fitted model (with the six disputed observations omitted) is shown on the log scale and back-transformed in Figure 11.3. The prediction SEs on the log scale are 0.079 for treatment combinations with two observations and 0.111 for treatment combinations with one observation, and the SEDs range from 0.111 to 0.157. The absorbances decrease across the concentrations, but it appears that preparation methods 1 and 3 show no difference in response between the first two concentrations, whereas method 2 shows a consistent decrease in absorbance across the full set.

11.5 Incorporating the Effects of Unplanned Factors

Another frequent cause of non-orthogonality is the occurrence of unplanned events within a designed experiment or observational study. For example, if pigeons graze a field experiment unevenly, then we can classify plots as wholly, partially or not grazed to quantify the damage done and then incorporate this new factor into the model to account for the effect of grazing on yield. Because the presence of such variables is (by definition) unplanned, they usually result in a non-orthogonal structure. For example, pigeons are unlikely to distribute their grazing evenly across treatments, and so the grazing factor is likely to be non-orthogonal to the treatment factor(s). We use the term **extraneous** for variables that are unrelated to the original design. In this section, we consider only extraneous variables that are qualitative (i.e. factors); quantitative extraneous variables (i.e. variates) are dealt with by a technique called analysis of covariance, which is described in Section 15.5.

Usually, the aim of analysis is to investigate whether treatments differ after the extraneous factors have been taken into account, and hence these factors should be added into the explanatory component before treatment terms. This leads to adjusted estimates of treatment effects, with the extent of the adjustment depending on the degree of non-orthogonality between the treatment and extraneous factors. The adjusted treatment effects should be more robust after the extraneous factors have been eliminated (corrected for), and including the extraneous factors usually reduces the estimate of background variation, and leads to more precise estimates and more sensitive tests.

It can also be useful to investigate whether there is any evidence of an interaction between the extraneous and treatment factors, but there may be practical difficulties in doing so. On the one hand, if only a small proportion of the full set of treatment × extraneous factor combinations is observed, evaluation of their interaction is based on a small amount of information and may be unreliable. On the other, if all of the combinations are present but unreplicated then fitting this interaction is uninformative (as each combination will be fitted exactly) and may leave insufficient ResDF for a reliable test. However, it is important to understand that omitting this interaction requires an implicit assumption that the interaction is zero.

EXAMPLE 11.4: PLANT HEIGHTS IN GLASSHOUSE*

A glasshouse experiment was done to investigate the effect of the dose of a growth regulator on plant height under controlled conditions. Six increasing doses (factor Dose) were each applied to four replicate plants in separate pots that were arranged according to a CRD in a grid layout consisting of four rows (factor Row) and six columns (factor Column) on a bench. Plant heights (cm, variate Height) were measured six weeks later. The data are listed in the experimental layout in Table 11.19 and are held in file HEIGHTS.DAT.

Preliminary analysis of the plant heights, using explanatory component [1] + Dose, revealed a trend in the plot of standardized residuals against fitted values (see Figure 11.4a). Further investigation showed that this was due to a strong pattern of increasing residual value across columns (see Figure 11.4b).

The glasshouse manager suggested that this pattern was real, as it could be explained by differential shading on one end of the bench. Columns of the design were therefore incorporated as an extraneous factor (called Column) in the analysis, with explanatory component [1] + Column + Dose. Table 11.20 shows the ANOVA tables for the models excluding and including this extraneous factor. There was strong evidence of differences between doses from the original model that ignored columns ($F_{5,18} = 4.03$). However, the variance ratio for the term Dose increased greatly once the effect of columns was eliminated ($F_{5,13} = 18.83$), as the shading effect was partly masking treatment differences. The predicted means for each dose before and after correction for column effects are shown with 95% CIs in Figure 11.5. When the column effects are ignored

TABLE 11.19

Layout and Heights (cm) for Plants in Pots in a Glasshouse Experiment (Example 11.4 and File HEIGHTS.DAT)

		Column of Layout					
		1	2	3	4	5	6
Row of	1	(1) 58.4	(3) 56.8	(4) 61.8	(4) 68.2	(5) 61.6	(3) 70.3
Layout	2	(3) 55.5	(4) 57.2	(6) 50.8	(6) 57.4	(1) 70.0	(5) 61.1
	3	(6) 49.9	(2) 60.9	(6) 50.7	(1) 71.3	(5) 61.3	(4) 70.0
	4	(2) 56.7	(3) 54.5	(2) 61.3	(2) 66.6	(5) 65.8	(1) 69.8

Note: The dose level (1–6) applied to each plant is given within parentheses.

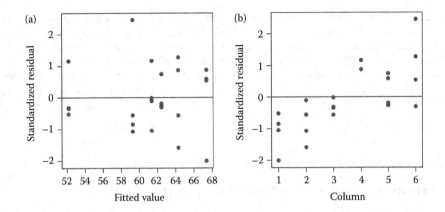

FIGURE 11.4
Plots of standardized residuals against (a) fitted values and (b) column positions using explanatory component [1] + Dose (Example 11.4).

TABLE 11.20

ANOVA Table for Plant Heights from a Glasshouse Experiment with Treatment Effects (Factor Dose) Fitted Either Ignoring or Eliminating the Extraneous Factor Column (Example 11.4)

	Ignoring Columns				Eliminating Columns				
Source	df	SS	MS	VR	P	df	SS	MS	VR	P
+ Column	—	—	—	—	—	5	618.9	123.8	33.49	< 0.001
+ Dose	5	536.1	107.2	4.03	0.012	5	348.1	69.6	18.83	< 0.001
Residual	18	478.8	26.6			13	48.1	3.7		
Total	23	1015.0				23	1015.0			

Note: SS = sum of squares, MS = mean square, VR = variance ratio, P = observed significance level.

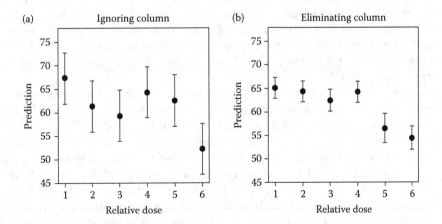

FIGURE 11.5
Predicted plant heights (cm) with 95% CIs (a) ignoring column (SED = 3.65, 18 ResDF) and (b) eliminating column effects using explanatory component [1] + Column + Dose (min SED = 1.14, max SED = 2.10, 13 ResDF) (Example 11.4).

(Figure 11.5a), it appears that only dose 6 restricts height. After adjusting for column effects (Figure 11.5b), prediction SEs (and hence CIs) are much smaller, and both doses 5 and 6 restrict height. The reduction in the estimate for dose 5 is large because this treatment occurred only in columns 5 and 6 on the bench, which had the largest responses. Clearly, any future experiment using this bench should use column as a blocking factor so that the shading can be accommodated in a more efficient manner and, ideally, as orthogonal to treatments.

11.6 Analysis Approaches for Non-Orthogonal Designs

Statistical software often includes several algorithms for the analysis of linear models. For example, most packages include algorithms to produce multi-stratum ANOVA or to fit a linear regression (Chapters 12 to 15), generalized linear model (GLM; Chapter 18) or linear mixed model (LMM; Chapter 16). Different algorithms present estimates in different formats, possibly using different parameterizations. But it is important to understand that, when they can specify the same model, two different algorithms using the same method will produce equivalent results. Alternative algorithms are typically provided in statistical software because different approaches can be used to take advantage of special cases: the most general algorithms can be inefficient for simpler models.

The simplest case occurs when a study consists of a set of unstructured units, so that the structural component of the model is not required. Any general algorithm for analysis of linear models should be able to process a study of this form. This is not the case when both explanatory and structural components are present, which ideally requires an algorithm that recognizes the separate roles of the two components of the model.

The provision of commands to produce a multi-stratum ANOVA is probably the most important requirement of software to analyse designed experiments with blocking or other structure. Unfortunately, the multi-stratum ANOVA can be defined only for orthogonal block structures, and where the treatment structure satisfies certain conditions of balance. In general terms, a block structure is orthogonal if all units at a given level of the hierarchy each contain the same number of units from a lower level. For example, in a RCBD, all blocks contain the same number of plots; hence, this is an orthogonal blocking structure. Similarly, the BIBD, LS and SP designs described in Chapter 9 all have orthogonal blocking structures (although the block and treatment structures are non-orthogonal for the BIBD, see Section 9.3).

For non-orthogonal block structures or unbalanced designs with a structural component, a more general procedure is required. Algorithms for linear mixed models (LMMs) can be used for this purpose, and these methods are introduced in Chapter 16. If facilities for LMMs are not available, then the explanatory and structural components must be combined into a single model. If most treatment comparisons are made within blocks, then it is often possible to get a good analysis by specification of the model terms in an order that mimics the multi-stratum analysis. However, this approach requires a good understanding of the experimental structure and loses information on treatment comparisons made between blocks; this strategy is known as the intra-block analysis. In the remainder of this section, we focus on strategies to produce an intra-block analysis that gives a reasonable approximation to the full analysis using both the explanatory and structural components.

11.6.1 A Simple Approach: The Intra-Block Analysis

In the simplest case, in which treatment effects are estimated within the lowest stratum, an intra-block analysis can be obtained by specification of blocking or other structural factors (and any extraneous factors) before treatment terms in the explanatory component. The structural term corresponding to the deviations (lowest stratum) should always be omitted in this context. For example, in a BIBD, the intra-block analysis can be obtained from

Explanatory component: [1] + Blocks + Treatments

The fitted model provides estimates of treatment effects adjusted for (eliminating) block effects. When a treatment term has deliberately been completely confounded with blocking or structure at some level, that treatment term should be fitted before the confounded structural term. For example, for the SP design presented in Example 9.2, the irrigation effects are confounded with whole plots within blocks, and so the combined explanatory component would be written as

Explanatory component: [1] + Block + Irrigation + Block.WholePlot + Species
                          + Irrigation.Species

This results in the same mean squares as in the multi-stratum ANOVA, but some care is required to obtain the correct variance ratios. The Irrigation mean square must be divided by the Block.WholePlot mean square, but the Species and Irrigation.Species mean squares must be divided by the ResMS. Following these principles, one can reconstruct the multi-stratum ANOVA table. Unfortunately, the confounding between the Irrigation and Block.WholePlot terms induces dependencies that make the parameterization, and hence the estimated Irrigation effects, difficult to interpret. For the same reason, it is difficult to accommodate pseudo-replication within the explanatory component (although this can be avoided by analysis of means of the pseudo-replicates in cases of equal replication).

In general, the intra-block analysis with a single model formula is sensible only when most of the treatment differences are estimated at the lowest level of the structure. In other cases, use of LMMs (see Chapter 16) is preferable. An example of intra-block analysis for a design with non-orthogonal blocks and treatments is presented below.

EXAMPLE 11.5: EFFECT OF TYPE AND SIZE OF CUTTING ON WILLOW YIELD*

A field experiment was designed to investigate whether the type of cutting planted affects the subsequent growth of willows. Cuttings of five different types (A–E, factor Type) were to be planted, and growth parameters would be measured over the following seasons, including yield at the end of the first year. At planting time, it was realized that the cuttings to be planted varied greatly in size, and that this might also have an effect on subsequent growth. Two options were considered here. Cutting size could be confounded with blocks, so that each block contained cuttings of the one size only. Alternatively, cutting size could be investigated as an extraneous factor, in addition to type. The second option was taken, and cuttings were classified as small (S), medium (M) or large (L, factor Size). Not all of the type × size combinations were available, and the total number of plots was fixed at 25. The design was based on a five-block RCBD with respect to cutting type, and the different sizes were allocated in as balanced a way as possible across blocks (factor Block) and cutting types. The yield (variate *Yield*) with allocation of size and type combinations to the five blocks is shown in Table 11.21 and stored in file CUTTINGS.DAT.

TABLE 11.21

Yield after First Year for Willows Grown from Cuttings of Different Types (A–E) and Size (S, M, L) in Five Randomized Blocks of Five Plots (Example 11.5 and File CUTTINGS.DAT)

Block	Plot	Type	Size	Yield	Block	Plot	Type	Size	Yield
1	1	D	M	15.16	4	1	E	M	23.84
1	2	E	L	18.31	4	2	D	L	21.30
1	3	A	M	23.94	4	3	B	M	24.51
1	4	B	S	24.37	4	4	A	S	25.77
1	5	C	L	12.04	4	5	C	M	18.34
2	1	D	L	15.30	5	1	A	S	23.01
2	2	A	L	19.81	5	2	C	L	14.74
2	3	E	M	16.45	5	3	E	M	21.67
2	4	B	S	18.45	5	4	D	S	17.30
2	5	C	M	17.28	5	5	B	M	16.63
3	1	B	M	24.56					
3	2	A	M	24.60					
3	3	C	S	25.11					
3	4	D	M	22.90					
3	5	E	L	25.71					

Each cutting type appears once in each block, and each size appears at least once in each block, although the size × type combinations are unequally replicated, and two combinations (S × E and L × B) were not available (Table 11.22).

Since the design is unbalanced, a multi-stratum ANOVA cannot be formed. However, the treatment information for the cutting type main effect could be retrieved from within-block comparisons (as all types occur once in each block) and so an approximate analysis should be acceptable here, as this is the main focus of interest. Block effects (factor Block) must be added into the model first, followed by the extraneous factor Size, so that cutting size can be eliminated before comparing cutting types (factor Type). As a preliminary model, we include the interaction between cutting size and type (term Size.Type) to give

Explanatory component: [1] + Block + Size + Type + Size.Type

In the resulting ANOVA table (Table 11.23), there is strong evidence of differences in yield between blocks ($F_{4,8}^B = 7.385$, $P = 0.009$), and a suggestion of a difference between

TABLE 11.22

Occurrence of Cutting Type × Size Combinations in the Willow Yield Trial (Example 11.5)

Cutting Size	Cutting Type					
	A	**B**	**C**	**D**	**E**	**Total**
S	2	2	1	1	0	6
M	2	3	2	2	3	12
L	1	0	2	2	2	7
Total	5	5	5	5	5	25

TABLE 11.23

Sequential ANOVA Table for the Willow Yield Trial (Example 11.5)

Change	df	Sum of Squares	Mean Square	Variance Ratio	P
+ Block	4	186.4298	46.6074	$F^B = 7.385$	0.009
+ Size	2	49.0181	24.5090	$F^S = 3.883$	0.066
+ Type	4	96.9158	24.2290	$F^T = 3.839$	0.050
+ Size.Type	6	21.0791	3.5132	$F^{S.T} = 0.557$	0.754
Residual	8	50.4896	6.3112		
Total	24	403.9324			

sizes ($F^S_{2,8} = 3.883$, $P = 0.066$), although this term was partially confounded with blocks and so may be masked by block differences. There was some evidence of a difference between cutting types ($F^T_{4,8} = 3.839$, $P = 0.050$). There was no evidence of an interaction between cutting size and type ($F^{S.T}_{6,8} = 0.557$, $P = 0.754$) and so this term was dropped and the model refitted, with all other terms being retained.

Predictions of yield for different cutting sizes can be calculated from the table of fitted values obtained from explanatory component [1] + Block + Size + Type. This gives a three-way table classified by Block, Size and Type, and marginal means can be taken for cutting type to give the predictions shown in Figure 11.6a.

Cutting types C and D produced the least yield, with larger yields produced by types A and E, and type B intermediate. For comparison, Figure 11.6b shows the predicted means obtained if the effect of cutting size is ignored, i.e. from explanatory component [1] + Block + Type. The two sets of predictions are very similar, partly because cutting sizes are reasonably balanced across the different types (so the analysis is close to orthogonal), and partly because the effects of cutting size are quite small. Predictions for cutting size are listed in Table 11.24, and it appears that first year yield is somewhat greater for smaller cuttings, with medium and large cuttings producing similar yield.

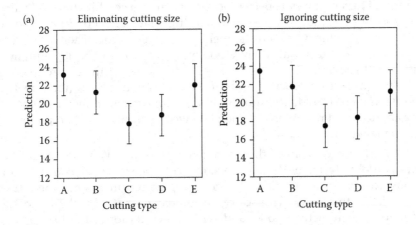

FIGURE 11.6

Predicted yields with 95% CIs (a) eliminating cutting size using explanatory component [1] + Block + Size + Type (min SED = 1.449, max SED = 1.538, 14 ResDF) and (b) ignoring cutting size (SED = 1.565, 16 ResDF) (Example 11.5).

TABLE 11.24

Predicted Yield with SE for Cutting Sizes from Explanatory
Component [1] + Block + Size + Type (Example 11.5)

	Size		
	Small	Medium	Large
Prediction	22.394	20.109	19.347
SE	0.9889	0.6667	0.9190

Note: Maximum LSD (5% significance level) = 3.0335.

EXERCISES

11.1* Thirty-two moths were assigned at random to separate flight mills, and the
distance (m) flown by each moth during one night was measured electroni-
cally. The species (A, B or C) and sex (F, M) of each moth was recorded. File
FLIGHT.DAT contains the mill number (*Mill*), with the species, sex and distance
flown (factors Sex, Species, variate *Distance*) for the moth in each mill. The
aim of the statistical analysis is to investigate whether there are any consistent differences among species or between sexes, and whether any difference
between sexes is consistent across species.

a. Write down an explanatory model for this experiment in terms of the two
explanatory factors (Sex and Species). Consider the replication of each of
the factor combinations and decide whether this structure is orthogonal.

b. Fit your model with the $\log_{10}$-transformed distances as your response variable. Use two different orders for adding terms into the model and explain
the differences in the corresponding sequential ANOVA tables.

c. Identify the best predictive model for these data. Interpret your model and
produce predictions with SE for the distance flown overnight by each sex
and species of moth.

11.2 Weeds within a crop can greatly decrease yield and there is interest in the
impact of different weed species, both alone and in combination. A RCBD with
two blocks of 30 plots was set up to investigate the effect of different densities
of barley and chickweed on the yield of a linseed crop. There were 29 treatments in total: a factorial combination of five densities of barley with five densities of chickweed (25 treatments), with duplicates of the control (no weeds),
plus two higher densities of each of the individual species (four treatments).
File DENSITY.DAT holds the unit numbers (*ID*), structural factors (Block, Plot),
the applied seed rate of barley and chickweed (variates *B, C*) and the resulting
grain yield (variate *Grain*).*

Create factor versions of the weed density variates. Write down an explanatory model in terms of these factors, identify the structural component of the
model for this trial, and fit the model using both components (the intra-block
analysis). Is there any evidence of an interaction between the weed species?
Identify the predictive model and write down its form. Produce predictions
with SE for each combination of weed seed densities present in the trial. (We
re-visit these data in Exercise 17.9.)

---

* Data from P. Lutman, Rothamsted Research.

11.3 A glasshouse trial was set up to investigate the profit associated with different methods of growing peppers (Mead et al., 2012, Chapter 7). The experiment continued over two years, and two blocks of six glasshouse compartments were used in each year (24 compartments in total), with one set of conditions applied to each compartment. The eight treatments were all combinations of standard (0) or enhanced (1) levels of heat, light and CO_2. All treatments were tested in the first year (four tested twice, four tested once) and the best five were tested again in the second year (with one treatment repeated in each block). The unit numbers (*ID*), structural factors (**Year, Block, DComp**) and treatment factors (**Heat, Light, CO_2**) are held in file PEPPERS.DAT, with a measure of profit (yield remaining after accounting for costs, variate *Profit*).

Analyse the profit, taking account of the experimental structure (years and blocks), as well as the three treatment factors. Identify a sensible predictive model and suggest which combination of the three factors should be used in practice to maximize profit. (We re-visit these data in Exercise 16.3.)

11.4 In Exercises 6.3 and 8.5 you analysed the score of potato scab from a CRD. Repeat your analysis and plot the residuals in field layout (as defined by the factors **Row** and **Col** provided in file SCAB.DAT). Identify any clear spatial trend and add suitable extraneous factors into the model to account for this. Compare your new model with the original, and comment on whether the increased complexity is justified.

11.5* A NIRS machine was used to measure the protein content of 35 accessions of wheat. Sets of six samples were analysed together in each run of the machine and the measurements were made in seven pairs of replicate runs. Each pair of runs used subsamples of seed from the same five accessions with a standard control sample (used in all runs). File NIRS.DAT holds the unit numbers (*ID*), structural factors (**Pair, Rep**), information on lines (factors **Type, Accession**) and protein measurements (variate *Protein*). Write down the explanatory and structural components of the model for this trial, and fit the model using both components (the intra-block analysis). Is there evidence of variation in protein content between the 35 accessions? What are the issues with the design of this experiment? Can you suggest a better design? (We re-visit these data in Exercise 16.3.)

11.6 Five pruning treatments were tested on apple trees (Pearce, 1965, Section 6.2). A balanced incomplete block design was used to allocate the five treatments (a–e) to four branches on each of 15 trees (60 branches in total). One of the outcomes measured was the length of shoots from the middle third of each branch, but this was only measured for treatments a, b and d. The shoot lengths (variate *Length*) are in file SHOOT.DAT with the unit numbers (*ID*), structural factors (**Tree, Branch**) and treatment factor (**Treatment**). Analyse these data, accounting for possible differences between trees as well as treatments. Can you identify which treatment produces the longest shoots? (We re-visit these data in Exercise 16.3.)

11.7 An experiment using a Latin square design was intended to compare the yield of six varieties of turnip, but three plots were damaged by vandals before harvest and their yield could not be obtained (Hand et al., 1994, Data Set 78). The yield (variate *FreshWt*, fresh weight in pounds per plot), plot numbers (**Plot**) and the design (**Row, Column**) and treatment (**Variety**) factors are held in file VANDAL.DAT. Write down a model for this trial that recognizes the structure of the Latin square design. Compare analysis of the yield by multi-stratum ANOVA with Healy–Westmacott estimation of missing values to an intra-block analysis excluding the missing plots. Do the two methods give the same conclusions in terms of variety comparisons? (We re-visit these data in Exercise 16.3.)

12

Models for a Single Variate: Simple Linear Regression

In Chapter 1, we described experimental and observational studies as scientific enquiries in which an outcome (or response) is investigated with the objective of understanding how it is affected by the experimental conditions. In this context, statistical models are used to quantify relationships between the response variable and one or more explanatory variable(s) that define the conditions. Two simple examples were illustrated in Section 1.3, where the single explanatory variable corresponded to either a qualitative variable (or factor, Example 1.1) or a quantitative variable (or variate, Example 1.2). In Chapter 4, we presented details of the analysis for data classified by a single explanatory factor, including the form of the underlying model, parameter estimation and statistical inference. This was mainly placed in the context of designed experiments. We now focus on the analysis of data where the single explanatory variable is quantitative, or a variate. This is usually known as **regression analysis**. However, the situation with either a qualitative or a quantitative explanatory variable results in the same basic form of linear model (Section 1.4). Both consist of a systematic component and a random component, with analysis based on the same underlying statistical theory to estimate parameters and predict from the fitted model. In this chapter, we are concerned only with models including a single explanatory variate. More broadly, regression analysis refers to the more general approach with any number of quantitative (and qualitative) explanatory variables. In Chapter 14, we shall extend the model to incorporate two or more explanatory variates (multiple regression), and in Chapter 15, we consider models that also include qualitative explanatory variables (factors), sometimes called regression with groups. In Chapter 16, we use linear mixed models to take account of the structure in the observations, including blocking and pseudo-replication.

Here, we begin by presenting the simple linear regression (SLR) model (Section 12.1) followed by parameter estimation (Section 12.2). ANOVA assesses whether the variation in the response that is associated with the explanatory variate is large in comparison to the background variation (Section 12.3), and also provides an estimate of this background variation used for inference on the model parameters (Section 12.4). A primary purpose of regression analysis is prediction of the response at given values of the explanatory variate (Section 12.5). Goodness-of-fit statistics can be used to assess the quality of the fitted model or to compare it against other potential models (Section 12.6). Uncertainty in the explanatory variable changes the interpretation of the model, and we discuss this issue in Section 12.7. We then show how to use replication to formally evaluate the fit of the model (Section 12.8). Finally, we describe two simple variations on the basic SLR model – the use of standardized explanatory variates and regression through the origin – and explain the process of inverse prediction (Section 12.9).

12.1 Defining the Model

The simplest model that can be used to describe the relationship between a response variable and a quantitative explanatory variable (or variate) takes the form of a straight line passing through the scatter of points arising when values of the response variable are plotted against the corresponding values of the explanatory variate. This type of model is known as a **simple linear regression** (SLR) and it can be represented mathematically as

$$y_i = \alpha + \beta x_i + e_i \,,\tag{12.1}$$

where y_i and x_i are the values of the response and the explanatory variates, respectively, for the ith observation. The quantity e_i represents a random deviation for the ith observation, and the subscript i ranges from 1 to N, where N is the total number of observations. The model is a straight line defined in terms of the model parameters α and β, as shown in Figure 12.1. Parameter α (often called the intercept or constant parameter) corresponds to the point at which the line intercepts the y-axis, and is the value of the straight line when the explanatory variate is equal to 0. Parameter β, the coefficient of the explanatory variate, is the slope (gradient) of the line, i.e. the change in the response produced by a unit change in the explanatory variate. The SLR is called *simple* because it contains a single explanatory variable and *linear* because the response is expressed in a linear form, i.e. as a sum of terms that each consists of a coefficient multiplied by an explanatory variable.

The deviation e_i represents the random stochastic or probabilistic element of the model, sometimes called the random noise or residual error. It can be visualized as the vertical displacement of the ith observation from the line (see Figure 12.1). In Section 4.1, we described the deviations as representing background variation about group means. In the context of regression analysis, background variation reflects the discrepancy in response between measurements from two units with the same value of the explanatory variate. However, here, we are fitting a structured model, in the form of a straight line, and if the pattern in the observations does not match the form of the model, then the deviations also

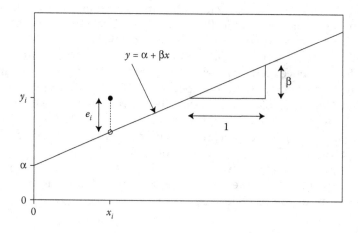

FIGURE 12.1

Representation of a SLR model, showing the line (—), with one observation (●) and its predicted value (o).

encompass the systematic difference between that pattern and the straight line. This is undesirable, as the model then does not capture the trend and the deviations no longer represent purely random variation. Therefore, you should always view the data before fitting the model, for example by using a scatter plot, to check that the underlying relationship could plausibly be a straight line. We consider the implications of this below and give methods for graphically examining the model fit in Chapter 13.

For the SLR model in Equation 12.1, the systematic component of the model is the straight line, $\alpha + \beta x_i$, and the random component is the deviation, e_i. Recall that in Chapter 4, we introduced a symbolic notation to represent linear models. This notation specifies the response variable and the systematic component of the model in terms of the explanatory variables. This reflects the form of model specification required for statistical software, although the details vary according to the package used. For SLR, the straight line relationship is the explanatory component of the model, and we write the model in symbolic form as

Response variable: *y*

Explanatory component: *[1] + x*

where the variate *y* contains the observed responses. The term *[1]* denotes a variate that takes value 1 everywhere, and is associated with the intercept parameter, α. The variate *x* contains the values of the explanatory variable and is associated with the slope parameter β.

In the context of designed experiments, we partitioned the systematic component into explanatory terms (associated with treatments applied) and structural terms (associated with the experimental structure, such as blocking or pseudo-replication). For regression analysis, the same partition applies, although the structural component can be omitted when no structure is present, as may be the case for observational studies. However, if the units are structured, then it is important that the structure is incorporated into the model. This extension of the model is not usually provided within software designed for regression analysis and we discuss this further in Section 15.3 and Chapter 16.

EXAMPLE 12.1A: DIPLOID WHEAT

Several morphological traits were measured for 190 seeds selected at random from a line of diploid wheat, *Triticum monococcum*, with the aim of identifying variables associated with differences in seed weight (Jing et al., 2007; Wheat Genetic Improvement Network (WGIN): www.wgin.org.uk). The variables measured were weight (mg), diameter (mm), length (mm), moisture content (%) and endosperm hardness (single-kernel characterization system index value). The data are in file TRITICUM.DAT which contains a variate *DSeed* to identify each seed in addition to variates *Weight, Diameter, Length, Moisture* and *Hardness*. A subset of the data is in Table 12.1 and the full set is given in Table A.1.

Seed size, as measured by length, is expected to be a major contributor to differences in seed weight, and so, we start by examining the relationship between seed weight and seed length, using the scatter plot presented in Figure 12.2.

The relationship between the two variates appears approximately linear; so, it makes sense to fit a SLR that relates the weight of the *i*th seed, *Weight_i*, to the length of the *i*th seed, *Length_i*, to investigate this relationship further. The statistical model is

$$Weight_i \;=\; \alpha + \beta\,Length_i + e_i \;,$$

TABLE 12.1

First Four and Last Four Observations of Seeds of Diploid Wheat from a Study to Identify Variables Associated with Variation in Seed Weight (Example 12.1A, Full Data in File TRITICUM.DAT and Table A.1)

Seed	Weight	Length	Diameter	Moisture	Hardness
1	30.15	3.27	2.09	10.27	−16.63
2	35.51	3.65	2.34	10.61	−8.27
3	29.16	3.36	2.15	10.27	−21.45
4	16.82	2.77	1.79	11.05	4.13
⋮	⋮	⋮	⋮	⋮	⋮
187	27.66	3.60	2.31	10.88	−22.68
188	26.54	3.58	2.29	10.49	3.30
189	30.90	3.17	2.03	10.37	−17.83
190	18.94	2.45	1.62	10.08	−7.06

Source: Data from H.-C. Jing and K. Hammond-Kosack, Rothamsted Research.

where e_i represents the deviation in the weight of the ith seed from the straight line relationship and i runs from 1 to 190 to represent the 190 observations. Parameter α represents the intercept, the expected seed weight for zero seed length. From simple biological arguments, we would expect the intercept to be zero, but this ignores aspects of the statistical modelling process that we discuss further in Example 12.1B. This model is written in symbolic form as

Response variable: *Weight*
Explanatory component: *[1] + Length*

where the variate *Weight* contains the observed seed weights and variate *Length* contains the corresponding seed lengths.

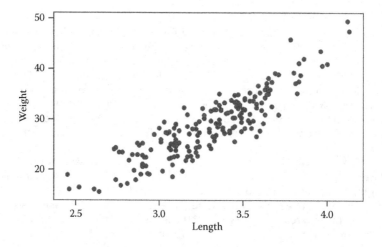

FIGURE 12.2
Scatter plot of weight (mg) versus length (mm) for 190 diploid wheat seeds (Examples 12.1A and 12.1E).

To fit the SLR model, some assumptions must be made about the deviations, e_i. These assumptions apply to any linear model, and so were presented in Section 4.1. Nevertheless, we repeat them here for completeness.

Assumption 1

$$E(e_i) = 0 \quad \text{for } i = 1 \ldots N .$$

The expected value (function E) of each deviation is assumed to be zero. This means that the population mean of the deviations is zero, which implies no systematic bias in the observations.   ∎

Assumption 2

$$\text{Var}(e_i) = \sigma^2 \quad \text{for } i = 1 \ldots N .$$

The variances (Var) of the deviations are the same for all units. This is also known as homoscedasticity, or homogeneity of variances.   ∎

Assumption 3

$$\text{Cov}(e_i, e_j) = 0 \quad \text{for all } i \neq j, \text{ and } i, j = 1 \ldots N .$$

The covariance (Cov) between deviations for two separate observations is zero, i.e. the deviations are independent.   ∎

Assumption 4

$$e_i \sim \text{Normal}(0, \sigma^2) .$$

The deviations follow a Normal distribution with mean 0 and variance σ^2.   ∎

In addition, we make an assumption on the explanatory variables:

Assumption 5

The values of the explanatory variables (factors or variates) are known without error.   ∎

Common violations of Assumptions 1 to 4 were described and discussed in Chapter 5. Here, we reiterate that Assumption 3 is most often violated when data are collected from the same source at different times, and that Assumption 4 is required to make any statistical inferences valid that rely on the Normal distribution. Assumption 5, which states that the explanatory variable is known or measured without error, is of particular importance for regression models. This assumption may be realistic in experiments where levels of the explanatory variate are controlled by the experimenter (e.g. fixed amounts of nitrogen fertilizer to be applied to field plots). However, in observational studies, values of explanatory variates are usually observed rather than under the control of an experimenter, and these observations are often prone to error. This does not invalidate the analysis, but it does change the interpretation of the fitted model and is discussed further in Section 12.7.

In Chapter 5, we introduced some diagnostic tools that can be used to check the plausibility of assumptions made about the deviations in models with a single qualitative explanatory variable. As the same assumptions about the deviations apply here, the same tools can be used to check the validity of models with a single quantitative explanatory variable (they are also appropriate for more complex regression models). These tools are revisited in Chapter 13. These tools use the residuals obtained from fitting the model, and for regression models, it is particularly important to use standardized residuals for this purpose (see Section 5.1.2). However, the assumption of a structured form for the response, here a straight line, means that the form of the model must also be checked. We refer to a mismatch between the observed pattern and fitted model as model misspecification, and some additional diagnostic tools to deal with such situations are described in Chapter 13.

12.2 Estimating the Model Parameters

In Section 4.2, we outlined the principle of least-squares estimation, the method that finds the best-fit model by minimizing the sum, across all observations, of the squares of the differences between the observed data and the fitted values. This principle can be used for any linear model, which includes the estimation of parameters for SLR. For the SLR model, the fitted values can be written as

$$\hat{y}_i = \hat{\alpha} + \hat{\beta} x_i \ .$$

As stated earlier, the hats (^) over y_i, α and β specify that they are estimates of population values for which the true values are not known. The simple residuals (see Section 5.1.1), which are estimates of the deviations, e_i, are computed as

$$\hat{e}_i = y_i - \hat{y}_i = y_i - (\hat{\alpha} + \hat{\beta} x_i) \ .$$

The quantity minimized to obtain the parameter estimates is the sum of these squared residuals, which for the SLR model is

$$\sum_{i=1}^{N} \hat{e}_i^2 = \sum_{i=1}^{N} (y_i - \hat{y}_i)^2 = \sum_{i=1}^{N} [y_i - (\hat{\alpha} + \hat{\beta} x_i)]^2 \ .$$

As in Section 4.2, when minimized, this quantity is known as the residual sum of squares, denoted as ResSS. Again, we do not present the mathematical details of the minimization process here, but the interested reader can find them in Section C.3.

We can write estimates of the model parameters, α and β, in terms of sums of squares and cross-products of the response and explanatory variates. The sum of squares for the response variable is the sum, over all observations, of the squares of the differences between the observed responses and their mean, written in mathematical terms as

$$SS_{yy} = \sum_{i=1}^{N} (y_i - \bar{y})^2 .$$

The sum of squares is simply equal to the unbiased sample variance multiplied by the degrees of freedom for this variance, $N - 1$ (see Section 2.1). We denote this sum of squares as SS_{yy}, where 'SS' denotes a sum of squares and the subscript identifies the relevant variate.

The sum of squares for the explanatory variate takes a similar form: the sum, over all observations, of the squares of the differences between the values of the explanatory variate and their mean, i.e.

$$SS_{xx} = \sum_{i=1}^{N} (x_i - \bar{x})^2 .$$

Again, the sum of squares for the explanatory variate equals its unbiased sample variance multiplied by $N - 1$.

Finally, the sum of cross-products between the response and explanatory variate is calculated as the difference between the observed response on each unit and the mean response, multiplied by the difference between the value of the explanatory variate on that same unit and the mean of the explanatory variate. These quantities are then summed over all observed units. This is mathematically written as

$$SS_{xy} = \sum_{i=1}^{N} (x_i - \bar{x})(y_i - \bar{y}) .$$

The symbol for the sum of cross-products, SS_{xy}, refers to the two variables used to form it, and the sum of cross-products is equal to the unbiased sample covariance between the two variates, s_{xy} (Section 2.5), multiplied by $N - 1$. Note that the sum of cross-products between a variate and itself is simply the sum of squares for that variate.

The least-squares estimate of the unknown population slope parameter, β, is equal to the sum of cross-products between the response and explanatory variate, divided by the sum of squares for the explanatory variate, i.e.

$$\hat{\beta} = \frac{SS_{xy}}{SS_{xx}} .$$

The estimate of the unknown population intercept parameter, α, can then be written in terms of the estimated slope and the sample means for the response and explanatory variates, thus

$$\hat{\alpha} = \bar{y} - \hat{\beta}\bar{x} .$$

It follows directly that the best fitting line passes through both sample means, as when $x_i = \bar{x}$, the fitted value is the mean response, $\hat{y}_i = \bar{y}$.

EXAMPLE 12.1B: DIPLOID WHEAT

We can now calculate parameter estimates for the SLR model for the diploid wheat data, which describe the variation in seed weight as a function of the explanatory variable seed length.

For these data, the sum of squares for the explanatory variate, seed length, takes the value $SS_{xx} = 19.2699$, and the sum of cross-products between seed weight and seed length is $SS_{xy} = 330.9297$. The mean length is $\bar{x} = 3.295$ mm and the mean weight is $\bar{y} = 28.658$ mg. The parameter estimates for the SLR model are therefore

$$\hat{\beta} = \frac{SS_{xy}}{SS_{xx}} = \frac{330.9297}{19.2699} = 17.173 ,$$

$$\hat{\alpha} = \bar{y} - \hat{\beta}\bar{x} = 28.658 - (17.173 \times 3.295) = -27.931 .$$

The fitted model can therefore be written as

$$\widehat{Weight}_i = -27.931 + 17.173 \, Length_i ,$$

where the hat over the variable name denotes the estimated fitted value. The units of the intercept and slope here are mg and mg/mm, respectively, and an increase of 1 mm in seed length is expected to produce an increase of 17.17 mg in seed weight. The intercept represents the estimated average weight for seeds of length zero (i.e. –27.93 mg). Biologically, this is a startling value for two reasons: we clearly cannot have negative seed weights, and we expect a seed with zero length to have zero weight. This means we need to check that the model is appropriate for the data, but it does not necessarily mean that the model is inappropriate. The fitted model represents the best fitting line over the range of observed seed lengths (in this case from 2.45 to 4.13 mm). If this line is not representative of the unseen relationship over the range from 0 to 2.45 mm, then it is possible for the predicted value of zero length to be inaccurate even if the model is a good representation of the observed data. In this example, a length of 0 mm corresponds to an extrapolation far outside the range of the observed data. We discuss the distinction between interpolation and extrapolation in Section 12.5.

The fitted model is shown in Figure 12.3 together with the observed data (and a 95% confidence interval [CI] for the fitted line, explained in Section 12.5).

You should always inspect the behaviour of the fitted model, particularly for more extreme values of the explanatory variate. Figure 12.3 suggests some curvature in the relationship as all observations are above the fitted line for the shortest and longest seeds (length < 2.6 or > 4.8 mm). A composite set of residual plots for this model, based on standardized residuals, is presented in Figure 12.4. The histogram and the Normal plot do not indicate strong departures from a Normal distribution for the residuals (see Section 5.2.3), and the absolute residual plot shows no evidence of variance heterogeneity. The fitted value plot suggests some trend in the residuals, with more negative residuals for intermediate weights and largely positive residuals for the lightest and heaviest seeds. This gives further evidence of some curvature in the relationship, although a straight line appears to be a reasonable overall approximation. We discuss the use of residual plots to investigate the fit of this model in more detail in Chapter 13. However, for now, we consider that the data are reasonably consistent with the assumptions underlying the linear model.

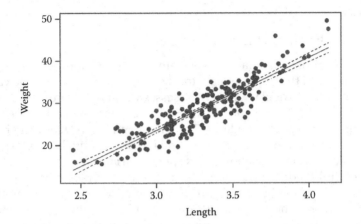

FIGURE 12.3
Scatter plot of weight versus length for 190 diploid wheat seeds together with the fitted straight line (—) and 95% CIs (– –) for the expected mean response (Example 12.1B).

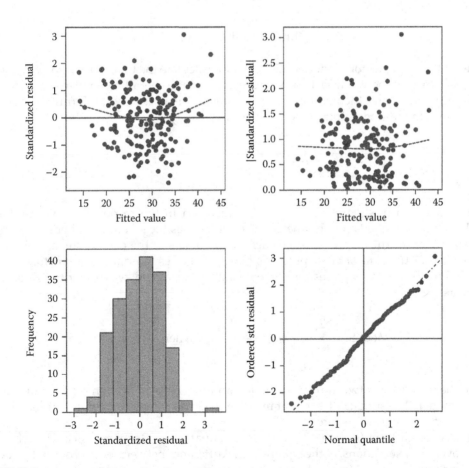

FIGURE 12.4
A composite set of residual plots after fitting a SLR model to the diploid wheat data (Example 12.1B).

12.3 Assessing the Importance of the Model

As for models containing qualitative explanatory variables (or factors), ANOVA can be used to assess how well the single explanatory variate model describes the variation in the observed response. In particular, ANOVA assesses whether the variation in the response that is explained by the explanatory variate is larger than the background variation. In more general linear regression models (Chapters 14 and 15), ANOVA can be used to further assess the relative importance of several quantitative or qualitative explanatory variables, or both, in explaining variation in a response variable.

In the case of SLR (and like the single-factor model of Chapter 4), we use ANOVA to partition the total variation of the response into the portion explained by the systematic component of the model (here, a straight line depending on the explanatory variate) and the residual, or unexplained, portion attributed to the random component of the model. As described in Section 4.3, ANOVA quantifies variation in terms of sums of squares. Hence, the total variation (TotSS) is partitioned into the variation due to the model, i.e. the regression line (ModSS, the model sum of squares) and the residual variation (ResSS, the residual sum of squares) so that

$$\text{TotSS} = \text{ModSS} + \text{ResSS} . \tag{12.2}$$

As with the single-factor model (Section 4.3), we calculate the total sum of squares by taking the difference between each observed value and the overall mean response, squaring these differences and then adding them together. This is exactly the form of the sum of squares for the response, introduced earlier, so

$$\text{TotSS} = \sum_{i=1}^{N} (y_i - \bar{y})^2 = \text{SS}_{yy} .$$

The model sum of squares represents the variation in the response accounted for, or explained by, the fitted straight line and is calculated as the sum, over all observations, of the square of the difference between each fitted value and the overall mean. This equals the square of the sum of cross-products between the response and explanatory variate, divided by the sum of squares for the explanatory variate; so, we can write this algebraically as

$$\text{ModSS} = \sum_{i=1}^{N} (\hat{y}_i - \bar{y})^2 = \frac{(\text{SS}_{xy})^2}{\text{SS}_{xx}}, \quad \text{where } \hat{y}_i = \hat{\alpha} + \hat{\beta} x_i .$$

Note the similarity between this expression and the TrtSS in Section 4.3.1 for the single-factor model; the estimated treatment mean in that case is replaced by the estimated fitted value from the regression here.

Finally, recall from Section 12.2 that the residual sum of squares is calculated as the sum, over all observations, of the square of the difference between each observed response and its associated fitted value. Alternatively, we can rearrange Equation 12.2 to show that the residual sum of squares is equal to the total sum of squares minus the model sum of squares, and we can therefore write this algebraically as

$$\text{ResSS} = \sum_{i=1}^{N} (y_i - \hat{y}_i)^2 = \text{SS}_{yy} - \frac{(\text{SS}_{xy})^2}{\text{SS}_{xx}} . \tag{12.3}$$

Once any two of the sums of squares are known, the third can be obtained by rearrangement of Equation 12.2.

As in the case of the single-factor model, we also need to calculate the amount of information associated with each sum of squares, quantified by the degrees of freedom. Again, there is a partition of the total degrees of freedom (TotDF) into a portion associated with the SLR (ModDF) and a portion associated with the residual variation (ResDF), such that

$$\text{TotDF} = \text{ModDF} + \text{ResDF} .$$

To calculate the df associated with each component, we use the same recipe developed in Section 4.3.1. The individual components in each sum of squares consist of fitted values from some model minus an adjustment. The df for a sum of squares counts the number of parameters required in the model used to calculate the fitted values, minus the number required to calculate the adjustments. In the TotSS, the values are the individual observations (requires N parameters) and the adjustment is the overall mean (requires one parameter). Hence, the total df is $\text{TotDF} = N - 1$. In the ModSS, the values are the fitted values from the SLR (requires two parameters) and the adjustment is again the overall mean (requires one parameter). Hence, the model degrees of freedom are $\text{ModDF} = 1$. And finally, in the ResSS, the values are the individual observations (requires N parameters) and the adjustment is the fitted values from the SLR (requires two parameters). Hence, the residual degrees of freedom are $N - 2$. Alternatively, the ResDF can be found by subtraction as

$$\text{ResDF} = \text{TotDF} - \text{ModDF} = N - 2 .$$

The model and residual sums of squares are then divided by their corresponding degrees of freedom to produce their respective mean squares. These are then on a common scale and so quantify the amount of variation associated with each component of the model. Of particular interest is the residual mean square (ResMS) which is an estimate of the background variability, denoted as s^2, and can be mathematically written as

$$s^2 = \text{ResMS} = \frac{\text{ResSS}}{N - 2} .$$

Intuitively, the ResMS quantifies the variation of all observations around the true regression line and – if the model is a good description of the data – it should arise from background variation alone. The model mean square (ModMS) arises from variation associated with the straight line relationship. If there is no linear dependence of the response variable on the explanatory variable, then the ModMS can arise only from chance background variation, and so should be of similar size to the ResMS, allowing for sampling variation. This concept is formalized by consideration of the expected values of these mean squares. The expected value of the residual mean square is the true background variation, σ^2, i.e.

$$E(\text{ResMS}) = \sigma^2 .$$

The expected value of the model mean square is equal to the true background variation, σ^2, plus the sum of squares for the explanatory variate, SS_{xx}, multiplied by the square of the true slope parameter, β, i.e.

$$E(ModMS) = \sigma^2 + \beta^2\, SS_{xx}\ .$$

If there is no linear dependence of the response variate on the explanatory variate, then the true slope parameter is zero ($\beta = 0$), and the second term in the expectation of ModMS also becomes zero, regardless of the values of the explanatory variate. Both mean squares will then have the same expected value. This is the basis for the use of an F-test in the context of ANOVA for a SLR, testing the null hypothesis of no linear dependence of the response variable on the explanatory variable, formally expressed as H_0: $\beta = 0$. This null hypothesis is compared against an alternative hypothesis of the presence of a linear dependence between these variables, namely H_1: $\beta \neq 0$, in which case the ModMS is expected to be larger than the ResMS.

We obtain the test statistic by dividing the model mean square by the residual mean square, and the quotient is the **variance ratio** or observed F-statistic, which we denote as F, i.e.

$$F = \frac{ModMS}{ResMS}\ .$$

If the null hypothesis is true, we expect the value of the variance ratio to be close to 1. If this ratio is larger, so that the variation associated with the regression line is greater than the background variation, then this gives evidence that the slope of the line is not zero.

More formally, under the null hypothesis, such a ratio of two independent mean squares has an F-distribution, and the amount of evidence can be quantified. Here, the F-distribution numerator df is 1 (ModDF) and the denominator df is $N - 2$ (ResDF). As in the previous chapters, for clarity, we usually specify the observed variance ratio with its df as subscripts, for example, $F_{1,N-2}$. If the observed statistic $F_{1,N-2}$ is larger than the $100(1 - \alpha_s)$th percentile of this F-distribution, denoted $F_{1,N-2}^{[\alpha_s]}$, then the null hypothesis is rejected at significance level α_s. Equivalently, an observed significance level, P, can be calculated as

$$P = \text{Prob}(F_{1,\,N-2} \geq F_{1,N-2})\ ,$$

where $F_{1,N-2}$ denotes a random variable with an F-distribution with 1 and $N - 2$ df. All the above calculations are conveniently summarized in an ANOVA table, as presented in Table 12.2.

EXAMPLE 12.1C: DIPLOID WHEAT

Consider again the diploid wheat seed data. Values of $SS_{xx} = 19.2699$ and $SS_{xy} = 330.9297$ were given in Example 12.1B, and the total sum of squares is $TotSS = SS_{yy} = 7294.4090$. The model sum of squares can be calculated from the sums of squares and cross-products as

$$ModSS = \frac{(SS_{xy})^2}{SS_{xx}} = \frac{(330.9297)^2}{19.2699} = 5683.1753\ .$$

TABLE 12.2

Structure of the ANOVA Table for a SLR

Source of Variation	df	Sum of Squares	Mean Square	Variance Ratio	P
Model	1	ModSS	ModMS = ModSS/1	F = ModMS/ResMS	$\text{Prob}(F_{1,N-2} > F)$
Residual	$N-2$	ResSS	ResMS = ResSS/$(N-2)$		
Total	$N-1$	TotSS			

TABLE 12.3

ANOVA Table for a SLR Model for Seed Weight with Explanatory Variate Seed Length (Example 12.1C)

Source of Variation	df	Sum of Squares	Mean Square	Variance Ratio	P
Model	1	5683.1753	5683.1753	663.117	< 0.001
Residual	188	1611.2338	8.5704		
Total	189	7294.4090			

By subtraction, we obtain the residual sum of squares as

$$\text{ResSS} = \text{TotSS} - \text{ModSS} = 7294.4090 - 5683.1753 = 1611.2338 .$$

All these values, together with their corresponding degrees of freedom, are combined to give the ANOVA table shown in Table 12.3.

The observed value of $F_{1,N-2} = 663.117$ is huge: the 0.1% critical value of the F-distribution with 1 and 188 df is $F_{1,188}^{[0.001]} = 11.176$; so, $P < 0.001$. Therefore, we have very strong evidence to reject the null hypothesis that the slope parameter is zero, and we conclude that there is a statistically significant linear relationship for seed weight in terms of seed length.

12.4 Properties of the Model Parameters

Having fitted the SLR model and obtained estimates of the model parameters, we can use statistical theory to make further inferences about their underlying, unknown values. If the deviations follow a Normal distribution (Assumption 4, Section 12.1), then the estimates of α and β also follow Normal distributions. In each case, the mean of the distribution is the unknown population parameter, and for this reason, the estimates are called unbiased. Their variances are functions of the explanatory variate, the number of observations and the unknown population variance σ^2. We estimate these variances by replacing σ^2 by its estimate s^2, the residual mean square (see Section 12.3), giving

$$\widehat{\text{Var}}(\hat{\alpha}) = s^2 \times \left(\frac{1}{N} + \frac{\bar{x}^2}{SS_{xx}} \right), \quad \widehat{\text{Var}}(\hat{\beta}) = s^2 \times \left(\frac{1}{SS_{xx}} \right).$$

The estimated standard error of a parameter estimate, $\widehat{\text{SE}}()$, is defined as the square root of its estimated variance.

It is sometimes of interest to test whether one of the parameters is equal to a specified value, usually zero. Hypotheses of this type can be evaluated by the one-sample t-test presented in Section 2.4.1. For example, to test the null hypothesis that the slope takes any pre-defined value c, i.e. H_0: $\beta = c$, we use the statistic

$$t_{N-2} = \frac{\hat{\beta} - c}{\widehat{SE}(\hat{\beta})} \, ,$$

which has a t-distribution with $N - 2$ degrees of freedom under the null hypothesis. An analogous test can be constructed for the intercept α.

The test of the null hypothesis that the slope parameter, β, equals zero, i.e. H_0: $\beta = 0$, against a two-sided alternative hypothesis, i.e. H_1: $\beta \neq 0$, is equivalent to the test of no linear dependence of the response variate on the explanatory variate against that of some linear dependence. Thus, the t-test is equivalent to the F-test obtained from the ANOVA table presented above, with $F_{1,N-2} = (t_{N-2})^2$.

The test of the null hypothesis that the intercept parameter, α, equals zero, i.e. H_0: $\alpha = 0$, against a two-sided alternative hypothesis, i.e. H_1: $\alpha \neq 0$, is used to determine if the model passes through the origin, i.e. that the expected value of the response variable is zero when the explanatory variate is zero. If the null hypothesis is accepted, then we might fit a model that contains only a slope parameter, although this model can have some undesirable properties which are discussed in Section 12.9.2.

We can calculate the $100(1 - \alpha_s)\%$ CI associated with these t-tests for the population parameters as

$$\left(\hat{\alpha} - t_{N-2}^{[\alpha_s/2]} \times \widehat{SE}(\hat{\alpha}), \quad \hat{\alpha} + t_{N-2}^{[\alpha_s/2]} \times \widehat{SE}(\hat{\alpha}) \right) ,$$

$$\left(\hat{\beta} - t_{N-2}^{[\alpha_s/2]} \times \widehat{SE}(\hat{\beta}), \quad \hat{\beta} + t_{N-2}^{[\alpha_s/2]} \times \widehat{SE}(\hat{\beta}) \right) ,$$

where $t_{N-2}^{[\alpha_s/2]}$ is the $100(1 - \alpha_s/2)$th percentile of a t-distribution with $N - 2$ degrees of freedom.

EXAMPLE 12.1D: DIPLOID WHEAT

Using the summary statistics and parameter estimates obtained in Examples 12.1B and 12.1C, we can calculate the estimated variances for the intercept and slope parameters as

$$\widehat{Var}(\hat{\alpha}) = s^2 \times \left(\frac{1}{N} + \frac{\bar{x}^2}{SS_{xx}} \right) = 8.5704 \times \left(\frac{1}{190} + \frac{3.295^2}{19.2699} \right) = 4.8743 \, ,$$

$$\widehat{Var}(\hat{\beta}) = s^2 \times \left(\frac{1}{SS_{xx}} \right) = 8.5704 \times \left(\frac{1}{19.2699} \right) = 0.4448 \, .$$

Hence, the estimated standard errors for $\hat{\alpha}$ and $\hat{\beta}$ are 2.2078 and 0.6669, respectively. For comparison with the ANOVA, we evaluate the evidence of linear dependence of weight on length using a t-test of the null hypothesis H_0: $\beta = 0$ against H_1: $\beta \neq 0$. The observed t-statistic is

$$t_{N-2} = \frac{\hat{\beta}}{\widehat{SE}(\hat{\beta})} = \frac{17.173}{0.667} = 25.751 \, .$$

TABLE 12.4

Parameter Estimates with Standard Errors (SEs), t-Statistics (t) and Observed Significance Levels (P) for a SLR Model for Seed Weight with Explanatory Variate Seed Length (Example 12.1D)

Term	Parameter	Estimate	SE	t	P
[1]	α	−27.931	2.2078	−12.651	< 0.001
Length	β	17.173	0.6669	25.751	< 0.001

The absolute value of this test statistic is compared with a critical value of the t-distribution with 188 df, for example, for a two-sided test with a significance level of 5%, this critical value is $t_{188}^{[0.025]} = 1.973$ and the null hypothesis is rejected. In fact, the test statistic is larger than the 0.1% critical value, $t_{188}^{[0.0005]} = 3.343$; so, we conclude that there is very strong evidence of a linear dependence of weight on length, in agreement with the ANOVA in Example 12.1C. As expected, the square of the observed t-statistic, $25.751^2 = 663.117$, is equal to the value of the F-statistic obtained in Example 12.1C. Properties of the parameter estimates, including their estimated standard errors, the t-statistics for testing the null hypothesis that each parameter is equal to zero and the associated significance levels are often summarized in a form similar to Table 12.4. Variations in this form are commonly produced by statistical software, with parameters implicitly identified via the associated explanatory variate.

Using the information accumulated so far, we can now obtain CIs for both model parameters. A 95% CI for the intercept α is obtained as

$$(-27.931 - (1.973 \times 2.208), -27.931 + (1.973 \times 2.208)) = (-32.286, -23.576),$$

and for the slope parameter, β,

$$(17.173 - (1.973 \times 0.667), 17.173 + (1.973 \times 0.667)) = (15.858, 18.489).$$

The CIs for the intercept and slope parameters often imply that there is a large set of possible fitted lines that are consistent with the data. It is therefore helpful to generate predictions and CIs for the fitted response rather than for individual parameters.

12.5 Using the Fitted Model to Predict Responses

Once a SLR model has been fitted, we can use the parameter estimates to predict the response for a given value of the explanatory variate. In the general case, prediction uses a fitted model to estimate the expected response for given values of all explanatory variables (Section 1.4). Here, we consider two different forms of prediction for a specified value of the explanatory variate, or prediction point, denoted x_{pred}. First, we predict the **expected mean response**, denoted $\mu(x_{\text{pred}})$, which is equal to the fitted response at x_{pred} based on the observed sample. For clarity, we use the notation $\hat{\mu}(x_{\text{pred}})$ for the estimate, which is calculated as

$$\hat{\mu}(x_{\text{pred}}) = \hat{\alpha} + \hat{\beta} x_{\text{pred}}.$$

Second, we predict the value of a **new individual response**, which we denote $\hat{y}_{\text{new}}(x_{\text{pred}})$. This prediction is equal to the fitted response at the prediction point, x_{pred}, plus the deviation associated with the new observation, denoted e_{new}, or

$$\hat{y}_{\text{new}}(x_{\text{pred}}) = \hat{\alpha} + \hat{\beta}x_{\text{pred}} + e_{\text{new}} \ .$$

The new deviation is unobserved and so unknown. If we assume it follows a Normal distribution with zero mean (in line with our assumptions on the deviations), then it is estimated at its expected value as zero. This prediction is therefore equal in value to the expected mean response given above. However, although the presence of the deviation does not affect the value of the prediction, it does influence its variance and standard error.

The estimated variance of the expected mean response at the prediction point x_{pred}, is written as

$$\widehat{\text{Var}}(\hat{\mu}(x_{\text{pred}})) = s^2 \times \left(\frac{1}{N} + \frac{(x_{\text{pred}} - \bar{x})^2}{\text{SS}_{xx}} \right) \ .$$

Again, here, the residual mean square, s^2, is used as an estimate of the unknown population variance, σ^2. The prediction variance takes its minimum value when x_{pred} is equal to the sample mean of the explanatory variate, $x_{\text{pred}} = \bar{x}$, at which point the uncertainty associated with the fitted line is minimized and the second term (within parentheses) is equal to zero. As x_{pred} moves away from the sample mean, this variance increases as uncertainty in the fitted response also increases.

The prediction for a new observation was written as the expected response plus a new deviation. The new deviation is independent of the fitted line and assumed to have variance equal to the population variance, estimated by s^2. The estimated variance of a prediction for a new observation is therefore equal to that for the expected mean response plus the estimated variance of the new deviation, which is

$$\widehat{\text{Var}}(\hat{y}_{\text{new}}(x_{\text{pred}})) = s^2 \times \left(1 + \frac{1}{N} + \frac{(x_{\text{pred}} - \bar{x})^2}{\text{SS}_{xx}} \right) \ .$$

Again, the minimum variance is obtained when $x_{\text{pred}} = \bar{x}$, and as x_{pred} moves away from $\bar{x}$, the variance increases. The additional variance of the new deviation means that this will always be greater than the variance of the expected mean response. For both types of prediction, the estimated SE is the square root of the estimated variance.

Finally, $100(1 - \alpha_s)\%$ CIs for the two types of prediction are obtained as

$$\left(\hat{\mu}(x_{\text{pred}}) - t_{N-2}^{[\alpha_s/2]} \times \widehat{\text{SE}}[\hat{\mu}(x_{\text{pred}})], \quad \hat{\mu}(x_{\text{pred}}) + t_{N-2}^{[\alpha_s/2]} \times \widehat{\text{SE}}[\hat{\mu}(x_{\text{pred}})] \right) ,$$

$$\left(\hat{y}_{\text{new}}(x_{\text{pred}}) - t_{N-2}^{[\alpha_s/2]} \times \widehat{\text{SE}}[\hat{y}_{\text{new}}(x_{\text{pred}})], \quad \hat{y}_{\text{new}}(x_{\text{pred}}) + t_{N-2}^{[\alpha_s/2]} \times \widehat{\text{SE}}[\hat{y}_{\text{new}}(x_{\text{pred}})] \right) ,$$

respectively, where $t_{N-2}^{[\alpha_s/2]}$ is the $100(1 - \alpha_s/2)$th percentile of the t-distribution with $N - 2$ degrees of freedom (the residual df). Figure 12.5 shows both types of CIs across the range of observed values of an explanatory variate x. The confidence limits have been joined to form envelopes, showing the CIs at each point on the fitted line. Since the estimated SEs of

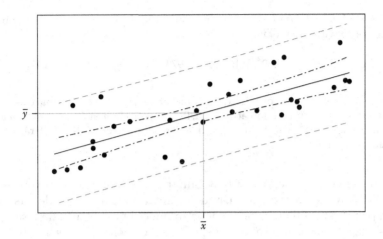

FIGURE 12.5
SLR prediction (——) with CIs for the mean response $\hat{\mu}(x)$ (– · –) and for prediction of a new observation $\hat{y}_{new}(x)$ (– –).

the predictions increase as the value of the explanatory variate moves away from its mean, the width of the CIs also increases on moving towards the ends of the range of the explanatory variate. Also, as a consequence of the difference between their variances, the CIs for the expected response are always smaller than those for a new observation.

EXAMPLE 12.1E: DIPLOID WHEAT

To get a more helpful measure of the uncertainty in the fitted line for the seed weight data, we calculate the 95% CIs for the expected mean responses for seeds with lengths equal to the smallest, mean and largest values of the sample. The shortest observed seed length is $x_{pred} = 2.45$ mm, with fitted response

$$\hat{\mu}(x_{pred} = 2.45) = -27.931 + (17.173 \times 2.45) = 14.144 .$$

This prediction has estimated variance

$$\widehat{\mathrm{Var}}(\hat{\mu}(x_{pred} = 2.45)) = s^2 \times \left(\frac{1}{N} + \frac{(x_{pred} - \bar{x})^2}{SS_{xx}} \right) = 8.569 \times \left(\frac{1}{190} + \frac{(2.45 - 3.295)^2}{19.268} \right) = 0.3628 ,$$

which corresponds to an estimated standard error of $\widehat{\mathrm{SE}}(\hat{\mu}(x_{pred} = 2.45)) = 0.6023$. The residual df is 188 (Table 12.3), and calculation of a 95% CI requires the 97.5th percentile of the t-distribution with 188 df, equal to $t_{188}^{[0.025]} = 1.973$. Hence, a 95% CI for the expected mean response at $x_{pred} = 2.45$ is calculated as

$$(14.144 - (1.973 \times 0.6023), 14.144 + (1.973 \times 0.6023)) = (12.955, 15.335) .$$

Similarly, the 95% CI for the expected mean response at the average seed length observed from our sample (i.e. $x_{pred} = \bar{x} = 3.30$ with $\widehat{\mathrm{SE}} = 0.2124$) is

$$(28.741 - (1.973 \times 0.2124), 28.741 + (1.973 \times 0.2124)) = (28.322, 29.160) .$$

Finally, the 95% CI for the expected mean response at the longest seed length observed (i.e. $x_{pred} = 4.13$ with $\widehat{SE} = 0.5959$) is

$$(42.995 - (1.973 \times 0.5959), 42.995 + (1.973 \times 0.5959)) = (41.819, 44.170) \, .$$

As expected, the CI for the predicted response at $x_{pred} = 3.30$ is much narrower (range 0.84) than that for the prediction at $x_{pred} = 2.45$ (range 2.37) or at $x_{pred} = 4.13$ (range 2.35). Figure 12.3 shows 95% CIs for the predicted response across the full range of the explanatory variate.

So far, we have made predictions only within the observed range of explanatory variate values. This is known as **interpolation**, and is valid as long as the model fits the observed data well (in particular, if there is no evidence of model misspecification; see Chapter 13). As we have seen, the uncertainty about the predicted values increases (i.e. the SE increases) for prediction points towards the ends of the observed range of the explanatory variate. This is inherited from our uncertainty about the slope of the underlying relationship: the fitted line must pass through the mean of both variates, so that this point is fixed. A small change in the slope can then have a larger impact at the ends of the observed range than close to the mean of the explanatory variate.

Of course, we can also make predictions outside the observed range of the explanatory variate, known as **extrapolation**. We should be careful when doing this, however, because in addition to the increasing uncertainty associated with the fitted line, we have no indication of whether the true relationship follows the form of the extrapolated line. This is illustrated in Figure 12.6. Here, the solid circles represent the data used to fit the SLR model indicated by the straight line. The open circles represent an additional sample for smaller values of the explanatory variate and illustrate the main danger of extrapolation: when considered over the extended range, the relationship is not a straight line, and so, predictions based on the fitted SLR give poor predictions for the smaller values of the explanatory variate. The same problems also occur for extrapolation in more complex models. Regression should always be regarded as an empirical descriptive technique: the fitted model describes the observed relationship between the response and explanatory

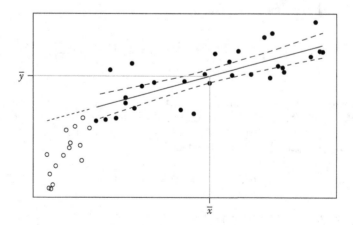

FIGURE 12.6
Fitted SLR model (—) with 95% CIs (– –) with data used to fit model (●), additional data not used to fit model (○) and extrapolation of fitted line (- - -).

variables and, in the absence of further information, this description can be valid only within the observed range of the explanatory variables.

The value of the estimated intercept parameter can be interpreted as a prediction of the response for a zero value of the explanatory variate, i.e. $x_{pred} = 0$. If zero is substantially outside the observed range of the explanatory variate, then the intercept may be a substantial extrapolation. We might then not be too concerned if the estimated intercept value seems to be biologically unrealistic: this merely implies that the fitted model is not valid over a wider range. As long as the fit seems plausible over the observed range, this is sufficient for the descriptive model to be regarded as adequate. In Section 12.9.1, we centre the explanatory variate to estimate the intercept (constant) parameter at a value within the observed range.

12.6 Summarizing the Fit of the Model

It is important to check the **goodness of fit** of a model. This is primarily required to ensure that the model provides a reasonable description of the observed data, and the graphical procedures described in Chapter 13 are a vital part of this process. However, it is also useful to have a simple numerical measure to compare different models for the same response variable. Goodness-of-fit statistics can be used to compare competing models based on different explanatory variates, or different transformations of the same explanatory variate (as we will see in Chapters 14 and 17). When values of the explanatory variate are replicated, we can formally test whether the fit of our model is acceptable, and this is discussed in Section 12.8.

Several goodness-of-fit statistics are available, each with advantages and disadvantages. The most common statistics are the coefficient of determination (R^2) and the adjusted coefficient of determination (adjusted R^2, or R^2_{adj}), sometimes expressed as the percentage variance accounted for. These statistics are described below. Other statistics are useful for regression models with several explanatory variates, and these are described in Section 14.8.

The **coefficient of determination**, here denoted as R^2, measures the proportion of variation in a data set that is accounted for by the fitted statistical model, calculated as the ratio of the model sum of squares to the total sum of squares, i.e.

$$R^2 = \frac{\text{ModSS}}{\text{TotSS}}.$$

Because of the relationship between the three sums of squares in Equation 12.2, we can rewrite this expression in terms of the residual and total sums of squares as

$$R^2 = \frac{\text{TotSS} - \text{ResSS}}{\text{TotSS}} = 1 - \frac{\text{ResSS}}{\text{TotSS}}.$$

The coefficient of determination can take any value between 0 and 1, with larger values indicating a closer fit of the model to the data. In the context of regression models with several explanatory variates, this statistic has the major disadvantage that it does not take account of the number of parameters estimated (see Section 14.8).

The **adjusted coefficient of determination**, denoted R^2_{adj} (or adjusted R^2) is an alternative goodness-of-fit statistic that takes into consideration both the number of observations, N, and the number of estimated parameters. This statistic is calculated as

$$R^2_{adj} = 1 - \frac{\text{ResMS}}{\text{TotMS}} = 1 - \frac{\text{ResSS}/(N-2)}{\text{TotSS}/(N-1)} = R^2 - \left(\frac{1-R^2}{N-2}\right),$$

where $\text{TotMS} = \text{TotSS}/(N-1)$. For SLR, there is a linear relationship between R^2 and R^2_{adj} that depends on the number of observations (N) and R^2_{adj} takes values between $-1/(N-2)$ (when $R^2 = 0$) and 1 (when $R^2 = 1$), with larger values indicating a better fit. Negative values imply an extremely poor fit, with the intercept-only model providing a better fit than the regression line. The adjusted statistic takes account of the number of parameters estimated in the model, and so it can guard against over-fitting, as will be discussed further in Section 14.8. Recall that ResMS is a measure of the variance not accounted for by the model and TotMS can be considered as a measure of the total variance in the data, and so, R^2_{adj} can be interpreted as the proportion of the variance accounted for by the model. When expressed as a percentage rather than a proportion (i.e. $100 \times R^2_{adj}$), this statistic is therefore sometimes called the **percentage variance accounted for**. Within this book, we usually report adjusted R^2 as a summary measure of fit for a model.

EXAMPLE 12.1F: DIPLOID WHEAT

The ANOVA for the SLR model for seed weight with explanatory variate seed length was shown in Table 12.3. The coefficient of determination for this model is

$$R^2 = \frac{\text{ModSS}}{\text{TotSS}} = \frac{5683.1753}{7294.4090} = 0.779 .$$

This is a fairly large value, suggesting a reasonable fit of the model to the data, which can be verified by checking Figure 12.3, which shows a strong positive linear relationship between seed weight and length. The adjusted coefficient of determination is

$$R^2_{adj} = 1 - \frac{\text{ResMS}}{\text{TotMS}} = 1 - \frac{8.5704}{38.5948} = 0.778 ,$$

which is very close to R^2 as $N - 2 = 188$ is very large and hence the adjustment $(1 - R^2)/(N - 2) = 0.001$ is small. The percentage variance in seed weight accounted for by the linear regression model using seed length as an explanatory variate is therefore 77.8%.

12.7 Consequences of Uncertainty in the Explanatory Variate

One of the basic assumptions of the linear model (presented in Section 12.1) is that the values of the explanatory variable(s) are known without error. In many cases, this assumption is not valid, as it is often impossible to ascertain the exact values of an explanatory variate. This problem is common in observational studies, in which values of explanatory variates are usually outside the control of the investigator. For example, in Example 12.1, all the variates are likely to be subject to some error in their measurements. It is important to appreciate the consequences of this for the fitted model: it does not make the model invalid, but it does change its interpretation.

When we fit a regression model, we think about fitting a model for the response in terms of the true value of the explanatory variate(s). In the SLR, uncertainty in the explanatory variate(s) acts to attenuate or dilute a regression relationship, so that the estimated slope tends to be smaller in size, and the estimate of background variability increases. This attenuation is less if the errors in the explanatory variate are small compared with the underlying variability of the (unobserved) true values. This is demonstrated in Figure 12.7 for Example 12.1, where the original SLR fit for seed weight as a function of length is shown in Figure 12.7a. In Figures 12.7b and c, the seed lengths have had random Normal errors with standard deviations of 0.08 and 0.16 mm added to them (25% and 50% of the sample standard deviation for seed length), and the model has been refitted (solid line). The original fit is shown by the dashed line, and you can see that as error in the explanatory variate increases, the spread of the observations also increases and the slope of the fitted line decreases.

When there are errors in the explanatory variate, the estimated regression line is therefore a biased estimate of the true relationship, i.e. the underlying relationship between the response and the true value of the explanatory variate. However, the estimated regression line gives a valid estimate of the relationship between the response and the explanatory variate as measured, i.e. with error. If the objective is prediction of the response for new observations of the explanatory variate, and if these new values will be drawn from the same population (with the same distribution of error on the explanatory variable), then the fitted line is appropriate. In Example 12.1, our fitted model will be valid for prediction of seed weight from length measurements of the same type. If our objective was prediction of seed length given seed weight, then the role of our two variables should be reversed and a different regression line would be obtained. If the objective is estimation of the relationship between the true values of both variables, then more advanced techniques are required (e.g. Carroll et al., 2006).

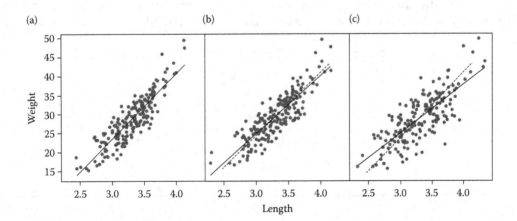

FIGURE 12.7
Seed weight plotted against (a) seed length, (b) seed length with Normal(0, 0.08²) errors added, (c) seed length with Normal(0, 0.16²) errors added, each with fitted SLR model (—) and original SLR fit (– –).

One way to deal with measurement error is to ensure that it is minimized when the data are collected. Technical replication (Section 3.1.1) can be useful in this context, with several independent measurements of the explanatory variate made for each observation, and this can reduce uncertainty due to both sampling variation and instrument or process variability. The mean of the technical replicates can then be used as the value of the explanatory variate.

Measurement error is less important in the case of a designed experiment, when the values of the explanatory variate have been pre-defined. However, these errors may still be present. For example, consider a field experiment with pre-defined levels of fertilizer application to be applied to plots: although operators will do their best to apply the required quantity, machinery is rarely precise enough to deliver this exactly. Similar problems of precision often occur, albeit on a much smaller scale, in laboratory experiments. As long as the errors can be regarded as random (rather than systematic), then no bias is introduced into estimates of the slope parameter, although the estimate of background variation will still be inflated.

12.8 Using Replication to Test Goodness of Fit

Replication is a basic principle of the statistical design of experiments and, as discussed in Chapters 3 and 4, involves the application of the same treatments to several independent experimental units. Differences between replicates with the same treatment give a direct estimate of background variability that arises from uncontrolled variation within the experimental process. For this reason, it is useful to have replication present wherever possible. In this section, we describe how to use replication to evaluate whether a SLR model gives an adequate representation of the observed data.

In the previous sections, we estimated the background variation directly from the residuals obtained after fitting a regression model. However, these residuals consist of two components that cannot be separated: variation of individual observations about the true but unknown trend (often called **pure error**); and systematic deviation of the fitted trend (here a straight line) from the true but unknown trend (commonly called **lack of fit**, and referred to earlier as model misspecification). When each of the values of an explanatory variate is replicated across different experimental units, these two components can be separated, as shown in Figure 12.8. The mean of replicate observations can be regarded as an estimate of the true

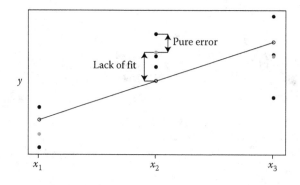

FIGURE 12.8
Partitioning the deviation into pure error and lack-of-fit components. Fitted SLR for response y (—) with observations (●), fitted values (○) and observed mean (◉) for each level of explanatory variate (x).

but unknown trend. Discrepancies between these mean values and the fitted regression line can then be used to assess whether the model shows any evidence of lack of fit, and variation between the replicate observations and their mean can be used to estimate pure error. This gives the basis for an objective evaluation of the adequacy of the fitted model.

EXAMPLE 12.2A: CROP TRANSECT BEETLE COUNTS*

A pilot study was done to investigate the pattern of an insect pest (beetles) entering a susceptible crop. It was suspected that the beetles entered the crop from the edge of the field and then progressed towards the centre. One field was surveyed periodically and, once the beetles were present in reasonable numbers, a transect was taken from the edge towards the centre of the field with samples taken at 2 m intervals. At each distance, beetle counts were made from four randomly selected plants, giving replicate measurements at each distance. The data are presented in Table 12.5 and can also be found in file TRANSECT.DAT. This file holds the distance into the crop in variate *Distance* and the corresponding beetle counts in variate *Count*.

Because of heterogeneity of variances, it is conventional to analyse these counts on the logarithmic scale (see Chapter 6). Figure 12.9 shows the $\log_{10}$-transformed counts, which suggest that a linear model on the log scale is plausible, but certainly should be checked for lack of fit. The variation between observations made on plants at the same distance represents pure error, and this variation is substantial.

TABLE 12.5

Beetle Counts from Transect Sampling, with Four Plants Sampled at Various Distances from the Edge of a Crop (Example 12.2A and File TRANSECT.DAT)

		Plant			
		1	2	3	4
Distance from the	0	21	33	25	16
edge of crop (m)	2	19	20	17	19
	4	8	10	8	8
	6	12	10	6	22
	8	10	6	9	11
	10	9	9	13	13

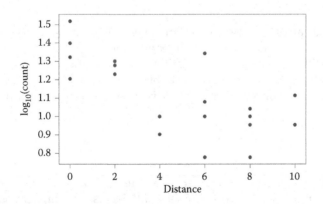

FIGURE 12.9

Logged beetle counts from replicate plants along a transect into the crop. Distances (m) are measured from the edge of the crop (Example 12.2A).

To explicitly identify the replicate observations, we must modify our notation to relabel the units. We shall refer to the distinct values of the explanatory variate as levels, which run from 1 to v. The kth level of the explanatory variate is represented as x_k, for $k = 1 \ldots v$, and the number of replicate observations for that level is denoted as n_k, with at least two replicates required for each level, i.e. $n_k > 1$. The response for the lth replicate observation of the kth level of the explanatory variate is represented as y_{kl}, $k = 1 \ldots v$, $l = 1 \ldots n_k$. The total number of observations is $N = n_1 + n_2 + \ldots + n_v$. The SLR model can be written with this new labelling as

$$y_{kl} = \alpha + \beta x_k + e_{kl} , \tag{12.4}$$

where e_{kl} is the deviation corresponding to observation y_{kl}. Note that because we have a single value of the explanatory variate for each level k, there is no need to have a second index for this variable. As usual, this model is written in symbolic form as

Explanatory component: *[1] + x*

where *[1]* is a variate taking value 1 in all units, associated with the intercept parameter α, and x holds the values of the explanatory variate associated with the slope parameter β. This model fits a straight line through the set of observations. To assess lack of fit, we want to partition the residual term into a term that fits a separate mean for each level of the explanatory variate plus deviations about this term. We can do this by fitting a factor, denoted facx, in the explanatory model. The factor facx is defined so that different levels of the factor correspond to distinct values of the explanatory variate, and can be interpreted as a factor version of the explanatory variate. We therefore add this factor to the model, in symbolic form, giving

Explanatory component: *[1] + x +* facx

In mathematical form, this can be written as

$$y_{kl} = \alpha + \beta x_k + \kappa_k + e_{kl}^* , \tag{12.5}$$

where κ_k represents an effect associated with level k of the explanatory variate, associated with factor facx, and e_{kl}^* is used to denote the deviations from this more complex model. In fact, we have partitioned the model deviations from Equation 12.4 into a lack-of-fit component that is common to each level of the explanatory variable, denoted as κ_k, and a pure error component comprising the separate individual deviations about this common component, denoted as e_{kl}^*, or

$$e_{kl} = \kappa_k + e_{kl}^* .$$

The above model (Equation 12.5) is over-parameterized, as we have $v + 2$ parameters to describe the pattern across v groups, and we shall discuss the implications of this later.

The ANOVA table for this model is Table 12.6. The model sum of squares, ModSS, is the same as from a SLR model. The sum of squares associated with the systematic component of the deviations, and factor facx, is called the **lack-of-fit sum of squares**, abbreviated as

TABLE 12.6

Structure of the ANOVA Table for a SLR with Residual SS Partitioned into Lack-of-Fit and Pure Error Components

Source of Variation	df	Sum of Squares	Mean Square	Variance Ratio	P
Model	1	ModSS	ModMS = ModSS/1	F^M = ModMS/PEMS	Prob($F_{1,\,N-v} > F^R$)
Residual					
Lack of fit	$v - 2$	LoFSS	LoFMS = LoFSS/($v - 2$)	F^L = LoFMS/PEMS	Prob($F_{v-2,\,N-v} > F^L$)
Pure error	$N - v$	PESS	PEMS = PESS/($N - v$)		
Total	$N - 1$	TotSS			

LoFSS. This sum of squares accumulates the squared differences between the fitted value from the SLR and the mean of the replicates at each value of the explanatory variate, and can be written as

$$\text{LoFSS} = \sum_{k=1}^{v} \sum_{l=1}^{n_k} \left(\bar{y}_{k\cdot} - \hat{y}_{kl} \right)^2 = \sum_{k=1}^{v} n_k \left(\bar{y}_{k\cdot} - \hat{\alpha} - \hat{\beta} x_k \right)^2 ,$$

where $\hat{y}_{kl}$ is the SLR fitted value for observation y_{kl} and $\bar{y}_{k\cdot}$ is the mean response for the explanatory variate value x_k or, in our previous notation,

$$\bar{y}_{k\cdot} = \frac{1}{n_k} \sum_{l=1}^{n_k} y_{kl} .$$

The LoFSS has $v - 2$ df, as v parameters are required to generate the mean values $\bar{y}_{k\cdot}$ and two parameters are required to generate the fitted values $\hat{y}_{kl}$. The remaining variation for this model arises from variation between replicates within each level of the explanatory variate, and is known as the **pure error sum of squares**, PESS, which can be calculated as

$$\text{PESS} = \sum_{k=1}^{v} \sum_{l=1}^{n_k} \left(y_{kl} - \bar{y}_{k\cdot} \right)^2 .$$

This sum of squares has $N - v$ df. As usual, mean squares are calculated by division of each sum of squares by its df. As the pure error mean square, denoted PEMS, is the best estimate of background variation, this quantity is used as the denominator for calculation of variance ratios.

The variance ratio F^M = ModMS/PEMS is used to test the model via the null hypothesis that there is no linear trend in the data, i.e. H_0: $\beta = 0$. Under this null hypothesis, this variance ratio has an F-distribution with 1 and $N - v$ df. If this variance ratio exceeds the chosen critical value of this F-distribution, it indicates the presence of significant linear trend. The second variance ratio, F^L = LoFMS/PEMS, is used to test for lack of fit via the null hypothesis that the deviations from the straight line, κ_k in Equation 12.5, are all zero, i.e. H_0: $\kappa_k = 0$, $k = 1 \ldots v$. Under this null hypothesis, the variance ratio has an F-distribution with $v - 2$ and $N - v$ df. If this variance ratio exceeds the chosen critical value of this F-distribution, it indicates that deviations of the group means from the fitted straight line are not all zero, and hence that we have some misspecification in the SLR model.

EXAMPLE 12.2B: CROP TRANSECT BEETLE COUNTS*

A SLR model for $\log_{10}$-transformed beetle counts, *logCount* = $\log_{10}$(*Count*), shows a statistically significant association with distance into the crop (F = 13.764, P = 0.001) and accounts for 35.7% of the variation (adjusted R^2 = 0.357). To investigate the lack of fit, we need to define a factor with six levels, one for each distance into the crop, and we call this factor fDist (also given in file TRANSECT.DAT). The model taking account of the lack of fit can be written as

Response variable: *logCount*
Explanatory component: *[1] + Distance* + fDist

This model accounts for 59.9% of the variation, which is much larger than for the SLR model, and the summary ANOVA table is shown in Table 12.7.

This table has partitioned the residual variation of the SLR (ResMS = 0.0259 with 22 df) into that associated with the fDist factor, which assesses lack of fit with 4 df, and the remainder, associated with variation within distances or pure error with 18 df. The pure error estimate of background variation (PEMS = 0.0161) is substantially smaller than the SLR estimate; so, the variance ratio for the explanatory variate *Distance* increases (F^M = 22.073, P < 0.001). The lack-of-fit variation associated with the factor fDist is also large compared with pure error (F^L = 4.321, P = 0.013), indicating significant deviations from the fitted line that require further investigation. Figure 12.10 shows the fitted line and group means.

The pattern of group means shows that there are more beetles within 2 m of the edge of the crop, and that the samples within the field (4–10 m from the edge) have smaller counts but do not continue to decrease with distance. The counts seem to fall into two groups rather than following the straight line required by linear regression, and this discrepancy is quantified by the lack-of-fit term. We can conclude that the SLR model is not suitable for these data and that further work is required to establish a better model.

We stated above that the model in Equation 12.5 is over-parameterized, and this arises because there are $v + 2$ parameters (α, β, κ_1 ... κ_v) used to fit means for v groups. Individual parameter estimates can therefore be difficult to interpret although the fitted values are equal to the observed group means. If there is no evidence of lack of fit, then the SLR model can be fitted to obtain the usual interpretable parameter estimates. If there is evidence of lack of fit, then the SLR model is not sufficient to explain the pattern. In this case, explanatory model *[1]* + facx, based on just the factor version of the explanatory variable, can be fitted to give an estimate of the response for each level to obtain

TABLE 12.7

ANOVA Table for $\log_{10}$(Beetle Counts) Testing for Lack of Fit (Example 12.2B)

Source of Variation	df	Sum of Squares	Mean Square	Variance Ratio	P
Model	1	0.3562	0.3562	F^M = 22.073	< 0.001
Residual					
Lack of fit	4	0.2789	0.0697	F^L = 4.321	0.013
Pure error	18	0.2905	0.0161		
Total	23	0.9256			

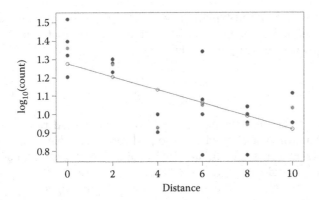

FIGURE 12.10
Data (•), fitted regression line (–o–) and group means (•) for logged beetle counts from transect sampling at distances (m) from the edge of the crop (Example 12.2B).

interpretable parameters. This factor-based model does not allow interpolation between levels in the same way as a regression model based on a variate. A better alternative would be to fit a regression model that follows the observed trend, perhaps a curved or non-linear model as discussed in Chapter 17.

Lack of fit can also be examined in more complex models, such as polynomial models (Section 17.1), multiple regression (Chapter 14) or regression with groups (Chapter 15). In each context, the technique for assessing lack of fit is the same: to add a factor version of the explanatory variable into the model everywhere that the variate version appears; this is illustrated in Example 18.4. Again, this should be done only where each value of the explanatory variable is replicated.

The advantage of using replication within the context of simple or multiple linear regression should now be clear: it allows for a quantitative assessment of model misspecification in addition to more subjective assessments based on residual plots. Unfortunately, it is possible to use replication only where the values of the explanatory variate are under the control of or can be chosen by the experimenter, which does not apply to most observational studies.

12.9 Variations on the Model

12.9.1 Centering and Scaling the Explanatory Variate

Centering is a simple transformation of a variate that subtracts the sample mean from all observations so that the transformed values have mean zero, i.e. they are centered about zero. The variance and standard deviation of the centered variate remain identical to that of the original variate. If zero is outside the range of the uncentered explanatory variate, this transformation can make the intercept parameter more easily interpretable (see Section 12.5). The **centered** model is written mathematically as

$$y_i = \alpha^* + \beta(x_i - \bar{x}) + e_i \ ,$$

which replaces the values of the explanatory variate with those same values after subtracting their sample mean, $\bar{x}$. This model can be fitted by the use of a new variate defined as $x_i^* = x_i - \bar{x}$ and by our rewriting the model as

$$y_i = \alpha^* + \beta x_i^* + e_i .$$

The intercept parameter, now relabelled as α^*, is the predicted response at value zero of the centered explanatory variate, which is equal to the sample mean of the uncentered explanatory variate. Hence, the estimated intercept parameter is equal to the sample mean of the response, $\alpha^* = \bar{y}$, since the fitted SLR line passes through the point $(\bar{y}, \bar{x})$ for any parameterization of the model. The estimated slope is unchanged.

As an alternative to centering, the explanatory variate may be **standardized** by subtraction of the sample mean from each observation, and then division of this by the sample standard deviation, s_x, as

$$x_i^{**} = \frac{x_i - \bar{x}}{s_x} .$$

The SLR model is then rewritten in terms of the standardized variate as

$$y_i = \alpha^* + \beta^* x_i^{**} + e_i .$$

The intercept here again represents the predicted response at the sample mean of the explanatory variate, and the slope parameter, relabelled as β^*, represents the change in the response for 1 unit change in the standardized explanatory variate, which is equal to a change of 1 standard deviation in the original explanatory variate. This form of the model is most useful when there are several explanatory variates (see Chapter 14) with very different scales: using standardized variates makes the slope coefficients directly comparable across different explanatory variates. For interpretation, it is often helpful to translate the slope parameter for the standardized explanatory variate back into the units of the original variate. This is done with the relationships

$$\hat{\beta} = s_x \hat{\beta}^* ; \quad \widehat{SE}(\hat{\beta}) = s_x \widehat{SE}(\hat{\beta}^*) ,$$

i.e. the estimated slope in terms of the original units is equal to the estimated slope for the standardized variate multiplied by the unbiased sample standard deviation for the original explanatory variate. The estimated standard error is similarly scaled by the unbiased sample standard deviation.

12.9.2 Regression through the Origin

We might also consider a restricted form of the SLR model with the intercept parameter set to zero, so that the response must be equal to zero when the explanatory variate is zero. This model may arise in two different ways. In some circumstances, it may be asserted from prior knowledge or expectations. This may be a reasonable biological assumption to make, for example in an early-growth experiment where one might expect zero biomass to correspond to zero shoot length. In other cases, the model may arise from interpretation of the results of a SLR, when the fitted model suggests that zero is a plausible value for the intercept parameter or, more specifically, when the null hypothesis that the intercept parameter is equal to zero, i.e. H_0: $\alpha = 0$, cannot be rejected. If zero is within, or close to, the range of the explanatory variate, and the overall pattern is clearly consistent with a zero intercept, then it may be sensible to omit the intercept parameter if it is not statistically significant. If zero is well outside the range, then we can only infer that the intercept of a SLR model should be zero if it is also reasonable to assume that a common linear relationship holds from zero up to the full observed range of the explanatory variate. Graphical diagnostics (see Chapter 13) should always be used to ensure that omitting the intercept has not introduced bias into the fitted model.

The new model, with $\alpha = 0$, is called **regression through the origin**, and takes the form

$$y_i = \beta x_i + e_i \, ,$$

where each term is described in Section 12.1. As with any other linear model, Assumptions 1 to 5 of Section 12.1 also apply here. We can write this model using symbolic notation by omitting the constant term and specifying the explanatory variate alone as

Explanatory component: x

To estimate the slope parameter in this model, we must define some new statistics. The uncorrected sums of squares for the response (USS_{yy}) and explanatory variable (USS_{xx}) and the uncorrected sum of cross-products (USS_{xy}) are defined as

$$\text{USS}_{yy} = \sum_{i=1}^{N} y_i^2; \quad \text{USS}_{xx} = \sum_{i=1}^{N} x_i^2; \quad \text{USS}_{xy} = \sum_{i=1}^{N} x_i y_i \, .$$

These quantities take a similar form to the sums of squares and cross-products defined earlier as SS_{yy}, SS_{xx} and SS_{xy}, but here we do not 'correct' the variables by subtraction of their sample means.

The least-squares estimate of the slope, β, for regression through the origin becomes

$$\hat{\beta} = \frac{\text{USS}_{xy}}{\text{USS}_{xx}} \, .$$

This takes a similar form to the slope estimate from the standard SLR model, but here, the uncorrected sums of squares are used in place of the corrected sums of squares.

The construction of the ANOVA table for this simplified model is also based on a partition of the total sum of squares as in Equation 12.2; however, the calculations are now based on uncorrected rather than corrected sums of squares. We modify the notation to reflect this change, with the uncorrected total (TotUSS), model (ModUSS) and residual (ResUSS) sums of squares defined as

$$\text{TotUSS} = \text{USS}_{yy}; \quad \text{ModUSS} = \frac{(\text{USS}_{xy})^2}{\text{USS}_{xx}}; \quad \text{ResUSS} = \text{TotUSS} - \text{ModUSS} \, .$$

TABLE 12.8

Structure of the ANOVA Table for Regression through the Origin

Source of Variation	df	Sum of Squares	Mean Square	Variance Ratio	P
Model	1	ModUSS	ModUMS = ModUSS/1	F = ModUMS/ResUMS	$\text{Prob}(F_{1,N-1} > F)$
Residual	$N-1$	ResUSS	ResUMS = ResUSS/$(N-1)$		
Total	N	TotUSS			

Since we have used uncorrected sums of squares, the degrees of freedom for each term must account for this. There is now no adjustment term present in the total or model sums of squares, so that their degrees of freedom are TotUDF = N and, as the regression model has one parameter, ModUDF = 1. The residual df can be obtained by subtraction as

$$\text{ResUDF} = \text{TotUDF} - \text{ModUDF} = N - 1 .$$

As usual, the sums of squares are divided by their degrees of freedom to form mean squares, and the structure of the ANOVA table is shown in Table 12.8.

The observed F-statistic, calculated as the model mean square divided by the residual mean square, can be used to evaluate the null hypothesis H_0: $\beta = 0$ against the alternative hypothesis H_1: $\beta \neq 0$. Under the null hypothesis, this statistic follows an F-distribution with 1 and $N-1$ df. The estimated slope parameter has an expected value equal to the unknown true value, β, with estimated variance

$$\widehat{\text{Var}}(\hat{\beta}) = s^2 \times \left(\frac{1}{\text{USS}_{xx}} \right) .$$

As before, the background variation is estimated from the residual mean square, with

$$s^2 = \text{ResUMS} = \text{ResUSS}/(N-1) .$$

Calculations of CIs for the estimated slope then follow the procedure in Section 12.4, but using ResUDF = $N-1$ to determine the appropriate t-distribution.

The fit of the regression through the origin can again be assessed by goodness-of-fit statistics. A modified version of the coefficient of determination, known as the **empirical coefficient of determination** and denoted as R^2_{emp}, is required and is defined as

$$R^2_{\text{emp}} = 1 - \frac{\text{ResUSS}}{\text{TotSS}} .$$

This statistic contrasts the uncorrected residual sum of squares to the corrected total sum of squares and can be compared with R^2 for models with an intercept, as both statistics have the same denominator. However, because the comparison is no longer related to a single partition of the total variation, R^2_{emp} can now take negative values.

EXAMPLE 12.3: AIR TEMPERATURE

Measurements of air temperature (°C) were made at approximately 9 a.m. on 100 days during 2006 ($N = 100$) with a standard glass mercury dry bulb thermometer and a new

TABLE 12.9

First Four and Last Four Measurements of Air Temperature
(°C) Made during 2006 Using a Standard Glass Mercury
Thermometer and a New Electronic Thermistor (Example
12.3, Full Data in File AIRTEMP.DAT and Table A.2)

Day Number	Mercury	Thermistor
1	5.3	5.3
7	6.6	5.5
8	8.9	8.7
13	6.9	6.7
⋮	⋮	⋮
344	10.6	10.4
349	4.9	4.0
351	−1.5	−2.4
353	0	−3.2

Source: Data from T. Scott and M. Glendining, Rothamsted Research.

electronic dry bulb thermistor probe. A subset of the data is presented in Table 12.9, with the full set shown in Table A.2 and in file AIRTEMP.DAT.

The aim of the analysis is to model the new thermistor measurements (response variate *Thermistor*) in terms of the standard mercury measurements (explanatory variate *Mercury*) to investigate the relationship between them. It is of interest to determine whether the measurements are equivalent, i.e. whether a line that passes through the origin (thermistor reads zero when mercury reads zero) with slope equal to 1 is a plausible model for these observations. The measurements range between −3.2°C and 28.4°C and the scatter of points does appear to pass through the origin (see Figure 12.11).

We start by fitting the SLR model (Equation 12.1), which accounts for 98.6% of the variation in the thermistor measurements (adjusted $R^2 = 0.986$), reflecting the very strong

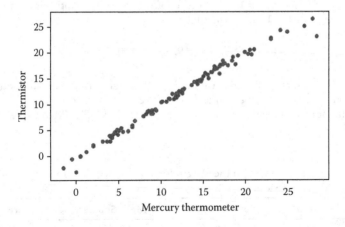

FIGURE 12.11
Scatter plot of air temperature measurements (°C) made by a standard glass mercury dry bulb thermometer and a new electronic dry bulb thermistor (Example 12.3).

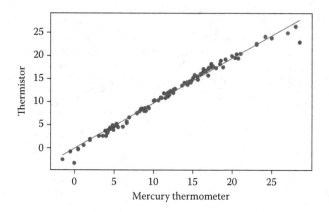

FIGURE 12.12

Air temperature measurements (°C) with fitted line from regression through the origin (Example 12.3).

linear relationship between the two sets of measurements. This model gives parameter estimates $\beta = 0.996$ (SE = 0.0120), $\hat{\alpha} = -0.262$ (SE = 0.1625) and the fitted model

$$\widehat{Thermistor}_i = -0.262 + 0.996\,Mercury_i .$$

First, we examine the intercept. The t-statistic for testing the null hypothesis H_0: $\alpha = 0$ is

$$t = \hat{\alpha}/\widehat{SE}(\hat{\alpha}) = -0.262/0.1625 = -1.613 ,$$

with 98 df. The observed significance level for this test is $P = 0.110$; so, there is no evidence that the estimated intercept is different from zero and it appears reasonable to drop this parameter and fit a regression through the origin. This model estimates the slope as $\beta = 0.979$ (SE = 0.0059). The fitted line is shown in Figure 12.12, and the associated ANOVA table is Table 12.10.

As in the SLR, there is strong evidence that the estimated slope is not equal to zero. However, here, we are more interested in whether the slope is equal to 1, corresponding to the null hypothesis H_0: $\beta = 1$. A t-statistic for testing this hypothesis against the two-sided alternative H_1: $\beta \neq 1$ can be calculated as

$$t = (\hat{\beta} - 1)/\widehat{SE}(\hat{\beta}) = (0.979 - 1)/0.0059 = -3.597 ,$$

with 99 df. The observed significance level for this test is $P < 0.001$; so, there is strong evidence that the slope is not equal to 1 and we reject H_0. Residual plots for this model based on standardized residuals are shown in Figure 12.13. The fitted values plot shows

TABLE 12.10

ANOVA Table for Regression through the Origin for Thermistor Readings with Mercury Readings as the Explanatory Variate (Example 12.3)

Source of Variation	df	Sum of Squares	Mean Square	Variance Ratio	P
Model	1	17,698.267	17,698.267	27,397.119	< 0.001
Residual	99	63.953	0.646		
Total	100	17,762.220			

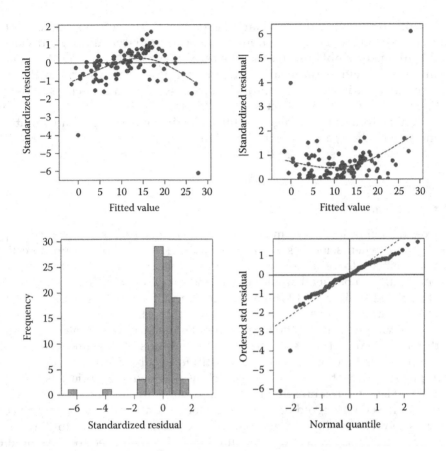

FIGURE 12.13
Composite set of residual plots based on standardized residuals from regression through the origin for air temperature measurements (Example 12.3).

a clear trend, with residuals tending to be negative for the lowest and highest temperatures, and positive for intermediate values. This suggests some non-linearity in the relationship which is also apparent in Figure 12.12 (and we discuss such patterns in the residual plots in more detail in Section 13.3). There are also two very large residuals, one at each end of the temperature range.

Putting all these results together suggests that there is not a 1:1 relationship between the two types of measurement, and that the thermistor measurements can markedly deviate from the traditional method for temperatures around or below 0°C or above 25°C.

Whether the thermistor can be used in practice as a substitute for the mercury readings may depend on the context. If the mercury measurement is regarded as a gold standard to be replicated, then the thermistor readings are clearly not adequate. On the other hand, if the required accuracy is less, and the likely range of use is within 5–20°C, then the small (although statistically significant) difference from the 1:1 relationship might not be of practical importance and thermistor readings may be acceptable (see Section 4.4 for a discussion of biological vs. statistical significance).

One case of regression through the origin where additional care may be required occurs when the origin represents some initial or control condition. Observations made at, or very close to, the origin may then show little or no variation. For example, consider a zero dose of inoculum in a disease-control experiment. The zero dose here acts both as a negative control (in the sense of Section 3.1) and as an initial condition, and should consistently result in no symptoms present. The zero background variation at the zero dose will be inconsistent with natural variation found for positive doses, and so, the assumption of a common variance across all deviations (Assumption 2, Section 12.1) does not hold (also see the discussion in Section 8.5). One solution is to exclude the initial/control condition (to avoid introducing bias into the estimate of background variation) and constrain the model to pass through the origin. Alternatively, if data are recorded as proportions or counts, then a GLM that accommodates such heterogeneity may be fitted (Chapter 18).

12.9.3 Calibration

The process of calibration, sometimes also called inverse regression or inverse prediction, is required when scientists wish to use a quick or easy procedure to estimate a quantity that is hard to measure directly. For example, in many laboratory procedures, a target molecule can be labelled with a dye, and then light absorbance by the dye can be directly related to the quantity in a sample. We will call the variable of interest the target and the variable to be measured the substitute variable. The calibration procedure uses known quantities of the target variable that span the range of interest (usually with replication) and measure the outcome in terms of the substitute variable. A regression model is then fitted with the substitute variable (which is subject to error) as the response and the target variable (which uses known quantities) as the explanatory variable. Here, we assume that the relationship is linear, but a similar procedure can also be followed for curved or non-linear models. Calibration will only be accurate if the regression relationship is a good fit, i.e. with adjusted R^2 close to 1, and with no evidence of model misspecification. The fitted model is then used to make predictions with confidence limits for the target variable given new measurements of the substitute variable. In non-mathematical terms, this process derives a range of plausible values for the target variable from the CIs for the fitted line, as shown in Figure 12.14 (also see Draper and Smith, 1998, Section 3.2).

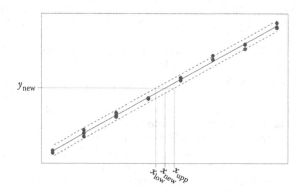

FIGURE 12.14
Inverse prediction for observed data (•) with fitted regression line (—) with 95% CIs for a new observation (– –). For new measurement y_{new}, the predicted value of the explanatory variable, x_{new}, occurs where the fitted line equals y_{new}, with 95% confidence limits, x_{low} and x_{upp}, obtained as the points at which the CIs equal y_{new}.

Suppose a calibration process leads to the fitted SLR model with predictions in the form

$$\hat{y}(x_{new}) = \hat{\alpha} + \hat{\beta} x_{new} \ ,$$

with estimated background variance s^2. If we have a new sample and take r independent measurements from it (technical replicates), then the mean of these values, denoted $\bar{y}_{new}$, gives us an estimate of the true value of the substitute variable, with associated error s^2/r. We assume that the process used to generate new measurements, and hence the associated errors, is the same one used to construct the calibration line. We can plug this new value into the prediction formula and rearrange it to get an estimate of the target variable as

$$\hat{x}_{new} = \frac{\bar{y}_{new} - \hat{\alpha}}{\hat{\beta}} \ ,$$

but we also need some measure of uncertainty in this estimate. The variance associated with a prediction for a mean of r observations at value x takes the form

$$\widehat{Var}(\bar{y}_{new}(x)) = s^2 \times \left(\frac{1}{r} + \frac{1}{N} + \frac{(x - \bar{x})^2}{SS_{xx}} \right) ,$$

and confidence limits can be formed from this value (as described in Section 12.5) for a range of values of x. Fieller's theorem shows that the values of x at which the $100(1 - \alpha_s)\%$ upper and lower confidence limits equal the value $\bar{y}_{new}$ give lower and upper $100(1 - \alpha_s)\%$ confidence limits for the estimate $\hat{x}_{new}$. These limits are shown in Figure 12.14 and their mathematical formula is

$$\hat{x}_{new} + \frac{g(\hat{x}_{new} - \bar{x})}{1 - g} \pm \frac{st_{N-2}^{[\alpha_s/2]}}{\hat{\beta}(1 - g)} \sqrt{\frac{(\hat{x}_{new} - \bar{x})^2}{SS_{xx}} + (1 - g)\left(\frac{1}{r} + \frac{1}{N}\right)}, \quad \text{where } g = \left(\frac{t_{N-2}^{[\alpha_s/2]}}{\hat{\beta}/\widehat{SE}(\hat{\beta})} \right)^2 .$$

These limits only exist when $g < 1$, which occurs when the t-test for the null hypothesis $H_0: \beta = 0$ exceeds the critical value $t_{N-2}^{[\alpha_s/2]}$ (see Section 12.4).

In building the calibration curve, it is important that the quantities of the target variable are known without error; if these values are also subject to uncertainty, then the fitted relationship will be subject to attenuation (see Section 12.7) and it would be better to regress the target on the substitute variable and directly predict from that relationship.

EXERCISES

12.1* An experiment was conducted to quantify the growth rate of transplanted cabbage plants. Forty cabbage plants were transplanted and four plants (chosen at random) were destructively sampled and the number of leaves present was counted on the day they were transplanted and 8, 14, 21, 28, 35, 37, 42, 44 and 46 days afterwards. File MEANCABBAGE.DAT contains sample numbers (*ID*), sample times (variate *Day*) and the mean number of leaves per plant at each sample (variate *NLeaves*). Fit a SLR and verify the estimates of the slope and intercept by calculating the relevant sums of squares. Plot the fitted model and comment on the quality of the fit. Give a 95% CI for the average growth rate over the period (as leaves per plant per day). (We re-visit these data in Exercises 13.2 and 18.2.)

12.2 The Rothamsted Insect Survey collects insects using 12.2 m suction traps at locations across the United Kingdom. As part of an investigation into long-term changes in the abundance of flying insects, indices of the total biomass collected per year (measured as wet weight) were created for 30 years from 1973 to 2002 for four locations (Shortall et al., 2009). The wet weights (variate *WetWeight*, g) collected from the Hereford trap in each year (variate *Year*) are held in file HEREFORD.DAT. Use a SLR to investigate whether there is evidence of any linear trend over time in the log-transformed wet weights, calculated as $\log_{10}(WetWeight + 0.5)$, and summarize the strength of the relationship. Use this model to predict the expected wet weight in 2010, and comment on the reliability of this prediction. Plot the fitted model and consider whether there are any aspects of the fit that you would wish to examine further. (We re-visit these data in Exercises 13.4 and 15.1.)[*]

12.3[*] A pilot study investigated whether measurements of leaf length and width made in the field could be used to accurately estimate leaf area. Twenty-five plants were chosen at random from a plot of a single variety and the length (cm) and width (cm) of the flag leaf on each plant was measured *in situ*. These leaves were then detached from the plants and their area was measured (cm^2) using imaging software. The data (variates *Leaf, Length, Width, Area*) are in file FLAGLEAF.DAT. Use SLR to explore the relationship between leaf area and its estimate constructed as length × width. Build and report a predictive model for leaf area, and critically assess its performance. In principle, we would expect leaves with zero length or width to have zero area; so, does it make sense to fit regression through the origin in this context?

12.4 An experiment was conducted to identify varieties of willow with high yields of dry matter. However, as accurate measurement of dry matter is time consuming, the use of a surrogate variable is desirable and several such variables were measured on a sample of 113 trees. File WILLOWSTEMS.DAT holds the values of dry matter (variate *DryMatter*) and several summary variables, including the length of the longest stem (variate *MaxLength*), which is the simplest to measure. Fit a SLR relating dry matter to the maximum stem length – could we reasonably use this as a surrogate variable? (We re-visit these data in Exercises 13.5 and 14.5.)[†]

12.5 The Rothamsted Insect Survey provided body mass (mg) and wing length (mm) measurements for a sample of moths from the *Noctuidae* family caught in a mercury-vapour trap at Rothamsted between 1999 and 2001 (Wood et al., 2009). These data are held in file NOCTUID.DAT and include unit numbers (*ID*), species name (factor *Species*), wing length (variate *WingLength*) and body mass (variate *Mass*) for each moth. The aim of this analysis is to predict body mass from wing length. Use SLR to investigate the relationship between $\log_{10}(Mass)$ and wing length. As wing lengths were measured to the nearest mm, and there are several observations at each distinct value of wing length, create a factor to test your SLR

[*] Data from R. Harrington and C. Shortall, Rothamsted Research.
[†] Data from I. Shield, Rothamsted Research.

for any evidence of lack of fit. What do you conclude? What other investigations, if any, would you like to make? (We re-visit these data in Exercise 15.2.)[*]

12.6 A microarray study investigated genes associated with the senescence of leaves. Forty-four plants were grown in a controlled environment and the seventh leaf was excised from four of these plants at 2-day intervals from 19 to 39 days after sowing (at the same point in the day/night cycle each time). The plants were allocated to sample dates at random, with a CRD design. Four subsamples (technical replicates) were taken from each leaf and allocated to separate microarrays. File SENESCENCE.DAT holds unit numbers (*ID*), design information (variate *Day*, factor BiolRep) and the expression value for three genes (variates *CATMA3A13560*, *CATMA2A31585* and *CATMA1A09000*) from each plant following normalization and combination of the values for the four technical replicates. Use SLR to predict the expression of gene *CATMA3A13560* over time. Is there any evidence of lack of fit to this relationship? (We re-visit these data in Exercises 13.1 and 17.2.)[†]

[*] Data from J. Chapman, Rothamsted Research.
[†] Data from V. Buchanan-Wollaston (PRESTA), University of Warwick.

13

Checking Model Fit

In Chapter 5, residual plots were used to investigate the assumptions underlying the linear model for the case of a single qualitative variable (factor). As discussed and briefly demonstrated in Chapter 12, all of these residual plots are also applicable to a model with a single quantitative variable or variate (i.e. SLR), as well as to the more complex regression models such as those described in Chapters 14 to 18. However, in models with quantitative explanatory variables, the additional question arises as to whether a linear trend is a good description of the relationship, and several diagnostic plots can be used to investigate this. Concepts such as influence and leverage can also help to examine the impact of individual points on the fitted line, and cross-validation techniques can quantify the predictive power of the model. In this chapter, we introduce these concepts, and review and introduce some techniques for checking the fit of regression models.

We start by considering the problem of model misspecification (Section 13.1) and formally define some residual plots that were presented in Chapter 12. We then review the different types of residuals first described in Section 5.1 and introduce two new types, prediction and deletion residuals, that are particularly helpful in the context of regression (Section 13.2). The use of residual plots in regression is then considered (Section 13.3), with particular reference to checking for model misspecification. The concepts of leverage and influence are then defined and discussed (Section 13.4). Finally, some simple cross-validation techniques are introduced (Section 13.5).

13.1 Checking the Form of the Model

The form of a SLR model asserts that the response changes as a straight line function of the explanatory variate. Any systematic deviation from this form implies that the SLR model is not appropriate, or that the model has been **misspecified**. The first step in any regression should therefore consist of plotting the observations, i.e. the values of the response variate against the values of the intended explanatory variate, to check whether a linear response is plausible, as in Figure 12.2. If the relationship is clearly curved, then transformation of the explanatory variate or a non-linear model should be considered (further details in Chapter 17). If the relationship appears linear, then the model fitting may proceed and we are then in a position to check the quality of our fitted model using numerical tools (such as goodness-of-fit statistics, Section 12.6) and diagnostic plots. The **fitted model plot**, in which the fitted model is superimposed on a plot of the observations, can be used to detect model misspecification: if the data follow the form described by the model, then the fitted line should reflect the trend in the observations across the full range of the explanatory variate. Examples are shown for Example 12.1 in Figure 12.3 and for Example 12.3 in Figure 12.12. In both, there is a suggestion that the model does not fit well at the ends of the range: in Example 12.1, the model appears to under-estimate the smallest and largest weights; in

Example 12.3, the model appears to over-estimate the smallest and largest temperatures. The **fitted value plot**, in which the standardized residuals are plotted against the fitted values (see Section 5.2 and below) can be used to investigate this pattern after removal of the overall trend. In the SLR model, this is equivalent to plotting the standardized residuals directly against the explanatory variate. This type of graph emphasizes deviations from the fitted model, as can be seen in Figures 12.4 and 12.13 for Examples 12.1 and 12.3, respectively. These graphs show clear curvature in the pattern of the residuals and hence some evidence of model misspecification. A more complex form of model might be investigated (see Chapter 17), especially if good prediction is required at the extremes of the explanatory variate. However, some common sense is also required. In both cases, the deviations from the fitted model are relatively small, and the model accounts for most of the variation in the response (with adjusted R^2 of 0.778 and 0.986 for Examples 12.1 and 12.3, respectively), and so, the SLR might be deemed adequate as a simple descriptive model. This is not the case in the next example.

EXAMPLE 13.1: ELISA ABSORBANCE READINGS

A set of eight ELISA readings were obtained for a series of increasing concentrations of a substrate. Here, the aim of the analysis is to describe the relationship between the absorbance reading and substrate concentration. The data are presented in Table 13.1 and can be found in file ELISA.DAT.

The absorbance is expected to be related to a power of the concentration and is hence approximately linearly related to the logarithm of the concentration. For convenience, we use the $\log_{10}$-transformation, with an offset (+1) included so that the background absorbance (zero concentration) can be included in the model. However, the relationship between absorbance and the $\log_{10}$-concentration is clearly not linear, as shown in Figure 13.1.

We already suspect that a SLR model is not appropriate for the data but, for the purposes of demonstration, we proceed with the analysis. The model in mathematical form is

$$Absorbance_i = \alpha + \beta \log_{10}(Conc_i + 1) + e_i ,$$

where the units are labelled with index $i = 1 \ldots 8$ with $Absorbance_i$ and $Conc_i$ being the observed absorbance and concentration for the ith observation. If variate *Absorbance*

TABLE 13.1

ELISA Readings (Absorbance) Obtained for Different Concentrations of a Substrate (Example 13.1 and File ELISA.DAT)

Concentration	Absorbance
0	0.100
0.5	0.678
1	1.107
2	1.609
4	1.958
8	2.202
16	2.414
32	2.485

Source: Data from Rothamsted Research.

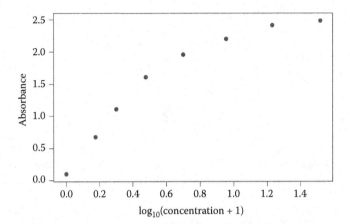

FIGURE 13.1
ELISA absorbance readings plotted against $\log_{10}$(concentration + 1) of a substrate (Example 13.1).

contains the ELISA absorbance readings, and variate *log10Conc* contains the $\log_{10}$-transformed concentrations, then this model can be written in symbolic form as

Response variable: *Absorbance*
Explanatory component: *[1] + log10Conc*

The fitted model accounts for 85.8% of the variation in the data (adjusted $R^2 = 0.858$), with the parameter estimates listed in Table 13.2. The slope parameter is clearly significantly different from zero ($t_6 = 6.575$, $P < 0.001$), indicating a strong linear trend in the absorbance response with changes in the $\log_{10}$-concentration. If we were to stop here in our investigation, we might think that this was a good representation of the data. But when we see the fitted model and fitted value plots in Figure 13.2, we realize that the straight line fits the data poorly. Although there is a trend present, in the sense that the absorbance reading increases with $\log_{10}$-concentration, there is also strong curvature in the relationship. Another model that accounts for the curvature must be sought for these data, and several possibilities are described in Chapter 17.

It can also be helpful to plot the observed response against the fitted values, particularly for the models with several explanatory variates considered in Chapter 14. Departures from the 1:1 line then indicate a poor fit of the model to the data. We might be tempted to assess this formally by fitting a regression through the origin to this representation, but the model is measured better by the goodness-of-fit statistics described in Sections 12.6 and 14.8. An alternative approach is cross-validation as described in Section 13.5.

TABLE 13.2

Parameter Estimates with Standard Errors (SEs), t-Statistics (t) and Observed Significance Levels (P) for a SLR Model for ELISA Absorbance Readings in Terms of $\log_{10}$-Concentration of Substrate (Example 13.1)

Term	Parameter	Estimate	SE	t	P
[1]	α	0.5458	0.19392	2.815	0.031
log10Conc	β	1.5284	0.23244	6.575	< 0.001

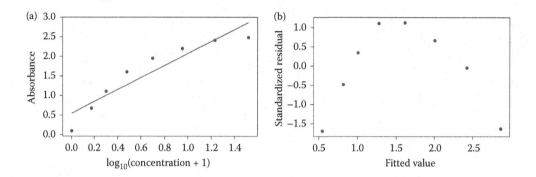

FIGURE 13.2
(a) Fitted model and (b) fitted values plot from SLR model for ELISA absorbance readings with $\log_{10}(\text{concentration} + 1)$ as the explanatory variate (Example 13.1).

In designed experiments, the values of the explanatory variable can be controlled and are often replicated. It is then possible to formally investigate model misspecification using the test for lack of fit introduced in Section 12.8.

13.2 More Ways of Estimating Deviations

In Section 5.1, we presented the simple and standardized residuals as estimates of the unknown deviations. These can be used to evaluate the validity of the underlying assumptions about the distribution of the deviations. We revisit these definitions here in the context of a SLR model, and we also introduce some new types of residuals that are useful for models with quantitative explanatory variables.

Recall from Section 5.1 that the **simple residuals**, $\hat{e}_i$, are defined as the difference between the observed responses and their fitted values. For the SLR model with

$$y_i = \alpha + \beta x_i + e_i \, ,$$

this takes the form

$$\hat{e}_i \ = \ y_i - \hat{y}_i \ = \ y_i - (\hat{\alpha} + \hat{\beta} x_i) \, ,$$

which is the difference between the response and the fitted straight line. The simple residuals from a SLR model do not have a common variance, as the variance of residual $\hat{e}_i$ depends on the value of the explanatory variate, x_i. For this reason, it is important to use **standardized residuals**, r_i, for regression models, which are defined as

$$r_i \ = \ \frac{\hat{e}_i}{\widehat{SE}(\hat{e}_i)} \, ,$$

where $\widehat{SE}(\hat{e}_i)$ is the estimated standard error of the ith simple residual. An explicit expression for this standard error is given in Section 13.4.2. These residuals are sometimes known as **internally Studentized residuals**. The standardized residuals are constructed to have a common variance equal to 1 (unit variance). Note that the standardized residuals apparently take the form of a t-statistic (an estimated quantity divided by its estimated SE, see Section 2.4) but – because the numerator and denominator are not independent – they do not follow a true t-distribution.

Both the simple and standardized residuals are estimated from the model fitted to the full set of data; this has the disadvantage that if an individual observation strongly influences the model fit, then its influence might not be detected. An alternative method based on prediction residuals overcomes this problem. We obtain the ith **prediction residual** by fitting the proposed model with the ith observation excluded, and then predict that observation from the new fitted model. This approach highlights responses that do not follow the general pattern of the model when it is fitted to the rest of the observations. Here we use the subscript (i) to denote quantities calculated with the ith observation omitted, and $\hat{y}_{(i)}$ denotes the prediction for the ith response from a model fitted excluding that observation. The prediction residual for the ith observation, denoted as $\hat{e}_{(i)}$, is defined as the difference between the predicted value, $\hat{y}_{(i)}$, and the response, y_i, i.e.

$$\hat{e}_{(i)} = y_i - \hat{y}_{(i)} .$$

We shall see an explicit formula for these residuals in Section 13.4.2. Since each observation is not involved in fitting the model that provides its predicted value, these residuals provide a valid measure of the predictive ability of a model. Further discussion and the use of these residuals in cross-validation methods is presented in Section 13.5. However, as for simple residuals, the prediction residuals do not have a common variance. We therefore define the **deletion residuals**, $r_{(i)}$, as a standardized version of the prediction residuals,

$$r_{(i)} = \frac{\hat{e}_{(i)}}{\widehat{SE}(\hat{e}_{(i)})} ,$$

where the estimated standard error $\widehat{SE}(\hat{e}_{(i)})$ is also calculated from a regression analysis that omits the ith observation. Deletion residuals are sometimes referred to as **externally Studentized residuals**, and these residuals do follow a true t-distribution. Hence, as a simple rule, observations with deletion residuals outside the range of ± 2 can be identified as potential outliers requiring further examination.

We have explained prediction and deletion residuals in terms of refitting the model to N subsets of the data obtained by excluding each observation in turn. In practice, all the quantities required can be calculated from the results of fitting the model to the full set. In particular, the deletion residuals can be directly computed from the standardized residuals as

$$r_{(i)} = r_i \sqrt{\left(\frac{N - p - 1}{N - p - r_i^2} \right)} ,$$

where r_i is the standardized residual, N is the sample size and p is the number of parameters in the model ($p = 2$ for SLR models). Other forms are given at the end of Section 13.4.2. The deletion residuals follow a t-distribution with $N - p - 1$ df, i.e. $N - 3$ df for a SLR model.

In general, we recommend the use of deletion residuals as they more readily identify outlying observations that have had a strong influence on the fitted model. Further discussion is presented in Section 13.4.

13.3 Using Graphical Tools to Check Assumptions

In Section 5.2, a composite set of residual plots was used to investigate the validity of the assumptions underlying a model with a single explanatory factor. In that context, the issue of model misspecification did not arise, because the model fitted an effect for each treatment or group. In that case, the residuals provide an untainted estimate of the model deviations. For SLR, if the model is misspecified then the residuals comprise a mixture of two components: one corresponding to the unknown model deviations and the other corresponding to the discrepancy between the model and the form of the response. For this reason, the residual plots can be used to assess properties of the deviations only if the model gives a good representation of the observed trend. When the form of the model is adequate, then the composite set of residual plots described in Section 5.2 can be used for models with quantitative explanatory variables, including the more complex regression models presented in Chapters 14, 15 and 17.

To recapitulate from Section 5.2, the residual plots can be used to check the assumptions that the deviations have equal variances (homogeneity of variances) and that they are consistent with observations from a Normal distribution. We usually check homogeneity of variance by plotting the standardized or deletion residuals, or their absolute values, against the fitted values from the model. In these graphs, variation is quantified as the vertical spread of the residuals: there should be no large change in this spread across the range of the fitted values (see Figure 5.2). If there is evidence of heterogeneity, then transformation of the response (Chapter 6) or a generalized linear model (Chapter 18) might be considered. The distribution of the residuals can be assessed with histograms and Normal probability plots (see Section 5.2.3). A histogram of residuals should show a symmetric, bell-shaped distribution, and the probability plot should yield an approximately straight line, with greater conformance to the expected shape being required for larger sample sizes.

EXAMPLE 13.2A: DIPLOID WHEAT

Recall that in Example 12.1, we fitted a SLR model to 190 diploid wheat seed weights with seed length as the explanatory variate. A subset of the data was presented in Table 12.1 and the complete data set can be found in file TRITICUM.DAT and Table A.1.

As noted above, the fitted model and fitted value plots suggested some evidence of model misspecification. However, the curvature in the data was small compared with the strong linear trend and so – with caution – we use the composite set of plots based on deletion residuals (Figure 13.3) to assess assumptions about the model deviations.

For this large data set, these graphs are very similar to those from the same model presented in Figure 12.4, which were plotted with standardized residuals. Variation of the residuals appears reasonably constant across the range of the fitted values, which accords with the assumption of homogeneity of variances. The histogram of the

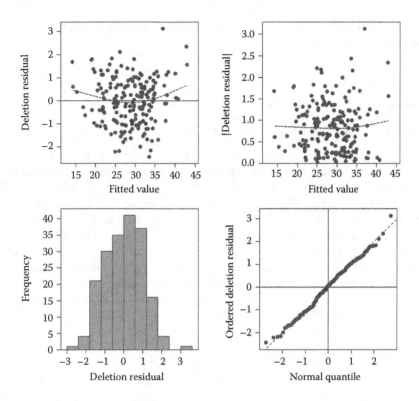

FIGURE 13.3

Composite set of residual plots, based on deletion residuals, for a SLR model with seed weight as the response and seed length as the explanatory variate (Example 13.2A).

residuals is symmetric and approximately bell shaped and the Normal plot is approximately a straight line, which together indicate the consistency of the residuals with a Normal distribution. The largest residual (corresponding to a fitted value of just under 37 mm) seems a little inconsistent with the rest of the distribution, although it does not stand out in the Normal plot. Weight and length measurements of this seed should perhaps be checked, but unless an error is found, it should be retained in the analysis (outliers were discussed in Section 5.4).

As stated in Section 13.1, in a SLR model, the fitted value plot may be substituted by one of the residuals against values of the explanatory variate, but this is not the case for the more complex regression models discussed in Chapter 14, which have several explanatory variates. In these models, the residuals can be plotted against each explanatory variate in turn to look for model misspecification. If the model fits well, then the residuals should be distributed homogeneously around zero without any systematic pattern.

The residuals can also be plotted against an additional explanatory variate that might help to explain the response, leading to a multiple regression model (see Chapter 14). If this graph shows a linear trend, then adding the new explanatory variate to the model might improve the fit. If the trend is non-linear, then transformation of the explanatory variable or a non-linear model (Chapter 17) might be required. Unfortunately, this graph gives a biased impression of the contribution that the new variate would make to the model, but an alternative graph, called an added variable plot, can be constructed to give an unbiased picture (see Section 14.6).

13.4 Looking for Influential Observations

When a SLR model is fitted, observations with more extreme values of the explanatory variate can have a large impact on the fitted line, which may make the resulting predictions unreliable. For example, if we drop one observation from the model and there is a large change in the estimated slope parameter, then we should be concerned about the robustness of the model. The basic concepts used to investigate this type of problem are leverage and influence. **Leverage** is a measure that identifies the more extreme values of the explanatory variate, which have the potential to be highly influential. However, leverage cannot quantify whether the observation has had a large impact on the fitted model, which is evaluated by its influence. The **influence** of an observation is a measure of the change in the fitted values that would occur if that observation was omitted. These concepts are illustrated for a SLR model in Figure 13.4.

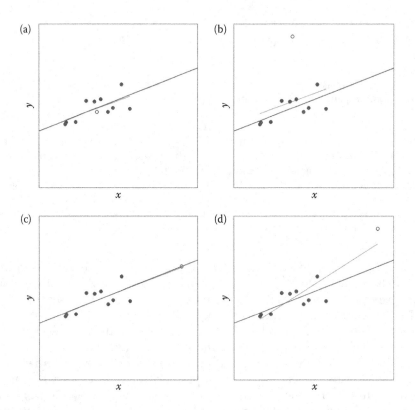

FIGURE 13.4
Leverage and influence of the highlighted point (∘): (a) small leverage and influence; (b) small leverage and large influence; (c) large leverage and small influence and (d) large leverage and influence. Black line represents the 'true' straight line relationship, and grey line represents the fitted model.

In each part of Figure 13.4, the black line represents the 'true' underlying straight line relationship while the grey line represents the fitted model. Each set of data has 10 observations in common (solid circles) with one additional highlighted observation (open circle) taking a different value in each plot. Figure 13.4a shows the highlighted point ($\circ$) with a value that gives it a small leverage (in the centre of the range of the explanatory variate) and with a small influence on the fitted line, which is then close to the true line. In contrast, the highlighted point in Figure 13.4b, while also having a small leverage, has exerted influence over the fit by increasing the value of the intercept (i.e. inducing a bias). In each of Figures 13.4c and d, the highlighted point has a large leverage. The highlighted point in Figure 13.4c has little influence on the fit, as it is consistent with the pattern in the rest of the data. In Figure 13.4d, the highlighted point is inconsistent with the rest of the data and has a large influence on the fitted line, causing changes in the estimates of both the intercept and slope parameters. Note that a point with large influence often appears inconsistent with the rest of the data, but might not appear as an outlier in residual plots if the fitted line is drawn towards it, as in Figure 13.4d. These examples demonstrate that although leverage, influence and outliers are often closely related, this is not always the case.

13.4.1 Measuring Potential Influence: Leverage

As described above, in a SLR, leverage quantifies the distance between the value of an explanatory variate for a given observation and the sample mean of that variate. If an observation is an outlier with respect to the explanatory variate (i.e. it has a particularly small or large value in comparison with the rest of the observations), then it is called a **leverage point**. Observations with large leverage can affect the fit of the model, but only if they are inconsistent with the overall trend. Therefore, a point with large leverage is not necessarily an influential point (e.g. Figure 13.4c). For this reason, leverage is most useful for assessing potential problems prior to analysis. For example, if an experimenter has control over values of the explanatory variate, then leverage can be assessed for different allocations (of value and replication) for the explanatory variate. However, the leverages give further insight into the form of the residuals discussed in the previous section, and into calculations of influence, and so we give further details here.

One common measure of leverage is called the **hat-value**. In a SLR model, the hat-values, also known as **leverages**, give a measure of the distance of the ith value of the explanatory variate, x_i, from its sample mean, $\bar{x}$, computed as

$$h_{ii} = \frac{1}{N} + \frac{(x_i - \bar{x})^2}{SS_{xx}},$$

where SS_{xx} is the sum of squares for the explanatory variate (as defined in Section 12.2) and N is the total number of observations. The name hat-value reflects the fact that these leverages are related to the fitted values ('y-hat'). In fact, the ith fitted value, $\hat{y}_i$, can be expressed as the sum of all N observed responses, identified as y_j, $j = 1 \ldots N$, multiplied by values h_{ij} defined as

$$h_{ij} = \frac{1}{N} + \frac{(x_i - \bar{x})(x_j - \bar{x})}{SS_{xx}}.$$

The fitted value for the ith observation can then be expressed as

$$\hat{y}_i = \sum_{j=1}^{N} h_{ij} y_j .$$

These weights capture the extent to which the jth observation affects the ith fitted value. If h_{ij} is large, then the jth observation has a substantial influence on the ith fitted value. The leverages, h_{ii}, therefore correspond to the influence that an individual observation has over its own fitted value. Note that the form of the weights, h_{ij}, becomes more complex for models containing several explanatory variates.

The leverage h_{ii} can take values between $1/N$ and 1 (i.e. $1/N \leq h_{ii} < 1$) and the sum of the leverages is always equal to the number of (independent) parameters in the model, p; their average is therefore p/N in general, and so $2/N$ for SLR models. Observations with large h_{ii} values are identified as having more leverage and, as a rule of thumb, values of $h_{ii} > 2 \times p/N$ are considered to be potential influential points. Leverages can be plotted against an explanatory variate or the fitted values.

EXAMPLE 13.2B: DIPLOID WHEAT

For the SLR model from Example 13.2A, the leverage threshold is $2 \times p/N = 2 \times 2/190 = 0.021$. Figure 13.5 shows the leverages plotted against the explanatory variate, with the threshold of 0.021 shown. This plot shows the quadratic relationship between the leverages and the explanatory variate in the SLR model, and that units with large leverages correspond to more extreme values of the explanatory variate. Clearly, most of the values are in the middle of the range with small leverage, but a few of the more extreme observations have leverages greater than the threshold of 0.021, with the maximum being 0.042. These observations are potentially influential points and should be further investigated by the influence measures described below.

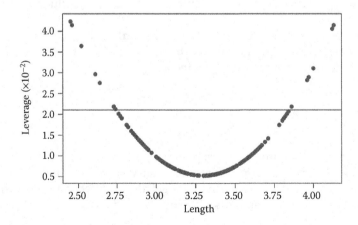

FIGURE 13.5
Leverages plotted against the explanatory variate (seed length, mm) for SLR with seed weight as the response (Example 13.2B). The horizontal line indicates leverage threshold (0.021).

13.4.2 The Relationship between Residuals and Leverages

We can write explicit expressions for the standardized, prediction and deletion residuals defined earlier in terms of the leverages, which clarify the relationships between these residuals. The estimated variance of the simple residual associated with the ith observation, $\hat{e}_i$, can be written in terms of its hat-value as

$$\widehat{\text{Var}}(\hat{e}_i) = s^2(1 - h_{ii}),$$

and the estimated standard error is the square root of this variance. It follows that uncertainty in the residual decreases as the leverage increases. This is because observations with very large leverage tend to get fitted more closely than observations with small leverage.

The standardized residual, r_i, is calculated as the simple residual divided by its estimated standard error, i.e.

$$r_i = \frac{\hat{e}_i}{\widehat{\text{SE}}(\hat{e}_i)} = \frac{\hat{e}_i}{\sqrt{s^2(1 - h_{ii})}}.$$

Since $\widehat{\text{SE}}(\hat{e}_i)$ is smaller for observations with large leverage, the standardized residuals of these observations tend to be slightly inflated relative to observations with smaller leverages. In practice, the range of leverages needs to be very large for this effect to become noticeable.

The prediction residuals, $\hat{e}_{(i)}$, can also be written more simply in terms of the simple residuals and the leverages as

$$\hat{e}_{(i)} = y_i - \hat{y}_{(i)} = \frac{\hat{e}_i}{1 - h_{ii}}.$$

So, we can obtain the prediction residual for the ith observation by re-scaling its simple residual by one minus its leverage. The variance of the prediction residual takes the form

$$\text{Var}(\hat{e}_{(i)}) = \frac{s_{(i)}^2}{1 - h_{ii}},$$

where $s_{(i)}^2$ is equal to the residual mean square (ResMS) obtained from a SLR with the ith observation omitted, and this variance can be directly calculated from the SLR results as

$$s_{(i)}^2 = s^2 \left(\frac{N - p - r_i^2}{N - p - 1} \right).$$

The deletion residual (Section 13.2) can then be expressed in terms of the other residuals as

$$r_{(i)} = \frac{\hat{e}_i}{\sqrt{s_{(i)}^2(1 - h_{ii})}} = \hat{e}_{(i)} \sqrt{\frac{(1 - h_{ii})}{s_{(i)}^2}} = r_i \sqrt{\frac{N - p - 1}{N - p - r_i^2}}.$$

13.4.3 Measuring the Actual Influence of Individual Observations

While leverage indicates the potential of individual observations to have a large impact on the fitted model, **influence statistics** measure the actual impact of each observation on the fitted model, and are hence generally more useful. Influence statistics help us to detect these individual observations that truly affect the fitted model, known as **influential points**. An influential point may affect one or more aspects of the fitted model. For example, in Figure 13.4b, the highlighted point affects the estimated intercept but not the slope; in Figure 13.4d, the highlighted point affects the estimates of both the intercept and slope. In both cases, the fitted model is changed, and in this section, we present some common influence statistics that measure the impact of individual observations on the overall fit of the model via changes in the fitted values.

Cook's statistic, D_i, measures the influence of an individual observation in terms of the change in the fitted values that would occur if that observation was omitted. For the ith observation, this statistic can be computed as

$$D_i = \left(\frac{r_i^2}{p} \right) \times \left(\frac{h_{ii}}{1 - h_{ii}} \right),$$

where r_i is the standardized residual, h_{ii} is the leverage for the ith observation, and p is the number of (independent) parameters in the model. Larger values of D_i correspond to observations with more influence on the fitted values. As a rule of thumb, values of $D_i > 1$ indicate influential points.

A modified form of Cook's statistic can be more useful for diagnostic plots, because its values can be used in half-Normal plots, where deviations in the tail of the distribution indicate the presence of potentially influential points. This modified statistic, C_i, is defined as

$$C_i = \sqrt{(N - p) \times \left(\frac{r_{(i)}^2}{p} \right) \times \left(\frac{h_{ii}}{1 - h_{ii}} \right)},$$

which uses the deletion residual rather than the standardized residual (Atkinson, 1984). A plot of the modified Cook's statistic against the explanatory variate(s) or fitted values can identify the location of influential points that exceed the threshold value of $C_i > 2\sqrt{(N - p)/N}$.

Figure 13.6 shows the modified Cook's statistics plotted against the fitted values and as a half-Normal plot (see Section 5.2.3) for the data of Figure 13.4c (large leverage, small influence) and Figure 13.4d (large leverage, large influence). When the highlighted observation has a small influence, all the C_i values fall below the threshold of $1.81 = 2\sqrt{(11 - 2)/11}$ (Figure 13.6a) and the half-Normal plot shows an approximately straight line (Figure 13.6b). When the highlighted observation has a large influence, it has a very large value of $C_i = 6.87$, substantially exceeding the threshold value (Figure 13.6c), and there is clear deviation from a straight line pattern in the half-Normal plot (Figure 13.6d).

Influential points might be outliers, but they do not necessarily appear as outliers in residual plots because the model fit has been adapted to accommodate them. In extreme cases, observations with both large leverage and large influence may be fitted almost exactly, giving a very small residual and possibly causing distortion elsewhere in the

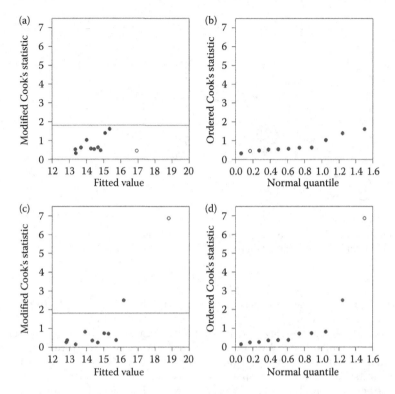

FIGURE 13.6
Modified Cook's statistics plotted against fitted values from the SLR in (a) Figure 13.4c and (c) Figure 13.4d; half-Normal plots of modified Cook's statistics from the SLR in (b) Figure 13.4c and (d) Figure 13.4d.

model. The investigation, and treatment, of influential points should be similar to that for other potential outliers, as discussed in Section 5.4. An important difference, in the context of regression models, is that we know that if we omit the influential observations, then the fitted model will change. Such influential observations should be checked for errors – in either the response or the explanatory variate or both – and corrected if necessary. If there is no evidence of any mistake, then the presence of the influential observations might indicate that the model is inadequate to explain the relationship: another explanatory variate might be required, or the shape of the relationship might be wrong. The influential observations should not be removed from the dataset without good reason, and any such action should be documented and reported. It may be helpful to consider the fit with and without any highly influential observations (possibly omitting potential outliers one at a time).

EXAMPLE 13.2C: DIPLOID WHEAT

The modified Cook's statistics for the SLR model of Example 13.2A are plotted against the fitted values and as a half-Normal plot in Figure 13.7.

There are $N = 190$ observations, and the SLR model has $p = 2$ parameters, giving a threshold value of $2\sqrt{(190-2)/190} = 1.99$. Using this threshold, we might identify six to eight influential points, and the plot of the modified Cook's statistics against the fitted values shows that most of the influential points are the seeds with the smallest and largest predicted lengths. The half-Normal plot appears curved rather than straight, and the eight largest values are not convincingly part of this trend. We know from our previous

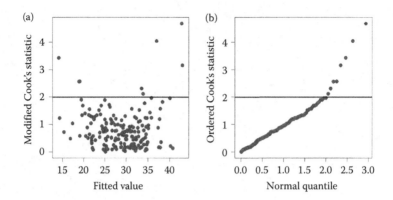

FIGURE 13.7
Modified Cook's statistics from SLR for seed weight with explanatory variate seed length: (a) plotted against fitted values and (b) a half-Normal plot (Example 13.2C).

analysis (Figure 13.3) that there is some doubt about the fit of the model in these more extreme regions of the explanatory variate (seed length); the influence statistics now suggest that these regions may also affect the overall fit of the model. However, if we omit the eight points whose modified Cook's statistic exceeds the threshold, then the change in the fitted line is small: the intercept decreases by -0.15 and the slope increases by 0.006 units; both changes are small compared with the standard errors of the estimated parameters (see Table 12.4). We can conclude that collectively, these influential points have only a small impact on the overall fit of the model.

13.5 Assessing the Predictive Ability of a Model: Cross-Validation

Cross-validation methods are used to assess the predictive ability of a model, and they can also provide an effective basis for choosing between competing models (see Section 14.9.3). Critical evaluation of a model is vital, as it enables limitations to be detected. This knowledge is especially important when the quality of real-life decisions depends on the reliability of predictions. For example, if a model is developed to predict contamination of grain via a sampling procedure, then the predictions must be accurate: if contamination is over-estimated, then good grain will be wasted; if it is under-estimated, then food quality might be compromised. To provide a realistic picture of its performance, a model should ideally be evaluated with an independent set of data, i.e. not the data on which the model was developed and fitted. The fitted model will tend to adapt to quirks in the original data that may not be representative of a wider population; so, the independent data should be representative of the population to which the model is going to be applied. The original data to which the model was fitted and the independent data used to test the model are commonly called the **training** and **validation** sets, respectively.

The cross-validation process consists of two steps. In the first step, a model is fitted to the training set. In the second step, this fitted model predicts the response for observations in the validation set. The differences between the observed and predicted values in the validation set, which we call **discrepancies**, may give some indication of ranges of the explanatory variable(s) for which predictions are reliable, and those for which they are unreliable. Note that the discrepancies are not the same as model residuals: residuals arise from the set of observations used to fit the model, whereas discrepancies arise from an independent set of observations not used in fitting the model. The overall predictive ability of the model may be quantified by summaries of these discrepancies in terms of statistics measuring bias or precision, which are described below. If the model is considered acceptable after the cross-validation process is complete, then a final model is often fitted from the combined training and validation sets.

In practice, it is often difficult or impractical to obtain further independent data and so, for the purposes of proper model validation, the original data may be split into two distinct subsets that form the training and validation sets. This cross-validation method is known as **data splitting**. At least half of the data will usually be allocated, at random, to the training set. The partitioning process has several drawbacks – both the quality of the fitted model and the results of the cross-validation may depend on the partition selected, particularly when the number of observations in either subset is small. The optimal partition depends on the context: a larger training set may produce a more robust model, but does not leave sufficient observations for reliable validation.

When there are too few data to be divided into two subsets, the **leave-1-out** cross-validation method can be used. Here, the training set (of size $N - 1$) contains all but one of the observations, which becomes the validation set (of size 1). This procedure is repeated for each observation in turn. Hence, the model is fitted N times: first, the model is fitted to the data with the first observation omitted and the response for the first observation is predicted; then the same procedure is followed for the second observation, and so on. In this case, the discrepancies are equal to the prediction residuals defined in Section 13.2. Another variant of this procedure, known as the **leave-k-out** cross-validation, splits the data into subsets of size k and uses these as validation sets with training sets of size $N - k$ (the remaining observations). Again, the model is fitted for each training set and predicts the response in each validation set. A third variant, known as **k-fold** cross-validation, splits the data into k subsets of approximately equal size and uses each subset in turn as the validation set.

In all these cross-validation methods, the predictive ability of the model is assessed on the discrepancies between the observed values and predictions made for the validation set. If these discrepancies are small, then the model is deemed to have good predictive ability. To define summary statistics that quantify the discrepancies, we need to identify the validation set separately from the training set. We illustrate the approach for the case of a SLR model, but the procedure is similar for more complex models. We denote the number of observations in the validation set as M and represent these observations as $Y_1 \ldots Y_M$, with $X_1 \ldots X_M$ as the associated values of the explanatory variate. From fitting a SLR model to the training set, we obtain parameter estimates $\hat{\alpha}$ and $\hat{\beta}$. These can be used to form predictions for the validation set as

$$\hat{Y}_i = \hat{\alpha} + \hat{\beta} X_i$$

for all observations, $i = 1 \ldots M$. The discrepancies, $Y_i - \hat{Y}_i$, measure the predictive ability of the model on the validation set. Plotting the discrepancies against the explanatory variate, X_i, might indicate specific ranges within which the model gives poor predictions.

Several statistics have been devised to summarize overall predictive ability. Here, we measure bias with the prediction bias statistic and we measure precision with the mean absolute difference and the square root of the mean square error. The **prediction bias** (PB) of the fitted model is estimated as the mean of the discrepancies across all observations in the validation set, written as

$$\mathrm{PB} = \frac{1}{M} \sum_{i=1}^{M} (Y_i - \hat{Y}_i) \, .$$

The PB takes a positive value if the predictions persistently under-estimate the observed response, and a negative value if they persistently over-estimate the observed response. If the predictions are unbiased, then the PB should be close to zero, but note that this can also occur if positive bias in one area is cancelled out by negative bias in another; so, the numerical summary should be interpreted alongside a graph of the discrepancies against the fitted values. This cancelling out cannot happen with the **mean absolute difference** (MAD), which is calculated as the mean of the absolute discrepancy values, i.e.

$$\mathrm{MAD} = \frac{1}{M} \sum_{i=1}^{M} |Y_i - \hat{Y}_i| \, .$$

The MAD is close to zero only if most of the discrepancies are small. A related measure, the **mean square error of prediction** (MSEP) is the mean of the squared discrepancies, written as

$$\mathrm{MSEP} = \frac{1}{M} \sum_{i=1}^{M} (Y_i - \hat{Y}_i)^2 \, .$$

This quantity is analogous to the ResMS from fitting the SLR model; the difference here is that estimation and prediction use different sets of data. If the predictive ability of the model is good, then the MSEP will be similar in size to the ResMS, on the assumption that the background variation is similar in the training and validation sets. The square root of the MSEP, known as the root mean square error (RMSE),

$$\mathrm{RMSE} = \sqrt{\mathrm{MSEP}} \, ,$$

is a common alternative to the MSEP. Both the MAD and RMSE take positive values, with large values indicating a model with poor predictive ability. Both statistics are easily interpreted as they are on the same scale as the observations; the major difference between them is that the RMSE gives more weight to large discrepancies. These statistics are especially useful for comparisons of different models. It can also help to express the PB, MAD and RMSE as a percentage relative to the average response, $\bar{Y}$, from the validation data set. For example, PB% $= 100 \times \mathrm{PB}/\bar{Y}$ represents the prediction bias as a percentage of the average response.

If you are concerned that the predictive ability of the model might change for specified subgroups, for example for large, medium or small values of the explanatory variate, then it may be appropriate to partition the validation set according to this criterion and to calculate and compare these summary statistics for each subgroup separately.

EXAMPLE 13.2D: DIPLOID WHEAT

In Examples 13.2A to 13.2C, we found some evidence of model misspecification in the SLR model that describes seed weight as a linear function of length. We can now examine the predictive ability of this model directly with cross-validation. To do this, we have split the data into two equally sized subsets: 95 observations were selected at random and allocated to the training set and the remaining 95 observations were allocated to the validation set. The SLR model was fitted to the training set, and accounted for 81.0% of the variance, with the parameter estimates shown in Table 13.3.

The parameter estimates accord with those from the full model (Table 12.4), although with a somewhat smaller intercept, steeper slope and larger standard errors, reflecting the reduced size of the training set.

The fitted model is shown in Figure 13.8a with the validation set, and Figure 13.8b shows the discrepancies between the observations in the validation set and their predicted values, plotted against the associated values of seed length, the explanatory variate.

The fitted model appears to run through the cloud of observations, but the discrepancies appear to show a general trend of decreasing value as seed length increases, with the exception of one very long seed. This suggests that the fitted slope is not quite following the trend in the validation set, although the general pattern is reasonable. The PB takes value –1.07, or –3.9% as a percentage of the mean seed weight ($\bar{Y} = 27.67$ in the validation set), indicating slight over-estimation of the response on average. The MAD and RMSE are 2.62 and 3.22, respectively, or 9.5% and 11.6% of the mean response. For comparison, the estimated background standard deviation from the training set was $s = 2.76$ and so, the average discrepancy in the validation sets is close to background variation in the training set. These results suggest no great problems with the SLR model. If we decided

TABLE 13.3

Parameter Estimates with Standard Errors (SEs), t-Statistics (t) and Observed Significance Levels (P) for a SLR Model for Seed Weight in Terms of Seed Length, Based on a Randomly Selected Training Set of 95 Observations (Example 13.2D)

Term	Parameter	Estimate	SE	t	P
[1]	α	−31.299	3.0537	−10.250	< 0.001
Length	β	18.358	0.9158	20.045	< 0.001

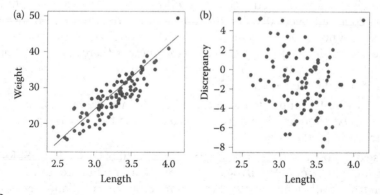

FIGURE 13.8

Cross-validation of SLR model for seed weight with seed length as the explanatory variate. (a) Validation set (95 observations) plotted with the SLR model fitted to the training set (remaining 95 observations) and (b) discrepancies plotted against the explanatory variate (Example 13.2D).

to fit a more complex model (e.g. Chapter 14), this type of cross-validation could be used to compare and evaluate different models.

In the case of leave-1-out cross-validation, the MSEP is known as the **predicted residual sum of squares** (PRESS), and a small value of the PRESS statistic indicates good predictive ability. Leave-1-out cross-validation is also closely related to the technique of jackknifing, which is used to obtain estimates of bias and precision for individual parameters or other model statistics. This technique is outside the scope of our account here, but you can find details in Efron and Tibshirani (1993).

EXERCISES

13.1 The microarray study introduced in Exercise 12.6 (data in file SENESCENCE.DAT) investigated gene expression associated with the senescence of leaves. Use SLR to predict the expression of gene *CATMA2A31585* over time, and use diagnostic plots and a formal test of lack of fit to assess the quality of this model. (We re-visit these data in Exercise 17.2.)

13.2* Now, consider the original data from the experiment described in Exercise 12.1. The numbers of leaves on each plant (variate *NLeaves*) are in file CABBAGE.DAT with unit numbers (*ID*) and sample dates (variate *Days*). Fit a SLR and use diagnostic plots to check the fit of the model. Would a transformation be appropriate here? If so, implement it and re-fit the SLR on your chosen scale. Plot the fitted model and check for any evidence of lack of fit. Give a 95% CI for the growth rate over the period (as leaves per day) and interpret this estimate. Can you reconcile this result with the one you gave in Exercise 12.1? (We re-visit these data in Exercise 18.2.)

13.3 Chickweed plants were sampled from a field trial to investigate whether the number of seeds produced could be related to the plant biomass, measured as dry weight (g). File CHICKWEED.DAT holds unit numbers (*ID*), the number of seeds (variate *NSeed*) and dry weights (variate *DryWt*) for 36 plants. Investigate the relationship between the variables, and use diagnostic plots to help decide whether you can use SLR to give a good description of this relationship. (We re-visit these data in Exercises 17.3 and 18.7.)*

13.4 In Exercise 12.2, you fitted a SLR to the log-transformed wet weight of flying insects collected over 30 years. Re-analyse these data without transformation, and use diagnostic plots to assess whether the model assumptions are better met on the untransformed or log scale. Check whether there is any sign of correlation in the errors between successive measurements on your chosen scale. (We re-visit these data in Exercise 15.1.)

13.5 In Exercise 12.4, you used SLR to establish whether the maximum stem length (variate *MaxLength* in file WILLOWSTEMS.DAT) could be used as a predictor of dry matter (variate *DryMatter*).

a. Use diagnostic plots to critically examine the fit of this SLR. Fit SLRs in terms of the other possible surrogate variables (variate *SumLength* is the sum of lengths of all stems, variate *SumDiam* is the sum of diameters of all stems and variate *LengthTop5* is the average length of the five longest stems) and

---

* Data from P. Lutman, Rothamsted Research.

investigate the quality of the fit in each case. Which variable is the best predictor of dry matter?

b. As there are 113 samples, there are sufficient data to compare these SLR models by cross-validation. Select 57 of the samples and re-fit the SLR for each explanatory variable from this subset, then assess their fit using the remaining 56 samples: calculate the prediction bias (PB), mean absolute difference (MAD) and root mean square error (RMSE). Which model has the best predictive properties? How did you select your samples, and was this method satisfactory?

(We re-visit these data in Exercise 14.5.)

13.6* Yield and a measure of disease were gathered from a field trial to try to establish a yield loss relationship. The unit numbers (*ID*), disease index (variate *Index*) and yield (variate *Yield*) are in file YIELDLOSS.DAT. Fit a SLR and evaluate the quality of this model. What happens if you exclude any highly influential observations from the model? What can you conclude about the reliability of the SLR?

13.7 The EXAMINE project (see Example 14.2) identified various measures of coldness during the winter as good predictors of the date of the first capture of various aphid species in suction traps. Here, we investigate the predictive ability of variable C60Day (the average temperature during the coldest 60-day period) to predict the date of the first capture of the aphid *Myzus persicae* (variable Mpe1st) at Long Ashton (in south-west England) over the periods 1970–1988 and 1993–2000. The data (variates *ID, Year, C60Day, Mpe1st*) are held in file LONGASHTON. DAT. Fit a SLR and use diagnostic plots to examine the fit. Is there any evidence of temporal correlation? Identify any influential observations and examine the impact of excluding them from the fitted model. What conclusions can you draw about the reliability of the SLR?*

---

* Data from R. Harrington, Rothamsted Research.

14

Models for Several Variates: Multiple Linear Regression

In Chapter 12, we introduced simple linear regression, which models a response variable as a straight line function of a single quantitative explanatory variable, or variate. In this chapter, we extend the concept to allow several variates in a **multiple linear regression** (MLR) model. This extension is analogous to the multi-factor model (Section 8.1) which investigates the simultaneous effects of several different qualitative variables, or factors, on the response. These extensions allow more realistic models, as usually several different explanatory variables might be associated with changes in the response, particularly in observational studies in which there is little or no control over the experimental conditions. In these circumstances, there can be strong correlation, or collinearity, between explanatory variates that can complicate the choice of which variates to include in a model. This chapter outlines the basic properties of MLR models and introduces methods for selection of explanatory variables.

As with any modelling exercise, the first step in building a MLR model is to explore the data, in this case to investigate the inter-relationships among the explanatory variates as well as those between the response and the individual explanatory variates (Section 14.1). The general form of the MLR model (Section 14.2) is an extension of the SLR model and, as in that model, parameter estimation is achieved by the method of least squares (Section 14.3). For a given set of explanatory variates, analysis of variance (ANOVA) is again used to estimate background variation, to assess whether the variability associated with the model is large compared with background variation, and to assess the contribution of individual explanatory variates to the model (Section 14.4). The estimate of background variation is used to make inferences on the model parameters, including predictions (Section 14.5). Prediction from the fitted model is often one of the main aims of a MLR, but accurate prediction requires a well-fitting model. We can investigate model misspecification by visualization of the contribution of individual explanatory variates to the model (Section 14.6). The choice of explanatory variates to include in a model can be complicated by the presence of correlation, or collinearity, between them and cases of very strong collinearity should be detected and avoided (Section 14.7). Additional goodness-of-fit statistics are available for MLR models (Section 14.8) and these can be used to compare models with different sets of explanatory variates. These statistics are utilized in various strategies for model selection (Section 14.9).

14.1 Visualizing Relationships between Variates

A MLR model aims to describe the relationship between a single response variable and two or more variates. Because correlation within the set of explanatory variates can affect the stability of a MLR model, we first inspect the data to detect which explanatory variates are clearly associated with the response, to see if these relationships are approximately linear, and to detect any strong correlations between pairs of potential explanatory variates.

As with SLR models, it is assumed that there is a straight line relationship between the response variate and each potential explanatory variate; however, in a MLR model, this relationship might hold only after one or more other explanatory variates has been taken into account, and so may not be immediately apparent. The presence of strong correlations between pairs of explanatory variates is an indication of **collinearity**, where the two variates are essentially measuring the same characteristic of the response, and which can create problems in interpretation and (in extreme cases) problems in fitting the model. These issues are discussed further in Section 14.7. Here, we investigate patterns of pairwise correlation between variates by constructing a correlation matrix (see Section 2.5) and by visualizing the underlying relationships in more detail with a scatter plot matrix.

Calculation of the correlation matrix for a response variate and set of potential explanatory variates summarizes all pairwise correlations within the set (see Section 2.5). A **scatter plot matrix** displays the pairwise scatter plots for the set of response and potential explanatory variates in the form of a matrix so that each row of plots has the same variate plotted on the y-axis and each column has the same variate plotted on the x-axis (see Figure 14.1).

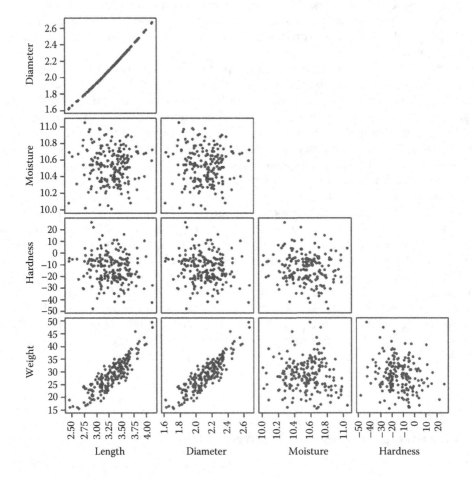

FIGURE 14.1
Scatter plot matrix for the diploid wheat seed data (Example 14.1).

This visual inspection of patterns of association also allows detection of curved relationships and of unusual, or outlying, observations.

EXAMPLE 14.1A: DIPLOID WHEAT

In Example 12.1, we described an experiment in which several morphological traits were measured on 190 seeds from a line of diploid wheat, *Triticum monococcum*. The traits measured on each seed were diameter (mm), length (mm), weight (mg), moisture content (%) and hardness index. The aim of the analysis was the identification of variables associated with differences in seed weight. The data can be found in file TRITICUM.DAT and Table A.1. We previously fitted a SLR model to describe seed weight as a linear function of seed length, and this model accounted for 77.8% of the variation in seed weight (adjusted $R^2 = 0.778$). Now, we want to know if we can improve this model by adding information from other explanatory variates.

A scatter plot matrix for the four explanatory variates (*Length, Diameter, Moisture, Hardness*) and the response variate (*Weight*) is shown in Figure 14.1.

Plots of seed weight against the explanatory variates are shown in the last row of the matrix. It is clear that weight is linearly associated with both length and diameter, but there is no obvious association of weight with either of moisture content or hardness. However, it is still possible that there is a relationship between weight and either hardness or moisture content after adjustment for length (or diameter), as this type of indirect relationship would not necessarily be visible here. There is a very strong association between length and diameter, indicating that these variables contain essentially the same information (are almost collinear). There are no other associations apparent within the set of explanatory variables. The correlation matrix (Table 14.1) corroborates these observations, and quantifies the strong correlation between diameter and length ($r = 0.999$) and between both of these variates and seed weight ($r = 0.883$ and 0.887, respectively).

14.2 Defining the Model

Having gained some insight into the structure of the data, we can start to build models. For the moment, we ignore the topic of model (or variable) selection, which is discussed in Section 14.9, and assume that we know which explanatory variates we wish to include in a MLR model.

The simplest MLR model is an obvious extension of the SLR model to relate a response variate to two explanatory variates. Where a SLR model fits a straight line in a two-dimen-

TABLE 14.1

Sample Correlations among Response (*Weight*) and Explanatory Variates, with Observed Significance Level in Parentheses, for the Diploid Wheat Study (Example 14.1A)

Length	—				
Diameter	0.999 (< 0.001)	—			
Moisture	−0.023 (0.748)	−0.021 (0.773)	—		
Hardness	−0.124 (0.088)	−0.125 (0.087)	−0.112 (0.125)	—	
Weight	0.883 (< 0.001)	0.887 (< 0.001)	−0.063 (0.390)	−0.207 (0.004)	—
	Length	Diameter	Moisture	Hardness	Weight

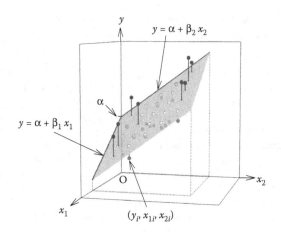

FIGURE 14.2

Plane ($y = \alpha + \beta_1 x_1 + \beta_2 x_2$) for a MLR with two explanatory variates x_1 and x_2. • observed values above the plane, ◒ observed values below the plane, ○ fitted values on the plane. Solid vertical lines represent the deviations, e_i, of the observations from the plane. O = origin (0, 0, 0).

sional space, a MLR model with two explanatory variates fits a plane in a three-dimensional space, as shown in Figure 14.2.

This model is represented mathematically as

$$y_i = \alpha + \beta_1 x_{1i} + \beta_2 x_{2i} + e_i \,, \tag{14.1}$$

where, as for the SLR model, the index i is used to identify the individual observations with labels from 1 to N (the number of observations). The value y_i is the ith observation of the response variate (symbolically denoted as variate y), x_{1i} and x_{2i} are the associated values of the first and second explanatory variates (denoted x_1 and x_2), and e_i is the deviation from the plane for the ith observation ($i = 1 \ldots N$). The intercept parameter, α, is the fitted response when both explanatory variates are zero, and parameters β_1 and β_2 are the slopes (gradients) of the response as the first and second explanatory variates vary, respectively. The standard assumptions presented in Section 12.1 all also apply to the MLR model.

We need also to extend the symbolic form of the explanatory component of the model. This component now includes the intercept and both explanatory variates as additive terms. So, the explanatory component of the model from Equation 14.1 is written symbolically as

Explanatory component: $[1] + x_1 + x_2$

where, as in Section 12.1, *[1]* denotes the variate taking value 1 everywhere, which is associated with the intercept parameter α, and the explanatory variates x_1 and x_2 are associated with the slope parameters β_1 and β_2, respectively.

This model represents a plane in three-dimensional space, as shown in Figure 14.2, defined in terms of the parameters α, β_1 and β_2. Parameter α is the intercept of the fitted plane with the y-axis at the values $x_1 = x_2 = 0$, and is also the predicted response at that point. Parameter β_1 is the change in value of the fitted plane for one unit increase in the first explanatory variate, x_1, with the value of the second explanatory variate, x_2, held constant. Similarly, parameter β_2 is the change in value of the fitted plane for one unit increase in the second explanatory variable, x_2, with the value of the first explanatory variable, x_1, held constant. This model is based on the assumption that the two explanatory variates affect the response variate independently, so that for a fixed value of the second explanatory variate, the relationship of the response variate with the first explanatory variate is a straight line: the intercept of this straight line varies according to the value of the second explanatory variate, but the slope remains constant. This can be seen mathematically by re-grouping the terms in Equation 14.1 as

$$y_i = (\alpha + \beta_2 x_{2i}) + \beta_1 x_{1i} + e_i .$$

With the value of the second explanatory variate held fixed, this is equivalent to a SLR model in terms of the first explanatory variate. Of course, a similar interpretation can be made for the second explanatory variate if the first is held fixed.

EXAMPLE 14.1B: DIPLOID WHEAT

From the analysis in Example 12.1, we already know that seed weight is strongly related to length. From biological arguments, we suspect that for a given length, the weight might also be affected by hardness of the seed, and so we add this second explanatory variate into the model. In mathematical form, the model can be written as

$$Weight_i = \alpha + \beta_1 \, Length_i + \beta_2 \, Hardness_i + e_i ,$$

where $Weight_i$, $Length_i$ and $Hardness_i$ are the weight, length and hardness index of the ith seed, respectively, and e_i is the deviation for that seed. As described above, α is the predicted seed weight for a seed of zero length and zero hardness. This is a substantial extrapolation beyond the range of the observed data and will probably not have biological meaning (see discussion at the end of Section 12.5). Parameter β_1 is the increase in seed weight for one unit increase in length (with hardness held fixed) and parameter β_2 is the increase in seed weight for one unit increase in hardness (with length held fixed). In symbolic form, this model is written as

Response variable: *Weight*
Explanatory component: *[1] + Length + Hardness*

where the variates *Weight*, *Length* and *Hardness* contain the values of seed weight, length and hardness index, respectively.

The MLR model can be extended from two to any number of explanatory variates, although simpler models – if plausible – are generally regarded as more desirable (see Section 14.9). The mathematical form of a general MLR model with q explanatory variates is

$$y_i = \alpha + \beta_1 x_{1i} + \ldots + \beta_l x_{li} + \ldots + \beta_q x_{qi} + e_i , \tag{14.2}$$

where y_i is the ith observation of the response, x_{li} is the associated value of the lth explanatory variate (denoted x_l in symbolic form, $l = 1 \ldots q$), and e_i is the deviation for the ith observation. Again, the standard assumptions presented in Section 12.1 all apply to this model. We use p to denote the total number of parameters in a MLR model. Here, there are $p = q + 1$ parameters (β_1 to β_q and α) to be estimated from the data. The fitted surface is now a hyper-plane in multiple dimensions and hard to envisage, but the parameters can still be interpreted as previously. Parameter α is the predicted response with $x_l = 0$ for all q explanatory variates. Parameter β_l is the change in value of the fitted plane for one unit increase in the *l*th explanatory variable with the values of all the other explanatory variates held constant.

14.3 Estimating the Model Parameters

Many of the principles introduced in the context of a SLR model apply directly also to a MLR model. Parameter estimates are obtained by the principle of least squares finding the parameter values that minimize the residual sum of squares, ResSS. For the general MLR model of Equation 14.2, the fitted values are written as

$$\hat{y}_i = \hat{\alpha} + \hat{\beta}_1 x_{1i} + \ldots + \hat{\beta}_l x_{li} + \ldots + \hat{\beta}_q x_{qi} \,, \tag{14.3}$$

where estimated values are again indicated by the hat (^) embellishment. The simple residuals are again the differences between the observed and fitted values

$$\hat{e}_i = y_i - \hat{y}_i.$$

The ResSS is the sum of the squares of these simple residuals and so can be written as

$$\mathrm{ResSS} = \sum_{i=1}^{N} (y_i - \hat{y}_i)^2 = \sum_{i=1}^{N} [y_i - (\hat{\alpha} + \hat{\beta}_1 x_{1i} + \ldots + \hat{\beta}_l x_{li} + \ldots + \hat{\beta}_q x_{qi})]^2 \,.$$

The estimated least squares parameters are those that minimize this quantity. We do not present the derivation of these parameter estimates, but details can be found in Rawlings et al. (1998). We also do not present the general form of the parameter estimates in the MLR model, as the expressions are often complex and difficult to interpret, and in practice we obtain these estimates from statistical software. However, for a MLR model with just two explanatory variates, the expressions are relatively easy to obtain, and give some insight into the adjustments made when more than one variate is present. As for the SLR model, estimates are based on the sums of squares and cross-products for the response and explanatory variates, introduced in Section 12.2. The sums of squares for the response and two explanatory variates are written as

$$\mathrm{SS}_{yy} = \sum_{i=1}^{N} (y_i - \bar{y})^2 \,; \quad \mathrm{SS}_{x_1 x_1} = \sum_{i=1}^{N} (x_{1i} - \bar{x}_1)^2 \,; \quad \mathrm{SS}_{x_2 x_2} = \sum_{i=1}^{N} (x_{2i} - \bar{x}_2)^2 \,,$$

where $\bar{y}$ is the mean of the response variate, and $\bar{x}_1$ and $\bar{x}_2$ are the means of the two explanatory variates, respectively. The sums of cross-products between the variates are written as

$$\mathrm{SS}_{x_1 y} = \sum_{i=1}^{N} (x_{1i} - \bar{x}_1)(y_i - \bar{y}) \,; \quad \mathrm{SS}_{x_2 y} = \sum_{i=1}^{N} (x_{2i} - \bar{x}_2)(y_i - \bar{y}) \,; \quad \mathrm{SS}_{x_1 x_2} = \sum_{i=1}^{N} (x_{1i} - \bar{x}_1)(x_{2i} - \bar{x}_2) \,.$$

Recall that the sums of squares and the sums of cross-products, as presented in Section 12.2, are simply scaled versions of unbiased sample variances and covariances, respectively. The two slope parameter estimates are then calculated as

$$\hat{\beta}_1 = \frac{(SS_{x_2x_2} \times SS_{x_1y}) - (SS_{x_1x_2} \times SS_{x_2y})}{(SS_{x_1x_1} \times SS_{x_2x_2}) - (SS_{x_1x_2})^2} , \quad \hat{\beta}_2 = \frac{(SS_{x_1x_1} \times SS_{x_2y}) - (SS_{x_1x_2} \times SS_{x_1y})}{(SS_{x_1x_1} \times SS_{x_2x_2}) - (SS_{x_1x_2})^2} ,$$

and the estimated intercept parameter can be written in terms of the estimated slopes and the response and explanatory variate means as

$$\hat{\alpha} = \bar{y} - \hat{\beta}_1 \bar{x}_1 - \hat{\beta}_2 \bar{x}_2 .$$

The estimated slopes for the explanatory variates ($\hat{\beta}_1$ and $\hat{\beta}_2$) usually differ from those obtained from two separate SLR models because of correlation between the two variates. However, if this correlation (and hence the sum of cross-products) is zero, i.e. if $SS_{x_1x_2} = 0$, then the estimated slopes are equal to those that would be obtained in the two separate SLR models, because the second term in both the numerator and denominator of the expressions for the slope parameter estimates becomes zero when $SS_{x_1x_2} = 0$. When the correlation between two explanatory variates is zero, we refer to them as orthogonal, and their effect on the response can be ascertained independently (as discussed for factor models in Section 11.1). When the correlation is not zero, the coefficient for one variate must be adjusted for the presence of the other in the model, with the adjustment depending on the covariance between them.

EXAMPLE 14.1C: DIPLOID WHEAT

We now estimate parameters in the MLR model of Example 14.1B for seed weight in terms of the explanatory variates length ($x_1 = Length$) and hardness index ($x_2 = Hardness$).

The sums of squares and cross-products for this set of variables can be calculated as $SS_{x_1x_1} = 19.2699$, $SS_{x_2x_2} = 29{,}721.0461$, $SS_{x_1y} = 330.9297$, $SS_{x_2y} = -3049.2033$ and $SS_{x_1x_2} = -94.0380$. The variate means are $\bar{y} = 28.658$, $\bar{x}_1 = 3.295$ and $\bar{x}_2 = 13.297$. Hence, the parameter estimates can be obtained as

$$\hat{\beta}_1 = \frac{(29721.05 \times 330.93) - (-94.04 \times -3049.20)}{(19.27 \times 29721.05) - (-94.04)^2} = 16.934 ,$$

$$\hat{\beta}_2 = \frac{(19.27 \times -3049.20) - (-94.04 \times 330.93)}{(19.27 \times 29721.05) - (-94.04)^2} = -0.049 ,$$

$$\hat{\alpha} = 28.66 - (16.93 \times 3.30) - (0.049 \times 13.30) = -27.795 .$$

The fitted MLR model can therefore be written as

$$\widehat{Weight}_i = -27.795 + 16.934 \times Length_i - 0.049 \times Hardness_i .$$

Hence an increase of 1 mm in seed length corresponds, on average, to an increase in seed weight of 16.93 mg for a fixed value of hardness. This value is a little different to that obtained in the SLR model of Example 12.1B (in which the estimated slope was

17.17) because of an adjustment for hardness index in the model. This adjustment is small because the correlation between these two variables ($r = -0.124$, Table 14.1) is also small. For a given seed length, an increase of one unit in hardness index corresponds to a decrease in seed weight, on average, of 0.049 mg.

14.4 Assessing the Importance of Individual Explanatory Variates

For a MLR model, the ANOVA is used to estimate the background variability which can then be used to make statistical inferences, to assess whether variation associated with the fitted model is larger than background variation, and to assess the contributions of individual explanatory variates to the model.

As with the SLR model, the summary ANOVA for a MLR model partitions the total sum of squares, TotSS, into a component due to the regression model, ModSS, and the residual variation, ResSS. The calculations for these quantities take the same generic form as those presented in Section 12.3, namely

$$\text{TotSS} = \sum_{i=1}^{N} (y_i - \bar{y})^2 \; ; \quad \text{ModSS} = \sum_{i=1}^{N} (\hat{y}_i - \bar{y})^2 \; ; \quad \text{ResSS} = \sum_{i=1}^{N} (y_i - \hat{y}_i)^2 \, .$$

Their form in terms of sums of squares and cross-products is now more complex and so is omitted here. The degrees of freedom are partitioned in a similar manner, following the same principles as for the SLR model. In the MLR model, the total df is still TotDF $= N - 1$. The number of parameters, p, is now equal to the number of explanatory variates plus one, i.e. $p = q + 1$. The model df is then ModDF $= q = p - 1$, and the residual df is ResDF $= N - p$. The form of the resulting ANOVA table is shown in Table 14.2.

An estimate of the background variation, s^2, is obtained from the residual mean square as

$$s^2 = \text{ResMS} = \frac{\text{ResSS}}{N - p} \, .$$

However, as for the SLR model, this quantity is a good estimate of the true background variation only if there is no model misspecification (see the discussion in Section 13.1). Diagnostic checks on the model fit and residuals should be made before conclusions are drawn; these checks are described in Section 14.6.

TABLE 14.2

Structure of the Summary ANOVA Table for a MLR Model with q Explanatory Variates and $p = q + 1$ Parameters

Source of Variation	df	Sum of Squares	Mean Square	Variance Ratio	P
Model	$p - 1$	ModSS	ModMS $=$ ModSS$/(p-1)$	F $=$ ModMS/ResMS	Prob($F_{p-1,N-p} >$ F)
Residual	$N - p$	ResSS	ResMS $=$ ResSS$/(N-p)$		
Total	$N - 1$	TotSS			

The F-test for the model, based on the variance ratio F = ModMS/ResMS, now relates to the null hypothesis that *all* of the slope coefficients for the explanatory variates are equal to zero, i.e. H_0: $\beta_l = 0$ for $l = 1 \ldots q$, and hence that there is no relationship between the response and the set of explanatory variates. The alternative hypothesis is that there is some relationship, and hence the regression coefficients $\beta_1 \ldots \beta_q$ are not all equal to zero. The F-test uses $p - 1$ numerator and $N - p$ denominator df, corresponding to the degrees of freedom for the model and residual mean squares, respectively. An observed value of F larger than the $100(1 - \alpha_s)$th quantile of this distribution gives evidence at significance level α_s of some relationship between the response and the set of explanatory variables. Alternatively, an observed significance level can be calculated as $P = \text{Prob}(F_{p-1, N-p} \geq F)$.

EXAMPLE 14.1D: DIPLOID WHEAT

The ANOVA table from the MLR model relating seed weight to length and hardness index is shown in Table 14.3. The observed variance ratio, F = 349.106, is larger than the 99.9th percentile of the F-distribution with 2 numerator and 187 denominator degrees of freedom ($P < 0.001$) giving strong evidence of a relationship between seed weight and this combination of explanatory variables. We should not be surprised as we had already established a strong association of seed weight with seed length in the SLR model.

A high significance ($P < 0.05$) for the F-test gives evidence of some association between the response variate and the set of explanatory variates, but does not indicate the contribution of individual explanatory variates to this association. These individual contributions can be evaluated by further partitioning the model variation (ModSS) into components associated with each of the explanatory variates, but when the explanatory variates are correlated, this partition is not unique. This leads us to the concept of a sequential ANOVA table, in which the sums of squares depend on the order in which explanatory variates are added into the model. We introduced this concept in Section 11.2 for models based on factors; here, we adapt it to regression models for explanatory variates.

14.4.1 Adding Terms into the Model: Sequential ANOVA and Incremental Sums of Squares

We can build a MLR model by starting with the intercept and adding each of the explanatory variates in turn, giving a sequence of sub-models that ends with the full model containing all of the explanatory variates. These sub-models can be used to form a **sequential ANOVA table**, which quantifies the change in the model sum of squares as explanatory variates are added into the model in a particular sequence.

To construct the sequential ANOVA table, we first need to define some quantities associated with the sequence of sub-models (previously defined for models with factors in

TABLE 14.3

Summary ANOVA Table for a MLR Model for Seed Weight with
Length and *Hardness* as Explanatory Variates (Example 14.1D)

Source of Variation	df	Sum of Squares	Mean Square	Variance Ratio	P
Model	2	5753.4738	2876.7369	349.106	< 0.001
Residual	187	1540.9352	8.2403		
Total	189	7294.4090			

Section 11.2). We identify the model sum of squares and df with a particular sub-model by explicitly specifying it within parentheses. For example, ModSS($[1] + x_1 + x_2$) is the model sum of squares associated with a model containing the intercept and explanatory variates x_1 and x_2, with degrees of freedom ModDF($[1] + x_1 + x_2$) = 2. This notation is not compact, but it has the advantage of being unambiguous. The model containing the intercept term alone is regarded as the initial or baseline model, and has zero sum of squares and df, i.e. ModSS($[1]$) = 0 and ModDF($[1]$) = 0; this is discussed further below.

The sequential ANOVA table is then derived from the set of model sums of squares and df. As each explanatory variate is added into the model, the increase in ModSS and ModDF is attributed to that variate. These changes in the sums of squares and degrees of freedom are called the **incremental** or **Type I sums of squares** and **df**, and both must always be greater than (or equal to) zero. These incremental quantities are labelled by both the explanatory variate added and the terms already in the model. For example, on adding the variate x_2 to a sub-model containing the intercept and variate x_1, we label the change as $+x_2 | [1] + x_1$, to be read as 'adding variate x_2 given that the intercept and variate x_1 are already in the model' or equivalently 'adding variate x_2 after accounting for (eliminating) the terms $[1] + x_1$'. The incremental sums of squares and df are denoted SS and DF, respectively. So, for example

$$SS(+x_2 | [1] + x_1) = \text{ModSS}([1] + x_1 + x_2) - \text{ModSS}([1] + x_1),$$
$$DF(+x_2 | [1] + x_1) = \text{ModDF}([1] + x_1 + x_2) - \text{ModDF}([1] + x_1).$$

Each quantity is calculated as a difference between the model containing all of the variates listed and the model containing only the variates listed after the '$|$' symbol, i.e. those in the previous sub-model. Again, this notation is somewhat cumbersome but unambiguous.

In the context of a sequential ANOVA, we use some abbreviations by considering the table as a whole. For example, instead of listing the change *and* the terms already present in the model, we can deduce the terms already in the model from previous lines in the ANOVA table and just indicate the additional term. Hence, we can use SS($+x_2$) and DF($+x_2$) to denote SS($+x_2 | M$) and DF($+x_2 | M$), respectively, where M is a list of terms added in previous lines of the ANOVA table. For example, Table 14.4 is the sequential ANOVA table for the MLR model obtained by our fitting first the variate x_1 and then the variate x_2. The first line of the table adds variate x_1 into a model containing the intercept only. The incremental sum of squares is then

$$SS(+x_1) = SS(+x_1 | [1]) = \text{ModSS}([1] + x_1) - \text{ModSS}([1]).$$

TABLE 14.4

Structure of the Sequential ANOVA Table for a MLR Model with Two Explanatory Variates, x_1 and x_2

Term Added	Incremental df	Incremental Sum of Squares	Incremental Mean Square	Variance Ratio
$+x_1$	DF($+x_1$) = 1	SS($+x_1$)	MS($+x_1$) = SS($+x_1$)/1	F^{x_1} = MS($+x_1$)/ResMS
$+x_2$	DF($+x_2$) = 1	SS($+x_2$)	MS($+x_2$) = SS($+x_2$)/1	F^{x_2} = MS($+x_2$)/ResMS
Residual	ResDF	ResSS	ResMS	
Total	$N-1$	TotSS		

We then add the variate x_2 into the model, with incremental sum of squares

$$SS(+x_2) = SS(+x_2 | [1] + x_1) = \text{ModSS}([1] + x_1 + x_2) - \text{ModSS}([1] + x_1) .$$

Mean squares are again calculated by division of the incremental sums of squares by the corresponding incremental df, and variance ratios for each explanatory variate are calculated with respect to the residual mean square, ResMS. The variance ratios can be used to test the null hypothesis that the response has no dependence on the explanatory variate being added, given that the explanatory variates previously included are also present in the model, i.e. that the coefficient for the explanatory variable added is equal to zero. For example, in Table 14.4, variance ratio F^{x_1} is used to test the hypothesis that the response has no association with variate x_1, given that the intercept is present in the model. The variance ratio F^{x_2} is used to test the hypothesis that the response has no association with variate x_2 given that both the intercept and variate x_1 are already in the model or, equivalently, whether adding variate x_2 has made any improvement to the SLR model containing x_1. Under the null hypothesis, each variance ratio in this sequential ANOVA table has an F-distribution with 1 numerator df and denominator df equal to ResDF.

As stated above, if the explanatory variates are correlated, then the values in the sequential ANOVA, and hence the incremental F-tests, depend on the order in which the variates are added into the model. This reflects the fact that the incremental F-tests are evaluating different hypotheses for different sequences of sub-models. In the example of two variates shown in Table 14.4, suppose that explanatory variate x_2 was fitted first, followed by variate x_1. In that case, the incremental F-test for explanatory variate x_2 gives evidence on whether the response is associated with that variate, given that the intercept is in the model. The incremental F-test for the second explanatory variate added (x_1) gives evidence on whether this variate leads to any improvement in the fit of a SLR model already containing x_2. These hypotheses are different from those tested in the original sequence of sub-models and so it is possible for different results to be obtained. This can lead to some ambiguity in choice of model as, for example, we may find that the incremental F-tests for both $x_1 | [1]$ and $x_2 | ([1] + x_1)$ are significant for model $[1] + x_1 + x_2$, but that only the test for $x_2 | [1]$ is significant when fitting $[1] + x_2 + x_1$ (where the order of terms defines the order in which the terms are added to the model). In general, when selecting a model, we aim for **parsimony**, i.e. using the fewest parameters possible to get an adequate description of the response variable. In this example, this principle would choose the model with only the intercept and explanatory variate x_2, because adding x_1 does not then improve the fit.

EXAMPLE 14.1E: DIPLOID WHEAT

The two incremental ANOVA tables from fitting the MLR model relating seed weight to length and hardness index are in Table 14.5. In this case, the conclusions are straightforward, as all incremental F-tests are significant, indicating that both variates are required in the model. Even though the correlation between the two explanatory variates is weak, $r = -0.124$, the values of the incremental F-tests for the different model orders are distinctly different.

There is some ambiguity in the process for testing incremental sums of squares that requires further explanation. For example, if we are considering a model containing just variate x_1, then the ResMS in the (sequential) ANOVA table is calculated having removed the effect of this term only. If we are considering a model with two explanatory variates, x_1 and x_2, and we add x_1 first, then we get the same incremental sum of squares for that variate,

TABLE 14.5

Sequential ANOVA Tables (Mean Squares Not Shown) from MLR for Seed Weight with
Explanatory Variates Length (*Length*) and Hardness Index (*Hardness*) (Example 14.1E)

Term	df	SS	VR	P	Term	df	SS	VR	P
+ *Length*	1	5683.18	689.68	< 0.001	+ *Hardness*	1	312.83	37.96	< 0.001
+ *Hardness*	1	70.30	8.53	0.004	+ *Length*	1	5440.64	660.25	< 0.001
Residual	187	1540.94			Residual	187	1540.94		
Total	189	7294.41			Total	189	7294.41		

Note: df = incremental df, SS = incremental sum of squares, VR = variance ratio.

but the ResMS in the resulting sequential ANOVA table is now calculated after removal of
both terms. The variance ratio for x_1 and the associated F-test, which depend on the ResMS,
will therefore differ between these two situations. Perhaps surprisingly, both tests are valid
even though they give different results, and each approach has its own advantages. First,
we consider the approach of adding one explanatory variate to the model at a time, forming
the sequential ANOVA table, and testing against the ResMS from the current model, i.e.
the ResMS calculated after estimation of only the terms in the current model. This has the
disadvantage that the approach may lack statistical power at early stages when important
explanatory variates have not yet been included in the model, because the ResMS will be
inflated and variance ratios will therefore be reduced in comparison with those for the
full model. The alternative approach is to use just one ResMS in the calculation of variance
ratios, taken from the ANOVA table for the full model including all explanatory variates.
This has the potential disadvantage that the ResMS may be based on relatively few df when
there are many potential explanatory variates present, so that the tests lack power. To illustrate this, we consider two contrasting situations: a designed experiment in which there are
a few pre-defined explanatory variables; and an observational study, where many potential explanatory variables might be available. In the first situation, the concept of the 'full
model', containing all explanatory variables of interest, is well defined, and in this context, it
makes sense to form a sequential ANOVA for the full model and to base variance ratios on
the ResMS from this full model. In the second situation, it is often not clear which explanatory variables are most likely to be relevant, and including the full set is unlikely to help. In
this context, adding one term to the model at a time, and using the ResMS from the current
model to form the variance ratios, appears more sensible. In practice, most examples fall
between these two extremes, and the use of common sense is required.

   Throughout this section, we have used a model containing the intercept term only as our
baseline model. This is the usual convention, but it requires some modification for regression through the origin (Section 12.9.2) when no intercept term is included in the model.
In this case, the definitions of model sum of squares and df used above are inappropriate,
and must be amended to the uncorrected versions defined in Section 12.9.2.

14.4.2 The Impact of Removing Model Terms: Marginal Sums of Squares

We can also take a somewhat different approach to this problem and, instead of progressively adding variates into a model, we can start with a model containing the full set of
explanatory variates and obtain **marginal F-tests** by removing each explanatory variate
from the model in turn. These marginal F-tests relate to the null hypothesis that the coefficient of the *l*th explanatory variate is zero given that the rest of the explanatory variates

TABLE 14.6

Form of Marginal F-Tests for Two Variates, x_1 and x_2, with Full Fitted Model $[1] + x_1 + x_2$

Term Dropped	Marginal df	Marginal Sum of Squares	Marginal Mean Square	Variance Ratio
$-x_1$	$\mathrm{DF}(-x_1) = 1$	$\mathrm{SS}(-x_1)$	$\mathrm{MS}(-x_1) = \mathrm{SS}(-x_1)/1$	$F^{x_1} = \mathrm{MS}(-x_1)/\mathrm{ResMS}$
$-x_2$	$\mathrm{DF}(-x_2) = 1$	$\mathrm{SS}(-x_2)$	$\mathrm{MS}(-x_2) = \mathrm{SS}(-x_2)/1$	$F^{x_2} = \mathrm{MS}(-x_2)/\mathrm{ResMS}$
Residual	ResDF	ResSS	ResMS	

are present in the model. The changes in the model sum of squares caused by dropping each variate in turn are often called the **marginal** or **Type III sums of squares**. We denote these marginal sums of squares and df by defining the variate to be dropped from the model and the model from which it is to be dropped, for example

$$\mathrm{SS}(-x_1 | [1] + x_1 + x_2) = \mathrm{ModSS}([1] + x_1 + x_2) - \mathrm{ModSS}([1] + x_2),$$
$$\mathrm{DF}(-x_1 | [1] + x_1 + x_2) = \mathrm{ModDF}([1] + x_1 + x_2) - \mathrm{ModDF}([1] + x_2).$$

The ' − ' sign indicates that the term is to be removed from the model following the '|' symbol. When the full model on the right-hand side is clear from context, we abbreviate the marginal sums of squares as $\mathrm{SS}(-x_1)$. The form of the marginal sums of squares and F-tests for a MLR model with two explanatory variates, x_1 and x_2, is shown in Table 14.6. In this table, $\mathrm{SS}(-x_1)$ is defined as above and $\mathrm{SS}(-x_2) = \mathrm{SS}(-x_2 | [1] + x_1 + x_2)$. Again, mean squares are obtained by division of the sums of squares by their df, and the ResMS is that taken from the full model specified.

EXAMPLE 14.1F: DIPLOID WHEAT

The marginal sums of squares and F-tests from the MLR model relating seed weight to length and hardness index are derived in Table 14.7. As expected, following the analysis of Example 14.1E, both marginal F-tests are significant.

Note that, for the last variate added to the model, the incremental (Type I) and marginal (Type III) sums of squares (and F-tests) are equal. In general, both the incremental and marginal sums of squares can help in deducing a suitable model. But when there are many explanatory variates, the situation becomes more complex as the number of different orders in which the variates can be fitted increases rapidly. One solution is the use of automatic methods for model selection and comparison, and these are described in Section 14.9 below. In addition to Types I and III, Type II and Type IV sums of squares have also been

TABLE 14.7

Marginal F-Tests from MLR for Seed Weight with Explanatory Variates Length (*Length*) and Hardness Index (*Hardness*) (Example 14.1F)

Term Dropped	Marginal df	Marginal Sum of Squares	Mean Square	Variance Ratio	P
− *Length*	1	5440.6436	5440.6436	660.249	< 0.001
− *Hardness*	1	70.2985	70.2985	8.531	0.004
Residual	187	1540.9352	8.2404		

defined. These are less commonly used and so are not considered here, but they are briefly described in Section 11.2.3.

F-tests can also be used in a more general situation to evaluate the effect of simultaneously adding or dropping a group of terms from a model (see Example 17.3B). These tests follow exactly the same procedures as outlined above, but they are rarely as useful as testing individual terms, and so we shall not give details here. For further information we recommend Rawlings et al. (1998).

14.5 Properties of the Model Parameters and Predicting Responses

As for the SLR model, in the MLR model, we can use statistical theory to obtain the sampling distribution of the parameter estimates and make statistical inferences about the true unknown parameters. If we can assume that the deviations follow a Normal distribution (Assumption 4, Section 12.1), then estimates of the model parameters also follow Normal distributions. These are unbiased estimates, so the mean of each distribution is the unknown population parameter. We represent the estimated variances of these parameters as $\widehat{\mathrm{Var}}(\hat{\alpha})$ and $\widehat{\mathrm{Var}}(\hat{\beta}_l)$, $l = 1 \ldots q$, for a MLR with q explanatory variates, and $\widehat{\mathrm{SE}}()$ denotes the corresponding estimated standard error. The formulae represented by this shorthand notation are complex and so omitted here – we rely on the calculations made by statistical software. However, note that the estimated variances use the estimate of background variation based on the residual mean square and so inherit the residual degrees of freedom.

The most common use of these distributions and variance estimates is in testing the null hypothesis that the parameter for a given explanatory variate equals zero, i.e. testing the null hypothesis H_0: $\beta_l = 0$ against the alternative hypothesis H_1: $\beta_l \neq 0$, for the lth explanatory variate, given that the remaining $q - 1$ variates are present in the model. This test statistic is calculated as

$$t_{N-p} = \frac{\hat{\beta}_l}{\widehat{\mathrm{SE}}(\hat{\beta}_l)} \, .$$

Under the null hypothesis, this test statistic follows a t-distribution with degrees of freedom equal to the residual df, $N - p$. In fact, this test is equivalent to the marginal F-test obtained by dropping the lth explanatory variate from the full model, and the marginal F-test statistic is equal to the square of the t-statistic. Most statistical software prints these t-statistics with the parameter estimates (including the intercept parameter). This enables a quick assessment of whether individual explanatory variates can be immediately omitted from the model without worsening the model fit. However, owing to collinearity, these tests can be misleading if more than one variate is dropped at a time and so a sequential approach is required. Automatic methods of model selection exist and these are discussed in Section 14.9.

A $100(1 - \alpha_s)\%$ confidence interval can be calculated for each regression parameter as

$$\left(\hat{\alpha} - t_{N-p}^{[\alpha_s/2]} \times \widehat{\mathrm{SE}}(\hat{\alpha}), \quad \hat{\alpha} + t_{N-p}^{[\alpha_s/2]} \times \widehat{\mathrm{SE}}(\hat{\alpha}) \right) ,$$

$$\left(\hat{\beta}_l - t_{N-p}^{[\alpha_s/2]} \times \widehat{\mathrm{SE}}(\hat{\beta}_l), \quad \hat{\beta}_l + t_{N-p}^{[\alpha_s/2]} \times \widehat{\mathrm{SE}}(\hat{\beta}_l) \right) ,$$

TABLE 14.8

Parameter Estimates with Standard Errors (SE), t-Statistics (t) and Observed Significance Levels (*P*) for a MLR Model for Seed Weight with Explanatory Variates Length (*Length*) and Hardness Index (*Hardness*) (Example 14.1G)

Term	Parameter	Estimate	SE	t	P
[1]	α	−27.795	2.1653	−12.836	< 0.001
Length	β_1	16.934	0.6590	25.695	< 0.001
Hardness	β_2	−0.049	0.0168	−2.921	0.004

where, as before, $t_{N-p}^{[\alpha_s/2]}$ is the $100(1 - \alpha_s/2)$th percentile of the t-distribution with degrees of freedom equal to the residual df, $N - p$.

EXAMPLE 14.1G: DIPLOID WHEAT

Table 14.8 shows the estimated parameters for a MLR model for seed weight in terms of length and hardness index, together with their estimated SE, t-statistic and the observed significance level (*P*) associated with testing the null hypotheses H_0: $\alpha = 0$ and H_0: $\beta_l = 0$, for $l = 1, 2$. The t-tests give evidence that the response depends on each of the explanatory variates when the other is present in the model, so neither explanatory variate can be removed without making the fit worse, which agrees with the results of Example 14.1F. We can verify that the squares of the t-statistics are equal to the marginal F-tests and give exactly the same observed significance level (Table 14.7).

As for a SLR model, we can use the fitted model to obtain predictions of the response for a given set of explanatory variate values, using the form given in Equation 14.3. An estimated standard error for the prediction can be calculated, but again we omit the formula here and obtain the values from statistical software. The concepts of interpolation and extrapolation carry over from the SLR model case, but the situation is now somewhat more complex because of the interplay between the explanatory variates. For a prediction to be considered as interpolation, it must lie within the range of values defined by the set of explanatory variates as a whole. For example, in the diploid wheat data (Figure 14.1), the observed seed lengths run from 2.5 to 4 mm and the observed hardness index lies between −50 and +20. However, if we consider the two-dimensional spread of values, there is only good coverage in the square defined by lengths in the range 2.75–3.75 mm and hardness index in the range −30 to 10. Predictions outside of this area should be considered as extrapolation, as there are too few data there to support the form of the model.

Confidence intervals for any prediction can be calculated, as shown in Section 12.5, to obtain an interval for either the expected mean response or for an individual new observation. Details can be found in Rawlings et al. (1998).

14.6 Investigating Model Misspecification

As for the SLR model, you should check the validity of the assumptions underlying the fitted model using the residuals. And, again, you should exclude the possibility of model misspecification before trying to interpret residual plots. This process is complicated by the presence of several explanatory variates, any of which may be subject to misspecification.

Here, we describe some graphical methods to verify the form of the model for individual explanatory variates. Once this has been done, the methods of Sections 5.2 and 13.3 can be used to check the properties of the standardized or deletion residuals. The methods of Section 13.4.3 can also be used to investigate the influence of individual observations with respect to the model as a whole.

The basic building block for all these methods is again the simple residual calculated as the difference between the observed and fitted values, i.e.

$$\hat{e}_i = y_i - \hat{y}_i = y_i - (\hat{\alpha} + \hat{\beta}_1 x_1 + \ldots + \hat{\beta}_q x_q).$$

Standardized, prediction and deletion residuals can be derived from the simple residuals as described in Section 13.2. Graphs of standardized or deletion residuals against each explanatory variate can indicate the presence of curvature in that particular relationship, but do not put this into the context of the overall model. To do this, we define a new type of residual, the partial residual. The **partial residual** for the ith observation on the lth explanatory variate ($l = 1 \ldots q$) is denoted by $\tilde{e}_{li}$ and calculated as the simple residual plus the contribution of the lth explanatory variate to the ith fitted value, i.e.

$$\tilde{e}_{li} = \hat{e}_i + \hat{\beta}_l x_{li}.$$

This gives a set of N partial residuals for each of the q explanatory variates. A scatter plot of the partial residuals against the values of the associated explanatory variate, known as a **partial residual plot**, should show a scatter of points around the fitted straight line representing the model. Any systematic deviation, such as substantial curvature in the scatter of points, may indicate misspecification for that explanatory variate, which might be dealt with by the methods described in Chapter 17.

EXAMPLE 14.1H: DIPLOID WHEAT

Figure 14.3 shows partial residual plots for the MLR model for seed weight with explanatory variates *Length* and *Hardness*, with the fitted component of the model ($\hat{\beta}_l x_{li}$, $l = 1, 2$) shown as a straight line in each case. As in the SLR model (Example 12.1), the

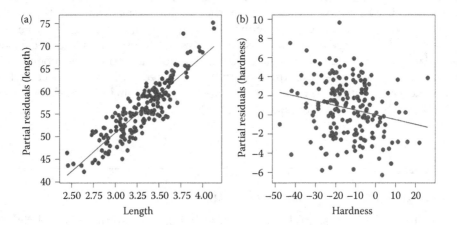

FIGURE 14.3
Partial residual plots for a MLR model for seed weight with explanatory variates (a) length (mm) and (b) hardness index, with fitted component of model (—) (Example 14.1H).

relationship with seed length is strong, with a slight hint of curvature at the ends of the range, where the partial residuals are all above the line representing the fitted model. However, this curvature is a small component of the overall trend. The relationship with hardness index is much noisier without any apparent curvature and, although statistically significant, the downwards trend might not be detected by eye.

The partial residual plot gives insight into the fit of an explanatory variate already included in the model. Residual plots can also help to investigate the potential role of an additional explanatory variate that has not yet been included in the model. Several naïve approaches are possible. Plotting the additional explanatory variate against the response (as in the initial exploratory data analysis) indicates whether a relationship exists in the absence of other explanatory variates, but it does not show whether a relationship exists after accounting for the variates already in the model. Plotting the standardized residuals against the additional variate does take some account of the current model, but the slope of this plot will not be equal to the estimated slope for the additional variable if it were added into the model (except in the special case where the additional variable is orthogonal to all of the variates in the current model). One solution is the **added variable plot**, in which the observed slope equals the estimated slope that would be obtained if the explanatory variate was added to the current model. The added variable plot is a scatter plot of the simple residuals from the current model against a set of adjusted values of the additional variate. The adjustment obtains the required slope in the plot. The adjusted values are the simple residuals obtained from fitting a MLR model with the additional explanatory variate as the response, and using the set of explanatory variates in the current model. In theory, added variable plots can be used to screen a set of potential additional variates for inclusion in a model, but in practice, these plots can be noisy and difficult to interpret. They should generally be used in combination with formal testing, for example with the incremental F-tests obtained by addition of each of the new variates in turn to the current model. Further details can be found in Atkinson (1985).

14.7 Dealing with Correlation among Explanatory Variates

The term **collinearity** is used to indicate linear dependencies, or strong correlations, between two or more explanatory variates. The simplest case of collinearity occurs when two explanatory variates are strongly correlated, either positively or negatively, and this can be detected from a pairwise correlation matrix for the full set of explanatory variates (e.g. Table 14.1). Perfect collinearity occurs when the correlation between two explanatory variates is exactly 1.0 or –1.0, which implies that once one of these variates is included as an explanatory term in a model, the other provides no additional information (as it can be predicted exactly from the first variate). Clearly, there is then no need to have both variates in a MLR model, and in fact there is no unique estimate of parameter values for perfectly collinear variates in a MLR model. The simple solution is to include only one of these variates. However, perfect collinearity is rarely found in practice; more often two variates will be strongly, but not perfectly, collinear. This can be seen in Figure 14.1, where seed length and diameter are strongly correlated. In these cases, the second variate may account for a small amount of additional variation once the first has been included. Unfortunately, the inclusion of strongly collinear variates in a MLR model has the effect of making their parameter estimates unstable and uncertain, which is reflected in large standard errors.

Multicollinearity is a more complex concept involving correlation among more than two variates. A set of variates is said to be perfectly multicollinear if any linear combination of the variates adds to a constant value. For example, suppose we weigh the above- and below-ground biomass of a plant separately, but then use the above-, below-ground and total biomass as explanatory variates: there is a perfect additive relationship between the three variables. Again, perfect multicollinearity is uncommon, but approximate multicollinearity can occur and makes parameter estimates unstable. Multicollinearity can be hard to detect, as it will often not be apparent from the pairwise correlations. **Variance inflation factors** (VIFs) can be used to detect (multi)collinearity within a set of explanatory variates. The VIF values are calculated for each explanatory variate ($x_1 \dots x_q$) as

$$\mathrm{VIF}_l = \frac{1}{1 - R_l^2}, \quad l = 1 \dots q,$$

where R_l^2 is the coefficient of determination (as defined in Section 12.5, but also see Section 14.8 below) obtained from a MLR model fitting variate x_l as the response in a model that includes all of the other explanatory variates. Values of the coefficient of determination R_l^2 close to 1 indicate the presence of multicollinearity and result in large values of the VIF. When the set of explanatory variates are approximately mutually orthogonal (i.e. all pairwise correlations are close to zero and R_l^2 is close to zero for $l = 1 \dots q$), then all the VIF values will be close to 1. The VIF can be interpreted as the inflation in the variance of the lth coefficient, β_l, compared to a situation in which the lth explanatory variate is orthogonal to the other explanatory variates. So, $\mathrm{VIF}_l = 10$ implies a 10-fold increase in parameter variance compared with this theoretical orthogonal scenario. Whether (multi) collinearity is problematic depends on the context, and O'Brien (2007) cautions against the unthinking use of thresholds (e.g. VIF > 10) to dictate that (multi)collinearity must be dealt with, for example by removal of one or more explanatory variates. We suggest that large VIF values (VIF > 10, or equivalently $R_l^2 > 0.9$) should prompt investigation of the multicollinearity, and consideration of whether it is either desirable or sensible to keep all the explanatory variates in the model. This is illustrated in the next example, and then discussed in more generality.

EXAMPLE 14.1I: DIPLOID WHEAT

We now model seed weight as a MLR model with four explanatory variates: seed length, hardness index, moisture content and diameter. The model is written in mathematical form as

$$Weight_i = \alpha + \beta_1 Length_i + \beta_2\, Hardness_i + \beta_3\, Moisture_i + \beta_4\, Diameter_i + e_i,$$

where α is the intercept parameter and β_1 to β_4 are the slope parameters for the four explanatory variates. In symbolic form, the model is written as

Response variable: *Weight*
Explanatory component: *[1] + Length + Hardness + Moisture + Diameter*

The R_l^2 and VIF_l values were obtained for each of the explanatory variates in turn by regression on the remaining explanatory variates, and the results are listed in Table 14.9.

TABLE 14.9

Coefficient of Determination (R^2) and Variance Inflation Factors (VIF) for Four Explanatory Variates (Example 14.1H)

Variate	R^2	VIF
Length	0.998	602.83
Hardness	0.029	1.03
Moisture	0.017	1.02
Diameter	0.998	602.79

The large VIF values of 602.8 for both diameter and length are expected because of the strong pairwise correlation ($r = 0.999$) between these two variates. The small VIF values (≈ 1.0) for moisture content and hardness index indicate that, in addition to them being uncorrelated with each of the other explanatory variates individually (Table 14.1), there is no linear combination of those variates that is related to either moisture content or hardness index. The impact of the collinearity between length and diameter can be seen by examination of the parameter estimates from the fitted model, i.e.

$$\widehat{Weight}_i = -12.97 - 40.16 \times Length_i - 0.052 \times Hardness_i - 1.81 \times Moisture_i$$
$$+ 90.94 \times Diameter_i,$$

also listed in Table 14.10. We can compare this to the fitted model obtained with only the first two variates, length and hardness (see Table 14.8), i.e.

$$\widehat{Weight}_i = -27.79 + 16.93 \times Length_i - 0.049 \times Hardness_i.$$

The most striking change is that the large positive coefficient for length in the model with two variates (+16.93) has changed to a negative coefficient (–40.16) in the model with four variates. This is surprising given the strong positive correlation of length with seed weight. The SE of this coefficient has also greatly increased (to be more than 20 times larger), as has the SE for the intercept. The coefficient for diameter is large and positive, as would be expected from the strong positive correlation between diameter and weight, but again with a large SE. The interplay between the coefficients for length and diameter may suggest that seeds with a larger diameter than expected for their length are also likely to be heavier. However, it is arguable that the small improvement to the fit on addition of diameter to a model already including length is not worth the difficulty in interpretation and the instability indicated by the large SEs of the parameters. On

TABLE 14.10

Parameter Estimates with Standard Errors (SE), t-Statistics (t) and Observed Significance Levels (P) for a MLR Model for Seed Weight with Four Explanatory Variates (Example 14.1H)

Term	Parameter	Estimate	SE	t	P
[1]	α	–12.970	10.1545	–1.277	0.051
Length	β_1	–40.155	14.5518	–2.759	0.002
Hardness	β_2	–0.052	0.0162	–3.238	< 0.001
Moisture	β_3	–1.805	0.9391	–1.922	0.014
Diameter	β_4	90.943	23.1755	3.924	< 0.001

balance here, we accept this argument and exclude diameter from the set of explanatory variates; we continue our search for a model in Section 14.9. In general, it should also be considered whether the conflict between the coefficients of length and diameter might be driven by a few seeds with atypical shapes, as this accommodation of a few outlying values might produce worse predictions for seeds with more typical shapes. This can be investigated using cross-validation; see Section 14.9.3.

Collinearity is often found in data sets with few observations, where there is a greater chance of **spurious correlation**. For example, a model may be used to predict insect counts in a field from temperature and humidity, but these variables have similar daily and annual cycles. If the study covers only warm dry days, then the temperature and humidity measurements will be strongly correlated. If some warm but wet days were also included, then some contrast between the variables will appear, the correlation decreases, and the separate role of the two variables might be identified. A similar situation occurs when measurements are obtained within only a small range of the explanatory variates – variates that have weak correlation over a wide range of circumstances may appear strongly correlated over a restricted range. Of course, the risk of spurious multicollinearity greatly increases as the number of explanatory variates increases. This emphasizes the importance of carefully 'designing' the observations to be collected in an observational study so that the correlations between potential explanatory variables are kept as small as is possible. We shall return to this matter briefly in Section 19.1.

Most statistical software produces warnings (usually based on the VIFs) if substantial collinearity is found. Collinearity may also be indicated by a significant overall F-test for the model when all marginal F-tests are not significant, or by large changes in parameter estimates and SEs when a new variate is added to the model (as illustrated in Example 14.1I). However, a change in estimated coefficients does not in itself indicate a problem. Consider the example of insect counts predicted by temperature and humidity introduced above, and suppose that insect counts tend to increase with temperature but decrease with humidity. In this case, we might find that humidity had a positive coefficient as a single explanatory variate in a SLR model because of the strong correlation between humidity and temperature. This could change to a negative coefficient in a MLR model that included temperature as an additional variate, as the response to humidity would now be modelled after accounting for temperature. The change in value has occurred because of collinearity, but the collinearity is not a problem here, as including both explanatory variates produces a more realistic model. This demonstrates the importance of understanding the biological context of the relationships modelled, and of using this knowledge when constructing a model, rather than just using statistical methods blindly.

Where collinearity occurs there are several possible approaches for dealing with it. If collinearity is present but not severe, and it is plausible that all of the correlated variates are introducing different biological information into the model, then all of the variates should be retained. If information is effectively duplicated across several variables, then some can be omitted with little loss of information. If the collinearity is very strong, then it is often better to drop variates progressively from the model: dropping either those with the largest VIF values, or those with large VIF values but least biological relevance. If you have few data, then making further observations over a wider range of the explanatory variates, in the hope of reducing the observed correlation between them, might help.

Finally, there are several other statistical techniques that can be used with the full subset of explanatory variates even when substantial collinearity is present. Techniques such as ridge regression and the lasso aim to minimize the residual sum of squares subject to a penalty on the size of the regression coefficients. This reduces the variance (and hence instability) of the regression coefficients, but at the cost of introducing bias, so that the expected values of the estimates no longer equal their true population values. These methods can result in a smaller prediction error (the expected squared difference between model predictions and the true underlying function), although choice of a suitable penalty introduces some further complexity into the analysis. Further details, with an intuitive comparison of these techniques, are given in Hastie, Tibshirani and Friedman (2001, Chapter 3).

14.8 Summarizing the Fit of the Model

Several goodness-of-fit statistics were introduced in Section 12.5 to summarize the fit of a SLR model. These statistics can also be calculated for MLR models and used to compare different MLR models. In addition, they can help to select which subset of variates should be included in a model, as described in Section 14.9.

The goodness-of-fit statistics presented in Section 12.5 can be used within the context of MLR models, after adjusting for the number of model parameters, p. These statistics are presented in Table 14.11 with some additional statistics useful for evaluation of MLR models.

The coefficient of determination (R^2) and adjusted coefficient of determination (R^2_{adj} or adjusted R^2) are defined as for the SLR model, although now the model sum of squares (ModSS) contains several model terms. For MLR models, the adjusted R^2 statistic is usually preferred to the coefficient of determination, as the latter always increases when a new variate is added to the model, even though there might be no real improvement in the model fit. The adjusted R^2 statistic takes account of the change in both the model sum of squares and the df through the residual mean square (ResMS) and can decrease if adding a new variate does not improve the model fit. Note that even if the adjusted R^2 statistic increases when a new variate is added, this might not correspond to a significant incremental F-test for the new term, and so is no substitute for a formal statistical test.

Other, more sophisticated, statistics are also available for comparing different models fitted to the same response variable. The information criteria, AIC (**Akaike information**

TABLE 14.11

Statistics Used to Assess Goodness of Fit in MLR Models

Statistic	Formula
Coefficient of determination (R^2)	$R^2 = \dfrac{\text{RegSS}}{\text{TotSS}} = 1 - \dfrac{\text{ResSS}}{\text{TotSS}}$
Adjusted coefficient of determination (R^2_{adj})	$R^2_{adj} = 1 - \dfrac{\text{ResMS}}{\text{TotMS}} = R^2 - \left(\dfrac{p-1}{N-p}\right)(1-R^2)$
Akaike information criterion (AIC)	$\text{AIC} = N \times \log_e(\text{ResSS}) + 2 \times p$
Schwarz Bayesian information criterion (SBC)	$\text{SBC} = N \times \log_e(\text{ResSS}) + \log_e(N) \times p$
Mallows' C_p	$C_p = \dfrac{\text{ResSS}_p}{\text{ResMS}_{\text{full}}} + (2 \times p) - N$

criterion) and SBC (**Schwarz Bayesian information criterion**), are widely used for model comparison. Both criteria multiply the natural logarithm of the ResSS by the number of observations, N, and then apply a penalty that depends on the number of parameters estimated, p. If the decrease in the ResSS on adding new explanatory variates into the model is large when set against the penalty for introducing those additional df into the explanatory model, then the values of the criteria decrease. Good models therefore correspond to small values of the information criteria. In the AIC, the penalty is simply twice the number of estimated parameters. In the SBC (sometimes also abbreviated as SBIC or BIC), the penalty is the number of estimated parameters, p, multiplied by $\log_e(N)$. The SBC penalty therefore takes account of both the number of parameters estimated and the number of observations. In practice, the SBC tends to give preference to simpler models than does the AIC. Both criteria can produce a ranking of competing models and can be useful for screening numerous possible models, as discussed in the next section. However, these criteria do not provide a formal test of difference in fit between competing models and small differences in criterion value may not indicate any meaningful difference in model fit.

The **Mallows' C_p** statistic corresponds to a situation with a total of m potential explanatory variates and is used to compare the fit of a sub-model containing q of these variates to the full model containing all of the m variates. In the Mallows' C_p formula, ResMS_{full} corresponds to the residual mean square of the full model (i.e. including all m explanatory variates) and ResSS_p is the residual sum of squares for a sub-model of interest that contains q explanatory variates ($q < m$) and hence $p = q + 1$ parameters. The value of C_p for the full model always equals $m + 1$, the total number of parameters in that model. Any sub-model containing q variates that has a similar value of the residual mean square (and hence similar precision) to the full model, will have a C_p value close to p. The Mallows' C_p statistic can also be used to screen competing models, and is best visualized by plotting values of C_p against p together with the line $C_p = p$, so that good models appear close to this 1:1 line. Again, this statistic does not provide a formal test of difference in fit between models.

The use of these statistics in model comparison is illustrated in Example 14.1J.

14.9 Selecting the Best Model

When there are many explanatory variates, subsets of them can be formed in many different ways giving numerous possible models. In this situation, one of the main aims of regression analysis is to choose a subset of explanatory variates that provide a good description of the response. Here, a good model is one that accounts for the maximum amount of variation in the response with the minimum possible number of parameters, following the principle of parsimony. The process of finding such a model is known as **model selection.**

Procedures for model selection fit a number of possible models, which are assessed by one or more summary statistics, usually the goodness-of-fit statistics defined in Section 14.8. If several models appear to perform equally well, each of these candidate models may be studied in detail to detect collinearity, misspecification, or departures from the underlying assumptions. The statistical significance of the estimated parameters should be checked, and any biological interpretation of the model parameters should be considered.

Ideally, candidate models should be selected from the set of all possible models. The number of possible models increases rapidly with the total number of explanatory variates, however. With q explanatory variates, there are 2^q different possible models (including the null model which contains the constant term only). For example, if we have 15 explanatory variates, then there are $2^{15} = 32{,}768$ possible models. The fitting and comparison of all possible models is often not computationally practical when there are many explanatory variates and, in these cases, automatic model selection strategies are usually employed. Some caution is required: if there are more explanatory variates than observations, then an exact fit to the data can be obtained, but this fit is completely uninformative with respect to the role of the explanatory variates in a larger study; this is an example of over-fitting (see also Section 17.1.2), where the model fits spurious detail at the expense of describing the larger-scale trends.

In fact, the practice of selecting models and estimating parameters on the same data is fraught with dangers, simply because models that over-fit often give better goodness-of-fit statistics. Their parameter estimates will be subject to bias resulting from the model selection procedure (selection bias), and our assessment of model fit will be over-optimistic (predicted errors will be too small). Fortunately, some of these problems can be reduced by cross-validation (introduced in Section 13.5). We return to this in Section 14.9.3 after we have described some common techniques for selecting models, and their associated problems, in more detail.

In the following example, we first illustrate model selection by fitting and comparing all possible subsets of explanatory variates (known as **all subsets selection**). In Section 14.9.1, we then discuss some of the automatic sequential procedures for selecting models that are useful for large numbers of explanatory variates.

EXAMPLE 14.1J: DIPLOID WHEAT

To find the best set of explanatory variates to describe seed weight, we fitted models with all subsets of the explanatory variates seed length, hardness index and moisture content. The variate diameter was excluded because of its collinearity with length, as discussed in Example 14.1I. A summary of the goodness-of-fit statistics obtained for the eight possible models is in Table 14.12.

The 'best' model according to each statistic is highlighted in bold type. For R^2 and adjusted R^2, larger values indicate better models, and both statistics have their largest values for the full model with all three explanatory variates. For the AIC and SBC statistics, smaller values indicate better models. The AIC takes its minimum value for the model with all three variates, but the SBC takes its minimum value for the model with

TABLE 14.12

Summary Statistics for MLR Models for Seed Weight: Goodness-of-Fit Statistics for All Possible Subset Models (Example 14.1J)

Explanatory Model	p	R^2 (×100)	R_{adj}^2 (×100)	AIC	SBC	C_p
[1]	1	0.00	0.00	1692.0	1695.3	704.6
[1] + Length	2	77.91	77.79	1407.1	1413.6	11.2
[1] + Hardness	2	4.29	3.78	1685.7	1692.2	668.4
[1] + Moisture	2	0.39	0.00	1693.3	1699.8	703.1
[1] + Length + Hardness	3	78.88	78.65	1400.6	**1410.4**	4.6
[1] + Length + Moisture	3	78.09	77.85	1407.6	1417.3	11.6
[1] + Hardness + Moisture	3	5.03	4.02	1686.2	1696.0	663.7
[1] + Length + Hardness + Moisture	4	**79.16**	**78.83**	**1400.0**	1413.0	**4.0**

Note: The 'best' model for each statistic is indicated in bold.

two variates: length and hardness index. On further inspection, this model is also close to the optimum values for AIC and both coefficients of determination. The Mallows' C_p statistic is relatively, but not convincingly, close to the target value of three for the model containing length and hardness index, but is much larger for all other one- and two-variate models.

Our candidate models for closer inspection are therefore the full model and the model containing length and hardness index. The simplest quantitative way of comparing these two models is to examine the incremental F-test for adding moisture content to a model already containing length and hardness index. The variance ratio for this test has value F = 2.57 with 1 and 186 df ($P = 0.111$). This suggests that there is no statistical improvement achieved by addition of moisture content to the simpler model. We have already found no evidence of misspecification for the model with length and hardness index as explanatory variates (Example 14.1H) and residual plots for this model accord with the assumptions regarding the deviations (Figure 14.4), although there is still a suggestion of underestimation of seed weight for very small and very large fitted values. The fitted model was presented in Example 14.1C, and the parameter estimates were listed in Table 14.8.

The final predictive model can be written as

$$\hat{\mu}(Length, Hardness) = -27.79 + 16.93 Length - 0.049 Hardness .$$

This model can be used to predict the potential gain in plant yield that might be expected for a given increase in seed length and decrease in hardness index, on the assumption

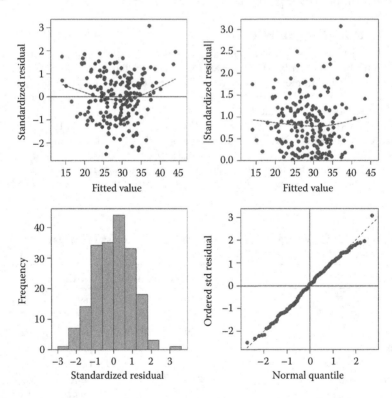

FIGURE 14.4
Composite set of residual plots based on standardized residuals for a MLR model for seed weight with explanatory variates length and hardness index (Example 14.1J).

that this could be achieved without affecting other aspects of the plant, such as the total number of seeds.

14.9.1 Strategies for Sequential Variable Selection

Automatic sequential selection strategies are often preferred when there are so many explanatory variates that fitting and comparing all sub-models becomes impractical. Automatic methods usually select a good set of explanatory variates, though not necessarily the 'best' set. They are designed to be computationally efficient and so to investigate a relatively small number of the possible sub-models, but still have the pitfalls of bias, over-fitting and over-optimism that we discuss in Section 14.9.2.

All of the sequential methods start with an initial or reference model, and at each step of the process they add or remove one explanatory variate. In all of the strategies presented here, the change at each step is evaluated via an F-test, and a threshold value must be chosen to control the process. It is easiest to explain the detail of these concepts in context, so below, we describe the three most common automatic selection methods: forward selection, backward elimination and stepwise regression.

Forward selection starts with the baseline model, containing the intercept term alone, and at each step adds the explanatory variate that gives the biggest improvement to the model fit, as measured by an incremental F-test, subject to this exceeding a threshold. This threshold can be defined as the minimum value of the incremental F-statistic that must be achieved in order for the variate to enter into the model, denoted F_{in}. If the F_{in} threshold value is chosen to be large, then the final model tends to contain fewer variates than for a smaller F_{in} threshold. The threshold can alternatively be defined in terms of the observed significance level of the incremental F-statistic, denoted SLE (significance level to enter). In this case, terms are added into the model only if the observed significance level is smaller than the SLE; the choice of a smaller SLE value leads to a final model with fewer terms.

Backward elimination starts with the full model, i.e the model containing all the explanatory variates, and at each step eliminates the variate that gives the smallest change to the model fit, as measured by a marginal F-test, subject to this being smaller than a threshold. This threshold can be defined as the maximum value of the marginal F-statistic that is allowed for a variate to be eliminated from the model, denoted F_{out}. If the F_{out} threshold value is chosen to be large, then the final model tends to contain fewer variates than for a smaller F_{out} threshold. The threshold can alternatively be defined in terms of the observed significance level for the marginal F-statistic, denoted SLS (significance level to stay). In this case, terms are eliminated from the model if the observed significance level is larger than the SLS; the choice of a smaller SLS value leads to a final model with fewer terms.

Both of these strategies can run into problems caused by multicollinearity among explanatory variates. In forward selection, a variable selected at an early step might not be required in the model once certain other variates have also been included. It might then be appropriate to remove it from the model, but this is not allowed within the forward selection framework. The reverse situation can occur with backward elimination, where variates removed at an early stage might later be used to improve the model, if this were allowed. The **stepwise** strategy incorporates such additional steps into the selection process. In its most general form, this procedure evaluates at each step the effect of dropping each of the explanatory variates currently in the model, and the effect of adding each explanatory variate currently excluded. Model fit is quantified by some goodness-of-fit statistic (often ResMS) and the step that gives the best value of this statistic will be taken, subject to the change passing the forward/backward threshold criterion. Many variants on this procedure are possible. For example, the variant **forward stepwise selection** starts as forward selection, but after each forward step, it switches to backward elimination, until no further variates can be dropped, and then forward selection is resumed. The variant **backward stepwise selection** uses the reverse procedure. In all cases, the final model is obtained when no further changes can be made. However, if the threshold F_{in} is smaller than threshold F_{out} (or SLE > SLS), then stepwise algorithms can get caught in a loop, adding and then dropping the same variable repeatedly, and so a maximum number of steps is often specified. Because of the switch between forward and backward steps, stepwise selection is more flexible than the single direction strategies and should therefore have a greater chance of selecting a good model for the data.

Clearly, the threshold values of F_{in} and F_{out} (or SLE and SLS) have a large influence on the model selected. The significance level associated with a particular F_{in} will obviously depend on the residual degrees of freedom, which decrease as new variates are added, so that fixing F_{in} results in the significance level increasing at each step. Conversely, fixing the significance level (SLE) results in the incremental F-statistic threshold increasing as more terms are added to the model. These changes will be small unless the residual df are small or the number of explanatory variates added is large. Conversely, the significance level associated with a fixed F_{out} will decrease as terms are dropped, and the marginal F-statistic threshold associated with a fixed significance level (SLS) will also decrease. But in both cases, remember that the residual mean square changes as terms are added or dropped and this may perturb the expected pattern.

For forward selection, it is often argued that the ResMS of the null model will be much larger than true background variation because it includes contributions from important explanatory variates not yet entered into the model. This will reduce the observed F-statistics so that a smaller threshold of F_{in} (or equivalently a larger value of SLE) is appropriate. A typical value of F_{in} is 2, and SLE values are commonly set around 0.15. These values are approximately equivalent for large data sets ($N > 100$), but the criterion SLE ≤ 0.15 is more stringent for models with fewer residual df. It can be argued that these thresholds should be tightened (i.e. F_{in} increases, SLE decreases) as more explanatory variates are included in the model and any inflation of the ResMS is reduced. Backward selection starts with all of the explanatory variates in the model and so the ResMS should not be inflated, although it may be an unreliable estimate if the residual df of the full model is very small; in this case, backward selection is not advised. A typical value of F_{out} for backward selection is 4, corresponding to a SLS value of 0.05 (for $N = 60$). An initial discrepancy between F_{in} and F_{out} (or SLE and SLS) values is therefore based on a sound rationale despite the fact that it can cause recursion in stepwise selection procedures.

The details of the selection procedures, such as use of F_{in} and F_{out} rather than SLE and SLS and default values for the thresholds, differ among statistical packages; you should therefore always check the documentation. These variations mean that packages may select different final models, again indicating the need for cautious use of such approaches. We suggest that automatic selection procedures are used as the first step in a modelling exercise, followed by comparison of the 'best' model(s) identified and taking account of the biological context. The analysis in the following example, which illustrates the use of stepwise selection, was done using GenStat.

EXAMPLE 14.2: APHID CATCH

The EXAMINE project collated data on aphid catches in suction traps across Europe (www.rothamsted.ac.uk/examine/) to investigate environmental and landscape influences on the timing of aphid flight and abundance. Here, we investigate the relationship

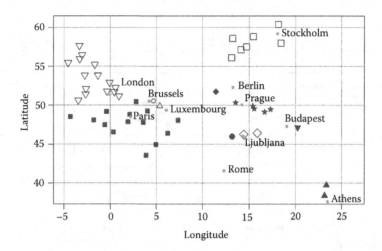

FIGURE 14.5

Approximate locations of 50 EXAMINE project aphid suction traps in 1995 in 11 European countries (with capital cities •): ○ Belgium (Brussels), ★ Czech Republic (Prague), ■ France (Paris), ◆ Germany (Berlin), ▲ Greece (Athens), ▼ Hungary (Budapest), • Italy (Rome), △ Luxembourg (Luxembourg), ◇ Slovenia (Ljubljana), □ Sweden (Stockholm), ▽ UK (London) (Example 14.2).

between the Julian day of the first catch of the aphid *Myzus persicae* at 50 locations (Figure 14.5) during 1995 and several geographical, meteorological and land-use variates. The three geographical variates were latitude, longitude and altitude of each trap. The 10 meteorological variates were monthly rainfall from October 1994 to May 1995, mean temperature for the coldest 30-day period at that trap site and mean temperature for the following 60-day period. These variates were chosen as those most likely to affect aphid flight dates, which were expected to be earlier in warmer, drier climates. The eight land-use variates gave the proportions of land in a circle of radius 75 km around the sampling site under different uses (coniferous, deciduous or mixed forest, grassland, arable land, inland waters, sea or urban). Note that these proportions do not sum to 1 for most sites as several land-use categories with overall small proportions have been omitted. The data set can be found in file EXAMINE.DAT and in Table A.3. Table 14.13 lists the explanatory variates and their symbolic names.

The Julian day of first catch ranges between 1 (1 January) and 205 (24 July) with mean 124.3 (4 May), lower quartile 100 (10 April) and upper quartile 146 (26 May). We first consider exploratory data analysis, as strong correlations between geographic and climate variables are likely, but a scatter plot matrix becomes impractical with 21 explanatory variates. A correlation matrix of the response variate (Julian day of first catch) and all explanatory variates can be scanned for instances of strong correlation. The response (date of first catch) shows a strong positive correlation ($r = 0.73$) with the trap site latitude and a strong negative correlation (-0.80) with the mean temperature in the 60 days after the coldest period. The strongest correlation (0.90) within the set of explanatory variates is between the mean temperature in the coldest 30-day period (denoted *C30Day*, see Table 14.13) and that in the following 60 days (denoted *F60Day*). This strong correlation is expected, but these variates are together intended to quantify the depth and length of the winter period, and so both will be retained for analysis. Not surprisingly, there is a negative relationship between latitude and *C30Day* (-0.48) or *F60Day* (-0.70) as winter temperature decreases as latitude increases. There are also strong positive correlations (0.70–0.76) between monthly rainfall in December, January and February. Finally, there is a strong positive correlation between the proportion of mixed and deciduous forest close to a trap site (0.72) and a positive correlation between monthly May rainfall and

TABLE 14.13

Explanatory Variates Available for Modelling Date of First Aphid Catch (Example 14.2)

Description	Name	Description	Name
Weather Variables		**Geographic Variables**	
Monthly rainfall:		Site latitude	*Latitude*
October	*OctRain*	Site longitude	*Longitude*
November	*NovRain*	Site altitude	*Altitude*
December	*DecRain*	**Land-Use Variables**	
January	*JanRain*	Proportion of area under:	
February	*FebRain*	Coniferous forest	*ConForest*
March	*MarRain*	Deciduous forest	*DecForest*
April	*AprRain*	Mixed forest	*MixForest*
May	*MayRain*	Grassland	*Grassland*
Mean temperature during:		Arable crops	*Arable*
Coldest 30-day period	*C30Day*	Inland water	*InlandWater*
Following 60-day period	*F60Day*	Sea	*Sea*
		Urban	*Urban*

the proportion of mixed forest (0.79). All other correlations are less than 0.70 in absolute value.

There are more than two million subsets of 21 explanatory variates, so testing all possible subsets is impractical. Instead, automatic model selection strategies were implemented with GenStat. Here, four selection strategies were used for the full set of 21 explanatory variates: forward selection, backward elimination and stepwise selection starting from either the null model (forward stepwise selection) or the full model (backward stepwise selection). Thresholds were chosen as $F_{in} = 2$ and $F_{out} = 4$. Hence, variates were added into the model if their incremental F-statistic was greater than 2 and were dropped if their marginal F-statistic was less than 4. The results for forward selection and forward stepwise selection were the same except for variate *MarRain* (monthly March rainfall) which the stepwise procedure cycled over adding and dropping (with $F = 2.45$). The models for backward elimination and backward stepwise selection were also the same except for variate *MarRain* which the stepwise procedure again cycled over adding and dropping (with $F = 2.15$). In both cases, the marginal F-tests for the *MarRain* variate were not significant ($P > 0.05$) and it was excluded. The models from the forward and backward strategies then accounted for 89.3% and 89.2% of the variation (adjusted R^2), respectively. The set of explanatory variables selected by forward stepwise selection were (in order, with the variate names as defined in Table 14.13)

F60Day, MayRain, OctRain, FebRain, Urban, DecForest, Longitude and *NovRain*.

The set of explanatory variables retained by backward stepwise selection were

FebRain, MayRain, OctRain, NovRain, C30Day, DecForest, Urban, Altitude, Latitude and *Longitude*.

To investigate differences in fit between these two models, we can use a scatter plot of the two sets of fitted values and calculate the correlation between them. In the scatter plot (not shown), the fitted values from the two models are very closely related, which

TABLE 14.14

Parameter Estimates (Standard Errors) and Observed Significance Levels
(*P*) for MLR Models for Julian Day of First Catch Obtained by Forward
Selection or Backward Elimination Strategies (Example 14.2)

Term	Forward Stepwise Selection		Backward Stepwise Elimination	
	Estimate (SE)	*P*	Estimate (SE)	*P*
[1]	238.4 (11.43)	< 0.001	−88.4 (52.01)	0.097
Latitude	—	—	4.68 (0.854)	< 0.001
Longitude	1.16 (0.415)	0.008	1.70 (0.494)	0.001
Altitude	—	—	0.05 (0.025)	0.047
OctRain	−0.46 (0.076)	< 0.001	−0.39 (0.079)	< 0.001
NovRain	0.23 (0.107)	0.036	0.28 (0.108)	0.013
FebRain	0.37 (0.090)	< 0.001	0.35 (0.095)	< 0.001
MayRain	−0.72 (0.087)	< 0.001	−0.63 (0.097)	< 0.001
C30Day	—	—	−5.45 (1.577)	0.001
F60Day	−14.74 (1.132)	< 0.001	—	—
DecForest	92.09 (28.387)	0.002	129.20 (34.911)	< 0.001
Urban	−194.76 (67.595)	0.006	−196.70 (68.107)	0.006

is reflected in the correlation coefficient of 0.995. The parameter estimates from both
models are shown in Table 14.14.

There is a set of seven explanatory variates common to both models, i.e.

Longitude, OctRain, NovRain, FebRain, MayRain, DecForest and *Urban,*

and the parameter estimates for these explanatory variates are broadly similar in
sign and size in the two fitted models. The backward method has retained *Latitude,
Altitude* and *C30Day* in place of *F60Day*, which was selected by the forward method.
Taking into account the observed correlations between these explanatory variates
(shown in Table 14.15) and their estimated coefficients, it seems plausible that the
combination of *Latitude, Altitude* and *C30Day* accounted for winter temperatures in
an equivalent manner to *F60Day*.

One strategy for finding a final model is to take a model consisting of the seven variates held in common across the two models, and then to make an exhaustive search on
adding subsets of the four variates that are in disagreement (*C30Day, F60Day, Altitude*
and *Latitude*). The results of this search are shown in Table 14.16, which evaluates the
adjusted R^2, AIC and SBC statistics for each model and obtains the observed significance level (*P*) associated with the marginal F-test for each of these four explanatory
variates in each model.

The best model in terms of all three criteria is the one obtained by addition of variate
F60Day only, the same model chosen by the forward selection method. The second best
model for adjusted R^2 contains the other three variates, i.e. the model selected by the
backward methods, but this is not the second best model for AIC or SBC, which instead
choose the model with *F60Day* and *Latitude*, in which the parameter for *Latitude* is
not significantly different from zero (*P* = 0.539). On balance, we prefer the simpler,
more parsimonious, model, in which only *F60Day* is added. Now, we are in position
to look more closely at the properties of this model. Partial residual plots (not shown)
suggest no evidence of model misspecification, and residual plots (Figure 14.6) show
no great cause for concern. The Cook's statistics (not shown) suggest the presence of
one influential observation, but omission of this observation has little impact on the

TABLE 14.15

Correlation Matrix for Response and Explanatory Variates Included in Candidate Models for Julian Day (*JDay*) of First Aphid Catch (Example 14.2)

	Longitude	Latitude	Altitude	OctRain	NovRain	FebRain	MayRain	DecForest	Urban	C30Day	F60Day	JDay
Longitude	—											
Latitude	-0.17	—										
Altitude	0.31	-0.27	—									
OctRain	-0.09	-0.40	0.05	—								
NovRain	-0.50	-0.06	-0.13	0.53	—							
FebRain	-0.55	0.05	0.08	0.50	0.67	—						
MayRain	0.20	-0.21	0.43	0.20	0.13	0.31	—					
DecForest	0.05	-0.36	0.59	0.13	0.03	0.25	0.62	—				
Urban	-0.28	0.09	-0.09	-0.06	-0.10	0.05	-0.24	-0.10	—			
C30Day	-0.60	-0.48	-0.28	0.36	0.40	0.33	-0.24	0.07	0.16	—		
F60Day	-0.39	-0.70	-0.28	0.31	0.24	0.09	-0.26	-0.02	0.12	0.90	—	
JDay	0.22	0.73	0.16	-0.50	-0.18	-0.08	-0.12	-0.08	-0.16	-0.69	-0.80	—

TABLE 14.16

Summary Statistics (R^2_{adj}, AIC and SBC) for Addition of All Possible Subsets of Variates *C30Day*, *F60Day*, *Altitude* and *Latitude* to a MLR Model with Seven Other Explanatory Variates, with Observed Significance of the Marginal F-Test for Each Explanatory Variate (Example 14.2)

p	Goodness-of-Fit Statistics			Observed Significance Level for Marginal F-Tests (P)			
	R^2_{adj}(×100)	AIC	SBC	C30Day	F60Day	Altitude	Latitude
8	46.23	549.68	564.97	—	—	—	—
9	**89.28**	**469.86**	**487.06**	—	**< 0.001**	—	—
9	82.20	495.20	512.41	—	—	—	< 0.001
9	81.74	496.47	513.68	< 0.001	—	—	—
9	46.77	549.96	567.17	—	—	0.238	—
10	89.11	471.38	490.50	—	< 0.001	—	0.539
10	89.01	471.83	490.95	0.877	< 0.001	—	—
10	89.01	471.85	490.97	—	< 0.001	0.941	—
10	88.29	475.04	494.16	< 0.001	—	—	< 0.001
10	86.19	483.25	502.37	—	—	0.001	< 0.001
10	81.29	498.44	517.56	< 0.001	—	0.880	—
11	89.15	471.92	492.95	0.001	—	0.047	< 0.001
11	88.96	472.82	493.85	—	0.002	0.513	0.374
11	88.96	472.82	493.86	0.513	0.071	—	0.380
11	88.73	473.82	494.85	0.886	< 0.001	0.959	—
12	88.99	473.37	496.31	0.297	0.522	0.297	0.175

Note: The selected model is indicated in bold.

parameter estimates. We therefore accept the forward selection model and move on to interpretation.

The fitted model (Table 14.14) suggests that the date of first catch of *M. persicae* is earlier within the year (has smaller fitted values) when the post-winter temperature (as measured by variate *F60Day*) is higher, for larger proportions of urban land use around the trap site, and when there is more rain in the previous October or in May of the same year. Conversely, the date of first catch is later (larger fitted values) when the trap site longitude is larger (further east within Europe), when the proportion of deciduous forest area around the trap site is greater, and when there is more rain in the previous November or in February of the same year. Of course, these variables do not vary independently, but a prediction of date of first catch can now be made for any site for which all these variables have been recorded. As the observations were all made in the same year, predictions cannot be made with confidence for other years without expansion of the study (as systematic differences between years would be expected), but this model should still provide information about the relative difference in flight dates for different environmental conditions.

As already noted, all automatic model selection strategies should be used with caution. There is usually no unambiguous 'best' model, because it will depend on the strategy used and the thresholds chosen. Similarly, even if all sub-models can be evaluated, the 'best' model may depend on the selection criterion used. It is often sensible to try several different approaches, as shown in Example 14.2, to select a few candidate models for further investigation. These should then be studied in detail with regard to the model fit, assumptions (by checking residuals and investigating outliers) and to their biological interpretation.

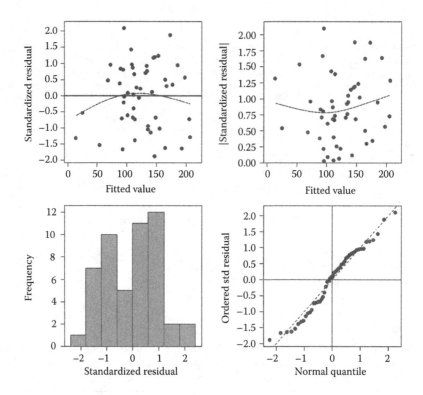

FIGURE 14.6
Composite set of residual plots based on standardized residuals for a MLR model for Julian day of first aphid catch with eight explanatory variates (Example 14.2).

14.9.2 Problems with Procedures for the Selection of Subsets of Variables

Above, we stated that procedures for model selection are subject to bias, over-fitting and over-optimism. In this subsection, we attempt to give some insight into these matters.

We start by considering the assumptions behind the incremental F-test for adding a term to a model, and how this relates to its role in forward selection. The distribution of the incremental F-test under the null hypothesis relates to a single pre-determined test. However, at the first step of the forward selection, the procedure calculates the incremental F-tests for each of the explanatory variates and chooses the largest to compare with the F_{in} threshold. Because we have deliberately chosen the largest of the set, this statistic will tend to be larger than we should expect under that F-distribution. The true significance level can therefore be much greater than the nominal value. We may therefore expect that some of the variables selected are actually unrelated to the response. This can lead to the phenomenon of over-fitting, where some of these extra variates accommodate random fluctuations at the expense of the overall trend. Including these extra explanatory variates in the model also reduces the ResMS and hence estimates of error – this makes estimates of uncertainty over-optimistic (too small). Similar considerations apply to all of the subset selection procedures in this section. For example, all subsets selection evaluates all subsets of a given size and chooses the best: again, chance variation can lead to the inclusion of some variables with no real underlying relationship with the response.

Miller (2002) suggests that one simple way to indicate whether uninformative variates have been included in a model is to introduce some new explanatory variates generated from random numbers, then to repeat the selection procedure including these new variates. The point at which these random variates start appearing in the selected model indicates the point at which no further useful information is being added, or where spurious information is being retained. Unfortunately, this approach requires that the number of random variates used is equal to the total number of explanatory variates, and so is impractical for small data sets with many explanatory variates. In other cases, it can give good insight into the possible reliability (or otherwise) of a selected model.

In addition to the inclusion of spurious variates, the process of model selection means that the coefficients of the selected variates tend to be biased. To demonstrate this, we consider two explanatory variates, both with the same underlying true correlation with the response. The variate that, by chance, appears more strongly correlated with the response in the observed sample is more likely to be selected, and will also tend to have a larger absolute regression coefficient (positive or negative) than expected. This is known as selection bias and should not be confused with the systematic bias arising from the omission of an important explanatory variate from the model. There is no way to avoid selection bias, except by selection of the model on one data set, then estimation of the model parameters from another, independent, set. However, the reduction in bias achieved by doing this might be outweighed by the increase in uncertainty caused by use of a smaller data set for inference.

These issues are intrinsic to the selection procedures and are discussed in detail by Miller (2002). In practice, they are difficult to avoid, but you should be aware of these problems and be properly sceptical about the results of any selection procedure. One method that can combat over-fitting is cross-validation, and its use in model selection is described in the next section.

14.9.3 Using Cross-Validation as a Tool for Model Selection

In Section 13.5, cross-validation was used as a tool to diagnose model fit; here we use it as a tool for model selection. In this situation, the purpose of cross-validation is to obtain an unbiased measure of the predictive ability of competing models, so the model with the best performance can be selected. In the simplest case, the data is partitioned into two parts: the training set and the validation set. The training set is used to estimate parameters for some candidate models. We then obtain predictions from each of the models for the units in the validation set. The predictive ability of the candidate models can be evaluated by calculation of statistics such as the mean square error of prediction (MSEP or RMSE), mean absolute difference (MAD) or prediction bias (PB), as defined in Section 13.5. The candidate model with the smallest value of MSEP, or some other combination of these statistics, is then selected. Because the validation set is independent of the training set, the MSEP gives an unbiased estimate of the squared error of prediction for the selected model. A model with too many explanatory variates, that over-fits the training set, is unlikely to give good predictions for the unrelated validation set; hence, this approach guards against over-fitting.

When the data set is too small to be divided into two separate subsets, k-fold cross-validation can be used instead. This variant divides the data into k subsets of (approximately) equal size. Each subset is used in turn as the validation set, with the remainder allocated to the training set. Again, the candidate models are fitted for each training set and then predict the response in each validation set. The evaluation statistics are calculated for each validation set, accumulated across sets, and then the model with best overall predictive ability is chosen. In this case, the training and validation sets are clearly not independent, but can still give a reasonable comparison of predictive performance.

14.9.4 Some Final Remarks on Procedures for Selecting Models

We have described above some of the perils of model selection. Despite these perils, model selection is useful for identification of important explanatory variates and obtaining predictions, so long as the results are treated with due scepticism. However, selection of a single model can be dangerous, as often several models, possibly containing different explanatory variates, can give a similar goodness of fit by adapting to different features of the data. You should remember that there is usually no 'true' model, and that we are usually seeking a descriptive model that gives a good prediction of the response. In general, this does not imply any causative relationship, and often we shall be unable either to identify or measure the underlying variables that actually cause the response. Hence, several different descriptions can perform equally well and cross-validation can be used to compare the predictive ability of different models. Alternatively, the technique of model averaging might be used to combine predictions from several different models (see Chapter 8 of Hastie et al., 2001, for a discussion of this topic).

We have described our strategies for model selection in terms of MLR, where the model consists of a set of explanatory variates. The methods apply to any linear models, including the models containing variates and factors and their interactions introduced in Chapter 15, where this topic is discussed further.

EXERCISES

14.1 A random sample of a vegetatively propagated family of 4-year-old loblolly pine trees was taken from a study located at Randolph County, Georgia, with the objective of describing average crown width (CW, m) in terms of explanatory variables that are simpler to measure, such as diameter at breast height (DBH, cm), total tree height (Ht, m) and height to live crown (HLC, m). Two additional crown variables were measured as the average from three randomly selected branches from each tree: branch diameter (DiamB, cm) and angle (AngB, degree). File CROWN.DAT contains unit numbers (*ID*) with these variates (*CW, DBH, Ht, HLC, DiamB, AngB*).[*]

   a. First consider the three simplest explanatory variates: *DBH, Ht* and *HLC*. Fit a SLR model with response crown width (*CW*) for each individual explanatory variate, and compare it to a MLR model including all three variates. Can you reconcile the results? Which subset of these three variates best describes crown width? Obtain residual plots to check the fit and write down and interpret your final predictive model.

   b. Now consider incorporating the two additional variables, *DiamB* and *AngB*, and repeat the subset selection process. Do you obtain the same result by considering these two additional variates with only those already selected in part (a) and, if so, discuss whether this will always be true or whether this strategy might sometimes fail?

14.2 Samples of foliage from plots of red pine were analyzed to establish whether foliar nutrients could predict growth (Bliss, 1970, Exercise 18.8). File FOLIAR.DAT contains the plot number (*Plot*) with the quantity (mg) of potassium (variate *K*) and calcium (variate *Ca*) found in foliar samples of given weight (variate *SampleWt*, g) together with the increase in height (variate *IncHt*, ft) and basal

---

[*] Data from FBRC, University of Florida.

area (variate *IncBA*, sq ft/acre) over a 5-year period. Construct biologically meaningful explanatory variates and fit MLR models for the increases in height and basal area. Comment on the fit of your models, examine them for any evidence of misspecification, and write down and interpret the best predictive model in each case. Give a 95% CI for the increase in height and basal area for a plot with 100 mg K and 20 mg Ca in a sample of weight 20 g.

14.3 A set of samples were processed to calibrate a near infrared reflectance (NIR) instrument for the measurement of the protein content of ground wheat (Fearn, 1983). File GROUND.DAT holds the sample number (*Sample*) and protein content (variate *Protein*, %) established by a standard method and measurements of the reflectance at six wavelengths (variates *L1–L6*). Can you find a stable MLR to predict protein content? Write down and interpret your final predictive model.

14.4 A forest biomass study combined data generated over several years from different research trials. In each study, several inventory plots were established and 2–12 slash pine trees from these plots were felled and components of biomass measured. The combined data set contains 174 trees from 50 inventory plots with a wide range of ages and sizes. For each tree, the total aerial biomass (TAB, kg), diameter at breast height (DBH, cm) and total height (Ht, m) were measured. Stand level variables were also obtained from each plot, including basal area (BA, m²/ha), total number of trees (N, trees/ha), quadratic diameter (QD, cm) and stand age (Age, years). The objective of the study is to construct a model that predicts total aerial biomass using the tree and stand variables. File SLASH.DAT contains unit numbers (*ID*) with plot numbers (*Plot*) and the explanatory variates (*TAB, Ht, DBH, BA, N, QD, Age*). The response *TAB* is usually log-transformed to obtain homogeneous variances. The generic model suggested in the literature takes the form

$$\log_e(TAB_i) = \alpha + \beta_1 \log_e(DBH_i) + \beta_2 \log_e(Ht_i) + \beta_3 \log_e(BA_i) + \beta_4 \log_e(N_i) + \beta_5 \log_e(QD_i) + \beta_6 \log_e(Age_i) + e_i ,$$

i.e. a MLR with six explanatory variates, with both the response and explanatory variates log-transformed. However, this set of explanatory variates often shows strong multicollinearity. Fit the MLR model described above, and critically evaluate it (e.g. investigate the collinearity, investigate misspecification using partial residual plots, plot the observed data against the fitted values). How robust is this model? Can you suggest a better model? (We re-visit these data in Exercise 16.6.)[*]

14.5 Exercises 12.4 and 13.5 used SLR to predict dry matter (variate *DryMatter*) in terms of one of four explanatory variates: *MaxLength* (length of the longest stem), *SumLength* (sum of lengths of all stems), *SumDiam* (sum of diameters of all stems) and *LengthTop5* (average length of the five longest stems). These variables are held in file WILLOWSTEMS.DAT. Investigate whether you can obtain better predictions of dry matter from a MLR model and check the fit of any candidate models. If you have more than one candidate model, compare their fit using cross-validation (as in Exercise 13.5b).

---

[*] Data from FBRC, University of Florida.

14.6   In Example 14.2, a MLR model was found for a set of 50 traps from the EXAMINE project in 1995 (data file EXAMINE.DAT).

   a.   Use residual plots to check the fit of this model (hint: look at the fitted values plot, a plot of the fitted versus observed values, Cook's statistics and partial residual plots). Is there any cause for concern?

   b.   Data from the same study in 1996 are held (in the same format) in file EXAMINE96.DAT. Evaluate the three candidate models found in Example 14.2 (i.e. the models from forward selection, backward selection and the final predictive model) by using cross-validation on the 1996 data. Which model performs best?

   c.   Perform model selection on the 1996 data to establish some candidate models. Check their fit using residual plots, as in part (a), and evaluate them by cross-validation on the 1995 data. Can you draw any conclusions by comparing the selected models across the 2 years?

   d.   Repeat the model selection exercise using the combined data set (held in file EXAMINE9596.DAT) and check the fit of candidate models using residual plots. Write down your final predictive model. How much confidence would you have in predicting for other years?

   e.   Describe the structure of the combined data set. Can you take account of this structure in regression analysis? (We explore this further in Exercise 16.7.)

15

Models for Variates and Factors

In the previous chapters, we developed models for one or more qualitative explanatory variables (factors; Chapters 4 to 11) and models for one or more quantitative explanatory variables (variates; Chapters 12 to 14). We now introduce models for a combination of qualitative and quantitative explanatory variables, i.e. one or more factors with one or more variates. Models for variates and factors arise in many situations, but they are simple extensions of the models discussed previously. We can think of them as either adding a variate to a model for factors, or vice versa. As an example of the first, consider a field trial set up as a CRD to study the effect of different types of fertilizers, where the linear model consists of a single factor to identify the response of each treatment group (fertilizer type). If differences in plant size between plots had been noticed (and measured) before the fertilizers were applied, the single factor model could be improved by incorporating an explanatory variate to quantify, and hence enable a correction for, the effect of initial plant size on final yield. This extension is known as **analysis of covariance** (ANCOVA), where an explanatory variate is used to account for underlying differences between experimental units. In the second case, we wish to incorporate information on groups into simple (or multiple) linear regression. The groups may arise from the application of different treatments to the experimental units (e.g. different varieties, or levels of water stress) or due to observed differences between experimental units (e.g. males and females of a species, or different soil types). Each group might exhibit a unique pattern of response, so the purpose of analysis is to investigate the differences, which might require separate intercept or slope parameters (or both) for each group. This process is often known as **regression with groups** or **parallel model analysis**.

In this chapter, we first focus on regression with groups for the case of a single factor and one explanatory variate. We start with an overview of the most common models (Section 15.1.1), and then give a detailed explanation of each model and the sequential analysis of variance (ANOVA) used for model selection (Section 15.1.2). We consider some variations, such as building the model from different sequences of sub-models (Section 15.1.3) or imposing constraints on the intercept parameters (Section 15.1.4). There are several ways to extend the model, and next we allow multiple variates, i.e. **multiple linear regression with groups** (Section 15.2). We then discuss regression with groups as a method for modelling linear trends within a structured designed experiment (Section 15.3) and explore the relationship between ANCOVA and regression with groups (Section 15.4). We can define more complex models, including both multiple factors and multiple variates, and in Section 15.5, we discuss the issues that then arise in model selection and prediction. Finally, we note that fitting a factor in a model is in fact equivalent to fitting a set of specially defined explanatory variates, called dummy variates, and this equivalence is explained in Section 15.6. Using this representation, we can write the model in matrix format, as used in mathematical statistical texts (Section 15.6.1).

15.1 Incorporating Groups into the Simple Linear Regression Model

The aim of simple linear regression with groups (SLR with groups) is to investigate whether the straight line relationship between the response and a single explanatory variate changes from group to group. This is done by a combined analysis of the whole data set which, as usual, aims to find the simplest model that accounts for the patterns observed. As in the previous chapters, ANOVA is used as the tool for model selection. As a motivating example, consider a controlled environment trial where a range of doses of fungal inoculum are applied to a set of plants from several varieties under conditions known to be conducive to infection. The aim of the experiment is to see how the number of lesions present after 1 week is related to the dose and whether this relationship changes across varieties, which may give a measure of variety resistance. If we can assume that the number of lesions increases linearly with the dose, then the most complex model for this experiment should allow separate lines (i.e. separate intercepts and separate slopes) for each variety (group). Observations consistent with this model are illustrated in Figure 15.1a. If ANOVA indicates no differences between the slopes of the regression lines across groups, then the model can be simplified, giving a set of parallel lines (with separate intercepts and a common slope); observations of this type are shown in Figure 15.1b. If there is no statistically significant difference in the average response levels, as indicated by the intercepts, then the model can

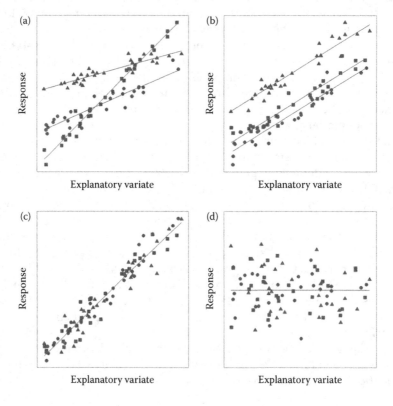

FIGURE 15.1
Data sets with three groups (●,■,▲) showing fitted lines (—) when the required model is (a) separate lines, (b) parallel lines, (c) single line and (d) null.

be simplified further, giving a single common line across groups, i.e. the SLR model (see Figure 15.1c). In this case, it is sensible to check for evidence of any association between the response and explanatory variate, and if there is not, then the response is best predicted by a constant value, as for the observations shown in Figure 15.1d.

15.1.1 An Overview of Possible Models

We start with an overview of the models outlined above, using a simple parameterization. As previously, we obtain least squares estimates for the parameters in the models. We do not derive these estimates here, leaving the calculations to statistical software. To define the models, we extend the notation for the SLR model to take account of the presence of groups. For convenience, we label the observations by the group to which they belong and then number the observations within each group. We therefore use y_{jk} to represent the kth observed response in the jth group, with the corresponding value of the explanatory variable denoted x_{jk}. The number of groups is denoted t, and so the index j runs from 1 to t groups ($j = 1 \ldots t$) and the number of observations in the jth group is denoted n_j, so the index k runs from 1 to n_j ($k = 1 \ldots n_j$). The total sample size, denoted N as previously, is the sum of the number of observations in each group, $N = n_1 + n_2 + \ldots + n_t$.

The most complex model for SLR with groups, which we call the **separate lines model**, allows a separate intercept and a separate slope for each of the t groups, and is written most simply as

$$y_{jk} = \alpha_j + \beta_j x_{jk} + e_{jk} , \tag{15.1}$$

where the parameters α_j and β_j represent the intercept and slope of the regression line for the jth group, and e_{jk} represents the random deviation for the kth observation in the jth group. The assumptions associated with the linear model listed in Sections 4.1 and 12.1 also apply here. In symbolic form, this model is written as

Explanatory component: grp + x.grp

where x holds the values of the explanatory variate and grp is a factor indicating the allocation of observations to groups. The grp term is associated with the individual group intercepts (α_j). The composite term containing the variate and factor, x.grp, fits a separate slope for each group (β_j).

EXAMPLE 15.1A: STAND DENSITY OF MIXED NOTHOFAGUS FOREST PLOTS

A survey was done of 41 plots containing natural stands of pure or mixed *Nothofagus* forest at the foot of the Andes. The resulting data can be found in file FOREST.DAT and are displayed in Table 15.1. The stands were classified into three types defined by the dominant species within the stand (factor Type) which was Coigue (type 1 with 13 plots), Rauli (type 2 with 9 plots) or Roble (type 3 with 19 plots). The variables recorded for each plot were the number of trees per hectare (stand density, variate SD) and the mean quadratic diameter in cm (variate QD).

The objective of the study was to model stand density as a function of quadratic diameter and to compare this relationship among the three types of stand. The usual model fitted to such data is a SLR model with both variables transformed to natural logarithms, and we follow this convention here. A scatter plot of the transformed variables (Figure 15.2) shows a negative relationship, with smaller log stand density corresponding to larger values of log quadratic diameter. On this log-log scale, the relationship

TABLE 15.1

Stand Density (Variate *SD*) and Mean Quadratic Diameter (Variate *QD*, cm) for 41 Plots of Mixed *Nothofagus* Forest Classified into Three Stand Types (Factor Type) According to the Dominant Species (Example 15.1A and File FOREST.DAT)

Type	SD	QD	Type	SD	QD	Type	SD	QD
Coigue	1780	22.11	Rauli	2970	13.02	Roble	3440	11.60
Coigue	980	30.50	Rauli	1500	14.84	Roble	1600	13.17
Coigue	3100	16.98	Rauli	4080	15.02	Roble	3100	9.48
Coigue	4120	12.69	Rauli	1600	15.44	Roble	1420	17.85
Coigue	2280	17.92	Rauli	2040	18.66	Roble	2060	15.85
Coigue	4760	15.19	Rauli	1960	18.02	Roble	2440	14.54
Coigue	4960	12.00	Rauli	2120	15.20	Roble	1720	16.20
Coigue	1520	19.51	Rauli	2160	19.60	Roble	1220	18.27
Coigue	1480	21.39	Rauli	2720	11.53	Roble	4080	8.88
Coigue	5560	10.87	Roble	3890	11.95	Roble	3440	11.65
Coigue	2000	23.94	Roble	1070	22.74	Roble	760	26.31
Coigue	2960	14.21	Roble	1720	14.41	Roble	3840	12.04
Coigue	3240	19.67	Roble	2920	12.58	Roble	1600	14.38
			Roble	2960	11.64	Roble	2320	12.60

Source: Data from Dra. Alicia Ortega Z., Universidad Austral de Chile.

is reasonably linear both overall and within groups, although the range of quadratic diameter values for Rauli plots (group 2) is much smaller than for the other groups.

The separate lines model can be written as

$$logSD_{jk} = \alpha_j + \beta_j \, logQD_{jk} + e_{jk} \, ,$$

where $logSD_{jk}$ and $logQD_{jk}$ are the natural logarithms of the stand density and mean quadratic diameter in the kth plot of the jth stand type, respectively ($j = 1 \ldots 3, k = 1 \ldots n_j$,

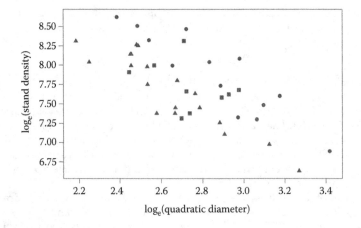

FIGURE 15.2

Logged stand density plotted against logged quadratic diameter (cm) for 41 plots classified by the dominant species (● Coigue/group 1; ■ Rauli/group 2; ▲ Roble/group 3) (Example 15.1A).

for $n_1 = 13$, $n_2 = 9$, $n_3 = 19$), α_j and β_j are the intercept and slope for the jth stand type, and e_{jk} are the random deviations. In symbolic form, this model is written as

Response variable: *logSD*
Explanatory component: Type + *logQD*.Type

where $logSD = \log_e(SD)$ and $logQD = \log_e(QD)$ are the $\log_e$-transformed variates. The explanatory terms Type and *logQD*.Type are associated with the separate intercepts and slopes for each stand type, respectively. The fitted model (adjusted $R^2 = 0.734$) is shown in Figure 15.3a, and the parameter estimates are given with their standard errors in Table 15.2.

The **parallel lines model** allows a separate intercept for each group, but imposes a common slope so that the fitted model consists of a set of parallel lines. This model can be written in its simplest form as

$$y_{jk} = \alpha_j + \beta x_{jk} + e_{jk}, \qquad (15.2)$$

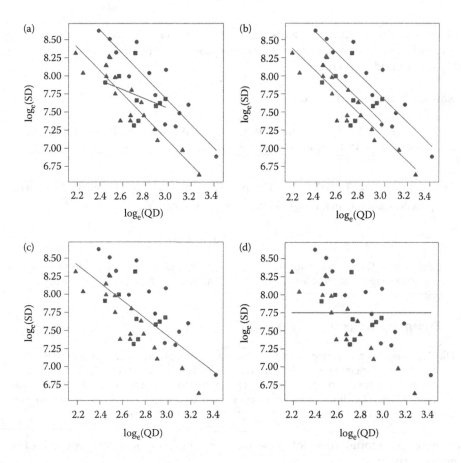

FIGURE 15.3
Logged stand density (SD) plotted against logged quadratic diameter (QD, cm): ● Coigue; ■ Rauli; ▲ Roble with fitted lines (—) generated by: (a) separate lines, (b) parallel lines, (c) single line and (d) null models (Example 15.1A).

TABLE 15.2

Models for Logged Stand Density in Terms of Stand Type (Factor Type) and Explanatory Variate $\log_e$(Quadratic Diameter) (Variate *logQD*) (Examples 15.1A to C)

Term	Parameter	Separate Lines Model (Example 15.1A) Estimate	SE	Parallel Lines Model (Example 15.1B) Estimate	SE	Single Line Model (Example 15.1C) Estimate	SE
[1]	α	—	—	—	—	11.115	0.5174
Type 1 (Coigue)	α_1	12.534	0.6734	12.270	0.4393	—	—
Type 2 (Rauli)	α_2	9.497	1.3727	11.926	0.4236	—	—
Type 3 (Roble)	α_3	11.949	0.5536	11.735	0.4041	—	—
logQD	β	—	—	−1.536	0.1516	−1.232	0.1884
logQD.Type 1	β_1	−1.628	0.2341	—	—	—	—
logQD.Type 2	β_2	−0.650	0.4999	—	—	—	—
logQD.Type 3	β_3	−1.617	0.2087	—	—	—	—

SE = standard error.

where the parameter α_j still represents the intercept of the regression line for the *j*th group, $j = 1 \ldots t$, and parameter β now represents the common slope of the parallel lines. In symbolic form, this model is written as

Explanatory component: grp + *x*

Here, the term grp is again associated with the individual group intercepts (α_j), and the term *x* is associated with the common slope parameter (β).

EXAMPLE 15.1B: STAND DENSITY OF MIXED NOTHOFAGUS FOREST PLOTS

The parallel lines model for logged stand density is written as

$$logSD_{jk} = \alpha_j + \beta \, logQD_{jk} + e_{jk} \, .$$

Parameter α_j is the intercept for the *j*th stand type, and β is the common slope of the decrease for logged mean quadratic diameter. In symbolic form, the explanatory model is written as

Explanatory component: Type + *logQD*

The Type term is associated with the separate intercepts for each stand type, and the term *logQD* is associated with the common slope. The fitted model (adjusted $R^2 = 0.723$) is shown in Figure 15.3b and the parameter estimates are given in Table 15.2. For this data set, the parallel lines model appears similar to the separate lines model, particularly for stands with Roble and Coigue as the dominant species.

The **single line model** does not allow for any difference between groups and is just a SLR model, written as

$$y_{jk} = \alpha + \beta x_{jk} + e_{jk} \, , \tag{15.3}$$

where the parameters α and β now represent the intercept and slope, respectively, of the common regression line. In symbolic form, this explanatory model is written as

Explanatory component: $[1] + x$

Recall from Chapter 12 that *[1]* represents a variate of length N that takes value 1 everywhere and is associated with the intercept parameter, α. As above, term x is associated with the common slope parameter, β.

EXAMPLE 15.1C: STAND DENSITY OF MIXED NOTHOFAGUS FOREST PLOTS

The single line model for logged stand density is written as

$$logSD_{jk} = \alpha + \beta \, logQD_{jk} + e_{jk}.$$

Parameter α is the intercept and β is the slope of the common line. In symbolic form, the explanatory model is written as

Explanatory component: $[1] + logQD$

The fitted model (adjusted $R^2 = 0.511$) is shown in Figure 15.3c and the parameter estimates are given in Table 15.2. For this data set, this model appears inappropriate, as most of the observations for Coigue-type stands appear above the fitted line, and most of those for Roble-type stands appear below the fitted line.

The **null model** does not allow for any difference between groups or for any relationship with the explanatory variate and is written as

$$y_{jk} = \alpha + e_{jk},$$

where parameter α now represents the overall population mean. In symbolic form, this model is written as a single term

Explanatory component: $[1]$

where the term *[1]* is associated with the parameter α.

EXAMPLE 15.1D: STAND DENSITY OF MIXED NOTHOFAGUS FOREST PLOTS

The null model is written as

$$logSD_{jk} = \alpha + e_{jk}.$$

Parameter α now represents the population mean logged stand density. The symbolic form was shown above. The fitted model, with adjusted $R^2 = 0$ and $\hat{\alpha} = 7.750$ (SE 0.0731) is shown in Figure 15.3d and is clearly inappropriate for this data set as it does not capture the clear negative correlation between the logged number of trees and the logged mean quadratic diameter.

We use ANOVA to determine the appropriate model for any data set objectively, by fitting the models described above in order of increasing complexity, i.e. the single line, parallel lines and separate lines models. There is no need to fit the null model as it is automatically used as the baseline for comparisons. This sequence of models can be fitted by progressive addition of terms into the model, but that procedure leads to a more complicated parameterization than used in this section. However, this parameterization is the default in most statistical software and so we explain it in some detail in the next section.

15.1.2 Defining and Choosing between the Models

We now look at the single line, parallel lines and separate lines models in more detail, using a more standard parameterization, and use the resulting sequential ANOVA to determine the most appropriate model for a given data set. As in the previous chapters (see Sections 11.2 and 14.4), each model is quantified in terms of its model sum of squares, ModSS, and df, ModDF. Recall that the model sum of squares (SS) is the sum of squared differences between the fitted values from the current model and those from the baseline (null) model. The model df is equal to the number of independent parameters required to fit the model minus one, where the adjustment accounts for the single parameter in the baseline model. A sequential ANOVA table is constructed from the incremental sums of squares and df derived from these model sums of squares and df (details below), and we can use F-tests from this ANOVA table to find an appropriate model for the observed responses.

15.1.2.1 Single Line Model

As described above, the single line model is a SLR model that ignores groups and takes the form shown in Equation 15.3, with symbolic form

Explanatory component: $[1] + x$

As in Section 14.4, we represent the SS for this model as $\text{ModSS}([1] + x)$ with $\text{ModDF}([1] + x) = 1$.

> **EXAMPLE 15.1E: STAND DENSITY OF MIXED NOTHOFAGUS FOREST PLOTS**
>
> For the stand density data, the parameter estimates are in Table 15.2 and the fitted model appears in Figure 15.3c. The model SS is equal to 4.583 with 1 df.

15.1.2.2 Parallel Lines Model

The parallel lines model represents the case where groups are constrained to have the same slope but allowed to have different intercepts. One simple form of this model is shown in Equation 15.2. However, it can also be considered as an extension of the single line model obtained by addition of the group factor into that model. Hence, the parallel lines model can also be written in symbolic form as

Explanatory component: $[1] + x + \text{grp}$

The variate *[1]* is still associated with a common intercept, and factor grp introduces a separate intercept for each group. In mathematical form, this model is written as

$$y_{jk} = \alpha + \beta x_{jk} + v_j + e_{jk} \, ,$$

where β is still the common slope of the regression lines across groups, α can be thought of as an overall intercept, and the parameters v_j ($j = 1 \ldots t$) can be thought of as group-specific deviations from the overall intercept. Unfortunately, this model is now over-parameterized as we have only t groups, but have $t + 1$ parameters that determine the intercepts of those t groups. Some form of constraint on the parameter estimates is therefore required. One possibility (used in Section 15.1.1) is the omission of the overall intercept term. The other possibility is to impose a constraint on the set of group intercepts, v_j ($j = 1 \ldots t$) and this is the choice usually made within statistical software. Here, we use first-level-zero constraints, which set $v_1 = 0$ and were previously used for factor models in Sections 4.5, 8.2.6 and 11.2.1. This parameterization changes the interpretation of the intercept and so we relabel that parameter. Some software packages, including SAS, use last-level-zero constraints which follow similar underlying principles but set $v_t = 0$ (see Section 4.5). Grouping the intercept terms together and relabelling parameters produces the model

$$y_{jk} = (\alpha_1 + v_j) + \beta x_{jk} + e_{jk}.$$

The intercept for the jth group is now $\alpha_1 + v_j$. Here, α_1 is still associated with a variate with value 1 everywhere (written symbolically as *[1]*), but because of the constraint $v_1 = 0$, parameter α_1 is now equal in value to the intercept for the first group. The parameter v_j is the difference between the intercepts for the jth and first groups (associated with the grouping factor, grp). The model SS is written as ModSS(*[1]* + x + grp), and this model estimates t intercepts and one slope parameter, hence ModDF(*[1]* + x + grp) = $(t + 1) - 1 = t$.

EXAMPLE 15.1F: STAND DENSITY OF MIXED NOTHOFAGUS FOREST PLOTS

Using first-level-zero parameterization, the parallel lines model for logged stand density is written as

$$logSD_{jk} = (\alpha_1 + v_j) + \beta logQD_{jk} + e_{jk},$$

with the explanatory component of the model written in symbolic form as

Explanatory component: *[1]* + *logQD* + Type

Parameter estimates for this model are $\hat{\beta} = -1.536$ (SE 0.1516), $\hat{\alpha}_1 = 12.270$ (SE 0.4393), $\hat{v}_2 = -0.344$ (SE 0.1083) and $\hat{v}_3 = -0.536$ (SE 0.0948), with $\hat{v}_1$ fixed equal to zero. The fitted models for the three stand types are therefore

Coigue (group 1): $\widehat{logSD}_{1k} = (\hat{\alpha}_1 + \hat{v}_1) + \hat{\beta} logQD_{1k} = 12.270 - 1.536 logQD_{1k}$

Rauli (group 2): $\widehat{logSD}_{2k} = (\hat{\alpha}_1 + \hat{v}_2) + \hat{\beta} logQD_{2k} = 11.926 - 1.536 logQD_{2k}$

Roble (group 3): $\widehat{logSD}_{3k} = (\hat{\alpha}_1 + \hat{v}_3) + \hat{\beta} logQD_{3k} = 11.735 - 1.536 logQD_{3k}$

This results in the same group intercepts as obtained in Example 15.1B and Table 15.2. The fitted parallel lines are shown with the data in Figure 15.3b, and the model SS is 6.524 with 3 df.

15.1.2.3 Separate Lines Model

This model allows separate intercepts and separate slopes for each of the t groups and was shown in simple form in Equation 15.1. This model can be considered as an extension of the parallel lines model, obtained by addition of a combined term formed from the explanatory variate and the groups factor. In symbolic form, the separate lines model is then written as

Explanatory component: $[1] + x + \text{grp} + x.\text{grp}$

Here, the term x is still associated with a component of slope held in common across groups, and the added term $x.\text{grp}$ specifies that a different slope for variate x is to be fitted for each of the t groups. In mathematical form, this model is written in full as

$$y_{jk} = \alpha + \beta x_{jk} + v_j + \eta_j x_{jk} + e_{jk} \,,$$

where parameters are defined as above, but with the introduction of η_j ($j = 1 \ldots t$) as group-specific deviations from the common slope, β. This model is now over-parameterized in terms of both the intercepts and the slopes, as we still have only t groups, but $t + 1$ parameters that determine the intercepts and $t + 1$ parameters that determine the slopes for those t groups. So, now we implement the first-level-zero parameterization for both intercepts and slopes. Grouping the intercept terms and the slope terms together, and again slightly relabelling the parameters produces the model

$$y_{jk} = (\alpha_1 + v_j) + (\beta_1 + \eta_j)x_{jk} + e_{jk} \,,$$

with constraints $v_1 = 0$ and $\eta_1 = 0$. As above, the intercept for the jth group is $\alpha_1 + v_j$, and the slope for the jth group is now $\beta_1 + \eta_j$. Here, β_1 is still associated with the explanatory variate (x), but because of the constraint $\eta_1 = 0$, β_1 is now equal in value to the slope for the first group. The parameter η_j (associated with term $x.\text{grp}$) equals the difference between the slopes for the jth and first groups. The model SS is written as $\text{ModSS}([1] + x + \text{grp} + x.\text{grp})$, and this model estimates t intercept and t slope parameters, with $\text{ModDF}([1] + x + \text{grp} + x.\text{grp}) = 2t - 1$.

EXAMPLE 15.1G: STAND DENSITY OF MIXED NOTHOFAGUS FOREST PLOTS

Using first-level-zero parameterization, we write the separate lines model for logged stand density as

$$logSD_{jk} = (\alpha_1 + v_j) + (\beta_1 + \eta_j)logQD_{jk} + e_{jk} \,,$$

with the explanatory component of the model written in symbolic form as

Explanatory component: $[1] + logQD + \text{Type} + logQD.\text{Type}$

Parameter estimates for this model are shown in Table 15.3, from which we can derive the fitted model for each stand type as

Coigue(group 1): $\widehat{logSD}_{1k} = (\hat{\alpha}_1 + \hat{v}_1) + (\hat{\beta}_1 + \hat{\eta}_1)x_{1k} = 12.534 - 1.628x_{1k} \,,$

TABLE 15.3

Parameter Estimates with Standard Errors (SE), t-Statistics (t) and Observed Significance Levels (*P*) for a Separate Lines Model for Logged Stand Density in Terms of Factor Type (1 = Coigue, 2 = Rauli, 3 = Roble) and Explanatory Variate *logQD* (Example 15.1G)

Term	Parameter	Estimate	SE	t	P
[1]	α_1	12.534	0.6734	18.615	< 0.001
logQD	β_1	−1.628	0.2341	−6.955	< 0.001
Type 1	ν_1	0.000	—	—	—
Type 2	ν_2	−3.037	1.5290	−1.986	0.055
Type 3	ν_3	−0.586	0.8717	−0.672	0.506
logQD.Type 1	η_1	0.000	—	—	—
logQD.Type 2	η_2	0.978	0.5520	1.772	0.085
logQD.Type 3	η_3	0.011	0.3136	0.036	0.972

$$\text{Rauli(group 2):} \quad \widehat{logSD}_{2k} = (\hat{\alpha}_1 + \hat{\nu}_2) + (\hat{\beta}_1 + \hat{\eta}_2)x_{2k} = 9.497 - 0.650x_{2k} \,,$$

$$\text{Roble(group 3):} \quad \widehat{logSD}_{3k} = (\hat{\alpha}_1 + \hat{\nu}_3) + (\hat{\beta}_1 + \hat{\eta}_3)x_{3k} = 11.949 - 1.617x_{3k} \,.$$

Again, these results match the group intercepts and slopes obtained in Example 15.1A and Table 15.2. The fitted separate lines are shown with the data in Figure 15.3a, and the model SS is 6.725 with 5 df.

15.1.2.4 Choosing between the Models: The Sequential ANOVA Table

We now have all of the ingredients required to build the sequential ANOVA table for this set of models, starting with the single line model, moving to the intermediate parallel lines model and then to the more complex separate lines model. Recall from Sections 11.2 and 14.4 that the incremental sums of squares and df in the sequential ANOVA table are calculated from the increase in the model SS and df as terms are added into the model. We calculate the incremental SS and df for the single line model by taking differences with the null, or baseline model, containing only the intercept term *[1]*, which has ModSS(*[1]*) = 0 and ModDF(*[1]*) = 0. Hence

$$\text{SS}(x|[1]) = \text{ModSS}([1] + x) - \text{ModSS}([1]) = \text{ModSS}([1] + x)$$

$$\text{DF}(x|[1]) = \text{ModDF}([1] + x) - \text{ModDF}([1]) = \text{ModDF}([1] + x) = 1$$

Similarly, on moving from the single line model to the parallel lines model, we obtain

$$\text{SS}(\text{grp}|[1] + x) = \text{ModSS}([1] + x + \text{grp}) - \text{ModSS}([1] + x)$$

$$\text{DF}(\text{grp}|[1] + x) = \text{ModDF}([1] + x + \text{grp}) - \text{ModDF}([1] + x) = t - 1$$

Finally, moving onto the separate lines model, we obtain

$$\text{SS}(x.\text{grp}|[1] + x + \text{grp}) = \text{ModSS}([1] + x + \text{grp} + x.\text{grp}) - \text{ModSS}([1] + x + \text{grp})$$

$$\text{DF}(x.\text{grp}|[1] + x + \text{grp}) = \text{ModDF}([1] + x + \text{grp} + x.\text{grp}) - \text{ModDF}([1] + x + \text{grp}) = t - 1$$

Recall also that within a sequential ANOVA table, we can abbreviate the incremental SS as SS(+term) to indicate that the term has been added into a model that already contains all of the terms in previous lines of the table. We follow a similar convention for the incremental DF. For this sequence of models, the incremental SS and DF can therefore be written as

$$SS(+x) = SS(x\,|\,[1]), \quad DF(+x) = 1$$

$$SS(+grp) = SS(grp\,|\,[1] + x), \quad DF(+grp) = t - 1$$

$$SS(+x.grp) = SS(x.grp\,|\,[1] + x + grp), \quad DF(+x.grp) = t - 1$$

We calculate mean squares as usual, by division of the incremental sums of squares by their df, and variance ratios by division of the mean squares by the residual mean square, ResMS, from the separate lines model. The sequential ANOVA table takes the form in Table 15.4.

As in SLR and MLR models, we should check for model misspecification before drawing conclusions. The residual plots described in Chapters 5 and 13 can be used for this check, and transformations may be used to stabilize variances if required (see Chapter 6). It may help to plot a separate graph for each group, especially when there are many observations within each group. If there is no suggestion of model misspecification, then we can proceed to use the sequential ANOVA to identify a parsimonious predictive model, i.e. the simplest model that describes the data well. As in Chapters 8 and 11, our full model is well defined, and so we start with the most complex model and progressively try to simplify it.

The variance ratio $F^{x.grp}$ is associated with adding the final term $x.grp$ into the parallel lines model to obtain the separate lines model. This tests the null hypothesis that the separate lines model gives no statistical improvement over the parallel lines model, i.e. H_0: $\eta_j = 0$ for $j = 1 \ldots t$, against the general alternative that this is not the case. If the null hypothesis is true, the variance ratio $F^{x.grp}$ has an F-distribution with $t - 1$ numerator and $N - 2t$ denominator df. If $F^{x.grp}$ is larger than the $100(1 - \alpha_s)$th percentile of this distribution, we reject this null hypothesis (at significance level α_s), conclude that we cannot simplify the separate lines model, and use this as our predictive model. In this case, the t separate slopes and t separate intercepts should be reported with their standard errors. If $F^{x.grp}$ is not significant, we move on to investigate the parallel lines model, using variance ratio F^{grp}.

The variance ratio F^{grp} is associated with addition of the factor term grp into the single line model to obtain the parallel lines model. This tests the null hypothesis that the parallel lines model gives no statistical improvement over the single line (i.e. SLR) model, i.e.

TABLE 15.4

Form of Sequential ANOVA Table for Regression with Groups Using Factor grp and Explanatory Variate x

Term Added	Incremental df	Incremental SS	Mean Square	Variance Ratio
$+x$	1	SS(+x)	MS(+x)	$F^x = MS(+x)/ResMS$
$+grp$	$t - 1$	SS(+grp)	MS(+grp)	$F^{grp} = MS(+grp)/ResMS$
$+x.grp$	$t - 1$	SS(+x.grp)	MS(+x.grp)	$F^{x.grp} = MS(+x.grp)/ResMS$
Residual	$N - 2t$	ResSS	ResMS	
Total	$N - 1$	TotSS		

SS = sum of squares.

H_0: $v_j = 0$ for $j = 1 \ldots t$, against the general alternative that this is not the case. If the null hypothesis is true, the variance ratio F^{grp} also has an F-distribution with $t - 1$ numerator and $N - 2t$ denominator df. If F^{grp} is larger than the $100(1 - \alpha_s)$th percentile of this distribution, we reject this null hypothesis and conclude that we cannot simplify the parallel lines model. In this case, the common slope and t separate intercepts should be reported with their standard errors. If F^{grp} is not significant, we move on to investigate the single line model, using variance ratio F^x.

The variance ratio F^x, associated with addition of the variate x into the null model to obtain the SLR model, is the test statistic for the slope in a SLR model (Section 12.3). If F^x is significant, then we conclude that the single line model adequately represents the data set, and the common slope and intercept should be reported with their standard errors. If F^x is not significant then we conclude there is no linear relationship between the response and the explanatory variate.

EXAMPLE 15.1H: STAND DENSITY OF MIXED NOTHOFAGUS FOREST PLOTS

The model SS and df from Examples 15.1E to G give the sequential ANOVA table in Table 15.5. We have already verified (Figure 15.2) that the relationship between the response and explanatory variate is approximately a straight line within each group and that there is no sign of model misspecification. A composite set of residual plots from the separate lines model is shown in Figure 15.4, with points labelled by groups, and shows no real cause for concern. We judge that the pattern in the absolute residual plot reflects the sparsity of observations for small or large fitted values rather than variance heterogeneity.

We therefore proceed to interpret the ANOVA table to establish our predictive model. The variance ratio for separate lines, $F^{lQD.T} = 1.724$ with 2 and 35 df ($P = 0.193$), gives no evidence that the slopes differ between types of plot. We therefore examine variance ratio F^T to test the null hypothesis that the intercepts are all equal. Here, $F^T = 16.629$ on 2 and 35 df ($P < 0.001$) giving strong evidence that separate intercepts are required, so we use the parallel lines model to describe the relationship between logged stand density and logged quadratic diameter. The equations of the fitted parallel lines were given in Example 15.1F. Logged stand density is smaller for larger values of logged quadratic diameter, and decreases at the same rate for all three stand types. For a given value of quadratic diameter, logged stand density for Rauli and Roble type stands is on average 0.344 and 0.536 units less, respectively, than for Coigue type stands. Figure 15.3b showed the data with the fitted parallel lines superimposed; this model appears to describe well the response within the observed range of logged quadratic diameter. In Exercise 15.4, we ask you to interpret this model on the original scale.

TABLE 15.5

Sequential ANOVA Table for Regression with Groups for Logged Stand Density with Factor Type and Variate *logQD* (Example 15.1H)

Term Added	Incremental df	Incremental SS	Mean Square	Variance Ratio	P
+ *logQD*	1	4.5833	4.5833	$F^{lQD} = 78.562$	< 0.001
+ Type	2	1.9403	0.9701	$F^T = 16.629$	< 0.001
+ *logQD*.Type	2	0.2011	0.1006	$F^{lQD.T} = 1.724$	0.193
Residual	35	2.0419	0.0583		
Total	40	8.7667			

SS = sum of squares.

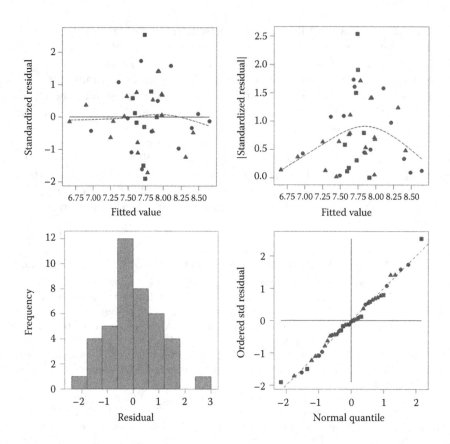

FIGURE 15.4
Composite set of residual plots based on standardized residuals (● Coigue; ■ Rauli; ▲ Roble) from the separate lines model for logged stand density (Example 15.1H).

In this section, we have outlined the basic procedure for identifying a good model for simple linear regression with groups. The main complication is in the different parameterizations that can be used. It is important to realize that although different parameterizations of a model result in different estimates of individual parameters, the fitted model is invariant to the parameterization, i.e. the same fitted model, and hence predictions, are obtained for any valid parameterization. Similarly, the sequential ANOVA table will have the same entries regardless of the model parameterization.

We have explained the first-level-zero parameterization in some detail because this type of parameterization is common in statistical packages, and is commonly misunderstood. Having explained the principles for obtaining the fitted model from the parameter estimates, the next challenge is in obtaining SEs for the amalgamated estimates of intercept and slope. Calculation of the SE of a sum (or difference) of estimates requires their covariances (see Example 15.1I and Section C.4), which might not be presented as standard output. One way to avoid this issue is to use the simple parameterization of Section 15.1.1 for the selected model, so that each intercept and slope is represented by a single parameter. Some models with several grouping factors cannot be represented in this manner, but most statistical software has facilities to calculate a SE for linear combinations of parameter estimates that can be used in this situation. Alternatively, if the interest is more in prediction of the expected response than in the components of the model, it may be more informative to produce predictions for specific values of the explanatory variates, and to quantify uncertainty with the prediction SEs or confidence intervals (CIs).

EXAMPLE 15.1I: STAND DENSITY OF MIXED NOTHOFAGUS FOREST PLOTS

Standard errors for the combined estimates of the intercepts in the parallel lines model (Example 15.1F) can be derived from the variance–covariance matrix for estimates of the unconstrained parameters, which takes the form

$$
\widehat{\mathrm{Var}}\begin{pmatrix}\hat{\alpha}_1\\\hat{\beta}\\\hat{v}_2\\\hat{v}_3\end{pmatrix}=\begin{pmatrix}0.1929 & & & \\-0.0658 & 0.0230 & & \\-0.0126 & 0.0028 & 0.0117 & \\-0.0193 & 0.0051 & 0.0053 & 0.0090\end{pmatrix}.
$$

For example, the variance of the intercept for Roble stands (group 3) can be expressed as

$$
\widehat{\mathrm{Var}}(\hat{\alpha}_1+\hat{v}_3)=\widehat{\mathrm{Var}}(\hat{\alpha}_1)+2\widehat{\mathrm{Cov}}(\hat{\alpha}_1,\hat{v}_3)+\widehat{\mathrm{Var}}(\hat{v}_3)
$$
$$
=0.1929+(2\times-0.0193)+0.0090
$$
$$
=0.1633.
$$

The estimated SE is then the square root of this value, with $\widehat{\mathrm{SE}}(\hat{\alpha}_1+\hat{v}_3)=0.4041$. A similar calculation gives $\widehat{\mathrm{SE}}(\hat{\alpha}_1+\hat{v}_2)=0.4236$, and $\widehat{\mathrm{SE}}(\hat{\alpha}_1+\hat{v}_1)=\widehat{\mathrm{SE}}(\hat{\alpha}_1)=0.4393$ since $\hat{v}_1=0$, matching the estimated SEs shown in Table 15.2.

In fitting the SLR with groups, we assume that the model deviations obey the assumptions stated in Sections 4.1 and 12.1. In particular, we assume that the deviations have a common variance and this implies that the variation is the same across all groups. You can check this assumption graphically by identifying groups, using different colours or symbols, in residual plots (see Section 5.2). It is not possible to use Bartlett's test (Section 5.3) here because the group sample variances are influenced by the values of the explanatory variate as well as by background variation. If variances differ between groups, and this heterogeneity cannot be corrected by applying a single transformation across all groups, then our assumptions no longer apply and conclusions from the ANOVA F-tests may be misleading. In this case, a weighted analysis might be appropriate but this is beyond the scope of this book (see Draper and Smith, 1998, or Montgomery et al., 2012).

As long as the residual df are not too small, then it is reasonable to perform model selection from the sequential ANOVA table as described above. If the residual df (ResDF) are small, then the estimate of background variation from the full model will be poor, giving low power for the analysis. In this case, if there is no evidence for the separate lines model, it is sensible to drop the x.grp term from the model and refit the parallel lines model to obtain a revised sequential ANOVA table. The incremental sums of squares and df for the model terms x and grp will be the same as those in the original table, but the combined term (x.grp) will now be merged with the residual to obtain a revised estimate of background variation. The variance ratios for terms x and grp are then calculated with respect to the revised residual mean square and will differ from those in the original table. This difference will usually be small if the ResDF in the original table were large (>30) or if the number of groups is small. Both sets of F-tests – from the original and revised tables – are valid, although their conclusions may differ. As a (somewhat arbitrary) rule of thumb, we suggest constructing the revised table when ResDF ≤ 10. In either case, once a model has been selected it is conventional to refit that model and to use its residual mean square as a basis for parameter SEs.

Finally, common sense is required in fitting these models. It is usually reasonable to compare behaviour across groups only if the range of the explanatory variate is similar across those groups. If there is not a strong overlap, there may be some ambiguity between differences in group intercepts and correlation with the explanatory variate. In some extreme cases, the sign of the estimated slopes can change depending on whether separate group intercepts are included in the model (e.g. see Figure 15.11c). Furthermore, comparison across groups is sensible only when there are sufficient observations within each group to give confidence in the conclusions.

15.1.3 An Alternative Sequence of Models

The sequence of models considered in the previous section started with the single line (SLR) model. We then added a factor to allow separate intercepts for each group, i.e. the parallel lines model, and finally added the interaction between the factor and variate to allow separate slopes for each group. Alternatively, we might start by adding the grouping factor into the null model giving a model of the form

$$y_{jk} = \alpha + v_j + e_{jk} \, .$$

This model consists of a set of parallel lines with zero slope. As discussed previously, this model is over-parameterized, as it has $t + 1$ parameters to describe only t intercepts and so we introduce a constraint. Again, we use the first-level-zero parameterization so the model takes the form

$$y_{jk} = \alpha_1 + v_j + e_{jk} \, ,$$

with $v_1 = 0$. As there is no explanatory variate in the model, the 'intercept' parameters can here be interpreted in terms of the population means for each group, where the parameter α_1 represents the population mean for group 1, and the effect v_j represents the difference between the population means for the jth and first groups. This model fits a separate population mean for each group and so is called the **separate groups model**, with symbolic form

    Explanatory component: *[1]* + grp

This is exactly the same model as used for a single explanatory factor in Chapter 4, with the model df equal to the number of groups minus one, i.e. $t - 1$.

Adding an explanatory variate into the separate groups model gets us back to the parallel lines model with a common slope and separate intercepts. We therefore have two routes to this model: we can either add the variate and then the factor into the model or vice versa. The combined term is then added to obtain the separate lines model. These two sequences of models are illustrated in Figure 15.5.

We now have two possible sequential ANOVA tables. Except in the balanced case, where observations within each group are made at the same values of the explanatory variate

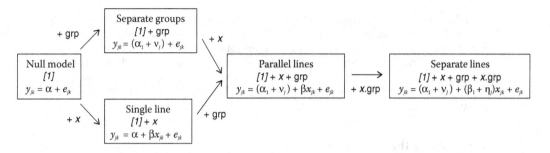

FIGURE 15.5
Two sequences of models for regression with groups.

and the two variables are orthogonal, the incremental sums of squares for adding the variate (*x*) and factor (**grp**) terms will differ between these two tables. The incremental SS and F-test for the combined term (*x*.**grp**) is the same in both sequences. As in Section 14.4, the aim of analysis is to identify the simplest model that describes the pattern in the data, and we suggest the following procedure. Starting with the separate lines model, we test whether simplification to the parallel lines model is permissible. If it is, we refit the model if the ResDF are small; otherwise we work from the two original sequential ANOVA tables. We then have two choices: we might drop either the grouping factor (to give the single line model) or the explanatory variate (to give the separate groups model). We assess the F-tests associated with both options. If both are significant, then we cannot simplify the parallel lines model. If the F-test for dropping the variate is significant, but that for dropping the grouping factor is not, then we drop the factor to obtain the single line model, and then test whether the variate should be retained in the model. If the F-test for dropping the grouping factor is significant, but that for dropping the variate is not, then we drop the variate to obtain the separate groups model, then test whether the groups should be retained in the model. If neither test is significant, then we drop the least significant term (i.e. that with the largest observed significance level) first, then test the other.

EXAMPLE 15.1J: STAND DENSITY OF MIXED NOTHOFAGUS FOREST PLOTS

For the stand composition data, the sequential ANOVA table obtained by addition of the Type factor into the model first is Table 15.6.

In Example 15.1H, we established that we did not require separate slopes for each group and therefore we start from the parallel lines model. The incremental F-test for

TABLE 15.6

Sequential ANOVA Table for Regression with Groups for Logged Stand Density with Factor Type and Variate *logQD*: Adding the Grouping Factor into the Model First (Example 15.1J)

Term Added	Incremental df	Incremental SS	Mean Square	Variance Ratio	P
+ Type	2	0.3025	0.1512	$F^T = 2.592$	0.089
+ *logQD*	1	6.2212	6.2212	$F^{QD} = 106.636$	< 0.001
+ Type.*logQD*	2	0.2011	0.1006	$F^{QD.T} = 1.724$	0.193
Residual	35	2.0419	0.0583		
Total	40	8.7667			

SS = sum of squares.

the *logQD* variate (after fitting the Type factor) is statistically significant ($F^{lQD}_{1,35} = 106.64$, $P < 0.001$, Table 15.6), indicating that the common slope in response to the logged quadratic diameter variate was non-zero. In the original ANOVA table, the incremental F-test for the Type factor (after fitting the *logQD* variate) was also highly significant ($F^{T}_{2,35} = 16.63$, $P < 0.001$, Table 15.5), indicating that separate intercepts were required. This confirms the parallel lines model cannot be further simplified and is therefore the most suitable model for these observations.

15.1.4 Constraining the Intercepts

Another model that might be considered relevant is one with a common intercept for all *t* groups but separate slopes for each group, known as the **common intercept model**. This model is based on the assumption that the straight line responses associated with the different groups all converge at a single point when $x = 0$. This model can be obtained by addition of the interaction directly to the single line model, and takes the symbolic form

Explanatory component: $[1] + x + x.\text{grp}$

Here, the term *[1]* corresponds directly to the common intercept, the variate *x* provides a common slope across all groups and *x*.grp specifies a different slope for each of the *t* groups. In mathematical form, and with first-level-zero parameterization, the resulting model is

$$y_{jk} = \alpha + (\beta_1 + \eta_j)x_{jk} + e_{jk} \ .$$

An example of this model is shown with artificial data in Figure 15.6a: the fitted lines diverge from a common intercept at $x = 0$.

Except in special circumstances, we advise against the use of this model. The arguments here are analogous to those against regression through the origin in Section 12.9.2. For this model to be useful, we first require that the origin ($x = 0$) is within (or close to) the range of the data, because we cannot objectively evaluate whether the common intercept

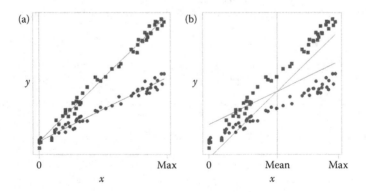

FIGURE 15.6

Non-invariance of the common intercept model: (a) fitted model (—) for two groups (•,■) and untransformed explanatory variate; (b) fitted model (—) with centered explanatory variate.

model gives a good fit unless some observations are present in this region. Second, the origin should have some absolute meaning within the biological context of the data. This is required because changing the position of the origin, for example, as when changing from the Celsius to Fahrenheit scale of temperature, will also change the required point of convergence. Figure 15.6b illustrates the fitted common intercept model when the position of the origin is changed by standardization of the explanatory variate: the fitted model is very different and now inappropriate for the observed response.

In general, we follow the principles of marginality (previously discussed in Sections 8.2.1, 8.3 and 11.2.2) also within the context of regression modelling. We consider a term as marginal to any term of which it is a sub-term; for example, terms A and B are both marginal to term A.B. The principle of marginality requires that for each term included in a model, all terms that are marginal to it, i.e. all sub-terms, should also be included. The only exception occurs when some sub-terms are not meaningful, for example, in nested models of form A/B, we do not require term B in the model before we fit A.B. We previously discussed this principle in the context of models containing only factors, but we apply it to all models. For example, the terms x and grp are both marginal to the term x.grp. The term *[1]* is regarded as marginal to all other terms. A model built by progressive addition of terms should only allow a term to be added if all of its sub-terms are already present, so we can add x.grp only to a model that already contains terms x and grp. Conversely, a term should not be dropped from a model if terms that it is marginal to are present in the model. For example, we should not drop term grp from a model that also contains term x.grp. Use of this principle with explanatory variates ensures that a model is invariant to changes of scale. The separate slopes with common intercept model discussed in this section disobeys the rule of marginality, as it does not include the sub-term grp, and so is not robust to change of scale.

15.2 Incorporating Groups into the Multiple Linear Regression Model

We now generalize the regression with groups model to allow several explanatory variates, incorporating groups into a multiple linear regression model. The aim of analysis stays the same, namely, to find as simple a model as possible that describes the response well. We first describe the general form of the model, and then use an example with two explanatory variates to illustrate the procedure of model selection in this relatively simple case.

For a model with q explanatory variates and a single factor with t groups, the most complex model allows a separate intercept for each group and a separate slope for each group for each explanatory variate. In the simplest parameterization, this model can be written in mathematical form as

$$y_{jk} = \alpha_j + \beta_{1j}x_{1jk} + \ldots + \beta_{lj}x_{ljk} + \ldots + \beta_{qj}x_{qjk} + e_{jk} , \qquad (15.4)$$

where, as previously, the units are labelled by index j indicating the group ($j = 1 \ldots t$) and index k labelling the observations within each group. The index l is used to identify the explanatory variates ($l = 1 \ldots q$) in the model. The model presented in Equation 15.4 has t intercepts (α_1 to α_t), and $q \times t$ slope parameters ($\beta_{11} \ldots \beta_{1t}$ to $\beta_{q1} \ldots \beta_{qt}$). Each group is associated with one intercept and q slope parameters, one for each explanatory variate. This model can be written in symbolic form as

Explanatory component: $grp + x_1.grp + \dots + x_l.grp + \dots + x_q.grp$

As before, grp is a factor (which allows a separate intercept for each group) and the term $x_l.grp$ is a combination of the grp factor and the lth explanatory variate, x_l, which allows that variate to have a separate slope for each group.

In practice, we use a somewhat more complex form of the model by progressively adding terms to the null model. With first-level-zero parameterization, this results in the mathematical form

$$y_{jk} = (\alpha_1 + \nu_j) + (\beta_{11} + \eta_{1j}) x_{1jk} + \dots + (\beta_{l1} + \eta_{lj}) x_{ljk} + \dots + (\beta_{t1} + \eta_{tj}) x_{qjk} + e_{jk} \ ,$$

with $\nu_1 = 0$ and $\eta_{l1} = 0$ for $l = 1 \dots t$. Then, α_1 is the intercept for the first group and ν_j is the difference between intercepts for the jth and the first groups. For the lth explanatory variate, β_{l1} is the slope for the first group and η_{lj} is the difference between the slopes for the jth and the first groups. In symbolic form, this model can be written as

Explanatory component: $[1] + x_1 + \dots + x_l + \dots + x_q + grp$
$\qquad\qquad\qquad\qquad + x_1.grp + \dots + x_l.grp + \dots + x_q.grp$

In fitting this more complex form of model, we need to be aware of several potential problems. First, there may be collinearity within the set of explanatory variates. This can be investigated with exploratory scatter plots and the methods described in Section 14.7. If variates are partially collinear, this can introduce ambiguity into the model selection process. If very strong collinearity is present, then the model may become unstable and one or more variates should be omitted. Second, in order for the separate lines model to be sensible, we now require a good overlap of values across groups for each of the explanatory variates as well as a reasonable number of observations in each group. Third, we may need to modify our strategy for model selection. With many explanatory variates, many different sequential ANOVA tables can be constructed by adding the variates into the model in different orders. If the full model, with separate slopes for each explanatory variate, makes biological sense and has a reasonable number of residual df then it is sensible to start from this model and use marginal F-tests to simplify it. When a term is dropped, the model is refitted and a new set of marginal F-tests calculated. An alternative strategy is to start from an intermediate model and to consider adding or dropping terms, refitting the model and recalculating the F-statistics each time the model is changed. In both cases, the principle of marginality should be respected: a term should be added only when all of its sub-terms are already present, and a term should not be dropped if it is a sub-term of another term still in the model. As previously, you should use diagnostic plots to check the fit of the model. We illustrate some of these principles in Example 15.2.

EXAMPLE 15.2: WEED SEED ABUNDANCE

An observational study was done to investigate whether the number of seeds produced by rye-grass could be related to plant characteristics. Between 17 and 24 samples were collected from each of four study sites (factor Site, with levels C, L, P and W). At each sample point within each site, the total number of seeds was counted and converted to number per m^2 (response variate *TotalSeed*), and the average head length (in mm, variate *HLength*) and average number of spikelets per head (variate *Spikelets*) on plants was recorded. The data are in Table 15.7 and file WEEDSEED.DAT. Before analysis, we transform the response variable using the log$_{10}$-transformation, i.e. *logSeed* = log$_{10}$(*TotalSeed*), to achieve homogeneity of variance of the residuals (see Chapter 6).

Preliminary exploration of the data indicates a fairly strong relationship between the two explanatory variates (sample correlation of $r = 0.67$, Figure 15.7). This implies that there may be some ambiguity as to which variate best explains differences in logged seed numbers. The range of both variates varies substantially across the four sites.

Plotting the log$_{10}$(number of seeds) against the number of spikelets or head length suggests a positive but noisy relationship in both cases (Figure 15.8), and that logged

TABLE 15.7

Observations of Total Seed Number per m^2 (Total) with Average Head Length (HL) and Number of Spikelets per Plant (Spikes) at Four Sites (Labelled C, L, P and W) (Example 15.2 and File WEEDSEED.DAT)

Site	HL	Spikes	Total	Site	HL	Spikes	Total	Site	HL	Spikes	Total
C	26.74	26.1	6232	L	26.21	21.1	13,320	P	33.76	26.4	120,937
C	23.77	28.7	6435	L	23.87	21.2	15,116	P	36.35	34.1	127,307
C	28.80	25.2	7022	L	27.58	21.1	15,243	P	35.98	28.7	137,416
C	29.89	30.1	10,700	L	23.57	19.6	16,830	P	31.22	27.5	161,070
C	31.01	29.1	11,524	L	28.33	24.7	18,856	P	33.96	25.4	162,154
C	26.00	26.3	12,814	L	24.39	26.1	20,930	P	32.76	27.7	173,734
C	31.25	21.4	13,093	L	20.95	21.7	24,200	P	29.47	27.7	181,971
C	33.48	27.7	14,991	L	30.03	24.6	24,369	P	32.35	25.9	190,408
C	24.86	28.0	15,137	L	26.92	26.0	24,944	P	35.48	30.8	215,477
C	25.71	28.0	16,162	L	27.47	23.1	25,097	P	34.53	30.5	245,200
C	26.79	27.9	16,956	L	26.46	27.0	28,136	P	30.59	27.2	246,758
C	31.19	27.8	16,962	L	28.19	23.1	31,434	P	31.75	29.2	300,595
C	30.79	26.8	17,234	L	27.81	21.8	33,256	W	20.09	19.8	59,321
C	33.17	28.7	17,409	L	27.26	21.6	34,690	W	21.92	22.6	59,960
C	30.74	26.1	17,414	L	31.30	24.2	38,623	W	18.21	18.0	62,700
C	31.62	25.5	18,828	L	27.20	21.5	54,260	W	22.74	20.4	66,096
C	30.22	28.1	22,611	L	30.38	25.9	58,827	W	26.92	24.9	78,618
C	31.59	28.6	23,690	P	28.28	24.2	35,042	W	23.00	22.1	80,400
C	29.72	26.1	25,108	P	31.50	25.1	42,312	W	24.91	25.9	84,607
C	31.23	27.6	26,121	P	38.44	27.5	60,867	W	19.99	19.2	90,436
C	37.12	30.0	28,161	P	33.99	25.1	62,047	W	20.80	21.0	93,800
C	33.12	27.5	28,637	P	29.31	23.4	65,286	W	23.85	24.3	105,700
C	25.22	27.2	31,439	P	29.16	24.6	80,327	W	20.35	23.1	106,321
L	19.66	23.5	7084	P	33.27	23.6	96,021	W	22.05	22.0	106,480
L	27.94	22.3	10,436	P	27.02	24.7	96,173	W	23.67	21.5	122,820
L	32.14	22.8	11,119	P	33.89	24.6	100,352	W	28.43	24.1	128,132
L	27.40	22.0	11,613	P	30.94	24.0	104,777	W	18.84	19.4	130,834
L	26.93	22.8	11,883	P	25.73	18.0	112,500	W	21.24	21.6	157,896
L	26.03	23.8	12,824	P	32.90	25.3	117,237	W	19.86	20.8	171,947

Source: Data from R. Alarcon-Reverte, Rothamsted Research.

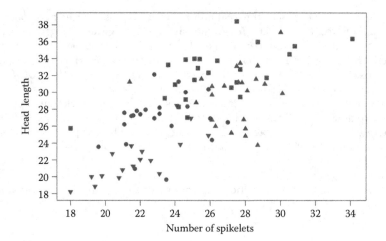

FIGURE 15.7
Average head length (mm) vs average number of spikelets per head at four sites: ▲ C (site 1), ● L (site 2), ■ P (site 3), ▼ W (site 4) (Example 15.2).

seed numbers may be more related to differences between sites than to either of the variates. For this reason, we fit factor Site as the first term in the model, followed by the two explanatory variates, and then the combined terms of factor Site with each of the two variates.

As there are between 17 and 24 observations at each site, it is a reasonable strategy to fit the full separate lines model and then seek to simplify it. The full model is expressed in symbolic form as

Response variable: *logSeed*
Explanatory component: *[1]* + Site + *HLength* + *Spikelets* + *HLength*.Site
                        + *Spikelets*.Site

This model can be written in mathematical form, with first-level-zero parameterization, as

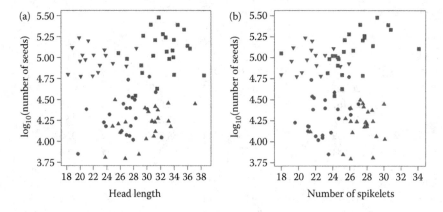

FIGURE 15.8
Total number of seeds per m² (log₁₀ scale) vs (a) average head length (mm) and (b) average number of spikelets per head at four sites: ▲ C, ● L, ■ P, ▼ W (Example 15.2).

$$logSeed_{jk} = (\alpha_1 + v_j) + (\beta_{11} + \eta_{1j})\ HLength_{jk} + (\beta_{21} + \eta_{2j})\ Spikelets_{jk} + e_{jk}\ ,$$

with $v_1 = 0$, $\eta_{11} = 0$ and $\eta_{21} = 0$. In this model, the response $logSeed_{jk}$ represents the kth observation at the jth site, with $j = 1 \ldots 4$, $k = 1 \ldots n_j$ and $n_1 = 23$, $n_2 = 23$, $n_3 = 24$ and $n_4 = 17$, and corresponding values of the explanatory variates indicated by $HLength_{jk}$ and $Spikelets_{jk}$. The overall intercept, α_1 (associated with term *[1]*), represents the intercept at the first site. The site-specific intercepts, v_j (associated with term Site), represent the difference in intercepts between the jth and the first sites. The overall slope for explanatory variate head length, β_{11} (associated with the term HLength), represents the slope with respect to head length at the first site. The site-specific slopes of head length, η_{1j} (associated with term HLength.Site), represent the difference between slopes at the jth and the first sites. Overall and site-specific slope parameters for the number of spikelets, β_{21} and η_{2j} (associated with terms Spikelets and Spikelets.Site), are interpreted similarly.

Table 15.8 shows two sequential ANOVA tables in abbreviated form, the first constructed by adding head length before number of spikelets, and the second using the other order. Recall that the estimated parameters are the same from the two fits, but the sequential ANOVA tables differ because of the partial collinearity between the two explanatory variates. The full model has a total of 12 parameters (one intercept and two slopes for each of the four sites) and can be interpreted as fitting separate planes in terms of the two explanatory variates for each site.

The next step in our analysis is the identification of a suitable model. We first consider the combined terms, Spikelets.Site and HLength.Site. We can obtain marginal F-tests for these terms from the sequential ANOVA tables in Table 15.8 and find that the F-statistic for dropping Spikelets.Site is less significant ($F_{3,75}^{Sp.S} = 0.65$, $P = 0.584$, Table 15.8a) and so drop this term. As there are 75 residual df in the full model, we do not refit the model, but can immediately consider whether we can then drop the term HLength.Site. Its F-statistic is not significant ($F_{3,75}^{HL.S} = 1.71$, $P = 0.172$, Table 15.8a) and so we also drop this term. There is therefore no evidence for separate slopes across sites for either of these variates.

We then consider whether we can drop either of the explanatory variates. We find that if we add the HLength variate after the Site factor and the Spikelets variate, there is no significant improvement to the model ($F_{1,75}^{HL} = 2.32$, $P = 0.132$, Table 15.8b). If we add the Spikelets variate after Site and HLength, then there is a small and borderline-significant improvement to the model ($F_{1,75}^{Sp} = 3.82$, $P = 0.054$, Table 15.8a). This indicates that HLength adds no information once Spikelets is in the model, but Spikelets may add

TABLE 15.8

Abbreviated Sequential ANOVA Tables for Separate Lines Model for Logged Seed Counts with Factor Site and Explanatory Variates Spikelets and HLength (Example 15.2)

(a)				(b)			
Term Added	Inc. df	Mean Square	P	Term Added	Inc. df	Mean Square	P
+ Site	3	4.406	< 0.001	+ Site	3	4.406	< 0.001
+ HLength	1	0.291	0.009	+ Spikelets	1	0.351	0.004
+ Spikelets	1	0.155	0.054	+ HLength	1	0.095	0.132
+ HLength.Site	3	0.070	0.172	+ Spikelets.Site	3	0.024	0.623
+ Spikelets.Site	3	0.027	0.584	+ HLength.Site	3	0.072	0.159
Residual	75	0.041		Residual	75	0.041	
Total	86	0.198		Total	86	0.198	

Inc. = incremental.

TABLE 15.9

Abbreviated Sequential ANOVA Tables for Separate Lines Model for Logged Seed Counts with Factor Site and Explanatory Variate *Spikelets* (Example 15.2)

(a) Term Added	Inc. df	Mean Square	P	(b) Term Added	Inc. df	Mean Square	P
+ Site	3	4.406	< 0.001	+ *Spikelets*	1	0.002	0.815
+ *Spikelets*	1	0.351	0.005	+ Site	3	4.522	< 0.001
Residual	82	0.042		Residual	82	0.042	
Total	86	0.198		Total	86	0.198	

Inc. = incremental.

a little information when *HLength* is in the model. We therefore drop the *HLength* variate from the model. This leaves the Site factor and *Spikelets* variate in the model; i.e. a parallel lines model in terms of *Spikelets*. We refit this model in both orders, giving the abbreviated sequential ANOVA tables in Table 15.9.

The incremental F-test for the *Spikelets* variate is not significant when it is fitted first (Table 15.9b), but is highly significant when fitted after factor Site (Table 15.9a). Factor Site is highly significant for both orders of fitting. The fitted model in Figure 15.9 shows the reason for this: differences in logged seed count between sites are so large that the relationship with number of spikelets can be detected only after we have corrected for this effect. This parallel lines model, with regression on *Spikelets*, is therefore our predictive model. Residual plots (not shown) indicate no conflict with the model assumptions, and plotting the residuals against the omitted *HLength* variate shows no evidence of any relationship. The predictive model accounts for 78.8% of the variation in the data (adjusted $R^2 = 0.788$), and can be written in simple form as

$$\hat{\mu}_j^*(Spikelets) = \hat{\alpha}_j + \hat{\beta}\,Spikelets\,,$$

where $\hat{\mu}_j^*(Spikelets)$ represents the prediction on the log scale at the jth site for the specified number of spikelets. The parameter estimates are listed in Table 15.10.

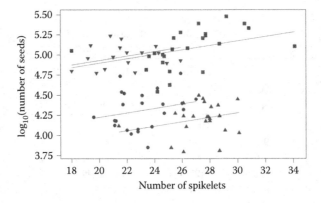

FIGURE 15.9

Logged number of seeds with predictive parallel lines model (—) in terms of average number of spikelets per head at each of four sites: ▲ C, ● L, ■ P, ▼ W (Example 15.2).

TABLE 15.10

Parameter Estimates with Standard Errors (SE), t-Statistics (t) and Observed Significance Levels (P) for a Parallel Lines Model for Logged Seed Counts in Terms of Variate *Spikelets* and Factor Site with Four Levels (Example 15.2)

Term	Parameter	Estimate	SE	t	P
Site 1	α_1	3.450	0.2646	13.040	< 0.001
Site 2	α_2	3.671	0.2249	16.324	< 0.001
Site 3	α_3	4.343	0.2547	17.048	< 0.001
Site 4	α_4	4.376	0.2142	20.432	< 0.001
Spikelets	β	0.0277	0.00955	2.895	0.005

Differences on the logarithm scale can be interpreted in terms of ratios on the original scale (Section 6.4), and so the predictive model can be interpreted as a multiplicative model, with

$$\hat{\mu}_j(Spikelets) = 10^{\hat{\mu}_j^*(Spikelets)} = 10^{(\hat{\alpha}_j + \hat{\beta}Spikelets)} = 10^{\hat{\alpha}_j} \times 10^{\hat{\beta}Spikelets} \ .$$

The difference between sites dominates the model. For a fixed number of spikelets, here denoted s, the $\log_{10}$-ratio of seed counts between any two sites (labelled i and j) can be written as

$$\log_{10}[\hat{\mu}_i(s)/\hat{\mu}_j(s)] = \hat{\alpha}_i + \hat{\beta}s - (\hat{\alpha}_j + \hat{\beta}s) = \hat{\alpha}_i - \hat{\alpha}_j \ .$$

We can therefore obtain a CI for the log-ratio in terms of a CI for the difference $\alpha_i - \alpha_j$, and then back-transform to get a CI for the ratio $\mu_i(s)/\mu_j(s)$. For example, consider sites 4 (W) and 1 (C). The estimated $\log_{10}$-ratio is

$$\log_{10}[\hat{\mu}_4(s)/\hat{\mu}_1(s)] = \hat{\alpha}_4 - \hat{\alpha}_1 = 4.376 - 3.450 = 0.926 \ ,$$

with SE 0.084. A 95% CI for the $\log_{10}$-ratio, $\log_{10}[\mu_4(s)/\mu_1(s)]$, can be calculated via the difference $\alpha_4 - \alpha_1$ as (0.759, 1.093). We back-transform to estimate the ratio μ_4/μ_1 as $10^{0.926} = 8.43$, so we expect 8.43 times as many seeds at a location in the fourth site as in the first site (for the same number of spikelets), with a 95% CI for this ratio equal to $(10^{0.759}, 10^{1.093}) = (5.74, 12.39)$.

Within a site, $\hat{\beta} = 0.0277$ (SE = 0.0096) represents the expected increase in logged seed count for an increase of one spikelet per plant. We can predict the relative change in seed count for an increase of one spikelet in terms of $\hat{\beta}$ using

$$\frac{\hat{\mu}_j(s+1)}{\hat{\mu}_j(s)} = \frac{10^{\hat{\alpha}_j} \times 10^{\hat{\beta}(s+1)}}{10^{\hat{\alpha}_j} \times 10^{\hat{\beta}s}} = 10^{\hat{\beta}} \ .$$

We can therefore predict that seed count will increase by a factor of $10^{0.0277} = 1.066$ (a 7% increase) for an increase of one spikelet per plant. We can calculate a 95% CI for $\hat{\beta}$ as (0.0086, 0.0468) and can back-transform this to give a CI for the relative change as $(10^{0.0086}, 10^{0.0468}) = (1.02, 1.11)$, corresponding to a 2–11% increase in seed count for plants with one additional spikelet.

15.3 Regression in Designed Experiments

Recall from Section 1.3 that we usually allow for both structural and explanatory components within our models. We have not yet allowed for a structural component within regression models because software for regression modelling does not generally support this (as discussed in Sections 11.6 and 12.1). It is possible to incorporate some types of structure within the explanatory component of the model, however; this is the intra-block analysis of Section 11.6.1 and, in the context of regression modelling, can be implemented as regression with groups. This approach can be used in studies where we wish to account for structure prior to estimation of the regression line. A common example is regression within a RCBD, where a fixed set of values of the explanatory variate are applied to experimental units within each block. The RCBD model allows for an additive difference between responses to the same treatment from different blocks which, in combination with a linear response to the explanatory variate, corresponds exactly to a parallel lines model. When replication is present, we can formally test for lack of fit to the common regression line (see Section 12.8). Recall that we regard structural terms as intrinsic to the model, and so we do not formally test or omit these terms. We also treat structural terms differently to explanatory terms in the predictive model as, although we wish to account for structure, we do not usually wish to include it in predictions. When a parallel lines model is appropriate, we would usually present the response averaged over structural variables as our predictive model, and we interpret this as the predicted response for the average conditions in the study. We illustrate these concepts in Example 15.3.

EXAMPLE 15.3: FORAGE MAIZE YIELDS

An experiment at Rothamsted Research in 1996 investigated the yield response of forage maize to nitrogen fertilizer. The experiment was designed as a RCBD with three blocks (factor Block) of four plots (factor Plot), with nitrogen fertilizer rates of 0, 70, 140 and 210 kg N (variate N). The whole crop forage yields from each plot (at 100% dry matter in tonnes/hectare, variate Yield) are shown in Table 15.11 and are held in file FORAGE.DAT. The aim of analysis is to model the yield as a function of applied nitrogen. We start by analysing the experiment as a RCBD, with model

TABLE 15.11

Whole Crop Forage Yield (Yield, t/ha) from a RCBD with Three Blocks and Four Nitrogen Fertilizer Rates (N = 0, 70, 140, 210 kg N) (Example 15.3)

Block 1			Block 2			Block 3		
Plot	N	Yield	Plot	N	Yield	Plot	N	Yield
1	0	10.42	1	70	11.62	1	70	11.13
2	140	12.21	2	0	11.98	2	210	12.57
3	210	12.85	3	210	12.81	3	0	9.82
4	70	12.22	4	140	12.67	4	140	10.92

Source: Data from P. Poulton, Rothamsted Research.

TABLE 15.12

Multi-Stratum ANOVA Table for Forage Yields from a RCBD
with Three Blocks (Factor **Block**) of Four Plots (Factor **Plot**) and
Four Nitrogen Treatments (Factor **FacN**) (Example 15.3)

Source of Variation	df	Sum of Squares	Mean Square	Variance Ratio	P
Block stratum					
Residual	2	2.8385	1.4192	4.399	0.067
Block.Plot stratum					
FacN	3	6.1434	2.0478	6.347	0.027
Residual	6	1.9359	0.3227		
Total	11	10.9178	0.9925		

Response variable: *Yield*
Explanatory component: FacN
Structural component: Block/Plot

where **FacN** is a factor with a separate level for each nitrogen application rate. The
resulting multi-stratum ANOVA is shown in Table 15.12.

The residual plots are adequate and a plot of the predicted means shows a linear trend
in applied nitrogen (Figure 15.10a). We could fit a linear polynomial contrast within the
multi-stratum ANOVA, but an intra-block analysis is appropriate for a RCBD and we
will take that option here. As a first step, we fit the parallel lines model and check for
lack of fit (as in Section 12.8) with

Explanatory component: Block + *N* + FacN

This fits the structural factor **Block** first, then adds explanatory variate *N* to obtain the
parallel lines model with a separate intercept for each block. To test whether the straight
line captures the pattern of response, we then add the factor version of the explanatory
variate (**FacN**) into the model, giving the sequential ANOVA in Table 15.13. The SS for
Block is the same as in the multi-stratum ANOVA, the SS for the nitrogen treatments
has been partitioned into variation due to the straight line (*N*) and deviations from it

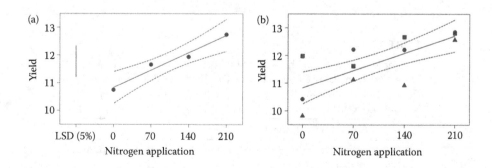

FIGURE 15.10
Predictive model (—) with 95% CI (– –) for forage maize yields (at 100% dry matter in tonnes/hectare) in terms of
nitrogen applied (kg) with (a) fitted treatment means (•) and LSD from multi-stratum ANOVA and (b) observed
data (• = block 1, ■ = block 2, ▲ = block 3) (Example 15.3).

TABLE 15.13

Sequential ANOVA for Forage Yields from a RCBD, Testing for Lack of Fit
to Parallel Lines Model (Example 15.3)

Term Added	Incremental df	Incremental SS	Mean Square	Variance Ratio	P
+ Block	2	2.8385	1.4192	4.399	0.067
+ *N*	1	5.9283	5.9283	$F^N = 18.374$	0.005
+ FacN	2	0.2150	0.1075	$F^{FacN} = 0.333$	0.729
Residual	6	1.9359	0.3227		
Total	11	10.9178	0.9925		

SS = sum of squares.

(FacN after eliminating *N*), and the residual SS remains unchanged. The straight line
accounts for most of the variation due to treatments and is highly significant ($F^N = 18.37$,
$P = 0.005$). There is no statistical evidence for lack of fit ($F^{FacN} = 0.33$, $P = 0.729$) and no
evidence of model misspecification in fitted model or residual plots when the lack-of-fit
term (FacN) is omitted. We therefore accept the parallel lines model as our predictive
model which, with first-level-zero parameterization, takes the form

$$\hat{\mu}_i(N) = \hat{\alpha}_1 + \widehat{Block}_i + \hat{\beta}N \ ,$$

where $\hat{\mu}_i(N)$ is the predicted yield (t/ha) in the *i*th block ($i = 1 \ldots 3$) for nitrogen application rate N ($0 \leq N \leq 210$), $\hat{\alpha}_1$ is the estimated intercept for the first block, $\widehat{Block}_i$ is
the difference in intercept between the *i*th and the first blocks, and $\hat{\beta}$ is the estimated
slope of the relationship with nitrogen application rate. These estimates are listed in
Table 15.14.

In this parallel lines model, we can summarize the overall performance by averaging across blocks to obtain the predictive model for average conditions within the
trial as

$$\hat{\mu}(N) = \frac{1}{3}\sum_{i=1}^{3} \hat{\mu}_i(N) = \hat{\alpha}^* + \hat{\beta}N \quad \text{where } \hat{\alpha}^* = \hat{\alpha}_1 + \frac{1}{3}\sum_{i=1}^{3} \widehat{Block}_i \ .$$

TABLE 15.14

Parameter Estimates with Standard Errors (SE), t-Statistics (t) and
Observed Significance Levels (*P*) for Parallel Lines Model for
Forage Yield from a RCBD with Three Blocks and Explanatory
Variate *N* (Example 15.3)

Term	Parameter	Estimate	SE	t	P
[1]	α_1	10.982	0.3279	33.487	< 0.001
Block 1	$Block_1$	0	—	—	—
Block 2	$Block_2$	0.345	0.3667	0.941	0.374
Block 3	$Block_3$	−0.815	0.3667	−2.223	0.057
N	β	0.0090	0.00191	4.696	0.002

The SE for the averaged intercept is calculated from the estimated variance–covariance matrix for the estimated intercepts (see Section C.4). Here, the averaged intercept is $\hat{\alpha}^* = 10.825$ (SE = 0.2505). The final predictive model therefore takes the form

$$\hat{\mu}(N) = 10.825 + 0.0090N .$$

This predictive model is shown with 95% confidence intervals and the observed data in Figure 15.10. It is clear that this predictive model gives a good fit to the original treatment means and passes through the centre of the observed data.

Where it is appropriate, the intra-block model is a valid alternative to multi-stratum ANOVA (see Sections 7.5 and 11.6). In principle, we prefer the multi-stratum ANOVA approach because it explicitly accounts for the different status of structural and explanatory terms in the model. In practice, it can be difficult to extract regression models (with SE) from multi-stratum ANOVA output, and so in these cases, we would accept the intra-block analysis. Where the intra-block model is not appropriate, the linear mixed models discussed in Chapter 16 give a more general and flexible approach to regression modelling when structure is present.

15.4 Analysis of Covariance: A Special Case of Regression with Groups

The analysis of covariance (ANCOVA, sometimes also known as ANOCOVA) is used to incorporate a limited form of regression with groups into the analysis of a designed experiment to adjust treatment comparisons for the presence of quantitative extraneous variables. This is separate from, and often in addition to, the inclusion of structure discussed in Section 15.3. In this context, a **covariate** is an explanatory variate thought to influence the response that was not taken into account in the design of the experiment. Adjusting for this variate ensures that comparisons between treatments are not biased by its presence and may reduce uncertainty in estimates of group differences. For example, consider a field experiment designed to compare yield response to drought. If it is thought that soil depth varies across the field, and that this may affect response to drought, then the soil depth can be measured within each experimental plot and used as a covariate to account for variation in drought response that is unrelated to the experimental treatments. However, the ANCOVA approach is sensible only if a parallel lines model is a good fit to the data and if the covariate values are unrelated to the groups. For example, consider the scenarios shown in Figure 15.11.

In Figure 15.11a, the effect of the covariate is the same across all three groups, and so the parallel lines model describes the response well. In this case, the differences among the three groups can be sensibly summarized by comparisons at any given value of the covariate; in practice comparisons are usually made at the covariate sample mean. In contrast, in Figure 15.11b, the effect of the covariate differs between the groups, so that differences among groups are highly dependent on the value of the covariate. In this case, we cannot summarize treatment differences without also taking into account the covariate value; here traditional ANCOVA, which is based implicitly on the assumption of parallel lines, is inappropriate and a separate lines model should be fitted and reported. In Figure 15.11c, although a parallel lines model appears appropriate, the range of the covariate differs

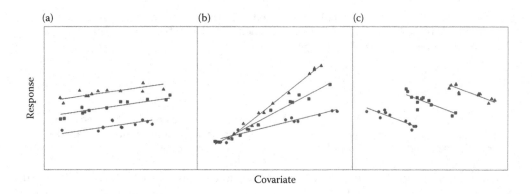

FIGURE 15.11
Adjusting treatment differences using a covariate with three groups (●,■,▲), showing predictive model (—): (a) common treatment difference for all covariate values; (b) treatment difference depends on covariate value; (c) covariate value dependent on group.

between the groups, and this can lead to problems of both estimation and interpretation, as differences in response can be attributed to either the covariate or the treatment groups. If the differences in the covariate values are intrinsic to the groups, i.e. if they would be replicated in other studies, then adjustment to a common covariate value might give a misleading impression of group differences. On the other hand, if we do want to adjust for the covariate but there is no overlap between the groups, then it is difficult to establish an appropriate value of the covariate at which to make comparisons, as this requires extrapolation for one or more groups. Fitting a separate groups model (Section 15.1.3) with the covariate as the response can be used to detect this situation.

Within the literature on analysis of designed experiments, a tradition of ANCOVA has developed that largely ignores the wider context of regression with groups. One reason for this is the use of multi-stratum ANOVA to account for structure, which can easily incorporate explanatory variates using a parallel lines model, but cannot easily accommodate a separate lines model. Conversely, some books consider ANCOVA to be nothing more than regression with groups, but this approach often misses the nuances associated with structure in designed experiments. We recommend a hybrid approach, which we now outline before presenting an example. Remember that the aim of ANCOVA is to present group comparisons adjusted for covariate effects.

The first step is to test whether the covariate is related to the groups, incorporating any structure into the analysis. This can be done with multi-stratum ANOVA for balanced designs, and with intra-block analysis (Section 11.6.1) or linear mixed models (Chapter 16) for unbalanced designs. If the covariate does differ systematically between groups, then you should ask why, what the implications for interpretation are, and whether ANCOVA is a sensible approach. If there is no such relationship, or if you decide to proceed anyway, then the next step is to consider whether a parallel lines model is plausible. If the number of observations within each group is reasonably large (>5), it might help to fit a separate lines model and to test explicitly for group differences in response to the covariate, again incorporating any structure. If the number of observations per group is small (≤5), then the fitted lines for individual groups are likely to be unreliable, so this step may be omitted. In either case, it usually helps to plot the response against the covariate with groups indicated (as in Figure 15.11) to give a visual assessment of the relationship, and of the extent of overlap in covariate values among groups. If the parallel lines model still seems plausible, it can then be fitted to obtain estimates of group differences (still incorporating any structure in the model). The covariate is always fitted first in the parallel lines model so that differences between groups are assessed after the covariate has been taken into account. The fitted model and residual plots should be assessed to check for any evidence of model misspecification before the results are accepted and interpreted.

EXAMPLE 15.4A: THOUSAND GRAIN WEIGHTS*

A field experiment was done to investigate the impact of growth regulator (+/–) on seed production for two varieties of oilseed rape (B or N) in a CRD with six replicates. Unfortunately, pigeons grazed some parts of the trial in early spring, and it was thought that this damage might affect plant growth and seed development. The extent of damage to each plot was recorded as percentage area grazed (variate *Damage*), to the nearest 10%, and the aim of analysis was to make treatment comparisons after accounting for bird damage, if possible. The response is thousand grain weight (abbreviated as TGW, held in variate *TGW*). The treatment combinations form the four factorial combinations of the two growth regulators and two varieties and are represented here by a single factor (Trt with groups labelled as +B, +N, –B, –N). The data are in Table 15.15 and file TGW.DAT.

Figure 15.12 plots variate *TGW* against variate *Damage*, with a unique symbol for each group. The range of pigeon damage clearly differs between treatments, with the second (+N) and third (–B) groups having large and intermediate amounts of damage, respectively.

The first model fitted was used to investigate the relationship between the covariate and the treatment groups formally, as

Response variate: *Damage*
Explanatory component: *[1]* + Trt

TABLE 15.15

Thousand Grain Weight (Variate *TGW*) and Pigeon Damage (% Plot Grazed, Variate *Damage*) for a Field Experiment Comparing Two Varieties N and B (Factor Variety) with (+) or without (–) Growth Regulator (Factor GR) (Example 15.4 and File TGW.DAT). Factor Trt Gives a Single Code for Each of the Four Treatment Combinations

Plot	GR	Variety	Trt	Damage	TGW	Plot	GR	Variety	Trt	Damage	TGW
1	+	N	+N	60	3.342	13	+	N	+N	60	3.150
2	–	B	–B	30	3.185	14	–	N	–N	60	3.436
3	–	N	–N	40	3.997	15	+	N	+N	50	3.793
4	+	N	+N	30	4.111	16	+	N	+N	40	3.937
5	+	B	+B	20	3.783	17	–	N	–N	40	3.901
6	–	B	–B	20	3.302	18	–	B	–B	30	3.357
7	–	N	–N	0	4.807	19	–	B	–B	30	3.562
8	–	N	–N	0	4.451	20	–	B	–B	30	3.338
9	–	B	–B	30	3.419	21	+	B	+B	0	3.749
10	+	B	+B	40	3.295	22	+	B	+B	30	3.138
11	+	B	+B	50	3.169	23	+	B	+B	0	3.756
12	+	N	+N	50	3.591	24	–	N	–N	0	5.019

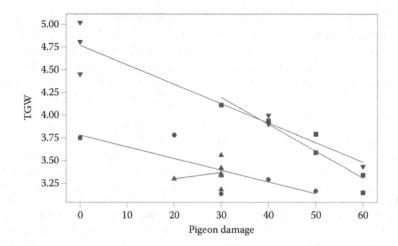

FIGURE 15.12
Thousand grain weight (TGW) with separate lines model (—) for pigeon damage (% plot grazed) with treatment groups (● +B, ■ +N, ▲ −B, ▼ −N) (Example 15.4A).

This model suggests there may be differences between treatment groups ($F_{3,20}$ = 2.642, P = 0.077), but it is not conclusive. A biological assessment of the matter concluded that the intention was to compare seed quality in the absence of pigeon grazing, so correction for such damage is appropriate in this context even if grazing did differ between treatments. The second model fitted was a separate lines model with the covariate and the treatment groups, specified as

Response variate: *TGW*
Explanatory component: *[1]* + *Damage* + Trt + *Damage*.Trt

The variance ratio $F_{3,16}^{D.Trt}$ = 2.710 (P = 0.080) suggested some differences in slope between treatments, and the fitted lines are shown in Figure 15.12. The slope for the third group (−B) is positive whereas those for the other treatments are negative, but this group has such a small range of grazing damage (five plots have 30% grazing and one plot 20% grazing) that this line cannot be considered reliable. A parallel lines model was therefore deemed plausible, and the effect of the treatments was tested. The incremental F-statistic for the factor Trt ($F_{3,16}^{Trt}$ = 28.874, P < 0.001) indicated large treatment differences after accounting for grazing damage. Estimates from the parallel lines model are shown in Table 15.16, and this predictive model is written in mathematical form, with first-level-zero parameterization, as

$$\hat{\mu}_j(Damage) = (\hat{\alpha}_1 + \hat{v}_j) + \hat{\beta}Damage ,$$

where $\hat{\mu}_j(Damage)$ represents the predicted TGW for the jth treatment group (j = 1 ... 4) with grazing damage *Damage*. Parameter $\hat{\alpha}_1$ is the estimated intercept for the first treatment (+B), $\hat{v}_j$ is the estimated difference in intercept for the jth treatment (relative to the first) and $\hat{\beta}$ is the estimated slope associated with grazing damage.

Predictions from this model evaluated at the average damage of 30.83% grazing are shown in Table 15.17a. The LSDs for these predictions (at significance level α_s = 0.05 with 19 df) range between 0.2301 and 0.2615. It is clear that groups 2 (+N) and 4 (−N) give greater seed weights than the other groups (+B, −B), with no significant differences within

TABLE 15.16

Parameter Estimates with Standard Errors (SE), t-Statistics (t)
and Observed Significance Levels (*P*) for Parallel Lines Model
for Thousand Grain Weight with Treatment Groups (Factor Trt)
Adjusted for Pigeon Grazing (Variate *Damage*) (Example 15.4A)

Term	Parameter	Estimate	SE	t	P
[1]	α_1	3.926	0.0954	41.132	< 0.001
Damage	β	−0.0190	0.00237	−8.018	< 0.001
Trt +B	v_1	0	—	—	—
Trt +N	v_2	0.648	0.1249	5.188	< 0.001
Trt −B	v_3	−0.026	0.1106	−0.235	0.817
Trt −N	v_4	0.787	0.1099	7.158	< 0.001

TABLE 15.17

Predicted Thousand Grain Weight with Standard Error (SE)
from Two Explanatory Models for Four Treatment Groups
with 30.83% Grazing Damage per Plot (Example 15.4A)

Treatment	(a) With Covariate		(b) Without Covariate	
	Prediction	SE	Prediction	SE
+B	3.339	0.0797	3.482	0.1586
+N	3.987	0.0881	3.654	0.1586
−B	3.313	0.0780	3.361	0.1586
−N	4.126	0.0797	4.268	0.1586

each of these two sets. Given our knowledge of factorial structures (Chapter 8), we should be able to clarify our inferences in terms of the underlying factors, variety and use of growth regulator, and we examine this further in Example 15.4B. For now, we compare the predictions from the ANCOVA (parallel lines model) with those from the separate groups model, which ignores the covariate (Table 15.17b). Predictions from the parallel lines model have been shifted according to the amount of grazing observed: the predicted TGW for the +N group, which was more heavily grazed, is adjusted upwards and the predicted TGW for the +B and −N groups, which were less heavily grazed, are adjusted downwards. The prediction SEs and SEDs are substantially smaller in the parallel lines model because the covariate has accounted for some of the variation among replicate plots.

Within the context of designed experiments, the traditional ANCOVA procedure differs from that given above in several ways. First, it is not usual to fit the separate lines model to investigate formally whether there is evidence against the parallel lines model. However, we find this step useful, and recommend it where there are sufficient observations within groups to make the analysis meaningful. Second, it is usual to present a single composite ANOVA table with F-tests for treatment groups after elimination of the covariate and for the covariate after elimination of the treatment groups. This ANOVA table is amalgamated from two different sequences of models: one with the covariate fitted first and the other with the factor(s) fitted first. In contrast, we have just used the sequential ANOVA table obtained by fitting the covariate first. Both approaches are correct, but we find our approach more straightforward. Another example of ANCOVA, this time for a structured experiment analysed by linear mixed models, is given in Example 16.2.

15.5 Complex Models with Factors and Variates

In general, we might have several explanatory factors and variates and the corresponding models become considerably more complex. The aims of analysis stay the same, namely, to find a predictive model that uses as few parameters as possible to describe the response. The automatic model selection techniques described in Section 14.9 can be used, but they must be modified somewhat, and these modifications are described in Section 15.5.1. Once we have identified a predictive model, then we need to decide which predictions to make, and this is discussed in Section 15.5.2.

15.5.1 Selecting the Predictive Model

In Chapters 8, 11, 14 and 15, we have used various different strategies for model selection, so here we try to set them out in a coherent framework. In all cases, we obey the principle of marginality when adding or dropping terms, and if our explanatory model contains terms associated with the structural component then we fit the structural terms first and do not test them (see Section 15.3). Our strategy will depend on whether the explanatory variables are orthogonal (see Section 11.1), and whether the full model is well defined. The full model consists of meaningful terms constructed from the set of explanatory variables and appropriate combinations. The full model is considered well defined if it was specified during the design phase of the study and has a reasonable number of residual df.

When the explanatory variables are orthogonal, we can use a single sequential ANOVA table for model selection. This usually only occurs in the context of a designed study, where the full model is well defined. We then start with the full model and progressively test and drop non-significant terms (respecting marginality) to identify the predictive model (e.g. Section 8.3).

When the explanatory variables are not orthogonal, there may be many sequential ANOVA tables. We will consider separately the cases when the full model is well defined and those where it is not.

If the full model is well defined, then we again fit the full model and progressively test and drop non-significant terms, respecting marginality. If there are few sequential ANOVA tables, then we might form all of them. If the residual df are large, then there is no need to refit the model if we can deduce all the required information from these initial tables. If there are many different sequential ANOVA tables, then forming them all is impractical and it will usually be easier to use marginal F-tests to progressively simplify the full model. At each step of the process, we then identify the terms that can be dropped (respecting marginality), and form a marginal F-test for each of these terms. We drop the least significant term, i.e. the one with the largest observed significance level (P) subject to $P > 0.05$, and then refit the model. We repeat this process until no further terms can be eliminated. This is the backward elimination procedure of Section 14.9.

If the full model is not well defined, then we must start from a simpler model and consider both adding and dropping terms as for the stepwise regression procedures described in Section 14.9, but in these more complex models we must now also respect marginality.

These latter two approaches suggest that the automatic selection procedures described in Section 14.9 are more widely useful, although the caveats stated there still apply and some modification is required to account for model terms involving explanatory factors. In particular, we now require automatic selection procedures to respect marginality when terms are added or dropped. Some caution is also required for model terms with more than 1 df. The numerator df for an F-test is always the change in model df obtained on adding or dropping the term. Since the critical value of an F-distribution decreases for larger numerator df, it is difficult to define a single threshold in terms of a critical value, especially where the df for the model terms cover a large range. In this case, it is usually more sensible to define thresholds in terms of the observed significance level, i.e. SLE or SLS (defined in Section 14.9).

EXAMPLE 15.4B: THOUSAND GRAIN WEIGHTS*

We now reanalyse the TGW data to recognize and exploit the crossed structure within the treatment groups. The four treatment groups (+B, +N, −B, −N) form a factorial set related to the factors Variety (with two levels, B and N) and GR, indicating the presence or absence of growth regulator (with two levels + and −). It is appropriate to replace the term Trt by a crossed structure Variety*GR that partitions the treatment effects into the main effect of variety, the main effect of growth regulator and their interaction (see Section 8.2). For a full analysis of this data set, we therefore repeat the procedure of Example 15.4A, but making this replacement throughout. First, we check whether the covariate is related to the treatments, using model

Response variate: *Damage*
Explanatory component: *[1]* + Variety*GR

The main effects are not significant with (by numerical coincidence) both $F_{1,20}^{GR} = 1.865$ and $F_{1,20}^{V} = 1.865$ (both $P = 0.187$), with the interaction of borderline significance ($F_{1,20}^{GR,V} = 4.197$, $P = 0.054$). In combination with the exploratory graphs in Figure 15.12, the reasoning given in Example 15.4A still stands, and we proceed with fitting the separate lines model for grain weight, specified as

Response variate: *TGW*
Explanatory component: *[1]* + *Damage* + Variety + GR + Variety.GR
                         + *Damage*.Variety + *Damage*.GR + *Damage*.Variety.GR

The covariate is fitted first, followed by terms associated with the treatment groups, with terms combining the covariate and the treatment groups at the end. Table 15.18 shows the observed significance levels of marginal F-tests from a sequence of models for this data, with this model labelled as Model 1.

All other terms are marginal to *Damage*.Variety.GR, so as a first step we can test only this term. We find it non-significant ($F_{1,16}^{D.V.GR} = 0.341$, $P = 0.567$) and so drop it and refit to obtain Model 2 of Table 15.18. In this ANCOVA setting, we are first interested in whether we can reduce to a parallel lines model, so we next examine the terms *Damage*.Variety and *Damage*.GR. We can omit term *Damage*.GR ($F_{1,17}^{D.GR} = 2.163$, $P = 0.160$) to obtain Model 3, but cannot then omit term *Damage*.Variety ($F_{1,18}^{D.V} = 5.581$, $P = 0.030$). Figure 15.12 suggests that the individual slopes for variety B are both less steep than those for variety N, so by using the crossed structure we can now detect this difference that was previously masked. We therefore retain the *Damage*.Variety term and can no longer regard this as a traditional ANCOVA. However, we can simplify the model further as term Variety.GR is eligible for testing but is not significant ($F_{1,18}^{V.GR} = 0.485$, $P = 0.495$). We drop the latter and refit the model (to obtain Model 4), and can then test term GR, which is not significant ($F_{1,19}^{GR} = 0.043$, $P = 0.839$). No further

TABLE 15.18

Observed Significance Level (*P*) for Marginal F-Tests in a Sequence of
Models for Thousand Grain Weights (Example 15.4B)

			P		
Term	Model 1	Model 2	Model 3	Model 4	Model5
[1]	—	—	—	—	—
Damage	—	—	—	—	—
Variety	—	—	—	—	—
GR	—	—	—	0.839	*
Variety.GR	—	0.806	0.495	*	*
*Damage.*Variety	—	0.015	0.030	0.020	0.015
*Damage.*GR	—	0.160	*	*	*
*Damage.*Variety.GR	0.567	*	*	*	*

Note: — = term in model but not eligible for testing, * = term omitted from model.

simplification is possible. The predictive model is therefore a separate lines model in
terms of explanatory variate *Damage* and the factor Variety, written as

Explanatory component: *[1]* + *Damage* + Variety + *Damage.*Variety

We can represent this predictive model in mathematical form, using first-level-zero
parameterization, as

$$\hat{\mu}_r(Damage) = (\hat{\alpha}_1 + \hat{v}_r) + (\hat{\beta}_1 + \hat{\eta}_r)Damage \ ,$$

Here, $\hat{\mu}_r(Damage)$ represents the predicted TGW for the *r*th variety (1 = B, 2 = N) with
grazing damage equal to *Damage*. Parameter $\hat{\alpha}_1 = 3.744$ (SE 0.1007) represents the intercept for variety B and $\hat{v}_2 = 1.051$ (SE 0.1345) is the difference in intercept for variety N
(recall $\hat{v}_1 = 0$). Similarly, parameter $\hat{\beta}_1 = -0.0125$ (SE 0.00344) represents the slope for
variety B, and $\hat{\eta}_2 = -0.0108$ (SE 0.00403) is the difference in slope for variety N, with
$\hat{\eta}_1 = 0$. The predictive model for the two varieties is shown in Figure 15.13 and can be
written as

Variety B: $\hat{\mu}_B(Damage) = 3.744 - 0.0125Damage$
Variety N: $\hat{\mu}_N(Damage) = 4.795 - 0.0233Damage$

We can conclude that there appears to be no effect of growth regulator on TGW, but
that TGW is decreased by pigeon damage, and that there is a strong varietal difference
which is also affected by the amount of pigeon damage. Although variety N always
had a larger TGW than variety B in this experiment, it is also more affected by pigeon
damage. For a 10% increase in plot damage, TGW is reduced by 0.125 units for variety
B, but by 0.233 units for variety N. This analysis suggests that the ANCOVA analysis of
Example 15.4A missed some of the nuances in the results by ignoring the structure of
the treatment groups.

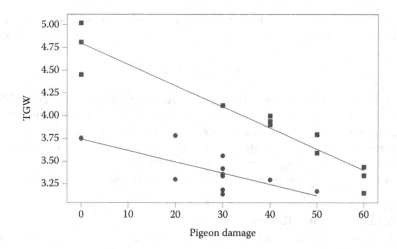

FIGURE 15.13
Thousand grain weight (TGW) with separate lines model (—) for pigeon damage (% plot grazed) with varieties B (•) and N (■) (Example 15.4B).

15.5.2 Evaluating the Response: Predictions from the Fitted Model

In Section 11.2.5, we described the process of prediction for models containing explanatory factors. All of the considerations discussed there also apply here, but now we must also incorporate explanatory variates into the process. We start with the predictive model. When all of the explanatory variables are factors, we form predicted values for each combination of the factor levels, and can then take marginal means (with care, as described in Section 11.2.5) to summarize differences between groups. When there are also variates in the model, the considerations are slightly different. It can be helpful to just present the fitted line for each group either in mathematical form (as at the end of Example 15.4B) or graphically with confidence intervals. Questions about differences in slope between groups are often best answered by directly testing these differences. Comparisons between groups can be more difficult. In a parallel lines model, comparisons between groups will be the same for any fixed value of the explanatory variates, and usually predictions are made at the sample mean of the variates so that the predictions are readily interpreted. In a separate lines model, differences between groups change according to the covariate value. Unless prediction is required at specific variate values, summary comparisons across groups are unlikely to be helpful and may be misleading. In either type of model, if the explanatory variates are distributed differently between groups, then it may be misleading to compare groups at a common value of the explanatory variates, as this may create predictions for combinations of variables that would never normally occur.

When making predictions from more complex models, you should therefore be careful to check that your predictions are meaningful and that any comparisons are interpretable.

15.6 The Connection between Factors and Variates

Up to this point, we have considered factors and variates as intrinsically different types of explanatory variables. In fact, a factor can be considered as a set of covariates with a particular structure, and in this section, we explain the connection in some detail. For simplicity, we start with the separate groups model of Section 15.3 written in mathematical form as

$$y_{jk} = \alpha_1 + v_j + e_{jk} .$$

(15.5)

Written in this form, we have several models; one for each group. To rewrite this as a single model in terms of the full set of parameters, we define a set of **dummy variates**. For a factor with t groups, we construct t dummy variates, labelled by subscript $l = 1 \ldots t$. The lth dummy variate corresponds to the lth group and its values are labelled by both the group, l, and the observation, jk, as d_{ljk}. The values of this dummy variate are 1 for observations belonging to the lth group, i.e. where $j = l$, and zero otherwise.

EXAMPLE 15.5A: CALCIUM POT TRIAL*

Consider the pot trial data of Examples 3.4 and 4.1. Four relative concentrations of calcium (A = 1, B = 5, C = 10, D = 20) were each applied to five individual plants growing in pots arranged as a CRD. At the end of the experiment, the total root length (cm) in each of the 20 pots was measured. The data and the set of dummy variates associated with the Calcium factor are in Table 15.19 and can be found in file CALCIUM2.DAT. The first variate, labelled d_1, takes value 1 for observations with treatment A ($j = 1$), and value 0 elsewhere. The remaining variates, d_2, d_3 and d_4, are formed similarly in relation to calcium treatments B, C and D, respectively.

TABLE 15.19

Calcium Pot Trial Data from Table 5.6 with Additional Labelling in Terms of Calcium Treatment (j) and Replicate (k) and Dummy Variates d_1–d_4 Corresponding to the Calcium Factor (Example 15.5A)

Pot	Calcium	Length	j	k	d_1	d_2	d_3	d_4
1	D	47	4	1	0	0	0	1
2	A	58	1	1	1	0	0	0
3	B	80	2	1	0	1	0	0
4	C	49	3	1	0	0	1	0
5	D	49	4	2	0	0	0	1
6	A	52	1	2	1	0	0	0
7	D	45	4	3	0	0	0	1
8	A	74	1	3	1	0	0	0
9	C	70	3	2	0	0	1	0
10	B	68	2	2	0	1	0	0
11	A	58	1	4	1	0	0	0
12	C	72	3	3	0	0	1	0
13	A	79	1	5	1	0	0	0
14	D	48	4	4	0	0	0	1
15	C	74	3	4	0	0	1	0
16	B	72	2	3	0	1	0	0
17	D	38	4	5	0	0	0	1
18	C	71	3	5	0	0	1	0
19	B	74	2	4	0	1	0	0
20	B	85	2	5	0	1	0	0

We can now rewrite the separate groups model in terms of the dummy variates as

$$y_{jk} = \alpha_1 + v_1 d_{1jk} + v_2 d_{2jk} + \ldots + v_t d_{tjk} + e_{jk} \,. \tag{15.6}$$

Because each unit belongs to exactly one group, only one of the values d_{1jk} to d_{tjk} is 1, with the values of the other dummy variates being 0. For example, for the first observation in group 2, we have $d_{221} = 1$ and $d_{2j1} = 0$ for $j \neq 2$. Hence

$$y_{21} = \alpha_1 + (v_1 \times 0) + (v_2 \times 1) + \ldots + (v_t \times 0) + e_{21}$$
$$= \alpha_1 + v_2 + e_{21} \,,$$

and so the model for the first observation in group 2 (or any other observation in any other group) is equivalent to the form given in Equation 15.5. In the new form of Equation 15.6, often called the **dummy variate representation**, the separate groups model looks like a MLR in terms of the dummy variates $d_1 \ldots d_t$ and can be written in symbolic form as

Explanatory component: $[1] + (d_1 + \ldots + d_t)$

We use parentheses () to emphasize the associations within the set of dummy variates that represent a factor. As explained previously, this model is over-parameterized. In this form, there is no information left after the first $t - 1$ dummy variates have been fitted, leading to last-level-zero constraints being imposed by default. We prefer to use first-level-zero constraints, so we impose $v_1 = 0$ and omit the first dummy variate from the model, giving symbolic form

Explanatory component: $[1] + (d_2 + \ldots + d_t)$

The main difference between a MLR and this model is that we add all of the dummy variates corresponding to a factor (those in parentheses) into the model as a group, to obtain a combined incremental sum of squares for the term, rather than adding the dummy variates individually.

EXAMPLE 15.5B: CALCIUM POT TRIAL*

The separate groups model for the calcium pot trial data can be specified with the dummy variates given in Table 15.19, with first-level-zero parameterization, as

Explanatory component: $[1] + (d_2 + d_3 + d_4)$

The model SS is 2462.95, giving the ANOVA table in Table 15.20 which matches the ANOVA table previously obtained (see Table 5.5).

The dummy variate representation can be used whenever a factor appears in a model. So, for example, the separate lines model for a factor grp with three groups (coded as dummy variates d_1, d_2 and d_3) can be written in symbolic form as either

Explanatory component: $[1] + x + \text{grp} + x.\text{grp}$

or equivalently, explicitly imposing first-level-zero constraints, as

Explanatory component: $[1] + x + (d_2 + d_3) + (x.d_2 + x.d_3)$

TABLE 15.20

Sequential ANOVA Table for the Separate Groups Model Fitted to the Calcium Pot Trial Data Using Dummy Variates d_2–d_4 (Example 15.5B)

Terms Added	Incremental df	Incremental SS	Mean Square	Variance Ratio	P
+ $(d_2 + d_3 + d_4)$	3	2462.95	820.98	10.753	< 0.001
Residual	16	1221.60	76.35		
Total	19	3684.55			

SS = sum of squares.

Again, terms within parentheses are added into the model as a group. Terms of the form $x.d_2$, composed of two variates, are equivalent to a single variate calculated by multiplication of the values of the contributing variates together for each observation. We now also have two ways to write our separate lines model in mathematical form. The following two forms are equivalent, and both use first-level-zero constraints, with $v_1 = 0$ and $\eta_1 = 0$:

$$y_{jk} = (\alpha_1 + v_j) + (\beta_1 + \eta_j)x_{jk} + e_{jk} \, ,$$

$$y_{jk} = (\alpha_1 + v_1 d_{1jk} + v_2 d_{2jk} + v_3 d_{3jk}) + (\beta_1 + \eta_1 d_{1jk} + \eta_2 d_{2jk} + \eta_3 d_{3jk})x_{jk} + e_{jk} \, .$$

Similarly to the symbolic form, composite terms of the form $d_{2jk}x_{jk}$ can be considered as a single value calculated by multiplication of the component values.

EXAMPLE 15.1K: STAND DENSITY OF MIXED NOTHOFAGUS FOREST PLOTS

The three dummy variates $(d_1, d_2$ and $d_3)$ required to represent the Type factor in the stand density data set are presented in Table 15.21 and can be found in file FOREST2.DAT. Calculation of the composite terms $logQD.d_1$, $logQD.d_2$ and $logQD.d_3$ are also shown in Table 15.21 as the product of the variates d_1, d_2 and d_3, respectively, with $logQD$.

The separate lines model for these data can be written in symbolic form either in terms of the Type factor as

Explanatory component: $[1] + logQD + \text{Type} + logQD.\text{Type}$

or using the dummy variates and first-level-zero constraints as

Explanatory component: $[1] + logQD + (d_2 + d_3) + logQD.(d_2 + d_3)$
  $= [1] + logQD + (d_2 + d_3) + (logQD.d_2 + logQD.d_3)$

Fitting the model in terms of factor Type gave the ANOVA table shown in Table 15.5. Using the dummy variates, and fitting the terms in parenthesis together gives the same sequential ANOVA table, as shown in Table 15.22. The parameter estimates are the same as those obtained in Table 15.3 in both cases, although now labelled by the dummy variates rather than by the factor levels.

The interpretation of factors as a set of dummy variates allows both types of explanatory variable to be considered within a single framework, which facilitates a unified

TABLE 15.21

Dummy Variates (d_1, d_2, d_3) and Their Products with Explanatory Variate *logQD* (*logQD.d$_1$*, *logQD.d$_2$*, *logQD.d$_3$*) for Three Groups (Factor Type) of Mixed *Nothofagus* Forest Calculated for Four Plots in Each Group (Example 15.1K)

Plot	Type	*logQD*	d_1	d_2	d_3	*logQD.d$_1$*	*logQD.d$_2$*	*logQD.d$_3$*
1	Rauli	2.57	0	1	0	0	2.57	0
2	Rauli	2.70	0	1	0	0	2.70	0
⋮	⋮	⋮	⋮	⋮	⋮	⋮	⋮	⋮
8	Rauli	2.98	0	1	0	0	2.98	0
9	Rauli	2.44	0	1	0	0	2.44	0
10	Roble	2.48	0	0	1	0	0	2.48
11	Roble	3.12	0	0	1	0	0	3.12
⋮	⋮	⋮	⋮	⋮	⋮	⋮	⋮	⋮
27	Roble	2.67	0	0	1	0	0	2.67
28	Roble	2.53	0	0	1	0	0	2.53
29	Coigue	3.10	1	0	0	3.10	0	0
30	Coigue	3.42	1	0	0	3.42	0	0
⋮	⋮	⋮	⋮	⋮	⋮	⋮	⋮	⋮
40	Coigue	2.65	1	0	0	2.65	0	0
41	Coigue	2.98	1	0	0	2.98	0	0

TABLE 15.22

Sequential ANOVA Table for Separate Lines Model for Logged Stand Density Using Explanatory Variate *logQD* and Dummy Variates d_2, d_3 to Represent Factor Type (Example 15.1K)

Term Added	Incremental df	Incremental SS	Mean Square	Variance Ratio	P
+ *logQD*	1	4.5833	4.5833	78.562	< 0.001
+ ($d_2 + d_3$)	2	1.9403	0.9701	16.629	< 0.001
+ *logQD*.($d_2 + d_3$)	2	0.2011	0.1006	1.724	0.193
Residual	35	2.0419	0.0583		
Total	40	8.7667			

SS = sum of squares.

mathematical treatment of linear models. The terms [1] and *[1]*, which both represent a vector with value 1 in all units, can be considered as equivalent. This framework allows statistical models to be written in matrix notation, and we give a very brief introduction to this topic below.

15.6.1 Rewriting the Model in Matrix Notation

So far, we have written our models in terms of individual observations. The use of dummy variates with matrix notation allows a succinct representation of the model for the whole set of observations simultaneously. Consider the separate groups model of Equation 15.6, and suppose we have six observations in each of the t groups. We can write the model for all observations, ordering by groups and then by observations within groups as

$$y_{11} = \alpha_1 + v_1 d_{111} + v_2 d_{211} + \ldots + v_t d_{t11} + e_{11}$$
$$y_{12} = \alpha_1 + v_1 d_{112} + v_2 d_{212} + \ldots + v_t d_{t12} + e_{12}$$
$$\vdots$$
$$y_{t5} = \alpha_1 + v_1 d_{1t5} + v_2 d_{2t5} + \ldots + v_t d_{tt5} + e_{t5}$$
$$y_{t6} = \alpha_1 + v_1 d_{1t6} + v_2 d_{2t6} + \ldots + v_t d_{tt6} + e_{t6}$$

(15.7)

We can then abbreviate this rather lengthy form using **matrix notation**. A matrix is simply a rectangular array of numbers, with rules for addition and multiplication that are explained in Section C.5. Our model can then be written as

$$\mathbf{y} = \mathbf{X}\boldsymbol{\tau} + \mathbf{e},$$

where $\mathbf{y}$ is a matrix with N rows and 1 column (a vector of length N) containing the observations, $\mathbf{X}$ is a matrix with N rows and $t + 1$ columns (an $N \times (t + 1)$ matrix) containing the known coefficients associated with the parameters, $\boldsymbol{\tau}$ is a matrix with $t + 1$ rows and 1 column (a vector of length $t + 1$) and $\mathbf{e}$ is a matrix with N rows and 1 column (a vector of length N) containing the deviations. These matrices are defined as follows:

$$\mathbf{y} = \begin{pmatrix} y_{11} \\ y_{12} \\ \vdots \\ y_{t5} \\ y_{t6} \end{pmatrix}; \quad \mathbf{X} = \begin{pmatrix} 1 & d_{111} & d_{211} & \cdots & d_{t11} \\ 1 & d_{112} & d_{212} & \cdots & d_{t12} \\ \vdots & \vdots & \vdots & & \vdots \\ 1 & d_{1t5} & d_{2t5} & \cdots & d_{tt5} \\ 1 & d_{1t6} & d_{2t6} & \cdots & d_{tt6} \end{pmatrix}; \quad \boldsymbol{\tau} = \begin{pmatrix} \alpha_1 \\ v_1 \\ v_2 \\ \vdots \\ v_t \end{pmatrix}; \quad \mathbf{e} = \begin{pmatrix} e_{11} \\ e_{12} \\ \vdots \\ e_{t5} \\ e_{t6} \end{pmatrix}.$$

The rules for matrix multiplication given in Section C.5 mean that this short form expands to give the full mathematical model in Equation 15.7. The matrix of coefficients, $\mathbf{X}$, is usually called the **design matrix** and its columns correspond to the explanatory variates. In this example, the first column corresponds to the overall constant term, denoted in our symbolic form as [1], and the 2nd to $(t + 1)$th columns contain the dummy variates for the t factor levels, denoted earlier as $d_1 \ldots d_t$. All of the elements in this case are therefore either 0 or 1. For Example 15.5A, with four treatment groups, the 2nd to 5th columns of the design matrix correspond to the values given in the last four columns of Table 15.19. Another simple example is the SLR model of Section 12.1, which takes the form

$$\mathbf{y} = \begin{pmatrix} y_1 \\ y_2 \\ \vdots \\ y_{N-1} \\ y_N \end{pmatrix}; \quad \mathbf{X} = \begin{pmatrix} 1 & x_1 \\ 1 & x_2 \\ \vdots & \vdots \\ 1 & x_{N-1} \\ 1 & x_N \end{pmatrix}; \quad \boldsymbol{\tau} = \begin{pmatrix} \alpha \\ \beta \end{pmatrix}; \quad \mathbf{e} = \begin{pmatrix} e_1 \\ e_2 \\ \vdots \\ e_{N-1} \\ e_N \end{pmatrix}.$$

We can use this matrix notation for models with any combination of factors and variates. Using this notation, parameter estimates and SEs can be written in a general form in terms of matrix operations on the design matrix, $\mathbf{X}$, and the response vector, $\mathbf{y}$, and this notation is therefore widely used in textbooks on mathematical statistics. We do not go into this further here but more details can be found in Mead et al. (2012) or Montgomery et al. (2012).

EXERCISES

15.1   The biomass data (wet weights in g) for all four sites in the study described in Exercise 12.2 (and Exercise 13.4) are held in file ALLSITES.DAT (variate *ID*, factor Site, variates *Year, WetWeight*). Use regression with groups on this combined data set to investigate whether the trend found at Hereford is the same as for the other three sites. Present a summary of the results from your analysis.

15.2   In Exercise 12.5, you fitted a SLR to the log-transformed body mass of a sample of moths with explanatory variate wing length and found evidence for lack of fit in the relationship. However, that SLR ignored the information on the species of each sample that was also recorded (in data file NOCTUID.DAT). Use this species information to investigate whether the relationship between log-transformed body mass and wing length is consistent across species. Test for lack of fit in your model and compare your results with those from Exercise 12.5. Can you reconcile the two analyses? Specify and interpret your final predictive model.

15.3   Many plant pathogens are dispersed through the crop by rain splash. To investigate the likely distance of travel, water drops of different sizes (weights) were dropped from various heights to give different velocities on impact. The average height of splash was measured for each combination of drop size and height. File SPLASH.DAT contains unit numbers (*ID*), the weight (variate *Weight*) and estimated terminal velocity (variate *Velocity*) for each run with the mean splash height (variate *MeanHt*). The aim of analysis is to predict splash height from drop velocity on impact. Form groups for the different weight classes and establish whether a common model across weight classes is appropriate. Would use of a common line lead to any erroneous conclusions? (We re-visit these data in Exercise 17.10.)[*]

15.4   In Example 15.1, we explored models to predict the density of stands of three types from sample measurements of quadratic diameter, with models based on the log-transform of both the response and explanatory variables. We identified a parallel lines model as being most suitable for this data (Examples 15.1F and H). Using the results of Section 6.4, rewrite this parallel lines model in terms of the stand density and interpret the difference between stand types on this scale.

15.5   A study was done to investigate the recovery of spring-applied fertilizer N in the harvested products of three arable crops (Macdonald et al., 1997): winter wheat, oilseed rape and potatoes. The recovery (% of applied N) was measured in potato tubers, rapeseed and wheat grain and here we investigate whether fertilizer recovery can be predicted by harvest index. The file RECOVERY.DAT contains sample numbers (*ID*), crop type (factor *Crop*), harvest index (variate *HIndex*) and fertilizer recovery rates (variate *Recovery*, %) from 8 plots of

---

[*] Data from Rothamsted Research.

wheat, 4 plots of potatoes and 12 plots of oilseed rape. Plot the data and discuss whether regression with groups is a sensible approach.[*]

15.6   As part of a study to quantify phosphorus (P) use efficiency in crops, data on the increase in Olsen P and increase in total P in 52 plots in long-term experiments across three sites with different soil types were compiled (Johnston et al., 2001). The file P.DAT contains index numbers (*Plot*), the site name (factor Site) and measurements of increase in Olsen P (variate *IncOlsenP*) and increase in total P (variate *IncTotalP*) for each plot. Investigate whether the increase in total P can predict the increase in Olsen P, and whether this relationship differs between soil types (sites). Find the simplest adequate model to describe these data, and write down and interpret your final predictive model.[†]

15.7   The Julian date of the last record of the aphid *Myzus ascalonicus* (shallot aphid) in the Insect Survey suction trap at Rothamsted was obtained for 1968 to 2005 (inclusive). Years could be classified as either early (last record < 210) or late (last record > 280). These groupings may be linked to the abundance of winged aphids in autumn, with early years corresponding to small (or absent) autumn migrations. Data file SHALLOT.DAT holds unit numbers (*ID*) with the year (*Year*), date of last observation (*JDate*) and classification as an early or late year (Group). Use regression with groups to establish whether there is any statistical evidence that the date of last observation (response *JDate*) is changing over time (explanatory variate *Year*) and whether any trend over time differs between early and late years (factor Group). Check for evidence of temporal correlation. Write down your predictive model and report your conclusions.[‡]

15.8   In Example 8.6, we analysed a designed experiment to investigate the affinity of a sugar transporter protein for a substrate within plant cells. We modelled the relationship between response $\log_e(Km)$ and the equivalent voltage using polynomial contrasts within ANOVA. The unit numbers (*ID*), structural factors (Rep, DUnit), input voltage (variate *Voltage*) and response (variate *Km*) are held in file VOLTAGE.DAT (Table 8.23). Refit the model as a linear regression, including replicates in the model. Is there any evidence of model misspecification? Check for lack of fit and verify that this gives the same results achieved in Example 8.6. Do you agree with the conclusions from our original analysis?

15.9   The impacts of several methods of forming ground cover in apple orchards were compared in a designed experiment (Pearce, 1983). The standard method (code O) was compared to five types of permanent crops (codes A–E). The experiment used four blocks of six trees, and treatments were allocated at random to trees within each block (a RCBD). The trees were old, and varied in productivity, and so their total yield (bushels) over the previous 4 years was provided as a covariate. The response was total yield (in pounds) over a 4-year period with the new treatments. File APPLE.DAT contains unit numbers (*ID*), the structural factors (Block, DPlot), treatment codes (factor Trt) and the crop from each tree before (variate *PrevCrop*) and during the experiment (variate *TotalCrop*). Investigate the impact of the treatments, taking both the design and the covari-

---

[*] Data from A. Macdonald, Rothamsted Research.
[†] Data from A.E. Johnston, Rothamsted Research.
[‡] Data from R. Harrington, Rothamsted Research.

ate into account. Is there any impact of including the covariate in the analysis? Which treatments would you recommend?

15.10 An experiment investigated the oxygen consumption of wireworm larvae at several temperatures (Bliss, 1970, Exercise 20.2). Consumption was expected to vary with larval size, so uniform batches of larvae of different sizes were tested and their mean weights were recorded. File OXYGEN.DAT contains unit numbers (*Unit*), the temperature group for each batch (factor Temperature), and the natural logarithms of mean bodyweight (variate *logBodyWt*, mg) and oxygen consumption per individual (variate *logConsumption*, mL/h). Is there any evidence of differences in oxygen consumption between temperature groups after taking body weight into account? What do you need to check before you can answer this question? Write down a predictive model for oxygen consumption at each temperature, and interpret the differences between temperatures. (We re-visit these data in Exercise 17.8.)

16

Incorporating Structure: Linear Mixed Models

In Chapters 7 to 9, we showed how to analyse data arising from designed experiments using multi-stratum ANOVA to take proper account of structure in the experimental units and that this approach led to appropriate estimates of parameter standard errors. However, multi-stratum ANOVA does not apply to unbalanced structures, and so its use is limited. In Chapter 11, we saw that combining the explanatory and structural components of the model – the so-called intra-block analysis – gives good results only for certain types of structure. We therefore need a more general approach to account for structure, and in this chapter, we introduce the class of linear mixed models. This class extends multi-stratum ANOVA to the cases of unbalanced and non-orthogonal structures, and extends regression models to include a structural component. We start with a short discussion of the need to include structure in models (Section 16.1) and then give a more formal definition of the linear mixed models that we use to achieve this (Section 16.2). We then describe methods for investigating the explanatory component of the model (Section 16.3) and aspects of the structural component (Sections 16.4 and 16.5) before considering prediction (Section 16.6) and model checking (Section 16.7). We analyse a data set in some detail to illustrate the concepts discussed in the previous sections (Section 16.8), and we explain some of the difficulties that can be encountered with this more general form of model (Section 16.9). Finally, we give a general overview of extensions to this class of models (Section 16.10).

16.1 Incorporating Structure

In Chapters 7 and 9, we analysed experimental studies with structure such as hierarchical blocking and pseudo-replication. We specified models using two separate components: the explanatory component was used to describe the relationship between the explanatory variables and the response, and the structural component was used to describe structure present in the observations. We argued that incorporation of the structure is required to generate the correct parameter SEs and df for hypothesis testing, and we achieved this with multi-stratum ANOVA.

In observational studies, structure is also often present and should be accounted for. For example, consider a large-scale ecological survey taken across fields growing several types of crop within designated farms in a region. The farms are not of interest in themselves, as they are intended to provide a representative sample, but systematic differences between farms are expected as a result of local management practices and so farms are regarded as a structural factor. Some explanatory variables might apply to whole farms, for example, type of farm, while others might be measured on individual fields, for example, crop (qualitative) or field area (quantitative). Incorporating the structure of the observations (in this case, Farm/Field) ensures that explanatory terms are compared to background variation in the correct level or stratum. Similarly, within the same study, several samples might be taken within each field to avoid bias due to small-scale variation. In the terminology of Section 3.1.1, the within-field samples are pseudo-replicates. Separation of within- and between-field variation (using the structural component Farm/Field/Sample) is required to assess accurately the precision of estimates for the effects of different crops and the relationship with field area.

Some statistical packages include algorithms for multi-stratum ANOVA which allow specification of both the explanatory and structural components of the model. However, algorithms for multi-stratum ANOVA require an orthogonal structure and balanced allocation of treatments (see Chapter 11) which are rarely present in observational studies and sometimes not in more complex designed studies. In Chapter 11, we demonstrated the principles of combining the two components of the model (the intra-block analysis), and showed that this approach is appropriate when most of the treatment information occurs at the lowest level of the structure, but can be problematic otherwise, particularly when treatments are applied at higher levels or when pseudo-replication is present (details in Section 11.6.1). In these cases, it will often be better to use a **linear mixed model** (LMM), specifying the model using two components, which are usually called the fixed and random models. For both experimental and observational studies, it is usually reasonable to allocate the terms of the structural component as random and the explanatory component terms as fixed. In general, the choice of which terms to classify as fixed and which as random depends on the aims of analysis, and we discuss this further in Section 16.5. A major advantage of LMMs is that the structure does not have to be balanced, and the model may contain any mixture of factors and variates. In the remainder of this chapter, we describe briefly the analysis of LMMs, illustrated with two examples.

16.2 An Introduction to Linear Mixed Models

A LMM is defined by the response, a **fixed model** and a **random model**. As stated above, here we equate the fixed model with the explanatory component, describing treatments or conditions that may affect the response, and equate the random model with the structural component, describing any structure present in the study.

EXAMPLE 16.1A: WEED COMPETITION EXPERIMENT

This experiment was introduced in Example 9.5, the layout was shown in Table 9.10 and the data are in file COMPETITION.DAT. This split-plot experiment investigated the competitive effects of weeds (factor Species), with and without irrigation (factor Irrigation), on grain yield of winter wheat (variate *Grain*), with four blocks (factor Block). Within each block, two irrigation regimes were applied to whole plots (factor WholePlot), each of which was split into four subplots (factor Subplot) in which the different weed species (no weeds, Am, Ga, Sm) were sown. This experiment has a nested structure with three strata: blocks, whole plots within blocks, and subplots within whole plots, i.e. Block/WholePlot/Subplot. The explanatory component was a two-way crossed structure, i.e. [1] + Irrigation*Species. Irrigation effects were estimated within the whole-plot stratum and species effects and the interaction were estimated within the subplot stratum. The LMM for this design translates directly from the explanatory and structural components as

> Response variable: *Grain*
> Fixed model: [1] + Irrigation*Species
> Random model: Block/WholePlot/Subplot

The assumptions behind the LMM differ slightly from those we have used previously. The fixed model is set up in exactly the same way as the explanatory component, usually with first-level-zero (or last-level-zero) parameterization (Section 11.2.1). The random model has a new set of assumptions, however. The effects associated with each random term are assumed to be a set of independent samples from a Normal distribution with a common variance, which is known as the **variance component** for that term. We have previously assumed that structural terms represent variation due to the physical structure of the experimental material or procedure: for continuous data it is then often reasonable to interpret these effects as samples from a Normal distribution. The model deviations become just one of these random terms, and the assumptions made for the deviations (Sections 4.1 and 12.1) also apply to each of the random terms. In addition, it is assumed that effects from different random terms are independent.

EXAMPLE 16.1B: WEED COMPETITION EXPERIMENT

The mathematical model for this split-plot experiment, with first-level-zero parameterization and a crossed treatment structure, can be written as

$$Grain_{ijk} = \mu_{11} + Block_i + Irrigation_j + (Block.WholePlot)_{ij} + Species_k$$
$$+ (Irrigation.Species)_{jk} + e_{ijk} , \tag{16.1}$$

where $Grain_{ijk}$ is the grain yield for the kth weed species ($k = 1 \ldots 4$; $1 =$ no weeds, $2 =$ Am, $3 =$ Ga, $4 =$ Sm) with the jth irrigation treatment ($j = 1, 2$; $1 =$ without, $2 =$ with irrigation) in the ith block, for $i = 1 \ldots 4$. The first-level-zero constraints impose $Irrigation_1 = 0$, $Species_1 = 0$ and $(Irrigation.Species)_{jk} = 0$ for $j = 1$ or $k = 1$. Parameter μ_{11} represents the population mean without irrigation or weeds, $Irrigation_2$ represents the effect of irrigation with no weeds, $Species_k$ ($k = 2 \ldots 4$) represents the effect of the kth weed species without irrigation, and $(Species.Irrigation)_{2k}$ is the effect of irrigation on the kth weed species relative to the effect of irrigation without weeds (see Section 11.2.1). The effects $Block_i$, $i = 1 \ldots 4$, are random block effects, assumed to be independent with common distribution $Block_i \sim$ Normal$(0, \sigma_b^2)$, the effects $(Block.WholePlot)_{ij}$ are random effects of whole plots within blocks, assumed to be independent with common distribution $Block.WholePlot_{ij} \sim$ Normal $(0, \sigma_w^2)$, and the deviations, e_{ijk}, are assumed to be independent with distribution $e_{ijk} \sim$ Normal$(0, \sigma^2)$, as mentioned previously. The variance components in this model are σ_b^2, σ_w^2 and σ^2.

The parameters of the LMM are the effects associated with the fixed terms and the variance components associated with the random terms. The random effects have a slightly different status, which is discussed further in Section 16.5. There is no requirement for a LMM to have a balanced structure, and so estimation by least squares is not always efficient. The usual alternative, maximum likelihood estimation, gives biased estimates of the variance components and so Patterson and Thompson (1971) introduced a method called restricted (or residual) maximum likelihood (REML) to estimate the variance components, and this is the approach we take. The method estimates the variance components by minimizing a quantity called the restricted (or residual) log-likelihood function (for more details, see Littell et al., 2006). The fixed effects are then estimated by the method of generalized least squares, conditional on the estimated values of the variance components. One advantage of the REML method is that it gives the same estimates of fixed effects and SEs as obtained from multi-stratum ANOVA when the structure is balanced and, where treatment information is divided across strata, estimates will be combined efficiently across strata into a single estimate.

16.3 Selecting the Best Fixed Model

The estimates of the fixed effects used with REML are often called BLUEs, which is an acronym for **best linear unbiased estimates**. The property of unbiasedness means that the expected value of the estimator is equal to the true parameter value. In this context, 'best' means that these estimates have minimum variance within the class of unbiased estimators, conditional on the variance components. In practice, we do not know the true values of the variance components and so substitute their REML estimates to obtain empirical BLUEs, often called eBLUEs.

EXAMPLE 16.1C: WEED COMPETITION EXPERIMENT

Fitting the split-plot experiment as a LMM with first-level-zero parameterization gives the estimates of fixed effects for terms Species and Irrigation.Species shown in Table 16.1. The estimate of the constant was $\hat{\mu}_{11} = 8.117$ (SE 0.4063) and the estimated effect of irrigation in the absence of weeds was $\overline{Irrigation}_2 = -0.935$ (SE 0.5344).

As described in Chapters 8 and 11, we usually wish to investigate the contribution of individual terms within the explanatory component (or fixed model) in explaining patterns of response. We need to take proper account of the structural component (or random model) and any non-orthogonality in the explanatory component (fixed model). Because it is not possible to construct a multi-stratum ANOVA table for a general unbalanced structure, in LMMs we take a slightly different approach and construct test statistics that account for the experimental structure.

Because of non-orthogonality, we still need to consider both incremental and marginal forms of these statistics; recall that incremental statistics reflect the change in fit on sequential addition of individual terms into the fixed model, and marginal statistics reflect the change on omission of individual terms from the full fixed model (see Sections 11.2 and 14.4). Recall that in a non-orthogonal structure, there may be many different sets of incremental and marginal statistics, corresponding to different orders of adding terms into or

TABLE 16.1

Estimated Fixed Effects with Standard Errors (SE) for Terms Species and Irrigation.Species with First-Level-Zero Parameterization in the Weed Competition Experiment (Example 16.1C)

Term	Parameter	– ($k = 1$)	Am ($k = 2$)	Ga ($k = 3$)	Sm ($k = 4$)	SE
Species	$Species_k$	0.000	−4.632	−1.437	−1.522	0.3613
Irrigation.Species	$(Irrigation.Species)_{2k}$	0.000	0.160	−1.670	−0.085	0.5109

dropping terms from the model (see Section 11.2). We again follow the principles of model selection discussed in Section 15.5.1. In particular, we respect the principle of marginality and add a term only if all marginal terms are already present in the model (e.g. add A.B in a crossed structure only if A and B are both present), and do not drop terms that are marginal to other terms present in the model (e.g. do not drop A or B if A.B is in the model). Here, we describe two types of test statistic in common usage in LMMs, Wald tests and approximate F-tests.

For a model term associated with a single effect, the **Wald statistic** is equivalent to the square of the t-statistic obtained by division of the estimated effect by its estimated standard error. When several effects are associated with a model term, the calculation is more complex. If the structure is orthogonal, then the Wald statistic is equivalent to the sum of squares for that term divided by the ResMS from the appropriate stratum. In the general unbalanced case, the marginal Wald statistic for a term is effectively the sum of squares of its estimated effects weighted by their estimated variance–covariance matrix. Under the null hypothesis of zero effects, on the assumption that the variance components are known, the Wald statistic has an approximate chi-squared distribution with df equal to the change in df when the term is added to the model (for an incremental test) or removed from the model (for a marginal test). This is a one-sided test, as estimates with either a large positive or negative value (with respect to their variance–covariance matrix) lead to a large positive value of the Wald statistic. As this distribution ignores the sampling variation associated with estimation of the variance components, it is analogous to the use of a Normal distribution rather than a t-distribution for the test for a single parameter estimate. We can see the impact of this approximation in the following example.

EXAMPLE 16.1D: WEED COMPETITION EXPERIMENT

This split-plot design is orthogonal, so there is a unique set of incremental Wald statistics, and these are shown in the third column of Table 16.2. These statistics can be verified in each case to be equal to the SS for the term divided by the ResMS from the appropriate stratum (see the ANOVA in Table 9.12). The observed significance levels for the Wald statistics (column 4 in Table 16.2) are smaller than those from the multi-stratum ANOVA table and, although the conclusions do not change, the strength of the evidence from the Wald tests appears greater. However, if the assumptions for the deviations are true, then the variance ratios from the multi-stratum ANOVA have an F-distribution, giving a known baseline for comparison and indicating that the Wald tests are over-confident.

Some caution is therefore required in the use of Wald tests, which tend to be too optimistic, i.e. to give false-positive results more often than would be expected. The reference

TABLE 16.2

Wald Statistics with Observed Significance Levels (P(Wald)) and Approximate F-Statistics with Estimated Denominator df (ddf) and Observed Significance Levels (P(F)) for the Weed Competition Experiment (Example 16.1D)

Term	df	Wald	P (Wald)	F	ddf	P (F)
+ Irrigation	1	9.480	0.002	9.480	3.0	0.054
+ Species	3	329.178	4.8×10^{-71}	109.726	18.0	9.3×10^{-12}
+ Irrigation.Species	3	16.747	8.0×10^{-4}	5.582	18.0	0.007

distribution for this test is known as an **asymptotic approximation**, which indicates that it holds only for large samples. In fact, the phrase 'large samples' is slightly misleading here, as the requirement is more specifically for the uncertainty in the variance–covariance matrix of the estimates to be small. This is difficult to check in a general situation, but in balanced situations requires that the ResDF should be large within strata where the fixed terms are tested.

To avoid this problem, various methods exist to convert Wald statistics into a form that has an approximate F-distribution, with denominator df that quantify uncertainty in the estimation of variances. The most popular method was developed by Kenward and Roger (1997, 2009). This method re-scales the Wald statistic so that it can be compared to an F-distribution with numerator df equal to those of the model term and an estimated denominator df. As with the Satterthwaite approximation (Section 9.2.3), the estimated denominator df will often be non-integer. For balanced designs, F-tests based on the Kenward–Roger method are identical to F-tests based on the variance ratios. This method is available in most software for LMMs, and these approximate F-tests should usually be preferred to the Wald tests.

EXAMPLE 16.1E: WEED COMPETITION EXPERIMENT

The fifth and sixth columns of Table 16.2 show the F-statistics and denominator df derived from the Kenward–Roger method. In this balanced case, the derived F-tests can be obtained by division of the Wald statistic for each model term by its df, and the estimated denominator df are equal to the ResDF from the appropriate stratum in the ANOVA table (Table 9.12). The resulting F-tests and observed significance levels (column 7 in Table 16.2) therefore exactly match those from the multi-stratum ANOVA table.

16.4 Interpreting the Random Model

The variance components associated with the random terms generate a variance–covariance matrix for the observations. The variance of an observation is equal to the sum of the variance components, and it can be derived from the algebraic form of the model: the variance of the fixed effects is zero, and the variance of each random effect equals its variance component. The covariance between any two observations depends on the random effects held in common across the observations, and it is the sum of the variance components for these common random effects. These calculations are illustrated in the following example.

EXAMPLE 16.1F: WEED COMPETITION EXPERIMENT

The estimated variance components for the weed competition experiment are in Table 16.3. The block variance component is smaller than the whole-plot and subplot variance components, which are similar in size.

To estimate the variance of a single observation from this experiment, we start with the model in Equation 16.1. The fixed terms do not contribute to the variance, and we have assumed that all random effects are independent (both within and across terms), so we do not need to account for covariances between random effects, which are all zero. The variance is thus equal to the sum of the variances of the random effects, which is the sum of the variance components, i.e.

TABLE 16.3

Estimated Variance Components for the Weed Competition Experiment (Example 16.1F)

Term	Parameter	Estimate	SE
Block	σ_b^2	0.0893	0.2732
Block.WholePlot	σ_w^2	0.3100	0.3072
Block.WholePlot.Subplot	σ^2	0.2610	0.0870

SE = estimated standard error.

$$
\begin{aligned}
\text{Var}(Grain_{ijk}) &= \text{Var}(Block_i + Block.WholePlot_{ij} + e_{ijk}) \\
&= \text{Var}(Block_i) + \text{Var}(Block.WholePlot_{ij}) + \text{Var}(e_{ijk}) \\
&= \hat{\sigma}_b^2 + \hat{\sigma}_w^2 + \hat{\sigma}^2 \\
&= 0.0893 + 0.3100 + 0.2610 \\
&= 0.6604 .
\end{aligned}
$$

This estimated variance is the same for all observations. The estimated covariance between observations from different subplots within the same whole plot (and hence the same block) can be derived similarly. Again, only the random terms contribute to the covariance and as we have assumed the random effects are independent, covariances between different effects are zero:

$$
\begin{aligned}
\text{Cov}(Grain_{ijk}, Grain_{ijl}) &= \text{Cov}(Block_i + Block.WholePlot_{ij} + e_{ijk}, Block_i + Block.WholePlot_{ij} + e_{ijl}) \\
&= \text{Var}(Block_i) + \text{Var}(Block.WholePlot_{ij}) \\
&= \hat{\sigma}_b^2 + \hat{\sigma}_w^2 \\
&= 0.3994 .
\end{aligned}
$$

The estimated covariance between observations from subplots in different whole plots within the same block is then equal to the estimated block variance component, 0.0893, and the covariance between observations from subplots in different blocks is zero. The covariance between observations therefore increases as their proximity within the hierarchical structure also increases.

Our original definition of the variance components, as variances of the random effects, required these variances to be positive. The interpretation of the variance structure in terms of variances and covariances between observations requires only that the total variance is positive and that the variance of any linear combination of observations is also positive (this property is known as positive-definiteness). In general, we use random terms to reflect structure and we expect units with random effects in common to be more similar than units without, and so variance components are usually expected to be positive. But occasionally circumstances arise when it is natural to allow variance components to take negative values. For example, in field experiments, blocks are laid out on areas of ground thought to be reasonably homogeneous with respect to fertility and other trends. If a mistake is made, then plots within the same block may be less alike than plots in different blocks, and this can be modelled only by using a negative variance component for blocks. A similar effect can occur if shelves in a CE cabinet are used as blocks to account for differences in lighting, but in fact a temperature gradient from the front to the back of the shelves has a much stronger effect. For these reasons, we prefer to allow variance components to be negative when required. Even if the true values of the variance components are positive, it is possible that they may be estimated as negative values due to sampling variability, particularly for terms with few levels. Some statistical packages always constrain estimates of variance components to remain positive, bounded below at zero (e.g. R function lmer), while others (e.g. GenStat and SAS PROC MIXED) give a choice on whether estimates should be constrained to remain positive or not. The default action differs between packages, so you should always check the documentation.

The presentation of estimated variance parameters also differs between statistical packages: SAS PROC MIXED and GenStat present the variance components, but R function lmer presents the square root of the variance components (labelled as standard deviations), arguing that these are easier to interpret as they are on the same scale as the observations. Standard errors of the variance component estimates are often provided, and the estimated variance components are often small compared with these SEs (e.g. Table 16.3), so it might be natural to think of dropping these terms from the model. We advise against this course for two distinct reasons. First, the SEs for variance components are reliable for testing only when there is a large amount of information contributing to the estimate; again, the SEs depend on an asymptotic approximation. A better approach is the use of **likelihood ratio tests** (LRTs); however, these tests are still the subject of research and their description is outside the scope of this book. Second, and more importantly in our context, the random terms have been included to describe the structure of the observations. This structure is a property of the data set and is used to obtain the denominator df for approximate F-tests: the removal of terms means that the random model no longer serves this purpose. There are contexts in which it is appropriate to try to simplify a random model, but this is not the case when it represents the structural component.

There is one situation in which it may be sensible to allocate part of the structural component as fixed rather than random terms. This situation occurs when there are few levels in a random term, so its variance component is poorly estimated, and when the explanatory terms vary at a lower level of the structure. For example, a RCBD with many treatments might have only two replicate blocks. The estimate of the block variance component is effectively based on only two agglomerated observations (related to the two block effects) and is unlikely to be reliable. In this design, all the treatment comparisons are made within blocks, and no information is lost by putting block as the first term into the explanatory component (the intra-block analysis of Section 11.6.1) or, equivalently in the context of a LMM, into the fixed model.

The remainder of this section explains the relationship between the variance components obtained from a REML analysis and the stratum variances obtained by multi-stratum ANOVA for a balanced set of data.

16.4.1 The Connection between the Linear Mixed Model and Multi-Stratum ANOVA

Estimates of variance components from REML are equivalent to those from multi-stratum ANOVA when the structure is balanced. Within the ANOVA table, the variance components are hidden contributors to the stratum variances, which are estimated by the stratum ResMS. For a balanced nested structure, the relationship between the stratum variances and the variance components is straightforward: each stratum variance is constructed as a weighted sum of variance components relating to random effects from that stratum and from all lower strata. The weight for each variance component is the number of observational units corresponding to a single random effect from that term. As an example, we consider the standard split-plot structure with m blocks, each with t_A whole plots, each of which in turn contains t_B subplots. We denote the stratum variances for blocks, whole plots and subplots respectively as ξ_b, ξ_w and ξ_s. These stratum variances are related to the variance components as follows:

$$\xi_b = \sigma^2 + t_B\,\sigma_w^2 + t_A\,t_B\,\sigma_b^2 \,,$$
$$\xi_w = \sigma^2 + t_B\,\sigma_w^2 \,,$$
$$\xi_s = \sigma^2 \,.$$

The block.wholeplot.subplot stratum is the lowest level, with random effects equal to the model deviations. Each deviation corresponds to a single observation, and so the weight for the subplot variance component (σ^2) is 1; this holds for all strata. In the block.wholeplot stratum, there are contributions from the whole-plot and subplot random effects. There are t_B observations within each whole plot, so this is the weight for the whole-plot variance component (σ_w^2), and again it applies to all higher strata. The block stratum contains contributions from the block, whole-plot and subplot random effects. Each block contains $t_A \times t_B$ observations and so this is the weight for the block variance component (σ_b^2).

EXAMPLE 16.1H: WEED COMPETITION EXPERIMENT

We can relate the estimated variance components in Table 16.3 to the estimated stratum variances obtained in Example 9.2. As in Section 9.2.3, we denote the estimates of stratum variances provided by the Block, Block.WholePlot and Block.WholePlot.Subplot residual mean squares as s_b^2, s_w^2 and s^2, respectively. To derive estimates of the stratum variances, we use the formula given above, i.e.

$$s_b^2 = \hat{\sigma}^2 + 4\hat{\sigma}_w^2 + 8\hat{\sigma}_b^2 = 0.2610 + (4 \times 0.3100) + (8 \times 0.0893) = 2.2158 \,,$$
$$s_w^2 = \hat{\sigma}^2 + 4\hat{\sigma}_w^2 = 0.2610 + (4 \times 0.3100) = 1.5012 \,,$$
$$s^2 = \hat{\sigma}^2 = 0.2610 \,.$$

As expected, these estimates match the stratum ResMSs shown in Table 9.12.

From the formulae for the stratum variances, we can deduce that whenever the stratum variance for a term is smaller than that for strata lower in the hierarchy, the variance component associated with that term must be negative. When statistical packages constrain variance components to remain positive, bounded below at zero, the variance component estimates may then not quite match those from the multi-stratum ANOVA table. Although the resulting differences are usually small, exact correspondence between multi-stratum ANOVA and REML is desirable and was one motivation for the development of the REML method. This is a strong argument for allowing negative estimates of variance components.

16.5 What about Random Effects?

We stated earlier that the parameters of the LMM are the fixed effects and the variance components, but we wrote down our models in terms of fixed and random effects. In this section, we discuss the status of the random effects.

The LMM can be written in two forms. The model written in terms of both the fixed and random effects, for example, Equation 16.1, is known as the conditional form since the response is conditional on the random effects. The marginal form of the model is obtained by integrating over the population of random effects. In the marginal form, the model is specified in terms of the expected value of the observations, determined by the fixed terms alone, and the variance–covariance matrix of the observations generated by the random terms (as described in Section 16.4). Estimation takes place in the marginal model, the parameters of which are the fixed effects and the variance components. However, we are still often interested in the values of the random effects, and so would like to estimate them. This is possible only when the variance component for the term is positive, as random effects cannot be defined with a negative variance. Because the random effects are not true parameters, we obtain predictors, rather than estimates, of their values that are called BLUPs, an acronym for **best linear unbiased predictors**. In this context, the adjective 'unbiased' can be slightly misleading, as it means that the expected value of a predictor is equal to the expected value of the population, which is zero. Given the (unknown) true value of a random effect, its BLUP is biased towards zero, a property known as **shrinkage**. The adjective 'best' here means that these predictors have minimum mean squared error (defined as variance plus squared bias), conditional on the variance components. Again, in practice, we do not know the true values of the variance components and so substitute their REML estimates to obtain empirical BLUPs, often called eBLUPs. The property of minimum mean squared error is attractive where accuracy in prediction is more important than unbiasedness, and is sometimes used as a justification for assigning terms to the random rather than the fixed model, particularly in the context of variety evaluation (see Smith et al., 2005, for discussion in this context).

All random effects, including the deviations, are estimated from a REML analysis as eBLUPs. This is different from multi-stratum ANOVA, which uses least-squares estimates for terms in the structural component, and so estimated effects for structural terms and residuals obtained from the two procedures often differ.

16.6 Predicting Responses

Prediction from LMMs follows the same basic principles laid out in Sections 11.2.5 and 15.5.2, but additional decisions must be made about the role of the random effects. Predictions are based on the selected model: we form a table of fitted values from this model classified by the explanatory variables (factors and variates) and then take marginal means to obtain the predictions required. In the case of LMMs, we must decide whether to make predictions conditional on the observed values of the random effects (known as **conditional or narrow-sense predictions**), or to make predictions with respect to the population of random effects (known as **marginal or broad-sense predictions**). All model terms are used to form conditional predictions, and so in this case the table of fitted values is classified by all the explanatory variables. For marginal predictions, each random term contributes its population mean value (zero) to the fitted values, and the table is classified only by variables that appear in the fixed model terms. Intermediate schemes are also possible, where predictions are conditional with respect to some random terms and marginal with respect to others. These options are discussed in some detail by Welham et al. (2004) and McLean et al. (1991). Marginal and conditional predictions take the same value for the LMMs considered here because the mean of the eBLUPs for random terms with independent effects and common variance is equal to the population mean of zero. Marginal predictions have larger SEs than conditional predictions because of the additional uncertainty associated with predicting for an unknown population rather than for an observed sample. Both types of prediction give the same SEDs for comparisons that do not directly involve random effects.

EXAMPLE 16.1I: WEED COMPETITION EXPERIMENT

Marginal predictions for the irrigation by species combinations ($\hat{\mu}_{jk}$ for the jth irrigation regime with the kth species) are formed by ignoring the structural component (i.e. random model) terms and forming predictions as

$$\hat{\mu}_{jk} = \hat{\mu}_{11} + \overline{Irrigation}_j + \overline{Species}_k + \overline{(Irrigation.Species)}_{jk} \; .$$

It is straightforward to verify that these predictions are equal to the treatment means given in Table 9.14, and they have a common SE equal to 0.4063. The multi-stratum ANOVA (Example 9.2) obtained a SE of 0.3778 for the same predictions, and the difference occurs because the ANOVA SE ignores contributions from the block and whole-plot effects. The SEDs are equal to 0.5344 for comparisons across irrigation regimes, and 0.3613 for within-irrigation regime comparisons. These are the same as the SEDs obtained from multi-stratum ANOVA because the contributions from structural (random) terms cancel when taking differences.

16.7 Checking Model Fit

In this more general context of LMMs, model checking becomes both more important and more complex. Model misspecification with respect to explanatory variates can be investigated with the techniques described in Sections 13.1 and 14.6. Assumptions regarding the random effects are investigated with the eBLUPs for each term (Section 16.5). The eBLUPs for the deviations are the equivalent of simple residuals and can be used in the residual plots described in Chapter 5 (see also Figure 16.3). Construction of the fitted values plot requires some thought, as fitted values can be defined either to include or to exclude eBLUPs associated with random terms (but always exclude the residual term). If random terms are included in the fitted values, then shrinkage can induce correlation between the residuals and the fitted values, so it is often better to exclude these terms. Histograms and Normal quantile plots of eBLUPs can be used to check the distributional assumptions for random terms.

There is no generally accepted analogue of the adjusted R^2 statistic to quantify the explanatory performance of LMMs. It is generally acceptable to state whether fixed terms show evidence of group differences (for factors) or linear trend (for variates), based on the outcome of approximate F-tests. One approach to calculating the percentage variance accounted for by the fixed model is based on the variance of an observation, as defined in Section 16.3. The baseline total variance can be calculated as the sum of the variance components when the constant term alone is included in the fixed model. This is compared with the sum of the variance components for the fixed model under consideration, and the percentage reduction measures the percentage variance accounted for by the fixed model. This statistic can be used to quantify the performance of different fixed models for a given structural component (random model). Note that the AIC and SBC described in Section 14.8 cannot be used to compare LMMs with different fixed terms when the variance parameters have been estimated by REML.

16.8 An Example

In this section, we analyse a real set of data in some detail to draw together and illustrate the ideas introduced in the previous sections.

EXAMPLE 16.2: WEED ABUNDANCE

During 2000–2003, data were collected from an extensive UK-wide field experiment, known as the Farm Scale Evaluations (FSEs), to determine the ecological effects of management regimes associated with either genetically modified (GM) herbicide-resistant or conventional crops. For each of four crops, the FSEs were designed with whole fields (blocks) split into two half-fields to which the treatments (factor **Treatment**, conventional or GM regime) were applied. Further information can be found in Case Study 19.1. Here, we consider only spring oilseed rape and analyse the counts of total weed abundance (variate *Weeds*) recorded in half-fields after the last herbicide application was made to the GM crop ('post-herbicide'; Heard et al., 2003). The seedbank in each half-field (variate *Seedbank*) was sampled before the crops were sown to provide a measure of initial seed densities. The aim of analysis here is to assess the impact of the two management regimes on weed abundance, taking into account the initial seedbank counts.

The trials used 62 fields during the spring seasons of 2000–2002 (factor **Year**, labelled chronologically as 1–3). The fields were located on 37 farms (factor **Farm**). Only one field per farm was used in each year, but some farms were studied in 2 or 3 years, with a different field used in each year (factor **Field**, numbered within farms as 1, 2 and 3). Half-fields (factor **DHalf**, labelled 1–2) are labelled systematically with respect to treatment although treatments were originally allocated to the halves at random (see Case Study 19.1). Two fields without seedbank counts, plus one further field where a zero seedbank count was regarded as suspect, were excluded from the analysis, leaving 59 fields (118 half-field data values). The data are held in file SOSR.DAT and shown in Table 16.4.

The weed and seedbank data are plotted as counts and on log-log axes in Figure 16.1. We should usually take a log-transform of the weed counts to accommodate variance heterogeneity; the log-log plot indicates that a linear relationship with seedbank counts is obtained if this variate is also log-transformed. These variates were therefore transformed to the $\log_{10}$ scale as *LogWeeds* = $\log_{10}$(*Weeds*) and *LogSeedbank* = $\log_{10}$(*Seedbank*).

We can now consider a preliminary model for the logged weed counts. The structure is hierarchical, with half-fields nested within fields, and fields nested within farms. Eighteen of the 37 farms have two or three separate fields used. In terms of these factors, we therefore write the structural component of the model as

Structural component: Farm/Field/DHalf

Since there is only one measurement per half-field, the half-field effects are the model deviations. The explanatory component of the model must account for year and treatment effects, and a crossed model is appropriate for these terms, i.e.

Explanatory component: [1] + Year*Treatment

TABLE 16.4

Weed and Seedbank Counts from Half-Fields under Conventional (C) or Genetically Modified (GM) Management Regimes in the FSE Study (Example 16.2 and File SOSR.DAT)

Farm	Field	Year	Weeds C	Weeds GM	Seedbank C	Seedbank GM	Farm	Field	Year	Weeds C	Weeds GM	Seedbank C	Seedbank GM
1	1	1	195	200	56	93	17	1	2	741	780	70	70
1	2	2	470	395	154	218	17	2	3	337	176	23	60
1	3	3	432	192	68	103	18	1	2	113	56	150	68
2	1	1	142	128	71	126	18	2	3	634	547	98	139
2	2	2	1625	180	60	117	19	1	2	1302	692	241	271
3	1	1	121	84	52	56	20	1	2	653	492	252	283
4	1	1	505	115	156	145	21	1	2	73	163	49	55
4	2	2	234	248	146	504	22	1	2	286	154	65	116
4	3	3	1266	1166	256	289	22	2	3	1040	324	158	51
5	1	2	54	125	311	73	23	1	2	487	288	100	153
5	2	3	68	406	49	237	23	2	3	702	1388	239	543
6	1	2	104	48	69	190	24	1	2	473	225	44	41
7	1	2	42	19	20	7	24	2	3	485	270	240	178
8	1	2	255	387	59	39	25	1	2	1631	7875	251	384
8	2	3	101	121	40	19	25	2	3	640	587	471	413
9	1	1	1815	381	133	128	26	1	2	358	25	241	216
9	2	2	403	461	182	120	26	2	3	198	46	50	149
9	3	3	817	1395	734	969	27	1	2	29	292	33	110
10	1	2	40	111	126	79	28	1	3	244	178	88	29
10	2	3	203	327	99	51	29	1	2	921	178	89	51
11	1	1	125	558	60	124	30	1	1	376	263	173	563
11	2	3	66	149	46	57	30	2	2	248	55	113	50
12	1	1	432	272	26	50	30	3	3	404	482	213	340
12	2	2	636	25	149	156	31	1	3	2103	367	394	530
12	3	3	356	51	102	55	32	1	3	354	233	72	66
13	1	2	449	56	62	61	33	1	3	403	142	136	56
14	1	2	2620	1743	302	260	34	1	3	261	310	25	85
15	1	2	314	602	487	152	35	1	3	2041	1176	389	693
16	1	3	708	571	85	167	36	1	2	171	677	50	88
							37	1	2	701	352	162	142

Source: Data from M. Heard, Centre for Ecology and Hydrology.

As a baseline, we first analyse the logged weed counts (*LogWeeds*) ignoring the initial seedbank counts. In addition, as we should ideally like to regard the initial seedbank counts as a covariate (see Section 15.4), we also use this model to check whether that covariate (*LogSeedbank*) is related to the explanatory terms.

To fit this as a LMM, the structural component becomes the random model and the explanatory component becomes the fixed model. Table 16.5 shows the estimated variance components for the two responses. Since the weed and seedbank counts are on different scales, we do not compare the values of the estimated variances for the two responses, but we do compare the pattern of relative sizes across strata.

In both cases, all three variance components are positive. This indicates some similarity across fields within farms, and across halves of the same field. This is expected, as weed management practices will differ between farms, and weed infestation often varies across

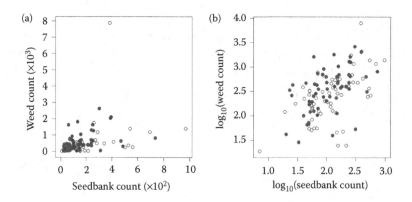

FIGURE 16.1
Weed counts for (•) conventional and (○) GM management regimes plotted against initial seedbank counts with both variables (a) untransformed and (b) transformed to logarithms (Example 16.2).

TABLE 16.5

Estimated Variance Components for Log_{10}-Transformed Weed and Seedbank Counts with Fixed Model [1] + Year*Treatment (Example 16.2)

Random Term	*LogWeeds*	*LogSeedbank*
Farm	0.0517	0.0521
Farm.Field	0.0628	0.0539
Farm.Field.DHalf (deviations)	0.1139	0.0443

fields within farms. For *LogWeeds*, the variation between half-fields is about twice that generated by farm and field effects, which are of similar sizes. For *LogSeedbank*, the three variance components are of similar sizes. This suggests that variation between half-fields is relatively smaller for the initial seedbank counts. Again, this might be expected where fields have been managed as single entities prior to the trial.

Table 16.6 shows the approximate F-tests for the fixed terms in these preliminary models. The Year and Treatment factors are orthogonal, so there is a unique table of incremental tests that can be used to investigate fixed terms. None of the fixed terms is significant for *LogSeedbank*. This matches our prior expectations as seedbanks were assessed before sowing and, given the relative consistency of seedbank counts within

TABLE 16.6

Incremental F-Statistics with Denominator df (ddf) and Observed Significance Level (*P*) for Log_{10}-Transformed Weed (*LogWeeds*) and Seedbank (*LogSeedbank*) Counts with Fixed Model [1] + Year*Treatment (Example 16.2)

Term	df	*LogWeeds*			*LogSeedbank*		
		F-Statistic	ddf	*P*	F-Statistic	ddf	*P*
+ Year	2	1.068	34.7	0.355	0.728	32.5	0.491
+ Treatment	1	5.415	56.0	0.024	1.413	56.0	0.240
+ Year.Treatment	2	0.048	56.0	0.953	1.576	56.0	0.216

whole fields, the possibility of an 'unlucky' random allocation is small. This analysis also confirms that there was no consistent difference in initial seedbank count across years or across treatments within years. For *LogWeeds*, the Year.Treatment interaction and Year terms are not significant, indicating no differences across years, but the Treatment term is significant, and indicates a consistent difference between the two management regimes, with the conventional treatment 0.145 units (SE 0.0719) larger than the GM treatment on the $\log_{10}$ scale.

To try to understand the denominator df used by the approximate F-tests, we can construct a dummy multi-stratum ANOVA table, as in Table 16.7. The structural component generates strata for farms (Farm), fields within farms (Farm.Field) and half-fields within fields (Farm.Field.DHalf). There are 37 farms, so the Farm stratum has a total of 36 df. There are 12 farms with two fields and five farms with three fields used, giving $12 + (5 \times 2) = 22$ df in total for the Farm.Field stratum. Finally, since each of the 59 fields has observations made on both halves, there are 59 df in total in the Farm.Field.DHalf stratum. Since treatments are applied to half-fields, the Treatment and Year.Treatment effects are estimated entirely within the Farm.Field.DHalf stratum, which removes three df (one for Treatment and two for Year.Treatment) and leaves 56 ResDF. This is the denominator df used by the F-tests for the Treatment and Year.Treatment terms, as we should expect. Although each farm uses a different field in each year, only five farms are present in all 3 years, and so effects for the Year term are estimated partly within and partly across farms, and the denominator df for this term is derived from both the Farm and Farm.Field strata. These denominator df then depend on both the allocation of information (which is the same for both responses) and on the relative values of the Farm and Farm.Field variance components (which differ), and so the denominator df for Year differ slightly for the two responses (estimated as 34.7 for *LogWeeds* and 32.5 for *LogSeedbank*).

Given the linear relationship between *LogWeeds* and *LogSeedbank* apparent in Figure 16.1, we might be able to improve our estimate of treatment effects by adjusting for the initial seedbank counts. Since most of the variation in seedbank counts occurred between rather than within fields, we expect that accounting for the initial seedbank will not have much impact on the estimated treatment effect (which is estimated from within-field comparisons), but we hope that accounting for this variation might increase

TABLE 16.7

Dummy Multi-Stratum ANOVA Table for the FSE Study with 59 Fields (Factor Field) on 37 Farms (Factor Farm) over 3 Years (Factor Year), and Two Treatments (Factor Treatment) Applied to Half-Fields within Fields (Factor DHalf) (Example 16.2)

Term	df
Farm stratum	
Year	2
Residual	34
Farm.Field stratum	
Year	2
Residual	20
Farm.Field.DHalf stratum	
Treatment	1
Year.Treatment	2
Residual	56
Total	117

the precision of the estimate (i.e. decrease its SE). We first check whether the relationship with the logged seedbank count is the same for both treatments (and across years) by fitting a model with separate slopes (see Section 15.4), leading to the fixed terms in a LMM as

Explanatory component: [1] + *LogSeedbank**Year*Treatment

The terms in this model are not orthogonal, and so there are many sets of incremental Wald tests. Here, we use marginal F-tests to select the predictive model. In the full model, we can test only the three-way term *LogSeedbank*.*Year*.*Treatment* and we find that it is not significant (Model 1 in Table 16.8, $F_{2,60.3}^{L.Y.T} = 0.262$, $P = 0.770$). We therefore drop this term and refit with all the remaining fixed and random terms. We can now test all of the terms containing two variables with the marginal F-tests shown for Model 2 in Table 16.8. None of these terms appears significant, so we drop the least significant of them first (*LogSeedbank*.*Year* with $F_{2,60.3}^{L.Y} = 0.075$, $P = 0.927$), refit the model, and find we can then drop each of the other two-variable terms in turn (*Year*.*Treatment* then *LogSeedbank*.*Treatment*). The relationship with initial seedbank counts is therefore consistent across both treatments and years, and treatment differences also appear consistent across years. This leaves only the single-variable terms in the model (Model 3 in Table 16.8). Marginal F-tests show no evidence of consistent differences between years ($F_{2,43.5}^{Y} = 0.698$, $P = 0.503$) and so the term *Year* can be dropped, but the remaining terms *LogSeedbank* and *Treatment* both have significant marginal F-tests, and form the predictive model. The full set of parameter estimates from this predictive model is shown in Table 16.9.

This predictive model can be written in algebraic form with first-level-zero parameterization as

$$\tilde{\mu}_{ijk}(LogSeedbank) = \hat{\mu}_1 + \widetilde{Farm_i} + \widetilde{Farm.Field_{ij}} + \widetilde{Treatment_k} + \hat{\beta} LogSeedbank .$$

Here, $\tilde{\mu}_{ijk}(LogSeedbank)$ is the predicted log_{10}-transformed weed count for a given value of the log_{10}-transformed seedbank count (*LogSeedbank*) with the kth treatment ($k = 1, 2$; $1 = C, 2 = GM$) in the jth field within the ith farm, for $i = 1 \ldots 37$, $j = 1, 2, 3$. We represent eBLUPs with a tilde (~) rather than a hat (^) embellishment, which we reserve for estimates of the model parameters. The intercept, $\hat{\mu}_1 = 1.297$(SE 0.2205), represents the prediction for the conventional (C) treatment for a zero value of *LogSeedbank*, which is

TABLE 16.8

Marginal F-Tests from Explanatory Models for Log_{10}-Transformed Weed Counts with Observed Significance Level (*P*) (Example 16.2)

Term	Model 1	Model 2	Model 3
[1]	—	—	—
L	—	—	$F_{1,85.3}^{L} = 31.797$ ($P < 0.001$)
Y	—	—	$F_{2,43.5}^{Y} = 0.698$ ($P = 0.503$)
T	—	—	$F_{1,57.8}^{T} = 7.969$ ($P = 0.007$)
Y.T	—	$F_{2,61.0}^{Y.T} = 0.103$ (P = 0.902)	*
L.Y	—	$F_{2,75.5}^{L.Y} = 0.075$ (P = 0.927)	*
L.T	—	$F_{1,63.7}^{L.T} = 0.081$ (P = 0.777)	*
L.Y.T	$F_{2,60.3}^{L.Y.T} = 0.262$ (P = 0.770)	*	*

Variable names: L = *LogSeedbank*, Y = Year, T = Treatment. — = term in model but not eligible for testing, * = term omitted from model.

TABLE 16.9

Parameter Estimates with Standard Error (SE) from the Final Linear Mixed Model for Log_{10}-Transformed Weed Counts (Example 16.2)

Parameter	Estimate	SE
Farm variance component	0.004	0.023
Farm.Field variance component	0.049	0.031
Residual variance	0.109	0.021
Constant	1.297	0.221
Treatment effect (C − GM)	0.173	0.061
Coefficient for *LogSeedbank*	0.612	0.106

outside of the observed range and so is an extrapolation. The estimate $\overline{Treatment_2}$ = 0.173 (SE 0.0611) indicates that the intercept for the C treatment is 0.173 units larger than that for the GM treatment. This difference is a little larger than that found in the initial analysis, with a smaller SE. In this parallel lines model, this difference between the treatments is the same for any value of *LogSeedbank*. The slope of the linear relationship between *LogWeeds* and *LogSeedbank* has estimate $\hat{\beta}$ = 0.612(SE 0.1055). We conclude that the number of weeds increases as the initial seedbank increases, and that the GM management system reduces the number of weeds. The predicted responses (omitting random effects) are shown with 95% CIs in Figure 16.2, together with the observations adjusted for Farm and Farm.Field eBLUPs. It is clear that the model follows the observed pattern reasonably well.

A composite set of residual plots is shown in Figure 16.3, with fitted values calculated excluding the random effects Farm and Farm.Field. The distribution of the residuals appears a little skewed to the left, and the fitted values plot shows a few large negative residuals for fitted values around 2.50; these can also be seen in Figure 16.2. These residuals correspond to half-fields with much smaller weed counts than would be expected from their initial seedbank counts. Of the four most negative residuals, two come from each treatment group. We suspect that these discrepancies are caused by patchiness of the weed populations.

The random farm effects have estimated variance of 0.0040, and the random field within farm effects have estimated variance 0.0494. The reduction in the estimated

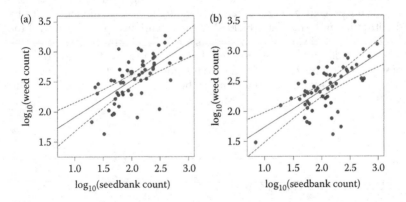

FIGURE 16.2

Fitted model (solid line) with 95% CI (dashed curved lines) for log_{10}(Weed count) in terms of log_{10}(Seedbank count) for (a) conventional or (b) GM management, with observations (•) adjusted for farm and field effects (Example 16.2).

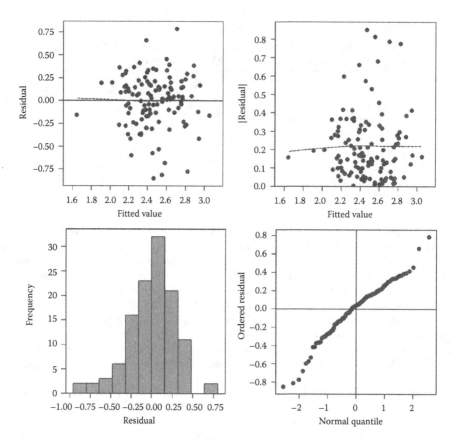

FIGURE 16.3
Set of composite residual plots from the predictive model for $\log_{10}$(weed count) (Example 16.2).

Farm variance component compared with the analysis without the covariate (Table 16.5) suggests that the initial seedbank accounts for most of the overall farm differences in weed count, which may in turn reflect differing farm management practices. To calculate the percentage variance accounted for by the final model, we fit a baseline model with random terms Farm/Field/DHalf, as mentioned previously, and the fixed term [1]. The estimated variance components are 0.0542 (Farm), 0.0589 (Farm.Field) and 0.1187 (Farm.Field.DHalf), giving total variance equal to 0.2318. From Table 16.9, the sum of variance components under the final model is equal to 0.1628. The percentage variance accounted for by the fixed terms is therefore calculated as $100 \times (0.2318 - 0.1628)/0.2328 = 30\%$.

16.9 Some Pitfalls and Dangers

Because LMMs are a more general class of model, allowing multiple random terms without any requirement for balance, the iterative algorithms used for estimation are also more general, with more possibility of failure. If such failures occur, there are several possible causes that should be considered.

The first possibility is that some of the variance components are not estimable. If the same, or equivalent, terms are put into both the fixed and random models, then no information is available for estimation of the variance component corresponding to the random term. Occurrence of this problem is not always obvious, as it is often possible to generate equivalent terms from combinations of different factors. A similar problem occurs if, in the terminology of balanced designs, there are zero residual df within any stratum, as the corresponding variance component then cannot be estimated. This usually indicates a lack of real replication for some combination of explanatory factors in the study. In either case, removal of the offending random term from the model also removes the corresponding stratum from the structure and the remaining variance components should be estimable. However, any explanatory terms that should be tested in that stratum will instead be tested in a lower stratum, and this should be reported as part of the analysis.

Problems may also occur for variance components that are estimable, but estimated as a negative value. As discussed in Section 16.4, some algorithms permit only positive estimates that are bounded below by zero, whereas others permit negative estimates; unfortunately both approaches have problems associated with them. Variance components that are fixed at a lower bound of zero are ignored by the Kenward–Roger method. The impact on approximate F-tests (Section 16.3) for explanatory variables tested at that level of the structure is equivalent to dropping the random term from the model, so that the resulting denominator df corresponds to a lower stratum. In some implementations, variance components are internally parameterized on a log scale which forces them to remain positive. This can lead to an apparent failure of convergence where the estimate should be zero (or negative), as zero estimates can never be reached on the log scale. This situation is easily detected if monitoring of the iterated estimates is examined. Finally, if negative estimates of variance components are allowed, occasional instability of the algorithm may result. The causes of this instability are usually due to difficulties in imposing positive definite constraints on the variance–covariance matrix as a whole during the estimation process. However, where negative estimates can be obtained, they can be used properly within the Kenward–Roger method.

Finally, we re-emphasize the role of the estimated variance components in the eBLUEs and eBLUPs. These estimates and predictors are usually treated as if the variance components were known, whereas in fact they are not. Uncertainty in the variance components leads to additional uncertainty in the eBLUEs and eBLUPs that is not accounted for in their SEs. An alternative approach that does account for this uncertainty is the use of Bayesian mixed models. One feature of this approach is the requirement for prior information (i.e. distributions) on the fixed effects and variance components; while uninformative priors for fixed effects are well established, the natural scale for priors on variance components is less clear (Gelman, 2006).

16.10 Extending the Model

In this chapter, we have briefly introduced some aspects of LMMs. This class of models can be applied in many different situations, but its flexibility can also be a weakness: it can be easy to fit an inappropriate model and to obtain misleading results. It is therefore vital to assess the model and results critically before proceeding to interpret them. Galwey (2006) and Littell et al. (2006) provide good introductions to LMMs.

We have discussed only the subset of LMMs in which the random effects for each term are independent with common variance; these are often called variance component models. In addition, we have insisted that the random model corresponds to the structural component of the model, with the fixed model corresponding to the explanatory terms. This approach can be generalized in several different ways, which we consider in turn.

There is rather more flexibility in the allocation of terms as fixed or random than our recipe of explanatory = fixed, structural = random acknowledges. There are several different grounds for assigning a term to the random model. First, if we believe that a set of random effects truly represents a sample from a (Normal) population then it is natural to assign them as random. This often applies to structure within experiments: those positions or locations or subsamples used are naturally regarded as a sample of the wider population that might have been used. In the context of field experiments repeated over several years, this reasoning often leads to treatment × year interactions being regarded as random. Second, if the aim of the experiment is to model variation across factor levels explicitly, then it is again natural to assign the factor as a random term. Finally, in Section 16.5, we noted the minimum mean squared error property of BLUPs, and remarked that this can be used as a motivation for fitting variety effects as random rather than fixed where the aim is accurate prediction of relative variety performance across a set of trials. This principle can be applied more widely, and again leads to explanatory terms being assigned to the random model. It is important to remember that adding terms into the random model changes the variance–covariance structure applied to the observations, and so may have an impact on SEs for other explanatory terms, and on the estimated denominator df for approximate F-tests.

Variance component models can be generalized if we allow more general variance–covariance models on the random effects, or on the deviations, and this leads to correlated error models. These models are widely used in the analysis of longitudinal data (repeated measurements), as they can model the correlation between successive measurements made on the same subject or unit, as well as allowing for changes in variance over time. A detailed review of this area is given by Verbeke and Molenberghs (2000). This type of model can also be used to account for spatial correlation in either experimental or observational settings, for example small-scale smooth trend across a field or glasshouse bench. Gilmour et al. (1997) give some detailed examples in the context of field experiments, but the same principles apply more widely. As these models become more complex, the dangers of misspecification and algorithmic problems also increase, so additional care and thought is required.

It is also possible to use smoothing splines, or penalized splines, to model non-linear responses within the framework of LMMs. These models do not have a pre-specified form; instead, the fitted response is determined by the observed trend in the data. The implementation within LMMs is facilitated by a coincidence in the form of the equations for estimating the spline for a given smoothness and those for estimating eBLUEs and eBLUPs in a specific LMM. The smoothness of the fitted curve is determined by the smoothing parameter, which is usually estimated via a variance component within the LMM context. You can find a good introduction to this topic in Ruppert et al. (2003).

Finally, extensions have been made to extend the class of LMMs to apply to non-Normal responses and non-linear models. The exact evaluation of the restricted log-likelihood function used to obtain REML estimates is much harder in these contexts, involving complex integrations, and so simpler approximate methods are often used. The main drawback to these methods is that the user might not know when the approximation is good enough to draw firm conclusions – this is currently an area of statistical research. Section

17.3 introduces the class of non-linear models, and these can be extended to include random effects through the class of non-linear mixed models (Pinheiro and Bates, 2000). Chapter 18 introduces the class of generalized linear models (GLMs) that can be used to model certain types of non-Normal responses, and this class can also be extended to include random effects (see e.g. Lee et al., 2006).

EXERCISES

16.1 In Exercise 9.8, you analysed a (slightly non-standard) split-plot design with multi-stratum ANOVA. Convert the explanatory and structural components for this experiment into a linear mixed model and fit this model. Obtain estimates of the variance components and verify that these match those obtained from the multi-stratum ANOVA. Use approximate F-tests to identify a predictive model and obtain predictions from this model and verify that the results match those from the multi-stratum ANOVA.

16.2* An experiment was done to establish conditions for infection of young brassica plants with a foliar disease. Plants were subjected to different temperatures and periods of leaf wetness after exposure to one of two isolates of the pathogen. The experiment used four CE cabinets, with temperatures (5°C, 10°C, 15°C, 20°C) allocated to cabinets and combinations of isolate (type 1 or 2) and leaf wetness (8, 16, 24, 48 or 72 h) allocated at random to trays within cabinets, with the four plants in each tray receiving the same treatment. (The original randomization has been lost so cabinets and trays are labelled systematically.) The experiment was done in four runs, with random allocation of temperatures to cabinets within each run. Four treatment combinations were omitted from the full factorial set (as pilot studies showed them to produce no infection) and temperatures 15°C and 20°C were omitted from the third and fourth runs, respectively. The unit numbers (*ID*), structural factors (Run, Cabinet, Tray, Plant) and treatment factors (Temp, Wetness, Isolate) are held with the response (variate *TotLesions*, count of total lesions) in file LESIONS.DAT. Write down the explanatory and structural components for the design of this experiment and translate these into a linear mixed model. Fit this model (using a transformation to account for variance heterogeneity if required) and use approximate F-tests to determine a suitable predictive model. What would you recommend as optimal conditions for infection in future experiments?

16.3 Exercises 11.3, 11.5, 11.6 and 11.7 comprised the intra-block analysis of a designed experiment by fitting block effects before treatment effects. We will now re-examine these experiments using mixed models to investigate when the intra-block analysis gives a good approximation to analysis by mixed models. Allocate the structural model terms as random and the explanatory terms as fixed and repeat the analysis. Obtain predictions for the target explanatory variable (stated below) with SE and SED and compare these to results from the intra-block analysis. Can you understand and explain any differences? Can you identify features of the data sets that make the use of mixed models advantageous?

   a. Identification of economic conditions for growing peppers in a glasshouse (Exercise 11.3). You must select a predictive model (using approximate F-tests) and produce predictions from the selected terms.

b. Measurement of protein by NIRS (Exercise 11.5). Obtain predictions for the full set of accessions.

c. Measurement of shoot growth for different pruning strategies (Exercise 11.6). Assess differences between the pruning strategies and produce predictions.

d. Estimation of variety differences from a vandalized experiment (Exercise 11.7). Obtain a set of variety predictions.

16.4 A dose–response experiment investigated the action of three insecticidal seed treatments on three clones of aphid. Eight doses (including a zero dose) of each insecticide were applied to batches of seed, and the (average) actual dose of each insecticide applied was recorded. Three plants were grown from each of these 24 treatments, and one plant with each treatment was allocated to each type of aphid clone. The experiment used six cages of 12 plants, with an unbalanced design (an alpha design) allocating the 72 treatment combinations to cages and plants. Adult aphids of the designated clone were introduced onto each plant and the number of nymphs present after 2 days was counted. The experiment was conducted in two runs, with each treatment combination present once in each run. File CAGE.DAT holds the unit numbers (*ID*), structural factors (Run, Cage, Plant) and treatment factors (Clone, Treatment, FDose) with the actual dose (variate *Dose*) and number of nymphs after 2 days (variate *Nymphs*).

Write down the structural and explanatory components of the model for this experiment in terms of the explanatory factors. Consider carefully whether dose should be crossed with the other factors or nested within treatment. Translate your model into a linear mixed model and fit it, checking the model assumptions. Investigate whether the response to dose is a linear function, and whether this relationship differs between insecticides or clones or both. Identify and present a predictive model to summarize the results of this experiment.[*]

16.5 The efficacies of six insecticidal treatments against aphids on vegetable brassicas were compared against an untreated control in a field trial. The trial comprised four complete replicates of seven plots, arranged as a grid with seven rows and four columns. The seven treatments were allocated to plots with a balanced row–column design, so that each treatment occurred once in every column (replicate) and in four different rows. Each plot was split into two, with two crops (cauliflower or savoy cabbage) allocated to the halves at random. Within each half-plot, 10 of the central (guarded) plants were sampled 14 days after the second spray application, and the number of peach–potato aphids on each plant was counted. The unit numbers (*ID*), structural factors (Row, Column, Plot, Halfplot, Plant), treatment factors (Insecticide, Crop) and response (variate *Aphids*) are held in file ROWCOL.DAT. There were 15 missing plants in this sample. Write down the structural and explanatory models and compare two methods of analysis: multi-stratum ANOVA using the Healy–Westmacott algorithm for estimation of missing values, and use of linear mixed models, with the missing observations omitted. Compare the results of the two analyses and discuss any differences.[†]

16.6 In Exercise 14.4, you constructed a model to predict tree total aerial biomass (*TAB*, data file SLASH.DAT). That analysis ignored the fact that several trees were

[*] Data from S. Foster, Rothamsted Research.
[†] Data from R. Collier, University of Warwick.

sampled from each inventory plot. We will now incorporate this structure into the model.

a. Write down the structural component of the model for these data. Consider which of the explanatory variables are assessed (predominantly) at each level of the structure. Discuss whether you could successfully model this structure using an intra-block analysis.

b. Using the structural component as random terms in a linear mixed model, fit all of the explanatory variables and perform backwards selection to identify a new predictive model. Interpret your model. Is there any evidence of correlation among measurements on trees from the same plot? What impact has accounting for this structure made to your predictive model?

16.7 In Exercise 14.6(d), you developed a MLR for the combined EXAMINE data set for years 1995 and 1996 (file EXAMINE9596.DAT). Now, we use linear mixed models to incorporate the crossed year × trap structure of this data set.

a. First, identify the level of the structure at which each explanatory variable shows variation, i.e. across years, across traps, or across year × trap combinations. How should this affect tests of the explanatory variables?

b. Fit a baseline linear mixed model with random terms Year*Trap and no fixed terms, and take note of the estimated variance components.

c. Add the terms you identified for the joint MLR in Exercise 14.6 into the fixed model. How do the estimated variance components change when these terms are added into the model? Use marginal approximate F-tests to decide whether all of the fixed terms are still required? How do these tests differ from those obtained in Exercise 14.6(d)? Write down your final model. What percentage of the variation does this final model account for?

16.8 New insect repellent compounds require testing in the field, and this process is complicated by large variations in insect abundance over both space and time. File MIDGE.DAT holds the results from a trial to test a potential repellent compound against the Scottish biting midge (variables *ID*, Day, Run, Tent, Volunteer, Treatment and *Total*). The trial used two tents (A and B) in different parts of the same location, with several runs during each evening of three consecutive days. During each run, one volunteer was allocated to each tent with either the test formula or a positive (known active compound) or negative (blank) control (the three treatments). The number of midges entering the tent was counted from 4–20 min after the start of the trial, and the total number was recorded. After an inter-run period of 20–30 min, the process was repeated with a different volunteer and compound in each tent. There were six volunteers available, and these were allocated to tents and compounds in as balanced a way as possible, but the resulting design is unbalanced.[*]

a. Write down the structure of this trial in terms of the Day, Run and Tent factors. Consider which factors are nested and which are crossed and identify the residual term. How much information is there at each level of this structure?

---

[*] Data from J. Pickett, Rothamsted Research, A.J. Mordue, Aberdeen University, and J. Logan, London School of Hygiene and Tropical Medicine.

b. Use tables of counts to investigate the allocation of volunteers and treatments to days, tents and runs. Examine the replication of volunteer × treatment combinations. Is it possible to get sensible estimates for all combinations? What happens if you fit an effect for an unreplicated combination?

c. Set up a linear mixed model with random term Day.Run and fixed terms Day*Tent + Volunteer + Treatment. Use your answers to parts (a) and (b) to justify this model. Fit this model and use diagnostic plots to check the assumptions. Refit with a transformation if necessary. When you are satisfied with the model fit, interpret the estimated variance components. Is there any evidence that the test treatment repels midges? Identify and interpret a predictive model for this trial.

d. Discuss what information from these results can be used to design future trials of this type. What principles would you recommend future designs should follow?

17

Models for Curved Relationships

The regression models discussed in the previous chapters fitted straight line relationships between a response variable and one or more explanatory variates. In many situations, this type of model adequately reflects the observed pattern, but sometimes a curved relationship is observed, or there might be a biological or physical reason for a curved relationship. Fitting a straight line will then produce an inadequate model and an alternative approach should be sought. In this chapter, we consider some simple techniques for fitting curved relationships in terms of one or more explanatory variates.

First, we describe approaches for a single explanatory variate that stay within the framework of linear regression (Section 17.1). The simplest approach to deal with curved relationships is transformation of the explanatory variate, so that a new transformed variate is used in place of the original (Section 17.1.1). A slightly different approach uses a combination of transformations of the explanatory variate together in a MLR model to create a curved relationship. This is often done with low-order polynomial functions (Section 17.1.2) or trigonometric functions (Section 17.1.3). We then extend these approaches to the case of two explanatory variates that act together (rather than independently) on the response, so that interaction between the explanatory variates is required to generate an appropriate curved surface (Section 17.2). Finally, non-linear regression is a more sophisticated approach that allows a wider range of models to be fitted (Section 17.3). However, this approach requires a different set of numerical and statistical techniques, which are mathematically and computationally more complex than those used in linear regression, and which we describe only briefly here.

17.1 Fitting Curved Functions by Transformation

We first consider transformations of a single explanatory variate as a means to produce curved relationships. A transformation of either the response or the explanatory variate changes the shape of the relationship. However, transformation of the response also changes the characteristics of the deviations, such as homogeneity of variance, as discussed in Section 6.1. For this reason, we use transformation of the response as a tool to find a scale that meets the underlying assumptions for the deviations, and we use transformation of the explanatory variate as a tool to manipulate the shape of the relationship with the response. In this section, we concentrate on the latter.

17.1.1 Simple Transformations of an Explanatory Variate

The aim of a simple transformation of the explanatory variate is to find a scale on which the relationship with the response becomes a straight line. The first step in the process is to plot the response against the potential transformation of the explanatory variate to see whether a straight line relationship is plausible on that scale. If it is, then the second step is to fit the model in terms of the transformed explanatory variate and to check for any signs of model misspecification (see Section 13.1). Alternative forms of transformation can be compared formally with goodness-of-fit statistics (such as adjusted R^2) and by graphical inspection of the different model fits or, if sufficient data are present, by cross-validation (Section 14.9.3).

The most common transformations of an explanatory variate (x) used in this context are the square root ($\sqrt{x} = x^{0.5}$), square (x^2), logarithm ($\log_e(x)$ or $\log_{10}(x)$), exponential ($\exp(x)$) and reciprocal ($1/x$) transformations. Typical shapes for these functions are shown in Figure 17.1. Trigonometric functions, for example, $\sin(x)$ or $\cos(x)$ or both, or other powers, for example, x^3, can also be used, and these are discussed in more detail in Sections 17.1.2 and 17.1.3. The modelling procedure entails calculation of the transformed variate, for example, $w = \sqrt{x}$, then a simple linear regression (SLR) model is fitted with the transformed variate w as the explanatory variate in the model.

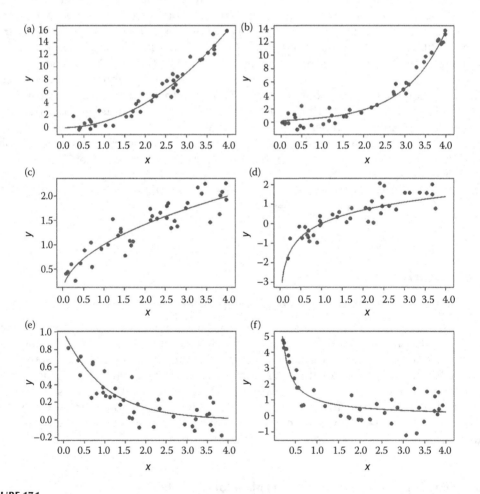

FIGURE 17.1
Typical shape of response (y) for simple functions of an explanatory variate (x): (a) $y = x^2$, (b) $y = 0.25e^x$, (c) $y = x^{0.5}$, (d) $y = \log_e(x)$, (e) $y = e^{-x}$, (f) $y = 1/x$. In each case, the line shows the underlying curve and the points show a sample of 40 observations taken from the underlying function plus Normal deviations with common variance.

Some of these functions are valid only for positive values of the explanatory variate ($x > 0$, e.g. for $w = \log_e(x)$, $\log_{10}(x)$ or $1/x$), and some are valid only for non-negative values ($x \geq 0$, e.g. for $w = \sqrt{x}$). If some of the values of the explanatory variate are outside the range allowed, for example, if $x_i = 0$ for a log transformation, then a pragmatic solution is to add a positive offset c to all the values of the explanatory variate, for example, $\log_e(x + c)$. The choice of a sensible offset value was discussed in Section 6.2 in the context of transformation of the response. Here, the aim is to find an offset such that all of the values of the explanatory variate fall within the allowed range and the relationship between the response and transformed explanatory variate is a straight line. Different values of the offset can have a large impact on the shape of the curve and should be evaluated both graphically and with goodness-of-fit statistics.

Predictions of the fitted line, together with standard errors (SEs) and confidence intervals (CIs), can be calculated in terms of the transformed variable (w) as for SLR (Section 12.5). These predictions, SEs and CIs also apply directly to the original explanatory variate (x); this is illustrated in Example 17.1.

EXAMPLE 17.1A: OLSEN P

The exhaustion land long-term field trial at Rothamsted Research has been used to investigate the relationship between crop yields and applications of soil fertilizer. The data in Figure 17.2a, Table 17.1 and file PHOSPHORUS.DAT are yields of spring barley from 20 plots in 1986 (variate *Yield*) with the available soil phosphorus content measured as Olsen P (variate *OlsenP*).

There is no suggestion of variance heterogeneity in yield, so there is no reason to transform the observed response. However, there is clear curvature in the relationship. An alternative plot of yield against the log-transformed Olsen P values (*LogOP* = $\log_{10}$(*OlsenP*), Figure 17.2b) appears to give a straight line. The transformed variate can therefore be used in a SLR model of the form

$$Yield_i = \alpha + \beta \, LogOP_i + e_i \, ,$$

where *Yield_i* is the yield, *LogOP_i* is the $\log_{10}$-transformed value of Olsen P and e_i is the deviation for the ith plot, $i = 1 \ldots 20$. The slope of the straight line is β, representing the

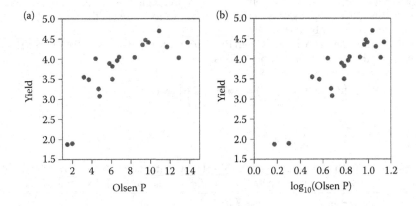

FIGURE 17.2
Yield plotted against (a) Olsen P phosphorus content, (b) $\log_{10}$(Olsen P) per plot for exhaustion land trial in 1986 (Example 17.1A).

TABLE 17.1

Yield and Olsen P Measurements from the Exhaustion Land Experiment at Rothamsted Research in 1986 (See Example 17.1A and File PHOSPHORUS.DAT)

Olsen P	Yield	Olsen P	Yield	Olsen P	Yield	Olsen P	Yield
6.8	4.05	10.9	4.70	3.7	3.49	4.8	3.08
5.8	3.89	9.5	4.47	2	1.90	12.9	4.03
3.2	3.55	6.1	3.50	9.2	4.35	13.8	4.41
1.5	1.88	6.1	3.82	9.8	4.41	8.4	4.04
11.7	4.30	4.4	4.01	6.6	3.96	4.7	3.26

Source: Data from P. Poulton, Rothamsted Research.

increase in yield for an increase of one unit of $\log_{10}$(Olsen P), and α is the intercept of this straight line. In symbolic form, the model can be written as

Response variable: *Yield*
Explanatory component: *[1] + LogOP*

This model accounts for 79.7% of the variation in the data (adjusted $R^2 = 0.797$) and the F-test from the ANOVA shows a strong association of yield with $\log_{10}$(Olsen P) ($F_{1,18} = 75.410$, $P < 0.001$). The parameter estimates are shown in Table 17.2, giving the predictive model

$$\hat{\mu}(LogOP) = 1.674 + 2.644\,LogOP \ .$$

We can rewrite this predictive model in terms of the untransformed explanatory variable as

$$\hat{\mu}(OlsenP) = 1.674 + 2.644 \log_{10}(OlsenP) \ ,$$

and the parameter SEs still apply on this scale. The intercept predicts the response when $\log_{10}(OlsenP) = 0$, corresponding to $OlsenP = 1$ on the original scale. This model gives an increase of 2.64 units in yield for one unit of increase in $\log_{10}(OlsenP)$. Since a one unit increase on the $\log_{10}$ scale is equivalent to a 10-fold increase on the original scale, this implies that a 10-fold increase in Olsen P would predict a 2.64 unit increase in yield. In practice, this model applies only across a sevenfold increase, as the Olsen P measurements range from 2 to 14 units.

Prediction at the mean value of Olsen P = 7.095, with $\log_{10}(OlsenP) = 0.8510$, can then be made, using the notation of Section 12.5, as

$$\hat{\mu}(OlsenP = 7.095) = 1.674 + (2.644 \times 0.8510) = 3.923 \ .$$

TABLE 17.2

Parameter Estimates with Standard Errors (SE), t-Statistics (t) and Observed Significance Levels (P) for the SLR for Yield with Explanatory Variate, *LogOP* = $\log_{10}$(*OlsenP*) (Example 17.1A)

Term	Parameter	Estimate	SE	t	P
[1]	α	1.674	0.2518	6.646	< 0.001
LogOP	β	2.644	0.3044	8.684	< 0.001

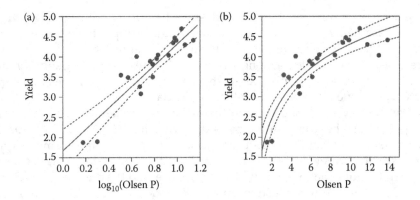

FIGURE 17.3
Yield with fitted model (—) and 95% CI (---) plotted against (a) $\log_{10}$(Olsen P) and (b) Olsen P values (Example 17.1A).

We can use the ResMS from the ANOVA ($s^2 = 0.1191$) and the results of Section 12.5 to form a 95% CI for this prediction as (3.756, 4.090). In Figure 17.3, the fitted line is plotted with 95% CI against both the log-transformed and the original Olsen P values. On the original scale, the fitted straight line becomes a curve, and the characteristic shape of the CI appears to change, although the width of the CI is the same at the equivalent points of the two x-axes.

Although the fitted model follows the overall trend in the data, there is a suggestion of model misspecification, as the fitted line lies above the observed yield at the extremes of the range. This is also clear in graphs of standardized residuals against the explanatory variate on either the transformed (Figure 17.4a) or original (Figure 17.4b) scale. Again, the same residuals are plotted in both parts of Figure 17.4, but the scale of the x-axis (and hence the trend line) has changed.

Finding an adequate transformation can be difficult and will not always be possible. In some cases, several models based on different transformations of the explanatory variate may appear plausible. The model chosen should give a good visual fit to the observations,

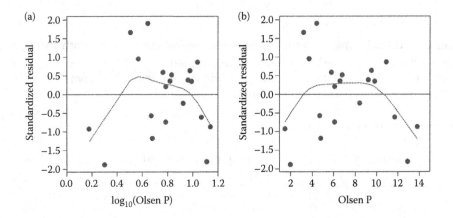

FIGURE 17.4
Residuals from fitted model with trend line (·····) plotted against (a) $\log_{10}$(Olsen P) and (b) Olsen P values (Example 17.1A).

show no evidence of misspecification, and should perform well with goodness-of-fit statistics. All else being equal, transformations with a simple biological interpretation should be preferred. However, remember that a biological interpretation is not essential for a purely descriptive model, as long as the results are interpreted appropriately. The more complex approaches described below should be considered if the relationship cannot be captured by a single transformed explanatory variate.

Simple transformation of one or more explanatory variates to achieve straight line relationships can also be useful in the context of MLR models (Chapter 14). Once a suitable transformation is identified, models are selected from the transformed explanatory variate(s). Take care, however, as the correlation between explanatory variates may distort the shape of individual relationships with the response, and the diagnostics of Section 14.6 can help to detect this. You can also extend MLR models by the addition of interactions between explanatory variates; this is discussed in Section 17.2.

17.1.2 Polynomial Models

It is often difficult to find a single transformation of an explanatory variate that can adequately describe a curved relationship. **Polynomial regression models** use several powers of the explanatory variate to introduce curvature into a relationship via a MLR model. Here, we consider only positive integer powers, i.e. x^q, where q is a whole number. The **order of a polynomial model** is equal to the highest power of the explanatory variate used. These models have the advantage that they are very flexible and can incorporate patterns of both increasing and decreasing response within a model, i.e. non-monotonic functions. Their major disadvantage is that high-order polynomials can lead to **over-fitting**, so that interpolation between observations can be unreliable. In addition, extrapolation may be unreliable even for low-order polynomials (see discussion later in this section).

The SLR model is the simplest case of a polynomial model, i.e. a straight line. Higher-order polynomial models are obtained by the addition of power transformations of the explanatory variate, such as x^2 or x^3, into the model. For example, a second-order polynomial, or **quadratic model**, includes the second power or square of the explanatory variate, and takes the form

$$y_i = \alpha + \beta_1 x_i + \beta_2 x_i^2 + e_i \ .$$

This model is fitted by the construction of a new explanatory variate with values equal to x_i^2, then by the fitting of a MLR model with two explanatory variates: the original and the squared values. So, if variate x contains the original values and x^2 contains the squared values, this MLR model is written in symbolic form as

Explanatory component: $[1] + x + x^2$

A polynomial model of order q has $p = q + 1$ parameters and can be written as

$$y_i = \alpha + \beta_1 x_i + \beta_2 x_i^2 + \ldots + \beta_{q-1} x_i^{q-1} + \beta_q x_i^q + e_i \ .$$

The sequential ANOVA table for a polynomial model of order q starts with the SLR model and then successively adds increasing powers of the explanatory variate into the model, giving incremental sums of squares and F-tests (see Section 14.4). Each power of the explanatory variate is associated with 1 df, and so a polynomial model of order q has q df. Predictions and CIs can be formed from the fitted model as for any MLR (Section 14.5).

EXAMPLE 17.1B: OLSEN P

Here, we consider polynomial regression as an alternative to the $\log_{10}$ transformation tried in Example 17.1A. We fit a cubic polynomial, written in symbolic form as

Response variable: *Yield*
Explanatory component: *[1] + OlsenP + (OlsenP)² + (OlsenP)³*

where the variate *Yield* holds the observations, *OlsenP* is the variate holding the Olsen P values and *(OlsenP)²* and *(OlsenP)³* represent variates holding the squared and cubed values of the *OlsenP* variate. Table 17.3 shows the sequential ANOVA table for this model, which accounts for 81.0% of the variation (adjusted $R^2 = 0.810$).

The incremental F-tests for the first two terms (linear and quadratic, $F_{1,16}^L = 62.063$, $F_{1,16}^Q = 20.559$, both $P < 0.001$) are significant, but the test for the cubic term is not ($F_{1,16}^C = 1.183$, $P = 0.293$). As the cubic term is added into the model last, its incremental F-test is also a marginal F-test and indicates that the fit of the model is not significantly worse if this term is dropped. We therefore drop this cubic term, and fit a quadratic model that accounts for 80.8% of the variation in the data (adjusted $R^2 = 0.808$), with all terms significant. The parameter estimates for the quadratic model are listed in Table 17.4.

Figure 17.5 shows the fitted quadratic and cubic polynomial models with 95% CIs. A difference in the fit of the two models appears for larger values of Olsen P, where the yield is stable: the quadratic model starts to move downwards, whereas the cubic model stays level. The 95% CIs are narrower in the centre of the range of the explanatory variate, and get much wider at the ends of the range (like the SLR models in Section 12.5). The CIs for the cubic model are wider than those for the quadratic model because the extra term reduces the ResSS only a small amount while introducing another estimated parameter with its associated uncertainty.

We check residual plots for evidence of model misspecification. Figure 17.6 shows standardized residuals from the quadratic and cubic models plotted against the explanatory variate. There is a suggestion of misspecification in both graphs, particularly at the smallest values of Olsen P.

Problems of collinearity can occur in polynomial models, particularly for higher-order models. For example, in Example 17.1B, large VIFs (> 100, see Section 14.7) are obtained when one fits the cubic polynomial. This can be avoided by the use of orthogonal polynomials

TABLE 17.3

Sequential ANOVA for a Cubic Polynomial Model for Yield with Explanatory Variate Olsen P (Example 17.1B)

Term Added	Incremental df	Incremental SS	Mean Square	Variance Ratio	P
+ *OlsenP*	1	6.9157	6.9157	$F^L = 62.063$	< 0.001
+ *(OlsenP)²*	1	2.2909	2.2909	$F^Q = 20.559$	< 0.001
+ *(OlsenP)³*	1	0.1318	0.1318	$F^C = 1.183$	0.293
Residual	16	1.7829	0.1114		
Total	19	11.1213	0.5853		

Note: SS = sum of squares.

TABLE 17.4

Parameter Estimates with Standard Errors (SE), t-Statistics (t) and Observed Significance Levels (*P*) for a Quadratic Polynomial Model for Yield with Explanatory Variate Olsen P (Example 17.1B)

Term	Parameter	Estimate	SE	t	P
[1]	α	1.283	0.3290	3.900	0.001
OlsenP	β_1	0.593	0.0964	6.156	< 0.001
(OlsenP)^2	β_2	−0.0279	0.00618	−4.510	< 0.001

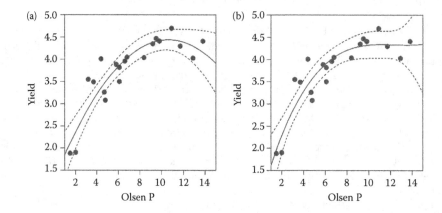

FIGURE 17.5

Observed yield and fitted curves (—) with 95% CI (---) for (a) quadratic and (b) cubic polynomial models (Example 17.1B).

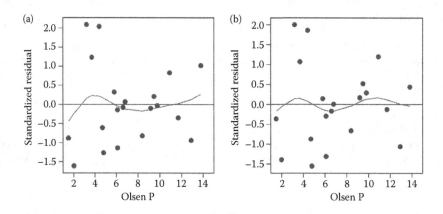

FIGURE 17.6

Residuals with trend line (·····) plotted against the explanatory variate for (a) quadratic and (b) cubic polynomial models (Example 17.1B).

rather than simple powers of the explanatory variate (see also Section 8.7). Orthogonal polynomials are constructed so that the *q*th function is of order *q* and is orthogonal to all of the lower-order functions. Figure 8.8 showed simple powers alongside the corresponding set of orthogonal polynomials. Orthogonal polynomials have zero pairwise correlations and so produce a stable model without collinearity problems. Their major disadvantage is that the estimated parameters are difficult to interpret, as the coefficients no longer relate to individual powers of the explanatory variate. We can deal with this problem by using orthogonal polynomials to establish the predictive model, and then refit it with simple powers of the explanatory variate to obtain interpretable parameter estimates. More information on orthogonal polynomials can be found in Bliss (1970). Alternatively, centering the original explanatory variate (see Section 12.9.1) before taking powers may reduce collinearity sufficiently to enable a stable model to be fitted (see Example 17.3B).

In general, polynomial models should be constructed sequentially by progressive addition of terms of higher orders, assisted by graphical inspection of the fit at each stage and sequential ANOVA tables. As in previous chapters, the aim is to produce a parsimonious model, i.e. a model of the lowest possible order that describes adequately the relationship between the response and explanatory variate. The model fitting process starts with a SLR model and higher-order powers of the explanatory variate are successively added. At each stage, lack of fit is tested if replicate observations are available (Section 12.8) and the fitted model and residual plots are examined visually for evidence of misspecification. If the fit appears inadequate, the next higher-order term can be added. If the fit appears good, then no further terms need to be added and the current model should be checked. The need for the highest-order term in the model should be verified by the use of a marginal F-test from the sequential ANOVA table. If this term is not statistically significant (e.g. $P > 0.05$), then it should be omitted, and a lower-order model will suffice. Once a suitable order for the polynomial has been established, all lower-order terms are retained in the model, even if not statistically significant. This strategy follows from our arguments on marginality (see Sections 8.3 and 15.5): we consider any lower-order power (x^k with $k < q$) to be marginal to a higher-order power (x^q). This also implies that we should fit lower powers of the explanatory variate before higher powers, as described above. Following this principle also ensures that the model can be translated to other scales if required, for example, from a model in terms of orthogonal polynomials to a model based on simple powers; an exact translation may not be possible if lower-order terms are omitted.

Polynomial models are essentially descriptive models, as there is rarely a biological interpretation for models of this form. An advantage of low-order polynomials is that they can flexibly adapt to follow the form of the relationship. However, as there is no constraint on the form of the curve outside the range of the explanatory variate, one should never extrapolate with these models.

As the order of the polynomial increases the residual sum of squares will decrease, and the fitted curve will pass closer to the observations. In fact, for a data set with k distinct values of the explanatory variate, a polynomial model of order $k - 1$ will pass through the mean at each value of the explanatory variate. If the observed values of the explanatory variate are unreplicated, then this curve fits each observation exactly. This perfect fit is counterproductive, as the model often becomes unreliable for interpolation as it attempts to accommodate detailed, and probably random, patterns in the relationship; this behaviour is known as over-fitting (see also Section 14.9). For example, Figure 17.7 shows a polynomial of order 8 fitted to the yield observations from Example 17.1. The fitted model has adapted to be much closer to the observations than the lower-order polynomials (Figure 17.5), but the interpolated model shows an unrealistic shape, particularly with respect to the sharp dips around Olsen P values of 1 and 13. This graph shows the importance of evaluating complex curved models at a dense set of explanatory variate values, as the full form of the curve (and any over-fitting) may not be apparent from the fit at the observed values of the explanatory variate.

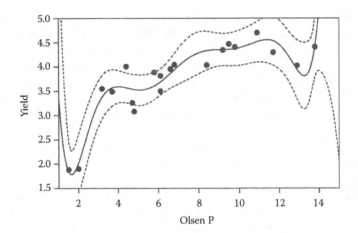

FIGURE 17.7
Observed yield from Example 17.1 with fitted polynomial model of order 8 (—) and 95% CI (– –).

Smoothing techniques provide an alternative to polynomial models, as they flexibly adapt to follow a curved relationship. These techniques constrain the roughness of the fitted curve and thus largely avoid problems with over-fitting. Regression splines can be implemented directly as a MLR model, while smoothing splines or locally weighted regression (loess) smoothers can be implemented as additive models with a penalized likelihood approach. These models are outside the scope of this book, but Ruppert et al. (2003) provide a good introduction.

17.1.3 Trigonometric Models for Periodic Patterns

Trigonometric regression models are MLR models used to describe periodic cycles, and are often used to model observations related to yearly or daily cycles, for example, mean monthly temperature as shown in Example 17.2. The period of the cycles, i.e. the number of time units corresponding to a full cycle, is assumed to be known and denoted as ω. Trigonometric regression models use sine and cosine transformations of the measurement times as explanatory variates. Recall that the sine and cosine functions are cyclic over time with a period of 2π radians. To convert our explanatory variate with period ω on to this scale, we use the transformations $\sin(2\pi t/\omega)$ and $\cos(2\pi t/\omega)$, where t is a variate of observed measurement times. For example, a simple trigonometric regression model for monthly data that exhibit yearly cycles, with $\omega = 12$, takes the form

$$y_i = \alpha + \beta_1 \sin\left(\frac{2\pi t_i}{12}\right) + \beta_2 \cos\left(\frac{2\pi t_i}{12}\right) + e_i \, , \tag{17.1}$$

where y_i is the ith observation made at time t_i with deviation e_i, and the unknown model parameters are α, β_1 and β_2. This is a MLR model with two explanatory variates, which are calculated as the sine and cosine functions of the observed times, t_i. This model can be converted into a more interpretable form as a single sine function, written as

$$y_i = \alpha + \gamma \sin\left(\frac{2\pi t_i}{12} - \theta\right) + e_i \, .$$

In this form, the parameters are the average response over a full cycle (α), the amplitude of the sine curve (equal to half of the range of the curve, γ) and the phase of the curve (θ). The phase is the lag behind the standard sine curve (which has its maximum at $\pi/2$ and minimum at $3\pi/2$ radians). The amplitude, γ, must always be non-negative ($\gamma \geq 0$). Using standard results for trigonometric functions, we can expand the sine function in the equation above to give

$$y_i = \alpha + \gamma \sin\left(\frac{2\pi t_i}{12}\right)\cos(\theta) - \gamma\cos\left(\frac{2\pi t_i}{12}\right)\sin(\theta) + e_i .$$

If we set $\beta_1 = \gamma\cos(\theta)$ and $\beta_2 = -\gamma\sin(\theta)$, then this is the model in Equation 17.1. We can therefore calculate the amplitude and phase of the fitted curve in terms of the original parameters as

$$\gamma = \sqrt{\beta_1^2 + \beta_2^2}, \quad \theta = \tan^{-1}(-\beta_2/\beta_1) .$$

This estimate of the phase is in radians and we can convert it to the scale of measurement by multiplying by $\omega/2\pi$. Although inference and SEs for estimates of the original parameters (α, β_1 and β_2) follow directly from properties of MLR models (Chapter 14), SEs for estimates of γ and θ are not straightforward, as these are non-linear functions of the original parameters. Approximate SEs can be calculated in statistical software by the delta method (see Casella and Berger, 2002).

EXAMPLE 17.2: ROTHAMSTED MONTHLY MEAN TEMPERATURE

The monthly mean temperatures at Rothamsted Experimental Station over the period 1891–1990 are listed in Table 17.5 and can be found in file TEMPERATURE.DAT. For this response (held in variate *Temperature*), we expect a yearly cycle and so trigonometric regression is appropriate.

The explanatory variate, *Month*, has values 1–12. To obtain cycles of period 12, i.e. equal to 1 year, we calculate explanatory variates *Sin* and *Cos* as

$$Sin = \sin(2\pi Month/12), \quad Cos = \cos(2\pi Month/12) .$$

The model can then be written in symbolic form as

Response variable: *Temperature*
Explanatory component: *[1] + Sin + Cos*

TABLE 17.5

Monthly Mean Temperatures (°C) at Rothamsted (UK) over the Period 1891–1990 (See Example 17.2 and File TEMPERATURE.DAT)

Month	Month	Temperature	Month	Month	Temperature
January	1	3.1	July	7	16.0
February	2	3.4	August	8	15.7
March	3	5.3	September	9	13.5
April	4	7.7	October	10	9.8
May	5	11.1	November	11	5.9
June	6	14.0	December	12	4.0

Source: Data from Rothamsted Research.

TABLE 17.6

Parameter Estimates with Standard Errors (SE), t-Statistics (t) and Observed Significance Levels (*P*) for a Trigonometric Regression for Monthly Mean Temperature at Rothamsted (Example 17.2)

Term	Parameter	Estimate	SE	t	P
[1]	α	9.12	0.139	65.432	< 0.001
Sin	β_1	−4.09	0.197	−20.722	< 0.001
Cos	β_2	−5.13	0.197	−26.007	< 0.001

The fitted model accounted for 99.0% of the variation in the data (adjusted $R^2 = 0.990$), and the parameter estimates are listed in Table 17.6.

The overall mean temperature is equal to the model intercept, which is estimated as 9.12°C. The amplitude of the fitted curve is calculated as

$$\hat{\gamma} = \sqrt{\hat{\beta}_1^2 + \hat{\beta}_2^2} = \sqrt{(-4.09)^2 + (-5.13)^2} = 6.56 \,,$$

hence the range of monthly temperatures is twice this value, equal to 13.12°C. Finally, we can consider the phase. We have

$$\hat{\theta} = \tan^{-1}(-\hat{\beta}_2/\hat{\beta}_1) = \tan^{-1}[5.13/(-4.09)] = \tan^{-1}(-1.255) \,,$$

and this has two solutions, $\hat{\theta} = 2.244$ and $\hat{\theta} = 5.385$ in the range $0 \leq \hat{\theta} < 2\pi$. The relationships $\hat{\beta}_1 = \hat{\gamma}\cos(\hat{\theta})$ and $\hat{\beta}_2 = -\hat{\gamma}\sin(\hat{\theta})$ tell us that $\cos(\hat{\theta}) < 0$ and $\sin(\hat{\theta}) > 0$, and hence $\pi/2 \leq \hat{\theta} < \pi$ radians (i.e. $1.57 \leq \hat{\theta} < 3.14$ radians) giving the solution $\hat{\theta} = 2.244$. This is then translated onto the scale of the time variate by division by 2π then multiplication by 12 to give the phase as 4.285 months. In the standard sine curve, the maximum and minimum occur at one-quarter and three-quarters of the period of the whole cycle, equivalent to three and nine months from the start of the yearly cycle here. In the fitted model, we therefore predict the maximum at $3 + 4.285 = 7.285$ months (between July and August) and the minimum at $9 + 4.285 = 13.285$ months, which because of the 12 month cycle is equivalent to 1.285 months into the year (between January and February). These features of the fitted curve can be verified in Figure 17.8, where the curve is extrapolated back to month 0 to demonstrate its periodicity. This graph suggests slight model misspecification at the extremes of the range, the fitted temperatures seem slightly too low at both the minimum and maximum points, but the fitted model describes the overall pattern well.

One difficulty with trigonometric regression is that the observations are often collected as time series or repeated measurements from the same unit, for example, monthly temperatures at a single site over several years. This often gives rise to serial correlation in the deviations, which contradicts the assumptions of independence underlying regression (see Section 12.1). Example 17.2 avoids this problem by using mean temperatures accumulated over 100 years; averaging over so many years dilutes the influence of serial correlations within years. Where strong serial correlation is present, methods for analysis of time series of longitudinal data that account for serial correlation should be used. More details about models for longitudinal data or repeated measurements can be found in Diggle et al. (2002). If the length of the cyclic period is unknown and has to be estimated, then this is no longer a linear model and the methods of Section 17.3 must be used.

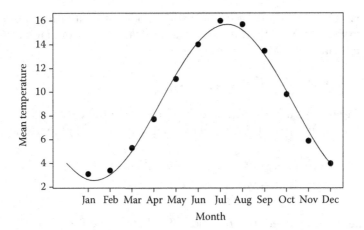

FIGURE 17.8
Mean monthly temperature (°C) at Rothamsted Experimental Station over the period 1891–1990 with fitted curve from trigonometric regression (Example 17.2).

17.2 Curved Surfaces as Functions of Two or More Variates

In Chapter 14, we considered MLR models that had several explanatory variates that acted independently. For example, in Example 14.1B, we modelled seed weight as a function of seed length and hardness. In these models, the change in the response due to one variate (e.g. length) is assumed to be the same regardless of the value of the other variate (e.g. hardness). For two explanatory variates, the resulting model can be represented as a plane in three-dimensional space (see Figure 14.2). This model is not always realistic, as the true three-dimensional surface might be curved rather than planar, which requires that the change in the response due to one variate depends on the value of the other variate. For example, the change in seed weight due to a change in length might also depend on the seed hardness. We can model some types of curvature by including an interaction between explanatory variates, and this is the subject of this section.

For simplicity, we start with the most basic MLR model based on two variates and written in the form

$$y_i = \alpha + \beta_1 x_{1i} + \beta_2 x_{2i} + e_i \, , \tag{17.2}$$

where y_i is the value of the ith observation, x_{1i} and x_{2i} are the corresponding values of the two explanatory variates and e_i is the deviation for that observation. The model parameters are the intercept α and the slopes, β_1 and β_2, respectively, for the two explanatory variates. We can introduce curvature into the model by including a new term, which combines the two variates, giving the model

$$y_i = \alpha + \beta_1 x_{1i} + \beta_2 x_{2i} + \beta_3 (x_{1i} \times x_{2i}) + e_i \, . \tag{17.3}$$

The extra term is equivalent to a new variate calculated by multiplication of the values of the two explanatory variates for each observation (e.g. $x_{3i} = x_{1i} \times x_{2i}$). In the spirit of crossed models (see Section 8.2), we might think of this model as containing the main effects of each explanatory variate plus their interaction. If x_1 is the first explanatory variate and x_2 is the second explanatory variate, we can write this model in symbolic form as

Explanatory component: $[1] + x_1 + x_2 + x_1.x_2$

The term $x_1.x_2$ represents a variate holding the product of the values of the two individual variates. We can also interpret this model by rewriting it in a slightly different form as

$$y_i = (\alpha + \beta_2 x_{2i}) + (\beta_1 + \beta_3 x_{2i}) x_{1i} + e_i$$
$$= \alpha^* + \beta^* x_{1i} + e_i . \tag{17.4}$$

Here, the model is considered as a function of values of the first explanatory variate, x_{1i}, for a specified constant value of the second variate, x_{2i}. In this form, we can see that this is like a SLR model in terms of the first explanatory variate where both the intercept (here $\alpha^* = \alpha + \beta_2 x_{2i}$) and slope (here $\beta^* = \beta_1 + \beta_3 x_{2i}$) depend on the value of the second explanatory variate. A similar interpretation in terms of the second variate can be formed by reversal of the roles of the two explanatory variates.

Including the combined term with both explanatory variates allows curvature in the fitted surface, but this curvature is of a specific form, so it is important to ensure that this matches the pattern seen in the data. We can check this by plotting residuals against both variates and by comparing the form of the observed and fitted surfaces using contour or surface plots. In the case of a designed experiment with replication, we can formally test for lack of fit (see Section 12.8) by fitting a factor version of the combined variates.

The sequential ANOVA table for the model of Equation 17.3 has three terms: one for each of the explanatory variates and one for the combined term, each with 1 df. The incremental sums of squares for the individual variates depend on the order in which they are fitted unless they are orthogonal (see Section 14.4). As usual, the aim is to find a parsimonious description of the response, so the simplest possible predictive model is sought. However, this process must again respect marginality, and both explanatory variates are marginal to the combined term. The individual explanatory variates should therefore be fitted before the combined term, and should not be dropped while the combined term is in the model.

EXAMPLE 17.3A: COTTON RESPONSE TO HERBICIDE AND INSECTICIDE

An experiment was done to evaluate the combined effects of five different doses of herbicide (0, 20, 40, 60 and 80 lb/acre) and five different doses of insecticide (0.0, 0.5, 1.0, 1.5 and 2.0 lb/acre) on the root growth of cotton plants in containers within a glasshouse. Four replicates of each treatment combination were arranged in a CRD. After three weeks, the dry root biomass (g/plant) was measured for each container. The treatment means are presented in Table 17.7 and in file COTTON.DAT. The residual mean square from the factorial model analysis of the raw data was 174 on 75 df.

In mathematical form, denoting the ith observation of biomass ($Biomass_i$) as a function of the herbicide ($Herbicide_i$) and insecticide ($Insecticide_i$) doses enables us to write a linear model with interaction (Equation 17.3) as

$$Biomass_i = \alpha + \beta_1 Herbicide_i + \beta_2 Insecticide_i + \beta_3 (Insecticide_i \times Herbicide_i) + e_i .$$

If the variates H and I contain the herbicide and insecticide doses, respectively, this model can be written in the symbolic form

TABLE 17.7

Dry Root Biomass (g/Plant) of 3-Week-Old Cotton Plants from an Experiment Evaluating the Effects of Different Amounts of Herbicide (H) and Insecticide (I) (Example 17.3 and File COTTON.DAT)

H	I	Weight	H	I	Weight	H	I	Weight
0	0	122.00	1	0	52.00	2	0	29.25
0	20	82.75	1	20	71.50	2	20	72.00
0	40	65.75	1	40	79.50	2	40	82.50
0	60	68.00	1	60	68.75	2	60	68.25
0	80	57.50	1	80	63.00	2	80	73.25
0.5	0	72.50	1.5	0	36.25			
0.5	20	84.75	1.5	20	80.50			
0.5	40	68.75	1.5	40	65.75			
0.5	60	70.00	1.5	60	77.25			
0.5	80	60.75	1.5	80	69.25			

Source: Data from Kuehl, R.O. 2000. *Design of Experiments: Statistical Principles of Research Design and Analysis* (2nd edition). Thomson Learning (Duxbury Press), Pacific Grove, California. 666 pp.

> Response variable: **Biomass**
> Explanatory component: *[1] + H + I + H.I*

where *H.I* can be calculated as the product of the variates *H* and *I*. The sequential ANOVA table for this model (Table 17.8) partitions variation between the treatment means into that accounted for by the explanatory terms and a remainder, which can be used to test lack of fit (see Section 12.8). The residual is calculated from variation between replicates and is an estimate of pure error, uncontaminated by lack of fit, and so we choose to use this residual for testing model terms.

We can first test the model as a whole by comparing the model mean square with the residual mean square ($F_{3,75} = 6.06$, $P < 0.001$) and this model accounts for 15.8% of the variation in the data (adjusted $R^2 = 0.158$). As we can partition the variation into pure error and treatment variation, we can also calculate the percentage of treatment variation accounted for by comparing the remainder mean square (157 with 21 df) with the treatment mean square (291 with 24 df) as

$$1 - \frac{\text{Remainder MS}}{\text{Treatment MS}} = 1 - \frac{3295/21}{(573 + 5 + 3102 + 3295)/24} = 0.460 .$$

TABLE 17.8

Sequential ANOVA Table for Cotton Root Biomass Model in Terms of Variates *H* (Herbicide Dose), *I* (Insecticide Dose) and the Combined Term *H.I* (Example 17.3A)

Change	Incremental df	Incremental SS	Mean Square	Variance Ratio	P
+*H*	1	573	573	$F^H = 3.29$	0.074
+*I*	1	5	5	$F^I = 0.03$	0.866
+*H.I*	1	3102	3102	$F^{H.I} = 17.83$	< 0.001
Remainder	21	3295	157	$F^{Rem} = 0.90$	0.589
Residual	75	13,050	174		
Total	99	20,025			

Note: SS = sum of squares.

TABLE 17.9

Parameter Estimates with Standard Errors (SE), t-Statistics (t) and Observed Significance Levels (*P*) for Cotton Root Biomass Model in Terms of Variates *H* (Herbicide Dose), *I* (Insecticide Dose) and Their Interaction *H.I* (Example 17.3A)

Term	Parameter	Estimate	SE	t	*P*
[1]	α	99.350	7.915	12.553	< 0.001
H	β_1	−29.050	6.4622	−4.495	< 0.001
I	β_2	−0.573	0.1616	−3.545	< 0.001
H.I	β_3	0.557	0.1319	4.223	< 0.001

This model therefore accounts for 46.0% of the treatment variation. We then consider the individual model terms. Because of the balanced allocation of treatments, variates *H* and *I* are orthogonal, and so we get the same incremental SS and tests for these terms fitted in either order. But we first examine the combined term, *H.I*, to see if we can simplify the model and find that the variance ratio ($F_{1,75}^{H.I} = 17.830$, $P < 0.001$) is highly significant and so we cannot. The predictive model therefore uses both variates and their combined term, and the estimated parameters are listed in Table 17.9. The remainder sum of squares gives no evidence of lack of fit ($F_{21,75}^{Rem} = 0.902$, $P = 0.589$).

In the form of Equation 17.4, with biomass as a function of herbicide dose for a given value of insecticide dose, the predictive model can be written as

$$\hat{\mu}(H,I) = (99.35 - 0.573I) + (-29.05 + 0.557I)H ,$$

where for brevity now *H* indicates the herbicide dose and *I* indicates insecticide dose applied. As the insecticide dose increases, the slope for herbicide increases and the intercept decreases. The fitted model is plotted with the observations in this form in Figure 17.9, and it can be seen that the slope is negative for small doses of insecticide and positive for larger doses.

Although a straight line seems a reasonable approximation to the shape for each fungicide dose, it is apparent that the fitted lines are not giving the best possible fit to the observations: the slope is clearly too gentle for the zero dose and too steep for the largest dose. Figure 17.10 shows contour plots for the observations and for the fitted model, which allows a visual comparison of the observed and fitted surfaces. Although there is some similarity across the two surfaces in terms of general trends, it is clear that the fitted model does not reproduce the observed trends well. This contradicts the non-significant test for lack of fit (based on F^{Rem}), which indicates that the discrepancy between the fitted values and treatment means is small compared with background variation. However, the presence of systematic (rather than random) discrepancies in the fitted model, as seen in Figure 17.9, suggests that some improvement in fit may be possible.

An interaction between variates introduces one type of curvature into the fitted surface, but more general forms will often be required. The simplest generalization is to extend the methodology used for polynomial models (Section 17.1.2) to two dimensions. A model of order *q* then contains all combinations of the explanatory variates with powers that sum to $\leq q$. A first-order model for two explanatory variates contains both individual variates (order 1) but not their interaction, which is of order 2; this model is a standard MLR (Equation 17.2). A second-order model adds the combined term and the squares of both variates. For convenience, we label the coefficients in these models by the powers of

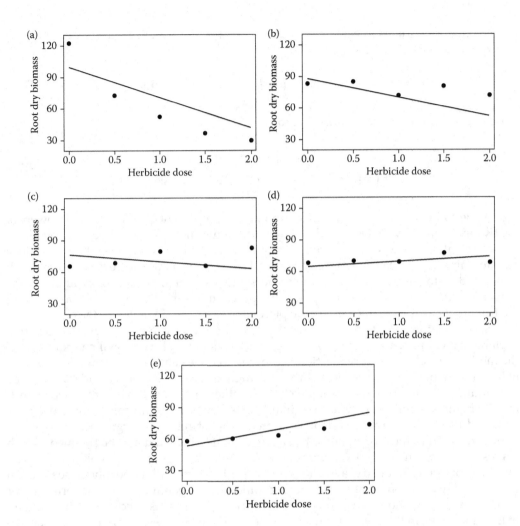

FIGURE 17.9

Observed cotton root dry biomass (g/plant) with predictive model in terms of herbicide dose (lb/acre), insecticide dose (a) 0, (b) 20, (c) 40, (d) 60, (e) 80 lb/acre, and their interaction (Example 17.3A).

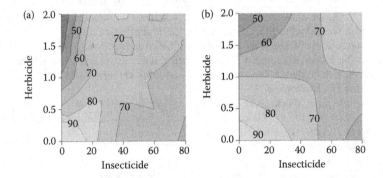

FIGURE 17.10

(a) Observed cotton root dry biomass (g/plant) and (b) predictive model for cotton root biomass in terms of herbicide dose (lb/acre), insecticide dose (lb/acre) and their interaction (Example 17.3A).

the explanatory variates in each term; hence, β_{11} is the slope associated with the product $x_{1i} \times x_{2i}$. The second-order model can be written as

$$y_i = \alpha + \beta_{10}x_{1i} + \beta_{01}x_{2i} + \beta_{20}x_{1i}^2 + \beta_{11}(x_{1i} \times x_{2i}) + \beta_{02}x_{2i}^2 + e_i \ . \tag{17.5}$$

Similarly, a third-order model takes the form

$$\begin{aligned} y_i = \alpha &+ \beta_{10}x_{1i} + \beta_{01}x_{2i} + \beta_{20}x_{1i}^2 + \beta_{11}(x_{1i} \times x_{2i}) + \beta_{02}x_{2i}^2 \\ &+ \beta_{30}x_{1i}^3 + \beta_{21}(x_{1i}^2 \times x_{2i}) + \beta_{12}(x_{1i} \times x_{2i}^2) + \beta_{03}x_{2i}^3 + e_i \ . \end{aligned} \tag{17.6}$$

Again, the components $x_{1i} \times x_{2i}$, $x_{1i}^2 \times x_{2i}$ and $x_{1i} \times x_{2i}^2$ represent combinations of powers of the two explanatory variates, calculated by multiplying the appropriate powers together.

The potential problems associated with polynomial models for a single explanatory variate, namely, collinearity between terms and over-fitting, may also be encountered with these models for several variates. The same solutions also apply, so collinearity can be reduced by the use of centered or orthogonal polynomials in place of simple powers, and the full fitted curve or surface should be plotted on a dense grid of values to check for any undesirable features. The previous model-building strategy can also be extended to two explanatory variates, so we start with a low-order model and use visual checks to see if the model fit is adequate. If replication is present then we can make a formal test for lack of fit. If the model is not adequate, then a set of higher-order terms can be added. Once you have found a suitable order, you should check whether the model can be simplified by testing the highest-order terms with marginal F-tests. The least significant term is dropped first, and then other terms retested. This process must respect marginality, so that if a term is retained in the model then all terms marginal to it should also be retained (which makes the model invariant to changes of scale). At each stage, a term is eligible for testing if it is not marginal to (i.e. a sub-term of) any other term still in the model (see Section 8.3.1). A sub-term is one that has all of its components in common with the term, so, for example, x_{1i}, x_{1i}^2, x_{2i} and $x_{1i} \times x_{2i}$ are all sub-terms of $x_{1i}^2 \times x_{2i}$. This process is demonstrated in Example 17.3B. In the predictive model, all terms eligible for testing should have statistically significant marginal F-tests.

EXAMPLE 17.3B: COTTON RESPONSE TO HERBICIDE AND INSECTICIDE

In Example 17.3A, we detected systematic discrepancies between the observed biomass and the predictive model with individual variates H (herbicide dose), I (insecticide dose) and their interaction, $H.I$. Here, we consider higher-order models to see if a better fit can be obtained. To avoid problems with collinearity, we centre each variate before calculating powers. The centered variates are calculated as $cH = H - 1$ and $cI = I - 40$, using variates defined in Example 17.3A.

For a second-order model, we need the main variates cH, cI, and the products $cH.cI = cH \times cI$, $(cH)^2 = cH \times cH$ and $(cI)^2 = cI \times cI$. The second-order model, from Equation 17.5, can then be written in symbolic form as

Response variable: *Biomass*
Explanatory component: *[1] + cH + cI + (cH)² + cH.cI + (cI)²*

This model accounts for 17.5% of the variation in the data (and 52.3% of the variation in the treatment means). Examination of the fitted curves and surface (as in Figures 17.9

and 17.10) shows that although this fitted model is closer to the observations, there are still clear systematic discrepancies visible. We therefore try a third-order model, as presented in Equation 17.6. This model requires four additional variates to be calculated, i.e. the cubes of the original variates and the products of the squared and linear terms, and can be written in symbolic form as

Explanatory component: $[1] + cH + cI + (cH)^2 + cH.cI + (cI)^2 + (cH)^3 + (cH)^2.cI$
$+ cH.(cI)^2 + (cI)^3$

Because the design is orthogonal, and we use the estimate of pure error as our residual mean square, we get a single sequential ANOVA table for this model shown in Table 17.10.

In the full model, we first test the cube terms and both the herbicide dose $(cH)^3$ $(F^{H3} = 0.03, P = 0.871)$ and insecticide dose $(cI)^3$ $(F^{I3} = 0.94, P = 0.336)$ terms can be removed from the model. Because the terms are orthogonal, we can immediately examine the third-order cross-product terms from the same table, and drop the product of the square of herbicide dose with insecticide dose $(cH)^2.cI$ $(F^{H2.I} = 1.36, P = 0.247)$. The square of herbicide dose $(cH)^2$ is then eligible for testing, and this term can also be removed from the model $(F^{H2} = 0.63, P = 0.432)$. No further terms can be removed. Parameter estimates in the final model are listed in Table 17.11, giving the predictive model as

$$\hat{\mu}(H,I) = 75.29 + 4.84(H - 1) - 0.016(I - 40) + 0.557(H - 1)(I - 40)$$
$$- 0.0070(I - 40)^2 - 0.01452(H - 1)(I - 40)^2 .$$

Because we have constructed the model in terms of centered variates, these centered variates must appear in the predictive model.

If we expand each term in full and gather together the coefficients for each combination of variables, the predictive model can be rewritten as

$$\hat{\mu}(H,I) = 105.34 - 40.66H - 1.17I + 1.718HI + 0.0071I^2 - 0.0145HI^2$$
$$= 105.34 - 1.17I + 0.0071I^2 + (-40.66 + 1.718I - 0.0145I^2)H .$$

TABLE 17.10

Sequential ANOVA Table for Third-Order Polynomial Model for Cotton Root Biomass Models in Terms of Centered Variates cH (Herbicide Dose) and cI (Insecticide Dose) (Example 17.3B)

Change	Incremental df	Incremental SS	Mean Square	Variance Ratio	P
+ cH	1	573	573	$F^H = 3.29$	0.074
+ cI	1	5	5	$F^I = 0.03$	0.866
+ $(cH)^2$	1	109	109	$F^{H2} = 0.63$	0.432
+ $cH.cI$	1	3102	3102	$F^{H.I} = 17.83$	< 0.001
+ $(cI)^2$	1	553	553	$F^{I2} = 3.18$	0.079
+ $(cH)^3$	1	5	5	$F^{H3} = 0.03$	0.871
+ $(cH)^2.cI$	1	237	237	$F^{H2.I} = 1.36$	0.247
+ $cH.(cI)^2$	1	1180	1180	$F^{H.I2} = 6.78$	0.011
+ $(cI)^3$	1	163	163	$F^{I3} = 0.94$	0.336
Remainder	15	1049	70	$F^{Rem} = 0.40$	0.975
Residual	75	13,050	174		
Total	99	20,025			

Note: SS = sum of squares.

TABLE 17.11

Parameter Estimates with Standard Errors (SE), t-Statistics (t) and Observed Significance Levels (P) for Cotton Root Biomass Predictive Model in Terms of Centered Variates cH ($=H-1$) and cI ($=I-40$) (Example 17.3B)

Term	Parameter	Estimate	SE	t	P
[1]	α	75.29	4.111	18.313	< 0.001
cH	β_{10}	4.844	5.814	0.833	0.407
cI	β_{01}	−0.016	0.0933	−0.169	0.866
$cH.cI$	β_{11}	0.557	0.1319	4.223	< 0.001
$(cI)^2$	β_{02}	−0.0070	0.00394	−1.783	0.079
$cH.(cI)^2$	β_{12}	−0.01452	0.005574	−2.604	0.011

In terms of herbicide (H), the predictive model is still a straight line for a given value of insecticide, but both the intercept and the slope vary as a quadratic function of insecticide dose. These straight lines are shown in Figure 17.11, and clearly provide a much better fit to the data than those from the simpler model shown in Figure 17.9. The shape of the fitted surface, shown as a contour plot in Figure 17.12, also appears a more reasonable fit to the observed surface. This model accounts for 23.2% of the total variation, and 71.7% of the treatment variation.

Given that we detected no formal evidence of lack of fit in our original model (linear plus interaction, Example 17.3A), we should check that our final model gives a quantifiable improvement in fit. For these two nested models, we can construct an F-test based on the change in the model sum of squares and df on adding the extra terms into the model (see end of Section 14.4). For the linear plus interaction model, we found $ModSS_1 = 3680$ with $ModDF_1 = 3$, compared to $ModSS_2 = 5414$ with $ModDF_2 = 5$ for our final predictive model. We compare this change to our estimate of background variation, i.e. ResMS = 174 on 75 df. The F-statistic is calculated as

$$\frac{(ModSS_2 - ModSS_1)/(ModDF_2 - ModDF_1)}{ResMS} = \frac{(5414 - 3680)/(5 - 3)}{174} = \frac{1734/2}{174} = 4.98 \,,$$

with 2 and 75 df, giving $P = 0.009$. There is thus strong evidence that the final model gives a better fit compared to the simpler model. This example demonstrates that the lack-of-fit test sometimes lacks power; it might be possible to improve a model even when the formal test for lack of fit is not statistically significant.

Example 17.3 was a designed experiment, and so a balanced set of combinations of the two variates had been used, which made the variates orthogonal and which greatly simplified the process of model selection. This is much less likely to occur in observational data, especially where variates are correlated. In general, many more observations are required to get a good spread of observations across two explanatory variates than for one, a situation known as the **curse of dimensionality**. Good coverage of the two-dimensional space spanned by two explanatory variates is essential if the model is to be robust across the full space, and you can check this by plotting the explanatory variates against each other. Predictions for regions with few observations should be treated as extrapolation and can be unreliable.

The models presented in this section, sometimes called **response surface** models, can be extended to three or more explanatory variates, but verification of the form of the model becomes much harder, because a full visual representation of the model requires four or

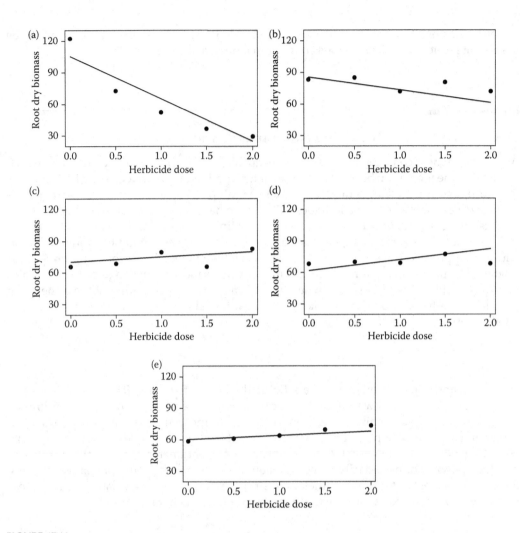

FIGURE 17.11
Observed cotton root dry biomass (g/plant) with predictive model in terms of herbicide dose (lb/acre) and insecticide dose (a) 0, (b) 20, (c) 40, (d) 60, (e) 80 lb/acre (Example 17.3B).

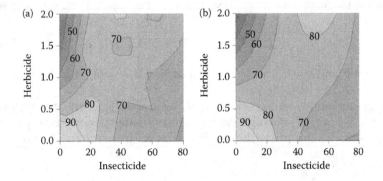

FIGURE 17.12
(a) Observed cotton root dry biomass (g/plant) and (b) predictive model for cotton root biomass in terms of herbicide dose (lb/acre) and insecticide dose (lb/acre) (Example 17.3B).

more dimensions. The curse of dimensionality also means that coverage is likely to be inadequate unless the observations come from a designed experiment.

17.3 Fitting Models Including Non-Linear Parameters

All of the models considered so far have been within the class of linear models, which means that the model can be written as a set of terms added together, each of which consists of an unknown coefficient (a model parameter) multiplied by a known value (an explanatory variable); for example, see Equation 17.2. As we have seen earlier in this chapter, these models can be used to fit curved as well as straight line relationships. However, these strategies provide only a limited range of models, and in some cases, a good fit cannot be found with this approach. The set of possible models can be widened by introducing **non-linear models**, i.e. models that cannot be written in linear form. An advantage of these models is that for some types of response, where there is a good understanding of the underlying process, a non-linear model with biologically meaningful parameters may be constructed.

One simple example of a non-linear model takes the form

$$y_i = \alpha + \beta x_i^{\theta} + e_i \,,$$

where α, β and θ are parameters to be estimated. If θ was fixed (e.g. $\theta = 2$), so the quantity x_i^{θ} was known, then this would be a linear model; it is the presence of θ as an unknown parameter that makes this model non-linear. Non-linear models can include several explanatory variates, although we consider only the case of a single explanatory variate here. In general, they may include several non-linear parameters, and so the number of parameters will not necessarily be one greater than the number of explanatory variates. For now, we label the set of p parameters in a non-linear model as $\gamma_1 \ldots \gamma_p$. Any non-linear model with a single explanatory variate can be written in general terms as

$$y_i = f(x_i, \gamma_1 \ldots \gamma_p) + e_i \,,$$

where y_i is the ith observation with value x_i of the explanatory variate and deviation e_i, and $f(x, \gamma_1 \ldots \gamma_p)$ gives the form of the non-linear function. For the example given above, we have

$$f(x_i, \gamma_1, \gamma_2, \gamma_3) = \gamma_1 + \gamma_2 x_i^{\gamma_3} \,,$$

where $\gamma_1 = \alpha$, $\gamma_2 = \beta$ and $\gamma_3 = \theta$. Our symbolic notation does not adapt easily to this framework, and so we do not use it here.

As for the linear models of earlier sections, we obtain least squares estimates for the parameters, but the process must be modified to estimate the non-linear parameters. The least squares estimates minimize the residual sum of squares (see Section 1.5), which for non-linear regression takes the form

$$\text{ResSS} = \min \sum_{i=1}^{N} (y_i - \hat{y}_i)^2 = \min \sum_{i=1}^{N} [y_i - f(x_i, \hat{\gamma}_1 \ldots \hat{\gamma}_p)]^2 \,.$$

Unlike the corresponding equations for linear models, these equations cannot be solved directly and an iterative algorithm is required to search numerically for the least squares estimates (see Seber and Wild, 1989, for further information). This algorithm might be unstable if it starts far from the solution: it might fail to converge, or it might appear to converge but at a point that does not give a global minimum of the ResSS (known as a local minimum). For this reason, you should provide good initial values for parameters whenever possible, for example, from previous related work or prior knowledge. Alternatively, several different sets of initial parameter values, for example, covering a regular grid, can be tried and the fitted model with the smallest ResSS is selected. Always plot the model with the observations to ensure that the fit is adequate. The assumptions presented in Sections 4.1 and 12.1 also apply to non-linear models, and residuals should be examined to check for model misspecification and the validity of the assumptions with the graphical diagnostic tools of Chapters 5 and 13.

Some of the most common non-linear curves are the exponential, logistic, Gompertz and inverse linear models and we discuss these briefly here, with some typical curve shapes shown in Figure 17.13.

The standard **exponential model** has the form

$$y_i = \alpha + \beta \exp(-\gamma x_i) + e_i \,. \tag{17.7}$$

Interpretation of individual parameters is not straightforward, but we might think of α as setting the level of the curve (analogous to an intercept), β as controlling the scaling (or effective range) of the curve and γ as controlling the curvature. The direction of this curve (and those considered below) varies according to the signs of the parameters β and γ. The

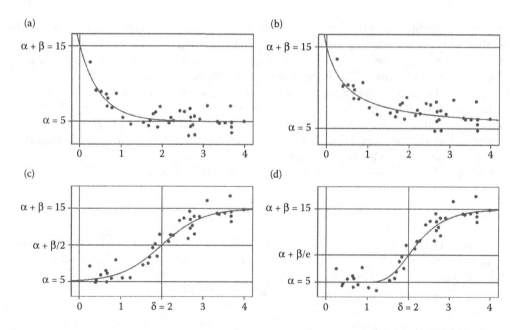

FIGURE 17.13
(a) Exponential curve with $\alpha = 5$, $\beta = 10$, $\gamma = 2$; (b) inverse linear curve with $\alpha = 5$, $\beta = 10$, $\gamma = 2$; (c) logistic curve with $\alpha = 5$, $\beta = 10$, $\gamma = 2$, $\delta = 2$; (d) Gompertz curve with $\alpha = 5$, $\beta = 10$, $\gamma = 2$, $\delta = 2$. Observations generated as function plus Normal deviations with common variance.

form for $\beta > 0$ and $\gamma > 0$ is shown in Figure 17.13a. Here, the value of the curve decreases as the explanatory variate increases, it crosses the y-axis at intercept value $\alpha + \beta$, and decreases towards a lower asymptote at value α. The rate of change decreases as the value of the explanatory variate increases. This curve is said to have a lower asymptote (limit) to the right (as the explanatory variate increases). An alternative to the exponential model is given by the **inverse linear model**, with form

$$ y_i = \alpha + \frac{\beta}{1 + \gamma x_i} + e_i . $$

For $\beta > 0$ and $\gamma > 0$ (Figure 17.13b), this curve also has a decreasing form with an intercept of $\alpha + \beta$ and a lower right asymptote of α, but is less sharply curved than the exponential model and approaches the asymptote more slowly. Varying the signs of parameters β and γ gives curves of different shapes. The exponential and inverse linear models both give decreasing functions with a lower right asymptote when $\beta > 0$ and $\gamma > 0$; increasing functions with a lower left asymptote with $\beta > 0$ and $\gamma < 0$; increasing functions with an upper right asymptote with $\beta < 0$ and $\gamma > 0$; and decreasing functions with an upper left asymptote when $\beta < 0$ and $\gamma < 0$. These models are often used for modelling growth curves or decay functions.

For S-shaped growth curves, we consider the logistic and Gompertz models. The **logistic model** takes the form

$$ y_i = \alpha + \frac{\beta}{1 + \exp[-\gamma(x_i - \delta)]} + e_i . $$

Again, we might think of α as setting the level of the curve, β as controlling the scale (or effective range) of the curve, γ controlling the curvature and the new parameter δ as defining the positioning of the curve with respect to values of the explanatory variate. A logistic model is shown in Figure 17.13c in the form with $\beta > 0$ and $\gamma > 0$. This curve has a lower left asymptote at value α as the explanatory variate decreases and an upper right asymptote at value $\alpha + \beta$ as the explanatory variate increases. The curvature is symmetric about δ, which is the value of the explanatory variate at which the slope of the curve is steepest (known as the inflexion point). At this point, the curve is at the midway point between the two asymptotes, and takes the value $\alpha + \beta/2$.

Another S-shaped curve is the **Gompertz model**, which is written as

$$ y_i = \alpha + \beta \exp\{-\exp[-\gamma(x_i - \delta)]\} + e_i , $$

and is shown in Figure 17.13d with $\beta > 0$ and $\gamma > 0$. This curve also has a lower left asymptote at value α and an upper right asymptote at $\alpha + \beta$, but is asymmetric about the value $x = \delta$. Again, the direction and shape of the logistic and Gompertz models can be manipulated by changing the signs and values of the parameters β and γ.

EXAMPLE 17.1C: OLSEN P

In this example, we compare the fit of the exponential and inverse linear models to those fitted in Examples 17.1A and B. We first consider the exponential model, which accounts for 83.1% of the variation (adjusted $R^2 = 0.831$) and gives fitted model

$$\hat{\mu}(OlsenP) = 4.405 - 4.412 \times \exp(-0.359 \times OlsenP) \, .$$

This model has $\beta < 0$ and $\gamma > 0$ and so is an increasing curve with an upper asymptote, as required by the shape of the response. The upper asymptote is at $\hat{\alpha} = 4.405$, but the intercept with the y-axis is not of interest as it is well outside of the range of the Olsen P measurements. The fitted curve is shown with 95% CIs in Figure 17.14a and clearly follows the pattern of response well.

In comparison, the inverse linear model accounted for 83.5% of the variation (adjusted $R^2 = 0.835$) and gives the fitted model

$$\hat{\mu}(OlsenP) = 4.962 - \frac{9.073}{1 + (1.208 \times OlsenP)} \, .$$

Within the range of the observations, this model has a very similar shape to that of the exponential model (Figure 17.14b). It has a slightly higher asymptote ($\hat{\alpha} = 4.962$) but decreases much more sharply below the smallest Olsen P measurement. The 95% CIs for these two non-linear models have quite different shapes, reflecting different sources of uncertainty in the two models. The inverse linear model shows much more uncertainty around the point of maximum curvature, whereas the exponential model shows more uncertainty moving towards the upper asymptote.

Table 17.12 summarizes the goodness-of-fit statistics for the models fitted in all parts of Example 17.1. Based on the adjusted R^2 and AIC statistics, the non-linear models fit better than all but the eighth-order polynomial model, which we previously dismissed on the grounds of over-fitting. The SBC, which penalizes the number of parameters more heavily, shows a clear preference for the non-linear models. There is little statistical difference in the fit of the two non-linear models, so either might reasonably be selected.

There are many variations and extensions of these models available in statistical software, as well as other types of non-linear models. In addition, most software allows user-defined non-linear functions to be fitted. Successful estimation of parameters in non-linear models depends on the amount of information available from the observations, as well as good initial values for the parameters. For example, the logistic or Gompertz curves

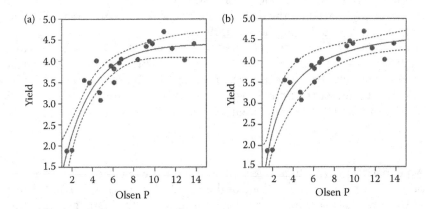

FIGURE 17.14
(a) Fitted exponential model (—) with 95% CI (---) and (b) fitted inverse linear model (—) with 95% CI (---) for yield in terms of Olsen P measurements (Example 17.1B).

TABLE 17.12

Summary Statistics for Models for Yield as a Function of Olsen P Measurements (Example 17.1A, 17.1B or 17.1C)

Example	Model	Number of Parameters	ResMS	R^2_{adj} (×100)	AIC	SBC
17.1A	SLR with $\log_{10}$(Olsen P)	2	0.1191	79.7	19.2	21.2
17.1B	Quadratic polynomial	3	0.1126	80.8	19.0	22.0
17.1B	Cubic polynomial	4	0.1114	81.0	19.6	23.5
17.1B	Eighth-order polynomial	9	0.0799	86.3	15.4	24.4
17.1C	Exponential model	3	0.0991	83.1	16.4	19.4
17.1C	Inverse linear model	3	0.0965	83.5	15.9	18.9

require some observations in the upper part of the curve to get a good estimate of the upper asymptote. Similarly, the precision of the curvature and position parameters (γ and δ) for these models becomes greater as the number of observations between the asymptotes increases. In general, it is desirable to have observations spread across the full range of the curve, and particularly at points where the slope of the curve changes. The parameterization used within non-linear models can influence the stability of the estimation procedure, and statistical packages use various different parameterizations. For example, the exponential model defined in Equation 17.7 can be written in an alternative form as

$$y_i = \alpha + \beta \varphi^{x_i} + e_i ,$$

where $\gamma = -\log_e \varphi$ and the other parameters retain their original interpretation. It can be helpful to try different parameterizations when problems with convergence are encountered, or to obtain parameters that have a biological interpretation.

Inference for non-linear models is not as straightforward as for linear models; in particular, SEs for parameter estimates and predictions are approximate, so that different statistical software can give somewhat different results. In most cases, approximate SEs and t-tests are reported in addition to ANOVA tables with approximate F-tests. Obtaining SEs for non-linear functions of parameters, for example, for $\gamma = -\log_e \varphi$, might also be necessary and is usually achieved by the delta method, sometimes called linearization (see e.g. Casella and Berger, 2002 or Seber and Wild, 1989).

Non-linear models can easily be extended to cases where different groups are present, and the approaches are analogous to those described in Chapter 15. The most general models allow all parameters to be separate among groups (e.g. a different asymptote for each group) and the most restrictive insist on common parameters across groups, with many intermediate models to be investigated.

EXERCISES

17.1* An experiment was done to establish the effectiveness of low doses of a fungicidal compound. Six fractions of the standard dose (1, 1/2, 1/4, 1/8, 1/16 and 1/32) were applied to individual leaves infected with a pathogen, and the number of colonies on each leaf were counted after a given period. Leaves without fungicide applied were included as a negative control. The design was a RCBD with three replicates, giving 21 leaves in total. File COLONIES.DAT holds the unit numbers (*ID*), structural factors (Rep, Leaf), the dose applied (variate *Dose*) and the number of colonies observed (variate *NColonies*).

    a. Find a transformation of the dose variable that gives an approximate linear relationship with the number of colonies. Fit a SLR using this transformed variate, including the design structure in your model. Check for lack of fit and state the predictive model.

    b. Find a non-linear model for the number of colonies that accounts for the structure of the design. Check for lack of fit. Compare this non-linear model with the transformation used in part (a) and state which model you prefer, with reasons.

17.2   The microarray study described in full in Exercise 12.6 investigated gene expression associated with senescence of leaves. File SENESCENCE.DAT holds design information (*ID*, variate *Day*, factor BiolRep) and the expression value for three genes (variates *CATMA3A13560*, *CATMA2A31585* and *CATMA1A09000*) from each plant following normalization.[*]

    a. Can you reasonably use polynomial regression to predict the expression of genes *CATMA2A31585* or *CATMA1A09000* over time? Over what range are your predictions reliable?

    b. Can you improve on these predictions by using non-linear models?

17.3   Exercise 13.3 analysed a set of chickweed plants from a field trial to investigate whether the number of seeds produced could be related to the plant biomass, measured as dry weight (g). There was evidence of variance heterogeneity, but the log-transformation required to stabilize the variance gave a curved relationship. Here, we try to find a model for that curved relationship, but now also include similar samples from several different experiments, carried out in different years and in different crops. File CWTRIALS.DAT holds unit numbers (*ID*) with a code for each trial (Trial), the year (factor Year) and crop type (factor Crop) as well as the number of seeds (variate *NSeed*) and dry weights (variate *DryWt*) for 193 plants. Find a transformation of the explanatory variate (*DryWt*) that linearizes the relationship with $\log_e(NSeed)$ and use regression with groups to establish whether the relationship differs between crops or years, or both. Identify a predictive model for the log-transformed number of seeds. Write down and interpret this predictive model. (We re-visit these data in Exercise 18.7.)

17.4   In Exercise 9.1, data from a field trial to investigate the effect of sulphur fertilizer on the yield of spring barley were analysed using ANOVA (data in file SULPHUR.DAT). Now use polynomial regression to model grain yield as a function of applied sulphur, accounting for the structure of the experimental design. Check for lack of fit. Write down the predictive model and give a 95% CI for grain yield with 25 kg S applied.

17.5   A microarray study was done to investigate the genes associated with infection of leaves by fungal pathogens. Ninety-six plants were grown in a controlled environment and the seventh leaf of each plant was excised at time zero and a mock inoculation was carried out (to give a baseline measurement). There were 24 sample points at 2-h intervals starting 2 h after the mock inoculation,

---

[*] Data from V. Buchanan-Wollaston (PRESTA), University of Warwick.

i.e. at 2, 4, 6 ... 48 h, and randomization was used to allocate four leaves to each sample point. Gene expression at the designated time was measured for each leaf. File BOTRYTIS.DAT holds unit numbers (*ID*) and structural factors (Hour, Leaf) with the expression values for one gene (variate *CATMA1A00045*). Plot the data. What do you notice about the pattern over time? Can you model this pattern using trigonometric regression? Is there any evidence of lack of fit to this relationship?[*]

17.6   An experiment was done to measure the response of yield to dose of nitrogen fertilizer. The design was a RCBD with four blocks of five treatments, corresponding to 0, 50, 100, 150 and 200 kg/ha of nitrogen applied and the response is plot yield. File FERTILIZER.DAT contains the unit numbers (*ID*), structural factors (Block, Plot), the amount of nitrogen applied (variate *N*) and the plot yields (variate *Yield*). Find a non-linear model to describe the response of yield to applied nitrogen. Check your model for misspecification and lack of fit. Write down and interpret the predictive model.

17.7   The yield response of Brussels sprout to applied nitrogen was investigated using a RCBD with three blocks of 13 plots. The treatments were 11 doses of nitrogen, between 0 and 250 kg/ha, with two replicates of 150 and 200 kg/ha per block. File SPROUTS.DAT contains the unit numbers (*ID*), structural factors (Rep, Plot), applied nitrogen (variate *Nitrogen*) and yield converted to tonnes per hectare (variate *Yield*). Plot the data and establish a predictive model to describe the pattern of response.[†]

17.8   Exercise 15.10 developed a model for oxygen consumption of wireworm larvae in terms of bodyweight and temperature groups. Now form a variate version of the temperature factor and investigate whether a surface can be developed in terms of temperature and bodyweight. Check your final predictive model for lack of fit and produce a visual representation of its fit. Write down and interpret your predictive model and comment on its usefulness.

17.9   Exercise 11.2 fitted a model for linseed yield in terms of barley and chickweed densities as factors. Fit a surface model for linseed yield using the two explanatory variables as variates. Use visual checks for model misspecification as well as formal tests for lack of fit. Write down your predictive model and state the range of values over which it can be considered reliable.

17.10   Exercise 15.3 established a separate lines model for splash heights in terms of velocity and weight classes. Now fit a surface in terms of the two explanatory variates. First, extract the estimates of intercept and slope for each weight class from the separate lines model and plot these against the weight values. Find a transformation of the weight variate that makes these two patterns into approximately straight lines – this is the transformation of weight to use in the surface model. Identify a parsimonious predictive model for your surface in terms of velocity and the transformed weight variate. Write down an equation for your predictive model and visualize it as a surface or contour plot. Does this give a better or worse fit than a surface constructed from the untransformed weight variate?

---

[*] Data from K. Denby (PRESTA), University of Warwick.
[†] Data from Horticulture Research International.

18

Models for Non-Normal Responses: Generalized Linear Models

We have seen in the previous chapters that linear models relate a response variable to one or more explanatory variables (factors or variates) and that inferences from these models rely on assumptions about the distribution of the response, often expressed in terms of properties of the model deviations (presented in Sections 4.1 and 12.1). Two of the most important assumptions are that the deviations, and hence the observed responses, have a common variance and follow a Normal distribution. These properties are required to make the F- and t-distributions valid for statistical inferences such as hypothesis testing and calculation of confidence intervals (CIs). In Chapter 5, we presented a set of diagnostic tools that can be used to check those assumptions. In Chapter 6, we then suggested transformation of the response variable to correct for heterogeneity of the variance and to make the distribution of the deviations approximate to a Normal distribution. However, for some types of response we expect, in advance of any statistical analysis, that their distributions will not be Normal, that their variances will be heterogeneous and that transformation might be unsatisfactory. Moreover, we can sometimes explicitly write the form of these distributions from knowledge of the underlying process(es) that generated the responses. Specifically, here we consider proportions that have been calculated from discrete counts (e.g. number of plants out of 20 affected by a disease) which are likely to have a Binomial distribution (Section 2.2.1), and responses that are generated as discrete counts (e.g. number of beetles caught in a pitfall trap during 24 h) which are likely to have a Poisson distribution (defined in Section 18.3). Responses with these probability distributions, and some others to be discussed later, can be analysed with generalized linear models (GLMs). This broad class of models allows the response to arise from one of several different probability distributions, extending the methods to situations other than the Normal distribution; however, this additional flexibility means that somewhat more complex estimation and inferential techniques are required.

In this chapter, we briefly introduce GLMs for Binomial and Poisson responses and describe the underlying models. We start with a general overview of the GLM model (Section 18.1). We consider a GLM for proportions with a Binomial distribution (Section 18.2), including some discussion about the detection and handling of over-dispersion (Section 18.2.2), checking model assumptions (Section 18.2.3) and the special case of binary responses (Section 18.2.7). We then introduce GLMs for discrete counts with a Poisson distribution (Section 18.3), and we end by describing briefly some other situations in which GLMs are used and some further extensions to these models (Section 18.4).

18.1 Introduction to Generalized Linear Models

A **generalized linear model** (GLM) extends the linear model framework to the situation where the responses have certain forms of non-Normal distributions, specifically distributions within the exponential family (Dobson, 1990) such as the Binomial and Poisson distributions. A Binomial distribution typically occurs where a fixed number of samples have been tested and the number passing (or failing) the test is counted. For example, in a survey of water sources we might take 20 separate samples from each source and count the number out of 20 with concentrations of mercury exceeding the maximum limit permitted for drinking water. A Poisson distribution typically occurs where responses consist of discrete counts. For example, we might count the number of viable seeds produced by individual plants to compare productivity across different varieties of a plant species. A GLM directly accounts for the particular characteristics of the distribution associated with a response and uses these characteristics in parameter estimation and inference. However, because we are now dealing with non-Normal distributions, we must modify the form of our models. Recall that in Section 1.3 we wrote our statistical model in the form

response = systematic component + random component ,

where the response was a numerical outcome, the systematic component was a mathematical function of one or more explanatory variables (factors, variates or both) and the random component (or model deviations) accounted for variation in the response not explained by the systematic component. Unfortunately, this partitioning of the model is specific to the Normal distribution and does not apply in a straightforward manner to non-Normal distributions. In general, it is more convenient to state the distribution of the response and to write the model in a different form as

E(response) = systematic component ,

i.e. the expected value of the response is equal to the systematic component of our model. In mathematical terms, we often write this as $E(y_i) = \mu_i$, where μ_i is the expected value of y_i, the response for the ith observation. The systematic component of the model is still a mathematical function of the explanatory variables, but now we allow a more complex form that involves a transformation. This transformation is used to account for boundaries on the range of possible values of the response variable, which should therefore also apply to the expected value. For example, for Poisson responses, the expected value must remain positive, while for Binomial responses where m tests have been made on each individual, the expected value must lie between 0 and m. The systematic component can then be expressed in general form as

g(systematic component) = linear function of explanatory variables ,

where the function g() is called the **link function** because it provides the link between the response and the explanatory variables. The linear function of the explanatory variables on the right-hand side of this equation, which may comprise any combination of factors and variates, is known as the **linear predictor**. Various different link functions can be used, but each distribution has a **canonical link** which has good mathematical properties and often works well in practice. For Binomial and Poisson responses, the logit and the log are the canonical link functions, respectively.

Once a model has been defined, the model parameters can be estimated. For non-Normal distributions, the simple method of least squares is no longer appropriate for parameter estimation. Instead, the principle of **maximum likelihood estimation** is used. We do not go into mathematical details here, but Dobson (1990) or Collett (2002) provide a good description. One of the consequences of this change is that, instead of obtaining exact SEs for parameter estimates, the estimated SEs become approximate, as does the calculation of CIs and hypothesis tests. These issues will be discussed in the following sections, where we consider the cases of Binomial and Poisson responses in more detail.

18.2 Analysis of Proportions Based on Counts: Binomial Responses

The Binomial distribution, which was introduced in Section 2.2.1, usually arises as the distribution of the number of successes out of a series of m independent binary tests (i.e. tests with only two possible outcomes: success or failure), where all tests have the same probability of success. In the context of a GLM, we have N Binomial responses, each of which is the result of a number of binary tests. The ith response consists of two pieces of information: the number of tests, denoted m_i, and the number of successes, denoted y_i. Note that the number y_i can take only integer values in the set 0, 1, 2, ... m_i, for $i = 1 ... N$. If only one test is made on each unit, so that $m_i = 1$, then we have **binary** observations that have only two possible values, zero or one. Many of the useful properties that apply to Binomial data fail in the case of binary data, and this is discussed in Section 18.2.7.

EXAMPLE 18.1A: DEMETHYLATION EXPERIMENT

This experiment is a pilot study intended to calibrate a scientific procedure. A demethylation agent is applied to plants: the agent has the effect of converting methylated nucleotides to non-methylated form, causing epigenetic changes that lead to abnormal phenotypes such as stunting and deformation (Amoah et al., 2008). The pilot study aimed to investigate the relationship between dose and the resulting proportion of plants with a normal phenotype. Seed was treated with the demethylation agent at six doses, including a zero control dose. Plants were grown in trays, each tray sown with seeds treated with the same dose of agent and each dose was replicated in four trays: two with 60 plants, and two with 100 plants. The trays were arranged as a CRD (Chapter 4). Table 18.1 lists the number of plants with a normal phenotype in each tray (*Normal_i*, $i = 1 ... 24$) with the number of plants per tray (*Total_i*). The data can also be

TABLE 18.1

Number of Normal Plants (Total Number of Plants) per Tray for Doses of Demethylation Agent (Example 18.1A and File DEMETHYLATION.DAT)

Dose					
0	0.01	0.1	0.5	1.0	1.5
59 (60)	58 (60)	54 (60)	4 (60)	3 (60)	3 (60)
58 (60)	59 (60)	53 (60)	11 (60)	2 (60)	3 (60)
99 (100)	98 (100)	88 (100)	14 (100)	2 (100)	1 (100)
98 (100)	99 (100)	87 (100)	15 (100)	1 (100)	3 (100)

Source: Data from S. Amoah, Rothamsted Research.

found in file DEMETHYLATION.DAT which contains explanatory variate *Dose*, response variate *Normal* and variate *Total* containing the number of plants for each tray, each identified using dummy index variate *DTray* (the original layout of trays was not recorded). Figure 18.1 shows the proportions of normal plants ($Prop_i = Normal_i/Total_i$) plotted against the dose applied. We can think of the agent acting on each seed independently, with the probability of producing a normal phenotype dependent on the dose applied. We therefore expect the number of plants with a normal phenotype in each tray to have a Binomial distribution.

Observations expected to follow a Binomial distribution with m_i tests can be denoted as $y_i \sim$ Binomial(m_i, p_i), where p_i is the underlying probability of success in each test, with $0 \le p_i \le 1$. The probability of observing a response y_i for the ith observation can then be written as

$$\text{Prob}(y_i\,;\,m_i,p_i) \;=\; \frac{m_i!}{y_i!\,(m_i - y_i)!}\,p_i^{\,y_i}(1 - p_i)^{m_i - y_i}, \quad \text{for } y_i = 0 \ldots m_i \,.$$

This probability depends on both the number of tests, m_i, which is a known value, and on the probability of success, p_i, an unknown parameter. We hypothesize that the probability of success may depend on explanatory variables, for example, in Example 18.1A, that the probability of obtaining a normal plant depends on the dose of the demethylation agent applied. If y_i follows a Binomial distribution, then its expected value and variance are, respectively

$$E(y_i) = \mu_i = m_i p_i \,; \quad \text{Var}(y_i) = m_i p_i(1 - p_i) \,.$$

The expected value is the product of the number of tests and the probability of success in each test. As the number of tests is fixed (once the data have been obtained), modelling the probability of success is equivalent to modelling the expected value. The variance is also a function of the number of tests and probability of success. This variance is small if p_i is

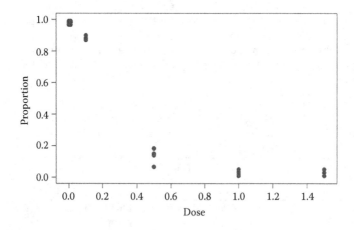

FIGURE 18.1
Proportion of normal plants per tray plotted against dose of demethylation agent (Example 18.1A).

close to either zero or one, and increases to its maximum at $p_i = 0.5$. This heterogeneity can be seen in Figure 18.1, where the observed proportions close to 0 and 1 show less variation than the proportions between 0.10 and 0.20 (for dose equal to 0.5). The variance can also be written in terms of the expected value as

$$\text{Var}(y_i) = \frac{\mu_i}{m_i}(m_i - \mu_i),$$

illustrating that the variance is a direct function of the expected value and the number of tests (m_i) for each unit.

18.2.1 Understanding and Defining the Model

To aid understanding, we introduce the GLM for Binomial responses with a single quantitative explanatory variable (variate) using notation like that in the previous chapters. Later, we shall write models for qualitative variables (factors) or a mixture of factors, variates and interactions. To make clear the distinction between the expected value of the data and its transformed value, we write $g(\mu_i) = \eta_i$, so η_i represents the expected value of the ith observation after transformation by the link function. Note that this usage of η, which is standard notation for GLMs, is somewhat different from that in previous chapters. The systematic component of a model with a single explanatory variate is then

$$\eta_i = g(\mu_i) = \alpha + \beta x_i,$$

so that, after transformation by the link function, the expected value of the ith observation is a straight line function of the explanatory variate, x_i. Recall from Chapter 12 that parameter α is the intercept of this straight line and parameter β is the slope. As stated earlier, the right-hand side of this equation is called the linear predictor, and so this is often referred to as the model on the **transformed** or **linear predictor scale**. For now, we concern ourselves with the **logit link** function, which is the canonical link for Binomial data, so that our model with a single explanatory variate can be written as

$$\eta_i = \log_e\left(\frac{\mu_i}{m_i - \mu_i}\right) = \alpha + \beta x_i.$$

We can rewrite the logit function in terms of the success probability, p_i, as

$$\log_e\left(\frac{\mu_i}{m_i - \mu_i}\right) = \log_e\left(\frac{m_i p_i}{m_i - m_i p_i}\right) = \log_e\left(\frac{p_i}{1 - p_i}\right) = \text{logit}(p_i).$$

This illustrates that the model above can equivalently be considered as

$$\eta_i = \text{logit}(p_i) = \alpha + \beta x_i, \tag{18.1}$$

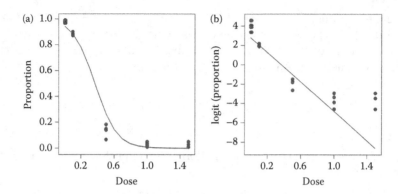

FIGURE 18.2
Observations (•) with fitted GLM (—, Binomial distribution and logit link) for explanatory variate *Dose* plotted on (a) natural and (b) logit scale (Example 18.1B).

i.e. the logit of the success probability is a linear function of the explanatory variate. The Binomial GLM with logit link is therefore often called **logistic regression**. The quantity $p_i/(1 - p_i)$ is known as the **odds** (in favour of success), so logit(p_i) is equivalent to the logarithm of the odds, or **log-odds**. Hence, another interpretation of this model is that the log-odds is a linear function of the explanatory variate. By rearranging Equation 18.1, we can write the model in terms of the success probability as

$$p_i = \frac{\exp(\eta_i)}{1 + \exp(\eta_i)} = \frac{\exp(\alpha + \beta x_i)}{1 + \exp(\alpha + \beta x_i)} \, . \tag{18.2}$$

This is often called the model on the back-transformed or **natural scale**. This model is a non-linear function of the explanatory variate x (see Figure 18.2a). Given estimates of the parameters, this formula can be used to predict the success probability for any value of the explanatory variate. If we multiply Equation 18.2 by m_i, then we can write this model equivalently in terms of the expected value as

$$\mu_i = m_i p_i = m_i \frac{\exp(\eta_i)}{1 + \exp(\eta_i)} \, .$$

Putting all of these properties together, we can interpret the Binomial GLM with logit link as a non-linear model that accounts for the Binomial distribution of the responses and its associated heterogeneity. To give a symbolic form for a GLM, we extend our previous definition to include the probability distribution and link function. This is illustrated in Example 18.1B.

As stated above, parameter estimation is achieved by the method of maximum likelihood, which is beyond the scope of this book. Here, we quote results obtained from GenStat rather than deriving estimates directly. Once parameter estimates have been obtained, the fitted model should be checked for misspecification. You can achieve this by plotting the observations and fitted values against the explanatory variable (see Section 13.1) to check that the fitted model follows the trend in the data.

EXAMPLE 18.1B: DEMETHYLATION EXPERIMENT

For the demethylation experiment introduced in Example 18.1A, we can fit a Binomial GLM with logit link for the number of normal plants in the ith tray (*Normal$_i$*) in terms of the dose applied to that tray (*Dose$_i$*). The model can be written as

$$Normal_i \sim Binomial(Total_i, p_i) ; \quad \eta_i = logit(p_i) = \alpha + \beta\,Dose_i \, ,$$

where p_i is the probability that *Dose$_i$* gives a normal phenotype, and η_i is its logit transformation. We can write the model in an extension of our symbolic form as

Response variable: *Normal*
Probability distribution: Binomial (Number of tests = *Total*)
Link function: logit
Explanatory component: *[1] + Dose*

We have now included some additional information. First, as usual, we define the response variable, which is the variate containing the number of successes per unit (*Normal*). Then we specify the probability distribution of the response, here the Binomial distribution. For this particular distribution, we must also define the number of tests performed for each observation (here, the number of plants per tray, *Total*). We then specify the link function, here the logit transformation. Finally, as usual, we give the explanatory component of the model in terms of the explanatory variables, here the intercept, *[1]*, and the explanatory variate, *Dose*.

We obtain the estimated parameters for this model from GenStat as $\hat{\alpha} = 2.793$ and $\hat{\beta} = -7.623$, giving the fitted model on the scale of the linear predictor, i.e. the logit scale, as

$$\hat{\eta}_i = 2.793 - 7.623 \times Dose_i \, .$$

On the natural scale, the fitted probability of a normal phenotype for the ith observation can be expressed as

$$\hat{p}_i = \frac{exp(\hat{\eta}_i)}{1 + exp(\hat{\eta}_i)} = \frac{exp(2.793 - 7.623 \times Dose_i)}{1 + exp(2.793 - 7.623 \times Dose_i)} \, .$$

Both forms are shown in Figure 18.2, with the observed proportion of normal plants (on the natural scale; Figure 18.2a) or the logit-transformed proportion (on the linear predictor scale; Figure 18.2b).

This model is clearly misspecified (see Section 13.1), as the fitted lines deviate from the trend in the plot. This is clearer on the scale of the linear predictor, where the trend in the data is evidently non-linear although the form of the GLM model demands a linear trend on this scale. This shortcoming can be tackled either with a different link function (see Section 18.2.7), with a polynomial function of the explanatory variate (see Section 17.1.2), or by transformation of the explanatory variable (as in Section 17.1.1), which is the route we take here.

We refit the model in terms of the log-transformed explanatory variate *logDose* = $\log_e$(*Dose* + 0.1). The offset of 0.1 (see Section 6.2.1) is required to deal with the zero (control) dose, and has been chosen pragmatically (by inspection, using trial and error) to give a reasonable straight line on the linear predictor scale. The revised model takes the form

$$\hat{\eta}_i = -3.188 - 3.148 \times logDose_i \, ,$$

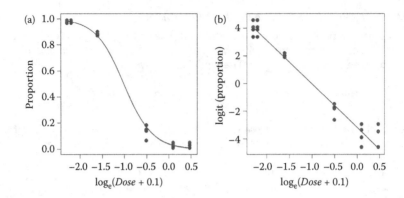

FIGURE 18.3
Observations (•) with fitted GLM (—, Binomial distribution and logit link) for explanatory variate $\log_e(Dose + 0.1)$ plotted on (a) natural and (b) logit scale (Example 18.1B).

where $logDose_i = \log_e(Dose_i + 0.1)$. On the natural scale, this gives

$$\hat{p}_i = \frac{\exp(\hat{\eta}_i)}{1 + \exp(\hat{\eta}_i)} = \frac{\exp(-3.188 - 3.148 \times logDose_i)}{1 + \exp(-3.188 - 3.148 \times logDose_i)} .$$

The fitted model is plotted on both the natural and linear predictor scales in Figure 18.3, which shows the fit to be much closer to the observed trend in the responses.

The estimated intercept ($\hat{\alpha} = -3.188$) and slope ($\hat{\beta} = -3.148$) parameters relate to the straight line fitted on the linear predictor, or logit, scale. The negative slope indicates that the proportion of normal plants is smaller for larger doses. We can write the predictive model as a continuous function of the original explanatory variable *Dose* as

$$\hat{\eta}(Dose) = -3.188 - 3.148 \times \log_e(Dose + 0.1).$$

From this formula, we can make predictions for any dose (staying within the observed range to avoid extrapolation). For example, for *Dose* = 0.3, with $\log_e(Dose + 0.1) = -0.92$, the predicted response on the logit scale is

$$\hat{\eta}(Dose = 0.3) = -3.188 - (3.148 \times \log_e(Dose + 0.1)) = -3.188 - (3.148 \times -0.92) = -0.304 .$$

We can back-transform this prediction to estimate the probability of getting a normal phenotype at this dose as

$$\hat{p}(Dose = 0.3) = \frac{\exp[\hat{\eta}(Dose = 0.3)]}{1 + \exp[\hat{\eta}(Dose = 0.3)]} = \frac{\exp(-0.304)}{1 + \exp(-0.304)} = \frac{0.738}{1.738} = 0.425 .$$

The estimated probability of our obtaining a normal plant after application of a dose of 0.3 units is therefore equal to 0.425, i.e. a 42.5% chance of obtaining a normal plant. We can translate this to an expected number of normal plants per tray by multiplying by the number of plants in the tray.

Having fitted a GLM that appears to give a reasonable description of the data in the fitted model plot, we must get some formal quantification of the fit and the uncertainty associated with the estimated parameters. These topics are discussed in the next sections.

18.2.2 Assessing the Importance of the Model and Individual Terms: The Analysis of Deviance

In GLMs, the fit of a model is quantified by calculation of the **deviance**, which is a measure of the discrepancy between the fitted model and the data. This comparison is made via a function, called the log-likelihood function, which takes into account both the link trans- formation and the underlying distribution and compares the fit of the proposed model against a perfect or **saturated model** that fits each observation exactly. For a Binomial distribution, the deviance for a model with fitted values $\hat{\mu}_i$ takes the form

$$D = \sum_{i=1}^{N} D_i^2 = 2 \sum_{i=1}^{N} \left[y_i \log_e \left(\frac{y_i}{\hat{\mu}_i} \right) + (m_i - y_i) \log_e \left(\frac{m_i - y_i}{m_i - \hat{\mu}_i} \right) \right]. \tag{18.3}$$

The fit of a model is usually summarized in an **analysis of deviance** (ANODEV) table. The ANODEV table starts with the total deviance of the observations, obtained as the deviance of a null or baseline model that assumes that the expected value is equal for all observa- tions, i.e. $\eta_i = \eta$ for $i = 1 \ldots N$. This total deviance is partitioned into the change in deviance that occurs when the explanatory component is fitted, here called the **model deviance** (ModDev), and a remainder, the **residual deviance** (ResDev), which is the change in devi- ance between the fitted and saturated models given by Equation 18.3. Each component of the total deviance has degrees of freedom associated with it, and those for the residual deviance are denoted ResDF. The ANODEV table is similar in spirit to the ANOVA table used to summarize the fit of a linear model (see Chapters 4 and 12) and takes the general form shown in Table 18.2 for a model with p (independent) parameters. For the model with an intercept and a single explanatory variate, we have $p = 2$. Because the components of the deviance generally increase as their degrees of freedom increase, it is helpful to divide the contributions by their degrees of freedom to get mean deviances that are on a common scale.

EXAMPLE 18.1C: DEMETHYLATION EXPERIMENT

In Example 18.1B, we modelled the number of normal plants (*Normal$_i$*) as a function of the log-dose of agent applied. The ANODEV table for this model is Table 18.3.

The model deviance represents differences between the null model (with one param- eter representing the overall mean on the logit scale) and the fitted model, here a regres- sion on logged dose. The change in deviance between these two models is 1874.77, with 1 df as one extra parameter has been added (the slope parameter). The residual devi- ance represents differences between the fitted model (in terms of logged dose) and the saturated model, which has an additional 22 parameters (to give 24 in total, one for each observation); the change in deviance here is much smaller. The total deviance represents

TABLE 18.2

ANODEV Table for a GLM with p Parameters and N Responses without Over-Dispersion

Source of Variation	df	Deviance	Mean Deviance	P (Chi-Squared)
Model	$p - 1$	ModDev	ModMDev = ModDev$/(p - 1)$	Prob(χ^2_{p-1} > ModDev)
Residual	$N - p$	ResDev	ResMDev = ResDev$/(N - p)$	
Total	$N - 1$	TotDev		

TABLE 18.3

ANODEV Table for the Demethylation Experiment with Explanatory Variate
$logDose = \log_e(Dose + 0.1)$ (Example 18.1C)

Source of Variation	df	Deviance	Mean Deviance	P (Chi-Squared)
Model	1	1874.772	1874.772	< 0.001
Residual	22	26.623	1.210	
Total	23	1901.395		

differences between the null and saturated models; the deviance and df from the model and residual contributions sum to the total values.

The appropriate method for assessment of the model depends on whether there is evidence of over-dispersion, and so we consider this issue next. The residual deviance incorporates systematic discrepancies between the model and the observed responses, variation between replicate observations (observations on independent experimental units with the same values of the explanatory variables), and sampling variation arising from the distribution of the data (here, the Binomial distribution). If there are no replicate observations and the fitted model provides an adequate description of the systematic trend, then only sampling variation contributes to the residual deviance. If this is true, then the residual deviance has an approximate chi-squared distribution (see Section 2.2.4) with df equal to the residual df. The null hypothesis that the model adequately describes the responses can therefore be rejected at significance level α_s if the residual deviance exceeds the $100(1 - \alpha_s)$th percentile of that chi-squared distribution. If this hypothesis is rejected, it indicates a poor fit of the model to the observations, which may happen for several reasons. First, the fitted model might not follow the observed patterns in the data (i.e. model misspecification), as illustrated in Figure 18.2. In this case, the explanatory variate(s) may be transformed to try and improve the fit (as in Example 18.1B), or an alternative link function might be considered. For example, the logit link function requires the shape of the curve (on the natural scale) to be symmetric around probability 0.5; one alternative, the complementary log–log link function, allows some asymmetry in this relationship and will give a better fit for some data sets. Second, the response may depend on explanatory variables that have not been included in the model; additional explanatory variables should be tested to see if they improve the model. Third, the assumed distribution might be incorrect. For example, the Binomial distribution requires that the individual tests that comprise each observation should be independent. If they are not, then the observed variation might not match that expected for the Binomial distribution, and this will be reflected in the residual deviance. Fourth, outliers or influential observations may have either distorted the model or inflated the residual deviance. These different circumstances can be investigated with the methods introduced in Chapter 13 and are discussed in Section 18.2.3. If replicate observations are present, then variation between replicates might inflate the residual deviance even if the model gives an adequate fit to the data, but the checks outlined above should still be made.

In general, the quality of the approximation to the chi-squared distribution for the residual deviance improves as the number of observations increases; this is known as an asymptotic approximation. For Binomial data, the approximation improves as both the number of observations, N, and the number of tests per observation, m_i, increase. However, the chi-squared approximation does not hold for binary data (i.e. $m_i = 1$), and the approach for this situation is discussed in Section 18.2.7.

If the residual deviance is larger than expected when compared with critical values of the appropriate chi-squared distribution, and if this cannot be dealt with by changing the model, then there is more variation present than can be accounted for by the assumed probability distribution. In this case, we say that the data show **over-dispersion**. The simplest way to deal with over-dispersion is by extending the model to scale the variance function. In a Binomial distribution, the scaled variance takes the form

$$\text{Var}(y_i) = \varphi\, m_i p_i (1 - p_i) = \varphi \frac{\mu_i}{m_i}(m_i - \mu_i).$$

The rationale for this approach is discussed by Collett (2002, Chapter 6). The parameter φ is a scaling factor, called the **dispersion parameter**, which is used to summarize the degree of over-dispersion present in the observations. Clearly, $\varphi = 1$ corresponds to the original model. This parameter can be estimated in several different ways. The **deviance estimate** of the dispersion is equal to the residual mean deviance (ResMDev), i.e.

$$\hat{\varphi} = \text{ResDev}/\text{ResDF}.$$

The **Pearson estimate** of the dispersion is equal to Pearson's chi-squared (goodness-of-fit) statistic divided by the residual df,

$$\hat{\varphi} = \frac{1}{\text{ResDF}} \sum_{i=1}^{N} \frac{(y_i - \hat{\mu}_i)^2}{\text{Var}(\hat{\mu}_i)} = \frac{1}{\text{ResDF}} \sum_{i=1}^{N} \frac{(y_i - m_i \hat{p}_i)^2}{m_i \hat{p}_i (1 - \hat{p}_i)}, \tag{18.4}$$

where $\text{Var}(\hat{\mu}_i)$ is the variance function associated with the probability distribution (with $\varphi = 1$), evaluated at the estimated expected value for the ith observation. The default method for estimation of the dispersion parameter varies between statistical packages. Either of these parameter estimates can be used to give a more realistic assessment of the contributions of explanatory variables in the ANODEV table, and to inflate the estimated SEs of parameters to reflect the observed variation. However, estimation of the dispersion parameter changes the way that contributions to the ANODEV table should be evaluated. We must therefore establish whether over-dispersion is present before attempting to interpret the ANODEV table.

EXAMPLE 18.1D: DEMETHYLATION EXPERIMENT

In the ANODEV table for the model with log-dose of the demethylation agent (Table 18.3), the residual deviance takes the value 26.62 on 22 df, with $P = 0.226$ when compared to the chi-squared distribution on 22 df. There is therefore no evidence of over-dispersion for this model.

18.2.2.1 Interpreting the ANODEV with No Over-Dispersion

If there is no over-dispersion present, then the model and residual deviance contributions approximately follow chi-squared distributions with degrees of freedom equal to the df for each contribution. We can use the model deviance to test whether the inclusion of the explanatory component has improved the fit when compared with the null model. The null hypothesis is that the response is not related to the explanatory component. For a model with a single explanatory variate, as in Example 18.1, the null hypothesis is equivalent to $H_0: \beta = 0$. If the model deviance, i.e. ModDev, is larger than the $100(1 - \alpha_s)$th percentile of the chi-squared distribution with degrees of freedom equal to the model df, then this null hypothesis can be rejected at significance level α_s, indicating that the explanatory component has improved the fit compared with the null model.

EXAMPLE 18.1E: DEMETHYLATION EXPERIMENT

In Example 18.1D, there was no evidence of over-dispersion for the model with log-dose of the demethylation agent. In the ANODEV table (Example 18.1C, Table 18.3), the model deviance represents the change on addition of the *logDose* explanatory variate into the model. This deviance takes the value 1874.77 with 1 df, with $P < 0.001$ when compared with the chi-squared distribution with 1 df. This test gives strong evidence that the proportion of normal phenotypes is related to the logged dose of the agent.

18.2.2.2 Interpreting the ANODEV with Over-Dispersion

If over-dispersion is present, then we expect all the components of deviance to be inflated, and so cannot compare them directly with a chi-squared distribution. Instead, we follow an approach similar to that taken in an ANOVA table (Chapters 4 and 12). The deviance contributions are divided by their degrees of freedom to get mean deviances that are on a common scale (analogous to the mean squares in ANOVA). The ratio of the model mean deviance (i.e. ModMDev) to the residual mean deviance (ResMDev) can then be used to assess whether the explanatory variable(s) have improved the fit compared with the null model. This introduces a new column of deviance ratios into the ANODEV table (see Table 18.4). Under the null hypothesis that the response is not related to the explanatory variable(s), the deviance ratio

$$F = \frac{ModMDev}{ResMDev}$$

has an approximate F-distribution, with numerator df equal to the model df (ModDF) and denominator df equal to the residual df (ResDF).

EXAMPLE 18.2A: LADYBIRD PREDATION

An experiment was done to investigate factors affecting predation by the Harlequin ladybird. Ladybirds of known sex (factor **Sex**, with levels 1 = female and 2 = male) were put individually into dishes containing six items of prey, which were either pea aphids or lacewing larvae (factor **Prey**, with levels 1 = aphid and 2 = lacewing). The experiment was designed as a RCBD with four rows (blocks) of four Petri dishes (one per treatment

TABLE 18.4

ANODEV Table for a GLM with p Parameters and N Responses with Over-Dispersion

Source of Variation	df	Deviance	Mean Deviance	Deviance Ratio	P (F)
Model	$p - 1$	ModDev	ModMDev	F = ModMDev/ResMDev	$\text{Prob}(F_{p-1,N-p} > F)$
Residual	$N - p$	ResDev	ResMDev		
Total	$N - 1$	TotDev			

combination) and was repeated on four occasions, although on one occasion only three rows could be completed, because of a shortage of lacewing larvae. The number of whole prey eaten after 60 min was counted within each dish (variate *Eaten*) and could be reasonably assumed to have a Binomial distribution. The final data set for analysis consisted of 60 observations (15 rows each with four treatment combinations), given in file PREY.DAT and in Table 18.5. For simplicity, here we do not distinguish between occasions and label the rows as 1 ... 15 (factor Row), combining the variation due to occasions and rows (within occasions) into a single term.

We wish to fit a GLM with Binomial distribution and logit link. Unfortunately, it is not possible to account properly for the structural component within the standard GLM framework, as there is no parallel to the multi-stratum ANOVA. As discussed in Section 15.3, we therefore have to either use a different method (see Section 18.4) or take an approximate approach by combining the explanatory and structural components. For a RCBD, treatment effects are estimated via within-block comparisons and an intra-block analysis allows us to exclude block (row) effects before we assess treatment terms, and so we take this approach. We use a two-way crossed structure (Section 8.2) to model the four treatments. This model can be written in mathematical form as

$$Eaten_{irs} \sim \text{Binomial}(6, p_{irs}); \quad \eta_{irs} = logit(p_{irs}) = \eta_{111} + Row_i + Sex_r + Prey_s + (Sex.Prey)_{rs} ,$$

where $Eaten_{irs}$ is the number of prey eaten in the ith row ($i = 1 ... 15$) by the rth sex ($r = 1$, 2 for 1 = female and 2 = male) with the sth prey type ($s = 1$, 2 for 1 = aphid, 2 = lacewing

TABLE 18.5

Number of Prey Eaten by the Harlequin Ladybird (Example 18.2A and File PREY.DAT) in an Experiment with 15 Rows of Four Dishes, Each Containing One Ladybird (Female, F, or Male, M) and Six Items of Prey (Pea Aphids, A, or Lacewing Larvae, L)

Row	Dish	Sex	Prey	Eaten	Row	Dish	Sex	Prey	Eaten	Row	Dish	Sex	Prey	Eaten
1	1	F	A	5	6	1	M	L	1	11	1	M	L	0
1	2	M	A	2	6	2	F	L	4	11	2	M	A	0
1	3	F	L	3	6	3	F	A	0	11	3	F	L	2
1	4	M	L	0	6	4	M	A	0	11	4	F	A	2
2	1	F	A	5	7	1	M	A	0	12	1	M	A	2
2	2	M	A	2	7	2	M	L	2	12	2	F	L	0
2	3	F	L	1	7	3	F	L	2	12	3	M	L	1
2	4	M	L	1	7	4	F	A	4	12	4	F	A	4
3	1	F	A	3	8	1	M	A	3	13	1	M	A	0
3	2	F	L	0	8	2	M	L	2	13	2	M	L	0
3	3	M	A	0	8	3	F	L	5	13	3	F	A	2
3	4	M	L	0	8	4	F	A	3	13	4	F	L	0
4	1	M	L	1	9	1	F	L	1	14	1	M	L	2
4	2	M	A	1	9	2	M	L	0	14	2	F	L	3
4	3	F	A	4	9	3	F	A	0	14	3	F	A	2
4	4	F	L	2	9	4	M	A	1	14	4	M	A	1
5	1	M	A	2	10	1	F	A	4	15	1	F	A	2
5	2	F	A	1	10	2	M	A	0	15	2	F	L	1
5	3	M	L	0	10	3	F	L	0	15	3	M	A	0
5	4	F	L	4	10	4	M	L	0	15	4	M	L	0

Source: Data from P. Wells, Rothamsted Research.

larvae), and p_{irs} is the probability that an item of prey in this category is eaten, with logit transformation η_{irs}. We use first-level-zero parameterization (see Section 11.2.1), so $Row_1 = 0$, $Sex_1 = 0$, $Prey_1 = 0$ and $(Sex.Prey)_{rs} = 0$ for $r = 1$ or $s = 1$. Then, η_{111} is the logit of the probability for the first level of all factors (i.e. females in the first row with aphids), Row_i is the relative effect of the ith row, Sex_2 is the difference in response between males and females (for aphid prey), $Prey_2$ is the difference in response between lacewing larvae and aphids (for females) and $(Sex.Prey)_{22}$ is the interaction effect, i.e. the additional difference for the combination of a male ladybird with lacewing larvae. The explanatory terms are the structural factor plus the two treatment factors and their interaction. On the logit scale, the model fits a separate effect for each sex × prey combination and allows a shift in the value for each row. This model is written with symbolic notation as

Response variable:	*Eaten*
Probability distribution:	Binomial (Number of tests = 6)
Link function:	logit
Explanatory component:	[1] + Row + Sex*Prey

The ANODEV table for this model is Table 18.6. The residual deviance is 69.66 with 42 df with $P = 0.005$ (compared to a chi-squared distribution with 42 df). There is therefore evidence of over-dispersion for this model. We first consider whether we can deal with this by changing the model. As this is a designed experiment where we have fitted effects for each row and each treatment combination, and there are no additional explanatory variables, we cannot identify any deficiency in the model that might be corrected. We might attribute the over-dispersion to variation between the behaviour of individual ladybirds, but we cannot usefully account for this within a simple model. We therefore include a dispersion parameter to model the over-dispersion, here estimated as the residual mean deviance,

$$\hat{\varphi} = \frac{\text{ResDev}}{\text{ResDF}} = \frac{69.659}{42} = 1.659 \ .$$

The model deviance represents the change in deviance when all of the explanatory terms are added into the model, with 17 df: 14 df for the 15 row effects (term **Row**) and 3 df for the crossed structure **Sex*Prey** (four treatment combinations). To assess whether this model explains any variation in the response, we use the ratio of the mean deviance for the model (4.214) with the residual mean deviance, to get 2.54 (= 4.214/1.659). We compare this deviance ratio to an F-distribution with 17 and 42 df, giving observed significance level $P = 0.007$, and so conclude that there is statistical evidence that the model explains some of the patterns in predation. We investigate the importance of individual model terms in Example 18.2B.

A good strategy for analysis is to fit an initial model with the dispersion parameter set equal to one, assess the quality of the fit (see Section 18.2.3) and, when the fit appears

TABLE 18.6

ANODEV Table for the Ladybird Predation Experiment (Example 18.2A)

Source of Variation	df	Deviance	Mean Deviance	Deviance Ratio	P (F)
Model	17	71.635	4.214	2.54	0.007
Residual	42	69.659	1.659		
Total	59	141.294			

adequate, to formally test whether the dispersion parameter is equal to one as shown above. Remember that this test is reliable only when the residual df is reasonably large and, for Binomial data, when the number of tests per observation (m_i) is not too small. If there is evidence that the dispersion parameter is larger than 1, then over-dispersion is present and the analysis should proceed accordingly.

Occasionally, the dispersion parameter might appear to be substantially less than 1 and then **under-dispersion** should be considered as a possibility. Under-dispersion occurs where, for a given distribution, we detect less variation than expected, with $\hat{\phi} < 1$. This is less common than over-dispersion and is often difficult to interpret or explain, and it is sensible to be wary in this situation. If the dispersion parameter is estimated as smaller than 1 when it is in fact equal to 1, then the significance of hypothesis tests will be inflated and estimated SEs will be too small. To avoid these problems, leave the dispersion parameter equal to 1 in cases of apparent under-dispersion.

18.2.2.3 The Sequential ANODEV Table

If the explanatory component consists of several different model terms, then we can calculate a set of incremental deviances (and df) from the change in deviance (and df) that occur on successive addition of individual terms into the model, producing a sequential ANODEV table analogous to the sequential ANOVA tables introduced in Sections 11.2.2 and 15.4.1. If there is no evidence of over-dispersion, then the incremental deviance is compared with a chi-squared distribution with df equal to the incremental df obtained on addition of the term into the model. If over-dispersion is present, then the deviance ratio for the term (incremental deviance divided by the incremental df, all divided by the residual mean deviance) is compared to an F-distribution, as illustrated in Example 18.2B. Because of the non-linear nature of the GLM, terms that would be orthogonal in a linear model (Section 11.1) will not be orthogonal in a GLM, i.e. the sequential deviance for a term in an ANODEV table depends on the order in which that term is added into the model, as illustrated in Example 18.2B. We can also construct a set of marginal deviances by calculating the change when a term is dropped from the model (c.f. Sections 11.2.3 and 15.4.2). In general, we follow the strategies for model selection outlined in Section 15.5.1 to obtain a predictive model.

EXAMPLE 18.2B: LADYBIRD PREDATION

Two sequential ANODEV tables for the ladybird predation experiment are shown in Table 18.7. In both sequences, we fit the structural factor Row first, followed by the explanatory crossed structure, which we fit as Sex*Prey in Table 18.7a and as Prey*Sex in Table 18.7b. First, we consider the former case. As we identified over-dispersion in Example 18.2A, individual model terms are assessed on their deviance ratios, with ResMDev = 1.659. Using similar notation for incremental deviances as that developed for incremental sums of squares earlier (Sections 11.2 and 15.4), we calculate the incremental deviance ratio for factor Sex as

$$F_{1,42}^{S} = \frac{\text{Dev}(+\text{Sex}|[1])/\text{df}(+\text{Sex}|[1])}{\text{ResMDev}} = \frac{(33.471/1)}{1.659} = 20.181 \, .$$

The numerator df for the F-statistic are the incremental df for the term added into the model and the denominator df are the residual df. Deviance ratios for other terms are calculated similarly from their incremental deviances and df.

TABLE 18.7

Two Sequential ANODEV Tables (Deviance Not Shown) for the Ladybird Predation Experiment with Explanatory Factors Row, Sex and Prey (Example 18.2B)

(a) Source of Variation	df	Mean Deviance	Deviance Ratio	P (F)	(b) Source of Variation	df	Mean Deviance	Deviance Ratio	P (F)
+ Row	14	2.34	$F^R = 1.41$	0.191	+ Row	14	2.34	$F^R = 1.41$	0.191
+ Sex	1	33.47	$F^S = 20.18$	< 0.001	+ Prey	1	4.62	$F^P = 2.79$	0.103
+ Prey	1	5.15	$F^P = 3.10$	0.085	+ Sex	1	34.00	$F^S = 20.50$	< 0.001
+ Sex.Prey	1	0.27	$F^{S.P} = 0.16$	0.690	+ Prey.Sex	1	0.27	$F^{P.S} = 0.16$	0.690
Residual	42	1.66			Residual	42	1.66		
Total	59				Total	59			

Our aim is to identify a parsimonious predictive model, so we progressively drop terms while respecting marginality (see Section 15.5). We therefore start by considering the interaction term, which is not significant ($F_{1,42}^{S.P} = 0.16, P = 0.690$). This test is the same in both sequential ANODEV tables because the term is fitted last in both cases. As the interaction is not significant, we can try to simplify the model further. We have many residual df (ResDF = 42) and there are only these two sequential ANODEV tables, so we can identify the predictive model from them. We therefore inspect the two main effects. In a linear model with this structure, factors Sex and Prey would be orthogonal, but Table 18.7 illustrates that this is not the case here, although the two tables are similar. We find that factor Sex is statistically significant whether it is fitted before or after factor Prey ($P < 0.001$ in both cases), and that factor Prey is not statistically significant in either sequence ($P \geq 0.085$). As factor Row represents a structural term, we do not consider removing it from the model (see Section 15.5). Our predictive model therefore takes the form

$$\text{Explanatory component:} \quad [1] + \text{Row} + \text{Sex}$$

Fitting this model leads to a residual mean deviance of 1.706 (with 44 df), and an observed F-statistic for the Sex main effect of $F_{1,44}^S = 19.617$ ($P < 0.001$). So this experiment gives strong evidence that the number of prey eaten by male and female ladybirds differ, but no evidence of any preference between the two prey types. We explore this difference between male and female ladybirds further in Examples 18.2D and 18.2E.

18.2.3 Checking the Model Fit and Assumptions

The first step of model checking consists of plotting the fitted model with the observed data. Figure 18.2 demonstrated that, for a model with a single explanatory variate, problems with model fit may be highlighted by plots of the fitted model on the scale of the linear predictor, where a straight line is expected. The residual plots described in Chapters 5 and 13 can also be used to give more information on the model fit, but the definition of the residuals needs to be extended for GLMs, and several methods are available. As previously, simple residuals can be defined as the difference between the observation, y_i, and its fitted value, i.e. $y_i - \hat{\mu}_i$. However, these residuals are subject to the same heterogeneity as the observations, and so are usually divided by the square root of the estimated variance of the distribution, $\text{Var}(\hat{\mu}_i)$, to give the set of **Pearson residuals**, defined for the ith observation as

$$\hat{e}_{Pi} = \frac{y_i - \hat{\mu}_i}{\sqrt{\text{Var}(\hat{\mu}_i)}} \ .$$

These residuals are called Pearson residuals as the sum of their squared values is equal to the Pearson goodness-of-fit statistic defined in Equation 18.4. Although we have now adjusted for heterogeneity of variance caused by the distribution of the observations, we still have to account for heterogeneity due to uncertainty in the predicted values (as in Section 13.2) and so we standardize the Pearson residuals by dividing them by their estimated SEs.

An alternative set of residuals are constructed as the square root of the contribution that each observation makes to the deviance (D_i defined in Equation 18.3) multiplied by the sign of the simple residuals; these are called the **deviance residuals**. For the Binomial distribution, the deviance residuals are calculated for the ith observation as

$$\hat{e}_{Di} = \text{sign}(y_i - \hat{\mu}_i)D_i = \text{sign}(y_i - \hat{\mu}_i)\left\{ 2y_i \log_e\left(\frac{y_i}{\hat{\mu}_i}\right) + 2(m_i - y_i)\log_e\left(\frac{m_i - y_i}{m_i - \hat{\mu}_i}\right) \right\}^{1/2} .$$

The sum of the squared values of these deviance residuals is equal to the residual deviance of the fitted model given in Equation 18.3. These residuals must also be standardized to give a common variance for diagnostic plots. Both the standardized deviance and standardized Pearson residuals can be generalized to prediction and deletion residuals via the same 'leave-one-out' technique used to derive these residuals in the Normal case (see Section 13.2).

Because the underlying probability distribution assumed for the observations is not Normal, we do not necessarily expect the residuals to conform to a Normal distribution. However, with a few exceptions, the standardized deviance residuals have been shown to give a reasonable approximation to a Normal distribution. For the case of a Binomial distribution, the exception is when the number of tests per observation, m_i, is small. In general, the distribution of the standardized Pearson residuals may be less close to a Normal distribution, and Collett (2002) shows some examples for Binomial data. The standardized deviance residuals can therefore be considered analogous to the standardized residuals discussed in Chapters 5 and 13, and are appropriate for use in the residual plots described in those chapters.

EXAMPLE 18.2C: LADYBIRD PREDATION

Figure 18.4 shows a composite set of residual plots with standardized deviance residuals from the predictive model that describes numbers of prey eaten in terms of the factors Row and Sex (see Example 18.2B). Six diagonal stripes can be seen in the fitted values plot, running downwards from the left-hand side to the right-hand side of the graph. These stripes correspond to the six distinct observed responses (0, 1 ... 5), and this type of pattern is likely to be found in any data set with a small number of discrete responses. There appears to be a little more variation in the centre of the range, but we judge the fitted values plot to be acceptable given the small Binomial total (six) per dish.

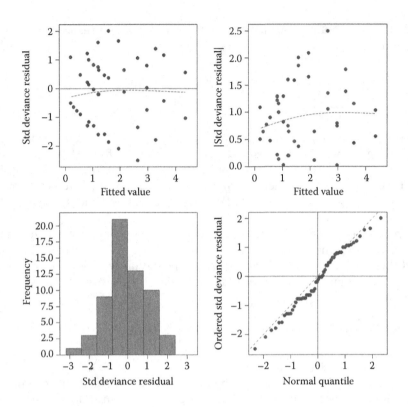

FIGURE 18.4
Composite set of residual plots based on standardized deviance residuals for the ladybird predation experiment (Example 18.2C).

The histogram and Normal probability plots suggest that the residuals give a reasonable approximation to a Normal distribution. These graphs therefore indicate no large discrepancies between the assumed model and the observed data.

18.2.4 Properties of the Model Parameters

As in the linear models seen previously, each parameter estimate in a GLM has an estimated SE that can be used for inference. The derivation of these SEs is beyond the scope of this book, but note that they are approximate and that they must include the multiplier $\sqrt{\hat{\phi}}$, i.e. the square root of the dispersion parameter, if this is estimated. If this multiplier is not used when over-dispersion is present, then the SEs under-estimate the uncertainty in the parameter values and this could lead to incorrect conclusions.

The decision on whether a term should be included in a model should be based on the sequential ANODEV table(s). A null hypothesis that a particular parameter is equal to zero can be tested by the parameter estimate divided by its SE, but remember that the interpretation and value of parameters associated with terms containing factors will depend on the parameterization of the model. Statistical software usually uses first- or last-level-zero constraints for GLMs (see Sections 4.5, 11.2 and 15.2 for further details). If there is no over-dispersion, then the ratio of a parameter to its SE has an approximate

Normal distribution. If the dispersion parameter is estimated, then this ratio has an approximate t-distribution with degrees of freedom equal to the residual df. If the absolute value of the ratio exceeds the $100(1 - \alpha_s/2)$th percentile of the appropriate distribution, then the null hypothesis that the parameter is equal to zero can be rejected at significance level α_s. The SEs can be used to construct approximate CIs for parameter values in the usual manner.

EXAMPLE 18.2D: LADYBIRD PREDATION

The predictive model fitted in Example 18.2B can be written in mathematical form with first-level-zero parameterization as

$$\hat{\eta}_{ir} = \text{logit}(\hat{p}_{ir}) = \hat{\eta}_{11} + \widehat{Row}_i + \widehat{Sex}_r ,$$

where $\hat{p}_{ir}$ is the predicted probability that an item of prey is eaten in the ith row ($i = 1 \ldots$ 15) for the rth sex ($r = 1, 2$; $1 = $ female and $2 = $ male) with logit transformation $\hat{\eta}_{ir}$. Then, $\hat{\eta}_{11}$ is the logit of the expected value for females in the first row, $\widehat{Row}_i$ is the relative effect of the ith row and $\widehat{Sex}_2$ is the difference in response between males and females on the logit scale. Table 18.8 shows the estimated parameters for this model with their estimated SEs.

The estimated effect of male ladybirds is $\widehat{Sex}_2 = -1.550$ (SE 0.3712), which indicates that males tended to eat less prey than females.

TABLE 18.8

Parameter Estimates (First-Level-Zero Parameterization) with Standard Errors (SE), t-Statistics (t) and Observed Significance Level (*P*), for the Ladybird Predation Experiment with Explanatory Factors Row (15 Levels) and Sex (Two Levels, 1 = Female, 2 = Male) (Example 18.2D)

Term	Parameter	Estimate	SE	t	P
[1]	η_{11}	0.386	0.6013	0.642	0.524
Row 1	Row_1	0	—	—	—
Row 2	Row_2	−0.200	0.8259	−0.242	0.810
Row 3	Row_3	−1.774	1.0117	−1.754	0.086
Row 4	Row_4	−0.408	0.8354	−0.488	0.628
Row 5	Row_5	−0.626	0.8496	−0.737	0.465
Row 6	Row_6	−1.121	0.8983	−1.247	0.219
Row 7	Row_7	−0.408	0.8356	−0.488	0.628
Row 8	Row_8	0.582	0.8171	0.712	0.480
Row 9	Row_9	−2.246	1.1303	−1.987	0.053
Row 10	Row_{10}	−1.417	0.9420	−1.504	0.140
Row 11	Row_{11}	−1.417	0.9414	−1.505	0.139
Row 12	Row_{12}	−0.626	0.8492	−0.737	0.465
Row 13	Row_{13}	−2.246	1.1387	−1.973	0.055
Row 14	Row_{14}	−0.408	0.8355	−0.488	0.628
Row 15	Row_{15}	−1.774	1.0119	−1.753	0.086
Sex 1	Sex_1	0	—	—	—
Sex 2	Sex_2	−1.550	0.3712	−4.176	< 0.001

18.2.5 Evaluating the Response to Explanatory Variables: Prediction

In general, examination of estimated parameters from the predictive model has limited scope, as it is usually the overall response to explanatory variables that is of interest. The presence of the link transformation makes prediction for GLMs more complex than for linear models, although the issues that arise are similar to those for the presentation of results following analysis of transformed data (see Section 6.3).

Prediction on the linear predictor scale is straightforward, as on this scale the model is linear and the estimated SE usually gives a good approximation of the uncertainty associated with the predicted value. Prediction for a specific combination of explanatory variables can be made on the linear predictor scale, then a CI can be generated from the Normal distribution (no over-dispersion) or t-distribution (over-dispersion present), and the prediction and its confidence limits can be back-transformed to the natural scale. While software will calculate SEs on the natural scale (via the delta method), these SEs tend to be much less accurate than those calculated on the linear predictor scale because they make an additional set of approximating assumptions. Back-transformed CIs therefore tend to give a better measure of uncertainty than these approximate SEs.

Further complications arise when averages over variables are required, or where the main objective of the study is comparison between groups, or both.

Averaging over variables is required for predictions for a subset of the explanatory variables. The usual procedure is to form predicted values for all combinations of the explanatory variables, i.e. at specified values of variates and all levels of factors. In a linear model, predictions for the variables of interest are then obtained as averages over the remaining variables (Section 15.5.2). In a GLM, we must also consider back-transformation to the natural scale and this leads to two possibilities, either averaging before back-transformation or averaging afterwards, and these two strategies will give different numerical results with different interpretations. This situation is discussed in the context of analysis of transformed data by Morris (1985) and illustrated in Example 18.2E. Averaging before back-transformation can be interpreted as making a prediction at an average value of the remaining variables. This gives individual predictions with SEs on the linear predictor scale, CIs can be formed for each prediction, and these CIs can be back-transformed to give a realistic measure of uncertainty on the natural scale. If instead the full set of predictions is back-transformed before averaging, this is analogous to predicting an average response on the natural scale for an experiment in which the predicted combination was applied with each combination of levels of the remaining variables. Unfortunately, only approximate SEs on the natural scale can be calculated for this type of prediction.

We now consider comparison between specific combinations of explanatory variables. Comparisons can easily be made on the linear predictor scale, with appropriate SEs, and so this is the scale on which you should test such comparisons. However, interpretation of comparisons on the natural scale can be difficult. We illustrate this problem using an experiment with a set of t treatment groups. We label the transform of the expected value for the jth group on the linear predictor scale as η_j, and are interested in the quantity $\eta_j - \eta_k$, with predicted value $\hat{\eta}_j - \hat{\eta}_k$. Ideally, as in the case of individual predictions, we should like to take a CI for this quantity and map it on to a meaningful quantity on the natural scale. This can be done for the log link function (see Section 18.3.1), since

$$\hat{\eta}_j - \hat{\eta}_k = \log_e(\hat{\mu}_j) - \log_e(\hat{\mu}_k) = \log_e\left(\frac{\hat{\mu}_j}{\hat{\mu}_k}\right),$$

so the comparison on the log scale is the log of the ratio of the predictions on the natural scale, and this ratio can be back-transformed and interpreted (see Example 18.3). For a logit link function, we find that

$$\hat{\eta}_j - \hat{\eta}_k = \text{logit}(\hat{p}_j) - \text{logit}(\hat{p}_k) = \log_e \left(\frac{\hat{p}_j/(1-\hat{p}_j)}{\hat{p}_k/(1-\hat{p}_k)} \right),$$

so the comparison on the logit scale is the log of the odds-ratio of the predictions on the natural scale. Unfortunately, the odds-ratio is rather less interpretable. In general, if the quantity of interest is the difference in expected values on the natural scale, i.e. $\hat{\mu}_j - \hat{\mu}_k$, then there is no real alternative to back-transforming predictions and using the approximate SE calculated on the natural scale. With link function g(), the difference is then estimated as

$$\hat{\mu}_i - \hat{\mu}_j = g^{-1}(\hat{\eta}_i) - g^{-1}(\hat{\eta}_j).$$

EXAMPLE 18.2E: LADYBIRD PREDATION

We established a predictive model in Example 18.2D, and now we want to understand how an estimated decrease for males of 1.55 units on the logit scale translates into number of prey eaten. Table 18.9 lists the full set of predictions and the back-transformed proportions. The predictions for male ladybirds are 1.55 units smaller than for female ladybirds in the same row on the logit scale, but the same difference varies between 0.10 (rows 9 and 13) and 0.37 (row 8) once back-transformed.

TABLE 18.9

Predictions for Ladybird Predation on Linear Predictor Scale and Back-Transformed as Probabilities for Each Sex in Each Row (Example 18.2D)

Row	Linear Predictor Scale (Logit)		Back-Transformed (Fitted Probability)	
	Female	Male	Female	Male
1	0.386	−1.164	0.595	0.238
2	0.186	−1.364	0.546	0.204
3	−1.388	−2.934	0.200	0.050
4	−0.021	−1.571	0.495	0.172
5	−0.240	−1.790	0.440	0.143
6	−0.734	−2.284	0.324	0.092
7	−0.021	−1.571	0.495	0.172
8	0.968	−0.582	0.725	0.359
9	−1.860	−3.410	0.135	0.032
10	−1.031	−2.581	0.263	0.070
11	−1.031	−2.581	0.263	0.070
12	−0.240	−1.790	0.440	0.143
13	−1.860	−3.410	0.135	0.032
14	−0.021	−1.571	0.495	0.172
15	−1.388	−2.938	0.200	0.050

To predict the difference in number of prey eaten between male and female ladybirds in an average row, we take the average of the predictions on the logit scale as

$$\hat{\eta}_{\cdot r} = \frac{1}{15}\sum_{i=1}^{15}\hat{\eta}_{ir}; \quad \hat{\eta}_{\cdot 1} = -0.553 \text{ (SE 0.2196)}, \quad \hat{\eta}_{\cdot 2} = -2.103 \text{ (SE 0.3128)}.$$

We can construct 95% CIs for these predictions on the logit scale as $(-0.996, -0.110)$ for females and as $(-2.733, -1.473)$ for males. We back-transform these estimates and CI, and estimate the probability of an item of prey in an average row being eaten by female ladybirds as 0.37 with 95% CI (0.27, 0.47), and by male ladybirds as 0.11 with 95% CI (0.06, 0.19). Note the asymmetry of the CI for the male ladybirds. Approximate SEs can be calculated directly for these back-transformed predictions as 0.051 and 0.030 for female and male ladybirds, respectively, so the approximation is better for the female than for the male ladybirds.

To predict the average difference in number of prey eaten between male and female ladybirds across the whole experiment, we take the average of the back-transformed predictions, as

$$\hat{p}_{\cdot r} = \frac{1}{15}\sum_{i=1}^{15}\hat{p}_{ir} \quad \text{for } r = 1, 2,$$

giving a predicted average proportion of prey eaten of 0.38 (approximate SE 0.044) for females and 0.13 (approximate SE 0.032) for males. In this example, these quantities differ only a little from those averaged on the linear predictor scale. In general, the appropriate scale for prediction will depend on the context of the study.

18.2.6 Aggregating Binomial Responses

It is not always clear how Binomial responses should be recorded. For example, consider an experiment looking at the prevalence of pests on different varieties within an orchard, where four individual branches are assessed as clean or infested on six trees of each variety. The investigator might wonder whether to record the results as binary scores (0 or 1) for each branch, as the number of infested branches per tree (out of 4), or as the number of infested branches per variety (out of 24)? As long as we fit the same explanatory component, we obtain the same parameter estimates at any of these scales, but we shall obtain a different residual deviance. As a rule of thumb, we suggest that the appropriate scale for analysis (and hence the minimum scale for recording measurements) is the smallest experimental unit present in the study (see Section 3.1), as this avoids the issues with binary data described in the next section. In our orchard example, this would be the individual tree, as the variety changes between but not within trees. The residual deviance then reflects expected tree to tree variation in the underlying susceptibility to disease in addition to Binomial sampling variation. A deviance larger than that expected for Binomial samples indicates that such variation is present, and this can be accounted for by the dispersion parameter φ.

In some circumstances, it can help to aggregate Binomial observations further, to give a single response for each of the study conditions, i.e. for each combination of explanatory variables, or group, present. It is appropriate to do this only when replicate observations are obtained under uniform conditions and no systematic differences between them are expected. The residual deviance can then be used to assess the fit, and any indication of over-dispersion indicates lack of fit in the model (as discussed in Section 18.2.2). This is useful only when the df associated with the model is smaller than the number of groups, and relies heavily on the assumption of a Binomial distribution to derive the sampling variance.

18.2.7 The Special Case of Binary Data

Binary responses, also known as a Bernoulli data, are a special case of Binomial data with only one test per observation (i.e. $m_i = 1$), so that the observations can take only the values 0 or 1. Analysis follows the same procedure as for other Binomial responses, but not all of the results discussed above are valid for binary data. In particular, the residual deviance does not give a reliable measure of over- or under-dispersion, and so the use of an estimated dispersion parameter is not recommended. The Pearson and deviance residuals are uninformative as they will not be distributed as an approximate Normal distribution, and a fitted values plot will often show strong patterns, even if the model is adequate.

From a practical point of view, it is better to avoid binary observations whenever possible, as they provide very little information per observation. One way of doing this is to take several independent replicate observations on each unit. For example, if the aim of an experiment is to assess disease incidence in a field trial then a binary assessment of each plot for presence or absence of the disease will be quick, but gives little information on the extent of infection (one plant infected per plot gives the same answer as all plants infected), and it can make it difficult to discriminate between treatments. If 10 (independent) plants per plot are individually assessed for presence of disease, then responses range from 0 to 10, giving some information on the extent as well as presence of disease, as well as a more tractable analysis. This is an example where sub-sampling within experimental units provides valuable extra information and, in this type of situation, data should always be considered as total counts within each unit rather than individual binary observations (see remarks in Section 18.2.6). In scenarios where binary data are unavoidable, replicate as much as possible to counteract the lack of information per observation.

18.2.8 Other Issues with Binomial Responses

In this chapter, we have described one common implementation of a Binomial GLM; however, many variations are possible. For example, some statistical software prefers the Pearson rather than the deviance estimate of the dispersion parameter and provides Pearson rather than deviance residuals. Similarly, the dispersion parameter might be fixed at 1 by default rather than estimated, or might be estimated but not used within the model for testing and inference unless this is explicitly requested.

The logit link is the canonical link for the Binomial distribution and widely used, particularly in medical applications, because of its interpretation in terms of odds-ratios, although in practice this may be difficult to explain to non-mathematicians. Historically, a method called probit analysis was used for dose–response studies (Finney, 1971), and the simplest probit analysis model is equivalent to a Binomial GLM with **probit link**. The probit function is the inverse of the cumulative distribution function for the Normal distribution, and can be interpreted in terms of a Normal tolerance distribution. Both the logit and probit functions are symmetric around probability $p = 0.5$ and usually give similar answers. Another option is the **complementary log–log link** function, $\log_e(-\log_e(1 - p))$, which has asymmetric curvature. In all cases, the fit of a model and residuals should always be checked graphically, as this may reveal an inappropriate link function.

It is common to find that the response needs to be modelled in terms of the logarithm of an explanatory variate, rather than in terms of the explanatory variate directly. This is interpreted by Collett (2002, Section 4.1) in terms of an asymmetric tolerance distribution, allowing a few individuals to have unusually high tolerances.

One limitation of regression within a Binomial GLM is that the success probability must tend to zero as the explanatory variate decreases to $-\infty$, and must increase to one as the explanatory variate increases to $+\infty$. If this is not the case, then a slightly more complex non-linear model is required, further details of these models are given in Collett (2002, Chapter 4) or Finney (1971).

One common use of logistic or probit regression in a dose–response context is the estimation of the dose required to achieve a certain response. For example, in pesticide studies the LD50, the dose required to kill 50% of a sample, is often used to compare compounds. This is different from a standard prediction in that we are trying to predict the value of the explanatory variate at which a certain response is obtained, rather than vice versa. This is an example of **calibration** (sometimes called **inverse prediction**), and was discussed for SLR in Section 12.9.3. Approximate SEs or CIs for this prediction, sometimes called fiducial limits, can be obtained from Fieller's theorem (Collett, 2002, Chapter 4). Note that use of an LD50 to compare compounds is sensible only if the responses can be fitted by a parallel lines model on the linear predictor scale; otherwise, a single value cannot capture the overall differences between the compounds.

In Chapter 6, we suggested a logit transformation to deal with proportion data where the numbers of trials m_i are reasonably large (> 20) and roughly equal across units, and the observed values are not too extreme (not too many observed proportions close to 0 or 1). This recommendation is justified as a Normal distribution can provide a reasonable approximation to the Binomial distribution under these conditions. This approach is particularly helpful when the experimental units are structured (e.g. a split-plot design), as this structure cannot always be accounted for easily in the GLM framework (as for regression, see Section 15.3). However, in all other cases, the use of the appropriate GLM is recommended.

Finally, think about the intended sampling scheme when Binomial data are to be collected. It is important that trials are independent, so there should be no competition for resources between the individuals assessed. The number of trials and number of observations should also be considered. Increasing the number of trials per observation also increases the precision of an individual observation, so very small numbers of trials should be avoided whenever possible, but increasing the number of observations may have more effect on the precision of the overall analysis.

18.3 Analysis of Count Data: Poisson Responses

A Poisson distribution arises as a count of the number of times a phenomenon occurs within a fixed interval of time or space. Examples of counts that may be modelled as a Poisson distribution include

- The number of bees arriving at a rape plant per minute
- The number of mutations in a given length of DNA after radiation is applied

- The number of pine trees per unit area of mixed forest
- The number of bacteria in a given volume of liquid

If an observation is Poisson-distributed, then it can take only non-negative integer values, $0, 1, 2 \ldots +\infty$. In theory, there should be no upper bound, but in practice, some physical upper bound can apply without invalidating the Poisson distribution assumption, so long as this limit is large enough in relation to the responses to avoid truncating the distribution. The Poisson distribution is defined by a single parameter, the mean μ. We write that an observation y_i is Poisson-distributed with expected value μ_i as $y_i \sim \text{Poisson}(\mu_i)$. The probability of obtaining a specific value y_i for the ith observation can be written in terms of its expected value as

$$\text{Prob}(y_i \, ; \mu_i) \; = \; \frac{\mu_i^{y_i} \, e^{-\mu_i}}{y_i!}, \quad \text{for } y_i = 0, 1, 2 \ldots +\infty \, .$$

The probability of observing a specific outcome, for example, $\text{Prob}(y_i = 0)$, depends only on the unknown parameter μ_i. If y_i follows a Poisson distribution, then both its expected value and variance are equal to the parameter μ_i, i.e.

$$E(y_i) = \mu_i; \quad \text{Var}(y_i) = \mu_i \, .$$

There is therefore a strong variance–mean relationship for this distribution. In the context of GLMs, we hypothesize that the expected value of the observation, μ_i, may depend on one or more explanatory variables.

EXAMPLE 18.3A: PEA APHID SURVEY

An ecological survey was done to investigate the co-occurrence of various insect predator and prey species. Here, we consider a subset of the data relating to one aphid species, the pea aphid, *Acyrthosiphon pisum*. In each of three fields, 15 randomly chosen triplets of adjacent bean plants were inspected and the number of pea aphids present on the three plants was recorded. The data are in Table 18.10, and file APHIDS.DAT contains explanatory factor Field (three levels) to identify the observations by field, factor Sample to label the 15 samples within each field, and response variate *AphidCount* which holds the total count of aphids at each sample point. The objective here is to determine whether infestation differed among the three fields.

The data are shown in Figure 18.5: they are discrete counts, and the variance between replicate observations for each field ($s_1^2 = 14.69$, $s_2^2 = 119.97$, $s_3^2 = 21.24$) appears to increase with the mean count ($\bar{y}_{1.} = 4.6$, $\bar{y}_{2.} = 15.4$, $\bar{y}_{3.} = 6.7$), although the variances are clearly much larger than the sample means in each case.

18.3.1 Understanding and Defining the Model

We recommend reading Section 18.2 before proceeding further as the analysis of Poisson responses using a GLM follows the same framework as the analysis of Binomial responses. The major difference between the two cases is in the form and interpretation of the model. Again, models can be written in terms of quantitative or qualitative variables, or both, but here we introduce the Poisson model using a single qualitative variable (factor) and later consider other cases. We label the units by groups ($j = 1 \ldots t$) and label observations within

TABLE 18.10

Counts of Pea Aphid from 15 Samples in Three Bean Fields (Example 18.3A and File APHIDS.DAT)

Field	Sample	Count	Field	Sample	Count	Field	Sample	Count
1	1	0	2	1	24	3	1	3
1	2	3	2	2	10	3	2	2
1	3	5	2	3	21	3	3	10
1	4	15	2	4	28	3	4	14
1	5	7	2	5	43	3	5	4
1	6	5	2	6	11	3	6	11
1	7	2	2	7	14	3	7	6
1	8	5	2	8	22	3	8	2
1	9	4	2	9	8	3	9	3
1	10	6	2	10	7	3	10	5
1	11	1	2	11	1	3	11	3
1	12	5	2	12	20	3	12	13
1	13	1	2	13	6	3	13	5
1	14	9	2	14	10	3	14	4
1	15	1	2	15	6	3	15	15

Source: Data from P. Wells, Rothamsted Research.

groups ($k = 1 \ldots n_j$), so that y_{jk} is the response for the kth observation in the jth group. The model with a single explanatory factor is then written as

$$E(y_{jk}) = \mu_j \quad \text{with } \eta_j = g(\mu_j) = \eta_1 + v_j ,$$

where g() is the link function, as described in Section 18.1. Each replicate observation in the jth group has a common expected value, namely, μ_j, and after transformation by the

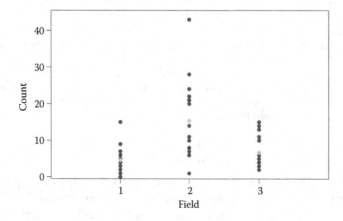

FIGURE 18.5
Counts of pea aphid (•) from each of three bean fields (Example 18.3A) with predicted field counts (•) from the fitted model (Example 18.3B).

link function, this takes the value η_j. The model uses first-level-zero parameterization (see Sections 4.5, 11.2 or 15.2 for details). The parameter η_1 represents the transformed value of the population mean for the first group, and v_j represents the difference between the jth and first group on the linear predictor scale, with constraint $v_1 = 0$.

The canonical link function for Poisson responses is the **log link**, and a model using this link function is called a **log-linear model**. The model can then be written as

$$\eta_j = \log_e(\mu_j) = \eta_1 + v_j \, ,$$

so the natural logarithm of the expected mean count changes according to the group it belongs to. We can rearrange this expression to write the model in terms of the expected counts as

$$\mu_j = \exp(\eta_j) = \exp(\eta_1 + v_j) = \exp(\eta_1) \times \exp(v_j) \, .$$

On the natural scale, this is a multiplicative model (see also Section 6.4), and the fitted values can take non-negative values only. The Poisson GLM with log link can therefore be considered as an exponential model that accounts for the Poisson distribution of the responses and their associated heterogeneity. This exponential model is not completely general (see Section 17.3), as it is constrained to have a lower asymptote of zero.

For counts held in response variate Y with groups labelled by the explanatory factor Group, this Poisson GLM can be represented in symbolic form as

Response variable: Y
Probability distribution: Poisson
Link function: log
Explanatory component: [1] + Group

As in Section 18.2, to fully specify the GLM, we need to give the probability distribution and link function in addition to the response variate and explanatory component of the model.

Parameter estimation is achieved by maximum likelihood estimation, and again results will be obtained directly from statistical software rather than being derived here. For data with a Poisson distribution, the deviance for a model with fitted values $\hat{\mu}_j$ takes the form

$$D = 2 \sum_{j=1}^{t} \sum_{k=1}^{n_j} \left[y_{jk} \log_e \left(\frac{y_{jk}}{\hat{\mu}_j} \right) - (y_{jk} - \hat{\mu}_j) \right] \, .$$

Once parameter estimates have been derived, you should use the procedures described in Section 18.2 to check the model fit before drawing any conclusions.

EXAMPLE 18.3B: PEA APHID SURVEY

We want to fit a model to the pea aphid data of Example 18.3A to investigate whether the expected count of this aphid differs among the three fields. Using the explanatory factor Field and the response variate *AphidCount* (see Example 18.3A), we can write the model in symbolic form as

Response variable:	*AphidCount*
Probability distribution:	Poisson
Link function:	log
Explanatory component:	[1] + Field

In mathematical form, this model is written with first-level-zero parameterization as

$$AphidCount_{jk} \sim \text{Poisson}(\mu_j), \quad \eta_j = \log_e(\mu_j) = \eta_1 + Field_j \,,$$

where $AphidCount_{jk}$ is the count for the kth observation in the jth field ($j = 1, 2, 3$) with expected value μ_j. Then, η_j is the log-transform of μ_j, and $Field_j$ is the difference on the log scale between the jth and the first fields. The estimated parameters are $\hat{\eta}_1 = 1.526$, $\widehat{Field_2} = 1.208$ and $\widehat{Field_3} = 0.371$, giving predicted values

$$\hat{\eta}_1 = 1.526, \quad \hat{\eta}_2 = 2.734, \quad \hat{\eta}_3 = 1.897 \,.$$

We can back-transform these values to estimate the expected number of aphids per sample in each field as

$$\hat{\mu}_1 = \exp(1.526) = 4.6, \quad \hat{\mu}_2 = \exp(2.734) = 15.4, \quad \hat{\mu}_3 = \exp(1.897) = 6.7 \,.$$

These predictions are equal to the mean counts for each field (see Example 18.3A) and Figure 18.5 shows these estimated field means with the observations.

18.3.2 Analysis of the Model

As described in detail in Section 18.2.2, the ANODEV table is formed by a partition of the total deviance into the change in deviance between the null model (overall mean) and the fitted model, and the change in deviance between the fitted model and the saturated model (where each observation is fitted exactly). If the residual deviance is larger than expected, then the fit of the model should be examined graphically to check for misspecification or outliers and addition of other explanatory variables should be considered. If these measures do not reduce the residual deviance to a value consistent with the expected chi-squared distribution, then, as with Binomial data, a dispersion parameter can be added to the model, so that

$$\text{Var}(y_i) = \varphi \, \mu_i \,.$$

The presence of an estimated dispersion parameter changes the interpretation of entries in the ANODEV table in the same manner as for Binomial responses (see Section 18.2.2.2), requiring the use of deviance ratios and tests based on the F-distribution. In practice, over-dispersion is usually present for count data. For models with several explanatory terms, sequential ANODEV tables or marginal tests can be used to identify the predictive model. Assessment of individual parameters and prediction follows as described in Section 18.2.4.

EXAMPLE 18.3C: PEA APHID SURVEY

Table 18.11 is the ANODEV for the model of Example 18.3B. Here, the residual deviance, ResDev, has value 191.475 with 42 df ($P < 0.001$ when compared with a chi-squared distribution with 42 df). So there is strong evidence of over-dispersion for this model, which fits with our preliminary observation that the within-field variances were much larger than the within-field means (Example 18.2A). We might speculate that this over-dispersion arises from variation in prevalence (patchiness) between different areas of each field and perhaps between plants. The deviance estimate of the dispersion parameter is equal to 4.559 and the model deviance ratio is calculated as $F = 52.728/4.559 = 11.566$. Compared with an F-distribution with 2 and 42 df, this deviance ratio is highly significant ($P < 0.001$). We therefore reject the null hypothesis and conclude that there are statistically significant differences in the mean count of pea aphids between fields.

Figure 18.6 shows the composite set of residual plots for these data based on standardized deviance residuals (Section 18.2.3). In these plots, the residuals appear somewhat skewed, but there is no strong evidence of variance heterogeneity and the Normal plots form approximately straight lines, so the model appears to give an adequate description of the data.

Further discussion with the investigator revealed that samples were taken along transects rather than from random positions in each field. In this case, one might suspect dependence between samples, with samples closer together on a transect being more strongly correlated than those further apart. We investigate dependence (Section 5.2.2) using an index plot of the standardized residuals against transect position (sample number) for each of the three fields separately (Figure 18.7a), and by plotting each residual against the residual for the previous sample on the same transect (Figure 18.7b). There is no evidence in either graph of correlation between successive observations.

We therefore accept the model and move on to interpretation. Our main interest is in quantifying differences between fields. With a log link function, we can use the property that differences on the log scale back-transform to give ratios on the natural scale (see Section 18.2.5). On the log scale, the estimated difference between the second and first fields is

$$\hat{\eta}_2 - \hat{\eta}_1 = \widehat{Field_2} = 1.208 \,,$$

with SE = 0.2929 and $P < 0.001$, indicating significantly larger counts in field 2. Since

$$\hat{\eta}_2 - \hat{\eta}_1 = \log_e(\hat{\mu}_2) - \log_e(\hat{\mu}_1) = \log_e(\hat{\mu}_2/\hat{\mu}_1) \,,$$

TABLE 18.11

ANODEV Table for the Pea Aphid Survey with Explanatory Factor Field (Example 18.3C)

Source of Variation	df	Deviance	Mean Deviance	P (Chi-Squared)
Field	2	105.456	52.728	< 0.001
Residual	42	191.475	4.559	
Total	44	296.931	6.748	

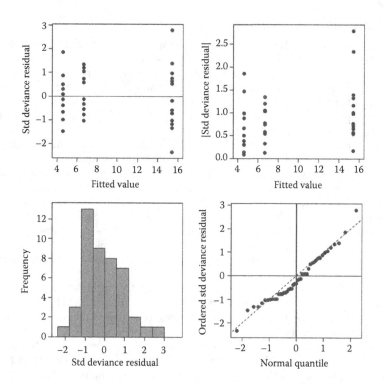

FIGURE 18.6
Composite set of residual plots based on standardized deviance residuals for the pea aphid survey (Example 18.3C).

the back-transformation of this difference gives us $\hat{\mu}_2/\hat{\mu}_1 = \exp(\hat{\eta}_2 - \hat{\eta}_1) = \exp(1.208) =$ 3.35. The expected count in field 2 is therefore estimated to be 335% of the expected count in field 1. We can construct a 95% CI for this quantity on the log scale as

$$\left[(\hat{\eta}_2 - \hat{\eta}_1) \pm t_{42}^{[0.025]} \times SE(\hat{\eta}_2 - \hat{\eta}_1) \right] = \left[1.208 \pm (2.108 \times 0.2929) \right] = (0.617, 1.799) .$$

When transformed back to the natural scale, the CI is $\exp(0.617, 1.799) = (1.85, 6.05)$, indicating that the ratio of expected counts between the fields may be smaller than 2 or as

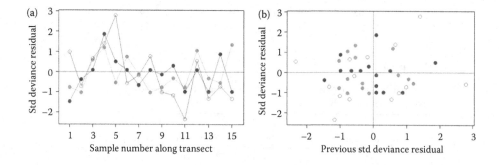

FIGURE 18.7
(a) Index plot of standardized deviance residuals against transect position within each of three fields and (b) plot of residuals against previous residuals (within transects) for the pea aphid survey (Example 18.3C). • field 1, ○ field 2, • field 3.

large as 6. If we follow a similar procedure for comparing the third field with the first field, we find

$$\hat{\eta}_3 - \hat{\eta}_1 = \widehat{Field}_3 = 0.371 \, ,$$

with SE = 0.334 and no evidence of a difference in expected count between these two fields ($P = 0.273$). Back-transformation estimates the ratio as 1.45 with 95% CI equal to (0.74, 2.84), confirming that a ratio of 1 is a plausible value. A similar calculation can be carried out for fields 2 and 3.

18.3.3 Analysing Poisson Responses with Several Explanatory Variables

In Example 18.3, we considered the case of Poisson responses with a single explanatory factor. We can also use log-linear models for variates or a mixture of factors and variates. Below, we demonstrate the modelling process for two explanatory variables, a factor and a variate.

EXAMPLE 18.4: CONIDIAL RELEASE EXPERIMENT

An experiment was set up with the primary aim of measuring aphid infection rates in response to differing doses of fungus. Aphids in inoculation chambers were subjected to conidia showers from sporulating cadavers from one of two different sources (a clone or a standard source) for one of eight time periods ranging from 0 to 80 min. Estimates of the conidial doses received by the aphids were obtained as counts of spores on slides placed in the chambers. Here, we investigate the relationship between the achieved dose (variate *Conidia*) and infection time (variate *Time*) for the two types of source (factor Source). Each time period and source combination was tested in each of two experimental runs (factor Run). Separate sources were used for each replicate of each time period and the observed counts are listed in Table 18.12.

The zero time period is a negative control: it should not be possible for any conidia to be released in no time, so this category just checks for contamination of slides, and the resulting zero counts verify that this was not present. We remove this category prior to analysis as it contains no information relating to the explanatory variable (see also discussion in Section 8.5). The data, excluding the zero time periods, can be found in file

TABLE 18.12

Number of Conidia Released by Different Sources over Eight Time Periods (Example 18.4 and File CONIDIA.DAT)

Time (min)	Source			
	Standard		Clone	
	Run 1	Run 2	Run 1	Run 2
0	0	0	0	0
5	6	71	8	44
10	71	223	173	209
15	157	426	165	383
20	568	1391	584	1188
25	883	1098	1296	627
40	1436	993	400	1628
80	3543	4295	4981	4302

Source: Data from J. Baverstock, Rothamsted Research.

CONIDIA.DAT. The aim of this analysis is to establish whether there is any difference in release rates between the two sources and this can be interpreted as a regression with groups (see also Section 15.1).

Preliminary investigation, plotting the log number of conidia against the time period, indicated a curved relationship in terms of time, but an approximate straight line relationship with a log transformation of time; hence, we construct the explanatory variate *logTime* = $\log_e$(*Time*). The experiment is set up as a RCBD, with runs as blocks and all the experimental conditions evaluated once within each run. As in Example 18.2, we incorporate the structural component (factor Run) in the explanatory component of the model to obtain an intra-block analysis, and fit the Run factor before the explanatory terms (see Section 15.3). The initial model fits separate lines for each source. In addition, as there are replicates for each treatment combination, we can formally investigate model misspecification with the lack-of-fit test described in Section 12.8, using a factor Period that has a separate level for each time period. The initial model can therefore be written in symbolic form as

Response variable:	*Conidia*
Probability distribution:	Poisson
Link function:	log
Explanatory component:	*[1]* + Run + *logTime* + Period + Source
	+ *logTime*.Source + Period.Source

The residual deviance of 2006.1 with 13 df for this model indicates substantial over-dispersion ($P < 0.001$). This cannot be explained in terms of outliers, misspecification or missing explanatory variables, and the residual plots (not shown) are adequate, so we use an estimated dispersion parameter, $\hat{\phi} = 2006.1/13 = 154.3$. We use marginal F-tests to identify the predictive model, respecting marginality, and the model selection process is shown in Table 18.13.

We start with the full model (Model 1 in Table 18.13) and examine the lack-of-fit term, Period.Source, which tests for deviations from the separate straight lines for each source. This term is not statistically significant ($P = 0.987$) and so we drop it. As we have few residual df here (ResDF = 13), we choose to refit the model excluding term Period.Source before proceeding to Model 2 in Table 18.13. Dropping a term does not change the other incremental deviances or mean deviances, but the dropped term is merged with the residual and so the residual deviance, residual df and deviance ratios all change. Because the mean deviance of the Period.Source term was substantially less than 1 and the residual df were small, the residual mean deviance for the revised model

TABLE 18.13

Observed Significance Level (*P*) for Marginal F-Tests in a Sequence of Models for the Conidial Release Experiment with Explanatory Variate *logTime* and Explanatory Factors Run, Period and Source (Example 18.4)

	P			
Term	Model 1	Model 2	Model 3	Model 4
Run	—	—	—	—
logTime	—	—	—	—
Period	—	0.038	0.034	0.028
Source	—	—	0.666	*
logTime.Source	—	0.432	*	*
Period.Source	0.987	*	*	*

Note: — = term in model but not eligible for testing, * = term omitted from model.

is much reduced (equal to 116.4 with 18 df). We can then examine terms *logTime*.Source (separate lines, $P = 0.432$) and Period (lack of fit to common line, $P = 0.038$). At this stage, we drop term *logTime*.Source and refit to get a parallel lines model with lack of fit (Model 3 in Table 18.13). We can then test terms Source (separate intercepts, $P = 0.666$) and Period (lack of fit, $P = 0.034$). There is therefore no need for separate intercepts, so we drop term Source, leaving the SLR with lack of fit (Model 4), which cannot be simplified further. This predictive model can be written in symbolic form as

Explanatory component: *[1]* + Run + *logTime* + Period

This fits a separate effect for each time period, and is equivalent to the simpler form

Explanatory component: *[1]* + Run + Period

We can write this model in mathematical form as

$$\log_e(\hat{\mu}_{ij}) = \hat{\eta}_{ij} = \hat{\eta}_{11} + \widehat{Run}_i + \widehat{Period}_j \ ,$$

where $\hat{\mu}_{ij}$ is the prediction of the expected value of counts in the jth time period for the ith run. To predict for an average run, we average over the runs to get

$$\hat{\eta}_{\bullet j} = \hat{\eta}_{11} + \frac{1}{2}\sum_{i=1}^{2} \widehat{Run}_i + \widehat{Period}_j \ .$$

To determine the extent and source of the lack of fit, we can compare these predictions with those obtained from a model excluding the lack-of-fit term, with explanatory component

Explanatory component: *[1]* + Run + *logTime*

This predictive model can be written in mathematical form in terms of continuous time (t) as

$$\hat{\eta}_i(t) = \hat{\alpha} + \widehat{Run}_i + \hat{\beta}\log_e(t) \ ,$$

and again, we can average this model over runs to predict for a typical run as

$$\hat{\eta}(t) = \hat{\alpha} + \frac{1}{2}\sum_{i=1}^{2} \widehat{Run}_i + \hat{\beta}\log_e(t) = \hat{\alpha}^* + \hat{\beta}\log_e(t) \ .$$

Figure 18.8 shows these predictions from both versions of the model on the natural and linear predictor scales.

There are two time periods, 20 and 40 min, where the counts appear inconsistent with the fitted line, either consistently larger (at 20 min) or smaller (at 40 min) than expected. Further investigation is required to determine whether this irregular behaviour is characteristic of the experimental system or an anomaly specific to this trial. In either case, since the fitted line broadly follows the observed trend, we can use it to indicate likely levels of conidial release to help design further experiments. We can transform the predictive model back to the natural scale as

$$\hat{\mu}(t) = \exp(\hat{\eta}(t)) = \exp(\hat{\alpha}^* + \hat{\beta}\log_e(t)) = \exp(\hat{\alpha}^*)\exp(\log_e(t^{\hat{\beta}})) = \hat{\lambda}t^{\hat{\beta}} \ ,$$

where $\hat{\lambda} = \exp(\hat{\alpha}^*)$. Our predictive GLM is therefore equivalent to a power model which is constrained to pass through the origin, i.e. $\mu(0) = 0$, while accounting for variance

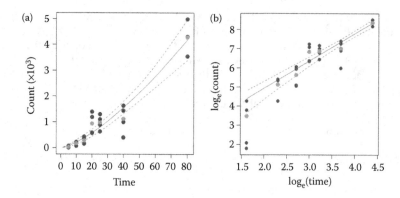

FIGURE 18.8

Observations (•) with predicted response from SLR (—) with 95% confidence intervals (---), and from model incorporating lack of fit (○) on (a) natural and (b) log scale for the conidial release experiment (Example 18.4).

heterogeneity and the strong variance–mean relationship inherent to the Poisson distribution. The slope coefficient (β) is estimated as 1.43 with 95% CI equal to (1.20, 1.65), so the power relationship is greater than linear ($\beta = 1$) but less than quadratic ($\beta = 2$).

18.3.4 Other Issues with Poisson Responses

The form of the variance–mean relationship in the Poisson model is quite restrictive and is not appropriate for all count data. The Negative Binomial probability distribution provides an extension that allows for some clustering in the responses by introducing another parameter into the model. This can be useful for zero-inflated Poisson responses, which occur when the responses resemble a Poisson distribution but with an unusually large number of zero counts. More sophisticated mixture models are also available in this context, and further details can be found in Ridout et al. (2001).

A GLM using the Poisson distribution with the log link function deals with discrete counts where the variance increases with the expected value. In Chapter 6, we suggested the logarithm transformation for data with this type of variance–mean relationship. If all of the expected counts are reasonably large (i.e. > 10), then a Normal approximation often provides a good approximation to the Poisson distribution. However, there is one important distinction between the GLM and transformation approaches in this case. As we saw in Chapter 6, the transformation approach leads to group population means being estimated by the group geometric means, as the means are taken after the logarithm transformation. In the GLM, the logarithm transformation is made on the expected value, so that (in simple cases) the estimated count for each group is the arithmetic mean. This is a major advantage of the GLM approach over transformation. The only disadvantage of the GLM approach is that it can be difficult to account properly for complex structure in the experimental units, where this is present.

18.4 Other Types of GLM and Extensions

In this chapter, we have considered Binomial and Poisson responses as being those most commonly encountered in biological research. Here, we describe two other common types of response that can be analysed using GLMs.

An extension to the case of Binomial proportions occurs when trials have more than two outcomes, which are ordered (**ordinal responses**). For example, instead of a plant being classified as healthy or infected, it might be classified as healthy or with slight, moderate or severe infection, giving four ordered outcomes instead of two. Models to deal with this situation are often called **ordinal regression**, and are related to logistic regression, but are beyond the scope of this book. Further details can be found in Agresti (2010).

Contingency tables summarize counts when each unit has been classified in terms of several factors. This type of data often arises from surveys. For example, a survey of farms might classify several weed species according to their growth habits, winter hardiness, and abundance in different types of crop, and the number of fields in each habit × hardiness × abundance × crop category forms a contingency table. The aim of analysis would be to establish any association between the classifying factors. For simple surveys, the table may be classified by just two factors, in which case the usual Pearson chi-squared test of association is appropriate (McClave and Sincich, 2012). For more complex surveys, a GLM can be used to investigate patterns of association. In this case, the responses have a **multi-nomial distribution**, but after conditioning on marginal totals, it can be shown that this is equivalent to fitting a GLM with a Poisson distribution and log link. A thorough overview of the area is given by Agresti (2007).

Finally, we note that the Normal distribution with the **identity link** function (i.e. no transformation) is a special case of a GLM. However, treating this case as a GLM leads to exactly the same analysis as discussed in the previous chapters, and so to avoid potential confusion we have not elaborated the connections here.

Since the GLM framework does not allow specification of a structural component within the model, we have used an intra-block analysis to deal with blocking structure in Examples 18.2 and 18.4. Other forms of analysis that explicitly account for a structural component, but which are beyond the scope of this book, include generalized linear mixed models (GLMMs, see Stroup, 2012), and hierarchical generalized linear models (HGLMs, see Lee et al., 2006).

EXERCISES

18.1 A series of experiments investigated the interactions between a fungus that infests aphids and broad bean plants. Here, we consider data from a trial in which germination of the fungal conidia was assessed on adult aphids. A batch of 50 aphids was exposed to fungal conidia then split into groups of 10 aphids which were allocated to five plants. Each plant was allocated to a sample time: 3, 6, 9, 12 or 24 h. At each time, 10 adult aphids were sampled from the designated plant and examined under a microscope to determine the total number of conidia present and the number that had germinated. The numbers of germinated and total conidia on aphids from each plant (variates *NGerm, Total*) are given with the sample time (variate *Time*) in file GERMINATION.DAT. Use a GLM with a Binomial distribution to investigate the pattern of germination over time, remembering to check for over-dispersion. Determine the predictive model and interpret the results.[*]

18.2 Exercise 13.2 (file CABBAGE.DAT) analysed the numbers of leaves on a set of cabbage plants as a function of days after transplantation. A log transformation was used to deal with variance heterogeneity. Repeat the analysis using a GLM

---

[*] Data from J. Baverstock, Rothamsted Research.

with a Poisson distribution for the untransformed response and a log link function. What is your estimate for the growth rate from this model? Is it comparable with that from your model from Exercise 13.2?

18.3* A pilot study investigated the period of leaf wetness required to successfully infect leaves with a foliar disease. Trays of four young plants with four leaves were sprayed with inoculum and then kept wet for a period of 16, 24, 48 or 72 h. The experiment used a CE cabinet with four shelves and was designed as a RCBD, with shelves used as blocks. File WETNESS.DAT holds the unit numbers (*ID*), structural factors (Shelf, Tray) with the wetness period (variate *Wetness*) and number of leaves infected (variate *NInf*, number out of 16). What distribution might you expect the number of infected leaves to follow? Use a suitable GLM to model the number of infected leaves in each tray, taking account of the design structure by including shelves in the model. Check for evidence of over-dispersion, check residual plots and carry out a formal test for lack of fit. Is there any evidence that wetness period affects the number of infected leaves? Predict the probability that a leaf is infected after 36 h of wetness, and give confidence limits for this prediction.

18.4 Example 12.2 analysed a set of insect counts from a transect sample and we used a log transformation to deal with variance heterogeneity. Repeat the analysis (the data are in file TRANSECT.DAT) using a suitable GLM and compare your results with the original analysis. Which analysis do you think is more appropriate?

18.5 The ecological survey described in Example 18.3 took several samples from each field surveyed, using the same transects and distances, but not necessarily the same plants in each sample. File APHIDS2.DAT contains data for the pea aphid collected from the next sample after the one analysed in Example 18.3. Repeat this analysis for the new sample. What conclusions do you draw? Can you extend your analysis to take account of the previous sample?

18.6 A greenhouse trial was undertaken to evaluate 63 families of loblolly pine for resistance to pine rust. The experiment was a RCBD with five replicates (blocks), and several seedlings from each family were tested in each replicate. Sets of seedlings were grown in trays, and each tray held 8–31 seedlings (median 17). File RUST.DAT holds information on the design (*ID*, factors Rep, DTray) and family allocation (factor Family) with the number of seedlings affected by rust and total number in each tray (variates *Rust, NSeedling*). Use a GLM to estimate the probability of rust occurring on a seedling for each family, after accounting for differences between replicates. Is there any evidence of differences in resistance among families? Identify the families where individual trees have less than 20% probability of being affected by rust.*

18.7 In Exercise 17.3, you analysed a set of field trials (data file CWTRIALS.DAT) to investigate whether the number of chickweed seeds produced by a plant could be related to its biomass, using a log transformation on both the response and dry weights. Repeat the analysis on the untransformed number of seeds using a suitable GLM. Does this model account for the variance heterogeneity?

---

* Data from FBRC, University of Florida.

18.8 Data from an agronomic trial is available to assess the effect of fungicide and a biological control agent on the incidence of white rot on onions. The trial was designed as a RCBD with five blocks of 12 plots. The 12 treatments were all combinations of three varieties with presence or absence of the fungicide (two levels) and the biological control agent (BCA, two levels). File BCA.DAT holds the unit numbers (*ID*), structural factors (Rep, Plot), treatment factors (Variety, Fungicide, BCA) and the total number of plants per plot (variate *Emerged*) and number with symptoms of white rot (variate *Disease*). Use a suitable GLM to identify a predictive model for both emergence and disease incidence. Note that the number of emerged plants is a small proportion of the seeds sown (which was not counted but was constant across plots) so is small compared to the unknown upper limit. What treatment would you recommend to maximize the number of unaffected plants for each variety?[*]

18.9 An investigation of response to insecticide used 28 cages of clones each produced from a single aphid. There were 14 cages of each type of clone (S and R) and a target dose of active compound was applied to each cage, with the actual dose recorded. After a given period, the number of moving aphids in each cage was counted, and the clones were classified according to presence of a marker suspected to affect tolerance of the compound. File CLONE.DAT contains unit numbers (*ID*), clone type (factor Clone), marker presence (factor Marker), and the logarithm of the dose applied (variate *LogDose*) with the number of moving aphids (variate *Moving*) and total aphids (variate *Total*) in each cage. Plot the data and comment on the structure of the groups (combinations of clones and marker types). Identify and write down a parsimonious predictive model to describe the data.[†]

18.10 A cage experiment was used to investigate the effect of three related insecticides on colonies of aphids with partial resistance to their common active compound. There were eight treatments: all combinations of the three insecticides or control (no insecticide) with two types of colony (susceptible or partially resistant). The experiment was organized as a RCBD with six blocks of eight cages, and one treatment combination was allocated to each cage in each block. A colony of the designated type was reared in each cage, and the number of live aphids was counted before the insecticide treatment was applied and then 2 and 6 days after application. Both births and deaths could occur within each cage between assessments. File REPEAT.DAT holds the structural factors (*ID, Block, Cage*), treatment factors (Insecticide, Clone) and responses (variates *Pre, Day2, Day6*). First, use a GLM to analyse the numbers before the insecticide treatment is applied. Should you take account of any differences in your analysis of the post-treatment numbers? How can you do this? How does this change the interpretation of the analysis?[‡]

18.11 The viability of carrot seed depends greatly on the conditions under which it is stored. Four batches of seed were stored in different conditions (labelled A–D). One hundred seeds were sampled from each batch: conditions A and B were sampled approximately every 60 days and conditions C and D were

---

[*] Data from J. Clarkson, University of Warwick.
[†] Data from S. Foster, Rothamsted Research.
[‡] Data from Horticulture Research International.

sampled approximately every 30 days, and the number of non-viable seeds was evaluated. File CARROT.DAT contains unit numbers (*ID*), the structural factors (Batch, Sample), explanatory variables (factor Condition, variate *Days*) and response (variate *Count*). Use a GLM to model the number of non-viable seeds over time in each condition and check the fit of the model carefully. Is there any evidence of model misspecification? Identify any features of the data that are incompatible with the GLM.[*]

---

[*] Data from D. Gray, Horticulture Research International.

19

Practical Design and Data Analysis for Real Studies

In the preface, we identified the aim of this book as being to provide an introductory, practical and illustrative guide to the design of experiments and subsequent data analysis in the biological and agricultural sciences. We have provided a brief overview of basic statistical concepts and terminology in Chapter 1, and ideas of summary statistics, probability distributions and simple statistical estimates and tests in Chapter 2. The bulk of the rest of the book has introduced and developed various statistical approaches associated with designing experiments and analysing the data generated (Chapters 3 to 11) or with analysing regression models (Chapters 12 to 15). We have tried to use common terminology across these sections to emphasize that the same form of model, the linear model, underlies all of these situations. We then described some more advanced techniques. Chapter 16 introduced linear mixed models that allow analysis of models with a structural component and any mixture of factors and variates in the explanatory component, with no requirement for a balanced structure. Chapter 17 extended the regression modelling approach to allow curved responses and non-linear models, and Chapter 18 introduced generalized linear models (GLMs) that allow analysis of models with any mixture of factors or variates in the explanatory component for data with certain types of non-Normal distribution. Throughout the book we have introduced real examples, either drawn from or inspired by our own experiences of working with scientists in research institutes and university departments. Our aim has been to show how the statistical approaches in this book can be used to address a range of real-life research problems across a number of application areas.

We hope that you have reached this final chapter of this book having worked through each of the preceding chapters and attempted some of the exercises. You should now have sufficient understanding of the various statistical concepts to enable you to apply what you have learnt to your own research. In this final chapter, we attempt to draw the various strands of this book together by introducing case studies that illustrate how to use this accumulated knowledge to develop appropriate designs for different experimental scenarios, and by discussing how to apply sensible analysis approaches for individual scientific problems.

We start with a summary of the various issues concerned with designing real studies (Section 19.1). We should consider the aims, hypotheses and treatments associated with a study separately from the available resources and constraints before allocating the treatments to the experimental material to construct an efficient design. During the design and planning stages, we also need to identify an analysis approach that enables us to address the aims of the study. In Section 19.2, we summarize and compare the various approaches introduced within the book, drawing out the similarities and differences between analysis methods for designed experiments and observational studies, and linking the analysis approach to the experimental aims. Finally, we discuss the information that needs to be presented when publishing the results of a study, including a description of the study design, data collection and analysis approaches, presentation of the results as provided by statistical software, and the interpretation of these results in the context of the scientific problem (Section 19.3).

19.1 Designing Real Studies

The basic principles that we need to consider when designing any experimental or observational study are always the same (as introduced in Chapter 3), but it is important to remember that almost every new study will be unique, in terms of either the questions to be asked or the resources that are available, or both. So, to develop an appropriate design for any new study it is important that we explore both of these components (questions to be asked, available resources) separately before finding the best way of combining them.

19.1.1 Aims, Objectives and Choice of Explanatory Structure

A sensible starting point is always to carefully consider the aims of the study (Section 3.1). In broad terms, these aims may be associated with identifying important differences in the response between treatments (combinations of selected levels of explanatory variables), understanding how the response from a biological system varies with changes in one or more explanatory variables, assessing how the response to one explanatory variable is affected by other explanatory variables, or simply in finding the combination of levels of explanatory variables that produces the best response. In most cases, it should be possible to re-express these aims and objectives in terms of testable hypotheses, which should then lead directly to the identification of the explanatory variables, the experimental treatments and the explanatory component of the model to include in the design of the study. Sometimes a study forms part of a larger research project, possibly being one of a sequence of studies or experiments, where information collected from previous studies should inform the design of this new study, or where we are gathering information that will be used to inform later studies. However, similar aims at different stages of a substantial research project may need to be addressed in different ways.

In Chapter 8, we discussed various ideas about extracting information from the explanatory component of the model to answer specific scientific questions. Consideration of the best approach should be included at the design stage of a study. Where there are multiple possible input variables, it is important to decide whether the inclusion of a factorial structure is useful; the possible benefits were described in Section 8.2.5.

At early stages in a project, the primary interest may be to identify those explanatory variables that have a major impact on the response (sometimes calling screening), rather than to determine the exact impact of each explanatory variable. An effective approach to this problem would be to use each explanatory variable at just two levels (low and high) within a multi-factorial arrangement. In industrial experimentation, specialized design approaches (e.g. Plackett–Burman designs, Plackett and Burman, 1946; see also Mead et al., 2012, Chapter 14) have been developed to provide a highly efficient approach for screening a large number of potential explanatory variables (although requiring the assumption that only the main effects of each explanatory variable are important). Once the most important explanatory variables have been identified, interest turns to the pattern of response across these key variables, using a factorial structure with those variables evaluated across a wider range of levels. Of course, this idea of screening explanatory variables using a small subset of their possible levels is really relevant only where the explanatory variables lie on some quantitative scale – for truly qualitative explanatory variables the concepts of low and high levels are meaningless (Section 1.3). For these variables, the numbers of levels should be identified from the aims and objectives (scope) of the study. Where both qualitative and quantitative explanatory variables are to be included, a pilot study with a few (combined) levels of the qualitative explanatory variable(s) might be used to screen for important quantitative explanatory variables. In any scenario, the convenience of this screening approach should be balanced against the possibility of missing important interactions.

For quantitative explanatory variables, it is necessary to select the number and spacing of the values (levels) to be used. This is easier in the context of designed experiments, where levels are directly under the control of the experimenter, but should also be considered in observational studies. Without prior knowledge to suggest otherwise, it is difficult to argue against having equally spaced levels. In a simple regression modelling context, the response is usually assumed to be linear but a low-order polynomial (Section 17.1.2), or polynomial contrast in the context of a multi-stratum ANOVA (Section 8.7), might also be appropriate. The number of levels needs to be sufficient to allow the fit of the selected polynomial model to be assessed; this requires at least three levels for a straight line model, four levels for a quadratic model, and higher-order polynomial models require additional levels. Replication can be used to give a direct test for lack of fit (Section 12.8). In the more general context of curved relationships (Chapters 17 and 18), it is important that levels cover the regions of greatest interest, which are often regions where the pattern of response changes most. It is always important that the selected levels span the full range of values relevant to the aims and objectives of the study. Where several quantitative explanatory variables are used, the ideas of factorial structures still apply, so observations should be selected to span both the range of interest for each individual explanatory variable, and, ideally, the combined ranges for all explanatory variables.

A final issue with regard to the choice of explanatory variables in a study is the need to include some sort of control or standard treatment (discussed in Section 8.5). Most trials use some sort of control treatment with known properties, to give assurance that the experiment has run as expected. In Section 8.5, we identified three different types of control: the positive control, the negative control and the standard. The inclusion of several controls will usually be relevant for studies concerned with the treatment of some detrimental activity, such as weed, disease or pest control in agricultural crops. The positive control provides the best possible response, and can provide a benchmark against which any new treatments can be compared. For example, in insecticide trials, a positive control might be some form of exclusion treatment that ensures that no pests infest the crop. By contrast, the negative control provides the worst-case scenario, and is often useful only in checking that some control of the detrimental activity is needed. For example, in insecticide trials, a negative control would be the lack of any chemical (or other) treatment, providing evidence of a pest infestation, and hence that the insecticide treatments are having a beneficial impact in controlling the pest. Finally, the standard or reference treatment can provide a known response or target value, so for an insecticide trial this might be the best commercially available product (or the most commonly used commercial product), and potential new products must perform at least as well as this standard treatment.

19.1.2 Resources, Experimental Units and Constraints

An important starting point when considering the resources to be used in a scientific study is to identify the experimental units. The choices made depend both on the aims of the study, and on the scientific methods used. For experimental studies, a useful definition was provided by Cox (1961), who stated that 'an experimental unit corresponds to the smallest division of the experimental material such that any two units may receive different treatments in the actual experiment'. In Section 3.1, we were slightly more precise, defining the experimental unit for each explanatory variable separately, since explanatory variables may be applied at different levels of the experimental material. A similar definition can be provided for observational studies, an experimental unit being defined, by analogy, as the smallest division of the biological material such that any two units may have different levels of the explanatory variable. Various examples of experimental units were listed in Section 3.1. It is also helpful to identify the measurement unit, the (biological) material on which each measurement is made. The measurement units may be the same as the experimental units for one or more explanatory variables, but often differ, as discussed in Section 3.1. The measurement unit almost always corresponds to the lowest level of experimental material. Where different experimental units are used for different explanatory variables, the ideas of multi-stratum designs should be used, such as the split-plot design introduced in Section 9.2.

Having identified the experimental and measurement units, the next step is to determine the maximum number of units that are available for the study. Often, this will be defined by the cost associated with using each unit (applying treatments, recording responses) and some constraint on the total funding available for the study. Other forms of constraint might include the amount of time taken to process each unit, or the physical space that is available within the experimental facility (e.g. glasshouse, controlled environment room or cabinet, incubator). It will sometimes be necessary to be able to complete the study within a certain period of time, or using some specific experimental facility. Where different sizes of experimental unit are required, resulting in a multi-stratum design, these issues need to be considered for each stratum in turn, including choices about the relative numbers of units at each level of the structure.

It is also important to identify any anticipated systematic sources of variability or structure within the experimental material. This might be caused by the way in which the experimental units have to be managed (e.g. a constraint on the number of units that can be processed within a certain period of time, or that can be contained within some physical space), or by the origin of the experimental material (e.g. plants raised from different batches of seed, or leaves on the same plant). Those units expected to have similar responses in the absence of any treatment should be grouped together into blocks, so that the systematic variation between these blocks can be separated from the background variation, hence increasing the precision of treatment comparisons by reducing the unit-to-unit background variability. This blocking is incorporated into the structural component of the model. In many situations, there will be multiple potential sources of variation, and so we may want to account for all sources in constructing the design. In Section 3.3, we discussed the distinction between nested and crossed structures; the presence of crossed structures naturally leads towards some form of row–column design (see Section 9.1), while nested structures with experimental units at several different levels suggest variations on the split-plot design (Section 9.2).

19.1.3 Matching the Treatments to the Resources

Having identified the combination of explanatory variables (treatment structure) to be included and the resources to be used for the study, the final step in constructing the design is to combine these two components. An important part of this process is determining the level of replication required to make likely the statistical detection of any treatment differences regarded as biologically important – i.e. to allow the demonstration of statistical significance for treatment differences large enough to be of biological interest. As discussed in Chapter 10, the amount of replication required depends on a number of (possibly competing) elements, including the explanatory model, the magnitude of the difference to be detected and the variability associated with the experimental units and/or measurement process. Where such information is available, a power analysis (Section 10.3) can be used to calculate the required replication.

Assuming that sufficient resources are available, the construction of the design then just depends on matching the treatment structure to the resources, taking account of any blocking or other structure required. Where no blocking or other structural constraints need to be accounted for, then a completely randomized design can be used (CRD; Chapter 4). In some circumstances, where blocking is used for administrative convenience rather than to account for unavoidable heterogeneity, there is flexibility to choose the block size to match the number of treatments, resulting in a randomized complete block design (RCBD; Chapter 7). More usually, sensible block sizes will not necessarily match the desired number of treatments, or several levels of structure may be present, and then some more complex design will be necessary. Some simple ideas were introduced in Chapters 9 and 11, but a wide range of design approaches are possible, as described in Mead et al. (2012).

In cases where there is little or no information available about the various sources of background variation, as at the start of a new project, it may be sensible to run a pilot study to gather information before embarking on any major experimentation. Such studies can be used to also provide some preliminary information about the key explanatory variables (e.g. to identify the range of a quantitative explanatory variable to be included). But unless these preliminary studies are done in a way that makes them compatible with the main experiments, they may represent a sub-optimal use of resources. One way to avoid this is to use an adaptive or sequential design approach (see, e.g. Mead et al., 2012, Chapter 20), where each stage of the experimental process provides information for the following stages, and each stage can be analysed separately or as part of the whole series.

One way of assessing how effectively the design of an experiment uses the available resources is to evaluate the division of the resources, as measured by the degrees of freedom, between and within strata where treatment comparisons are made. As discussed in Section 10.2, a reasonable 'rule of thumb' is that there should be between 10 and 20 residual degrees of freedom in each stratum of the design where treatment comparisons are made; this ensures a reasonable estimate of background variability is obtained. Having too few (< 10) residual degrees of freedom in a stratum may result in low power for detecting treatment differences in that stratum; so increasing the replication at that level of the design might be sensible. Having too many (> 20) residual degrees of freedom in a stratum gives no real advantage, and may imply that replication (in that stratum) can be reduced or that the opportunity to answer additional questions, through the inclusion of further treatment factors, should be taken. In balanced designs with a nested structure, such as the split-plot design (Section 9.2), there are a fixed number of smaller units nested within each larger unit, and so changing the replication of larger units also changes the total number of smaller units. In these designs, the residual df are larger (sometimes much larger) for the lower strata, and this imbalance of information is one of the disadvantages of nested designs.

19.1.4 Designs for Series of Studies and for Studies with Multiple Phases

In many cases, an individual study forms part of a larger series. This is almost always the case for field trials, where results for a single trial in a single year can be notoriously unrepresentative. For example, official guidance for studies concerned with the development and testing of new crop protection products (e.g. European and Mediterranean Plant Protection Organization; http://www.eppo.int/) indicates that experiments should be repeated across multiple sites, at which pest presence or intensity or timing might vary, as well as in multiple years, in which environmental conditions might impact on the effect of different treatments. Ideally, similar designs will be possible for each site × year combination, but sometimes there will be different constraints for each trial, requiring different designs. Similar issues occur in crop breeding programmes, as new varieties are required to perform well across a wide range of environments. In this context, limited seed in early generation trials may produce constraints, so that not all potential lines can be trialled in all environments within the same year, and new lines will be introduced in following years with less promising lines dropped from the programme.

For most series of studies, it is important that data from each separate study (e.g. at a single site in a single year) can be analysed on its own, so that the individual characteristics of each study can be determined before combining the data. The simplest design for a set of studies would use the same set of treatments and same design for each study. In this case, if the background variation is similar across the studies, then a combined analysis is straightforward within the ANOVA framework, incorporating study as a high-level structural component within which the common design is nested, and allowing for an interaction between the within-study explanatory component and study (i.e. the possibility of different treatment effects in the different studies). If the background variation differs, or if the set of treatments is common but a different design is used for each study, then a combined analysis is possible but must account for the individual study designs and allow for different levels of background variation; this can be achieved using linear mixed models (Chapter 16). Where a different set of treatments is used in each study, then some overlap – a subset of common treatments – must be present if the trials are to be analysed together. A combined analysis in these circumstances relies heavily on the assumption that no study × treatment interaction exists, particularly for the comparison of treatments not tested within the same trial. If several common treatments are present, then this assumption can be tested (to a limited extent) within that set of common treatments, but this assumption cannot be tested at all where only one common treatment is used. Where a set of studies cannot all use the same treatments, we therefore strongly advise the use of a large overlap, with the common set preferably including treatments that span the full range of responses. Again, analysis of the combined set of studies usually requires the use of linear mixed models.

Careful thought about design is also needed in the increasingly common context of two- or multi-phase studies. These often occur where a crop is grown in the field and then harvested and processed in the laboratory or to produce some food product. Treatments may be applied in both the field and laboratory phases, and the design must account for structure in both the field and the laboratory, as well as ensuring that suitable harvest samples can be obtained for the later processing phase. For example, a study to examine factors influencing bread-making quality of wheat might use a RCBD with several varieties and fertilizer regimes in the field, then split the harvest from each plot into four sub-samples, each to be used for a different variation in the bread-making process. The four samples from each plot are processed together, with each plot processed on a separate day. The experimental structure must account for blocking in the field and processing time in the laboratory, as well as all treatment effects. Such studies might then involve further phases, for example, the taste testing of the resulting bread by several people, each giving a subjective score. In designing multi-phase studies, it is important to be aware of all constraints during each phase of the study, to ensure that the effects of each treatment (and interactions) can be extracted in the analysis, and to identify the structure within each phase. It can be useful to confound structure between phases, for example, using blocks in the field as blocks in the laboratory, and it may be necessary to allow separate analyses to be made at the end of each phase. A useful approach and overview has been developed by Brien and Bailey (2006).

Multi-phase trials are now common in the use of high-throughput technologies developed in the 'omics revolution, where the impact of different treatments on the levels of gene, protein or metabolome expression is measured. Typically, these studies involve an experimental phase (in the field or, more usually, controlled environment) during which treatments are applied, and plant material from each experimental unit is then harvested and processed to produce one or more samples used for the 'omics phase. In this context, there are often severe cost constraints on the total number of samples that can be used, and many technologies can process only small numbers of samples simultaneously. Given that substantial costs may be involved in obtaining expression readings for a single sample, it is important to ensure that the experimental phase is well designed, taking account of any constraints in the 'omics phase. Common issues at the 'omics phase include the allocation of experimental treatments to small blocks, the balance between different types of replication, and the assessment of response along a time course. Case Study 19.2 discusses the allocation of experimental treatments to small blocks in the context of two-channel microarray gene expression studies. Most 'omics studies include both biological replication (samples from different biological organisms) and technical replication (several sub-samples prepared from each unit in the experimental phase to allow for variation during the sample processing and measurement phases). Technical replication is particularly important if variability in the sample processing or measurement stages is large, whereas biological replication is important to ensure that results are not specific to one organism or sample. For time course studies, involving samples collected over time, it is important to identify whether the data form a cross-sectional study (samples collected from different organisms at each time point), or a longitudinal study (samples collected repeatedly from the same organisms over time). A longitudinal study may provide more precise comparisons across time points, but the analysis must account for correlation between samples taken from the same organism.

19.1.5 Design Case Studies

As previously noted, each study is unique, and so it is impossible to provide a generic recipe for how to design any study. However, to illustrate some of the issues identified above, we present three case studies from our own experiences.

Case Study 19.1: Designing a Large-Scale, Multi-Site Field Experiment

Spring 2000 marked the beginning of an extensive ecological experiment, known as the Farm Scale Evaluations (FSEs). The aim of the study was to compare the effects of two treatments on various indicators of farmland biodiversity, including both plants and invertebrates. The two treatments represented the composite effects of management practices associated with genetically modified herbicide-tolerant (GMHT) and conventional crop varieties. Simultaneous experiments were to be carried out for beet, maize, and spring and winter oilseed rape crops. The null hypothesis for each crop was that there was no difference between the two treatments in abundance and diversity of the chosen indicators. The two-tailed alternative hypothesis was that the treatments differed, in either a positive or negative direction.

The FSEs, carried out by a consortium of UK research institutes, were to form the largest, and most highly scrutinized, ecological study of its kind to date, costing in the region of £6 million (Clark et al., 2006). Undertaking an experiment of such magnitude required much planning in practical, biological and statistical terms. The project began in 1999 with a pilot study to develop sampling protocols and to inform a statistical study to determine an appropriate design for the experiment. This statistical study examined choice of design structures, the power of any potential design to detect treatment differences of a given size, and choice of subsequent analysis approaches for the data collected. Full details can be found in Perry et al. (2003) and Rothery et al. (2003); here we focus on four specific design issues: choice of experimental unit, allocation of treatments to units, estimation of sample size and sampling strategy.

As it was important that the results of the FSEs were representative of commercial British agriculture as a whole (e.g. farm location, farming intensity, weather conditions, soil types), farms throughout Britain, especially in those areas where the chosen crops were typically grown, were to be selected to take part in the study. The most pertinent design issue was then the definition of the experimental unit. Two choices were considered (Figure 19.1): half fields within whole fields (i.e. a RCBD with fields as blocks and half-fields as experimental units), or whole fields within farms (i.e. a RCBD with farms as blocks and fields as experimental units). There were many biological considerations (e.g. mobility of insects and their behaviour at different spatial scales) but these had to be balanced with statistical considerations. The primary statistical argument for the half-field option was the potential reduction in residual variation that might be achieved due to two half-fields being more similar to each other (e.g. in soil type, surrounding habitat, previous management) than two paired whole fields. Limited data from previous studies, coupled with the small amount of data from the pilot study, suggested that half-fields were indeed likely to be less variable than paired whole fields. Other more practical issues, such as the availability of whole fields and ease of sampling, were also contributing factors. The final decision was made to use half-fields as the experimental unit. Only one field per farm was used in each year, but some farms were sampled in 2 or 3 years, with a different field used in each year.

The boundary line used to split any field into two was first determined by assessing the many factors that might influence the variability of wildlife within the field; the optimal choice being the line that divided the field into two halves as close to identical as possible with respect to these non-treatment influences on biodiversity. An example of the detailed plans drawn up to inform this decision for each field is shown

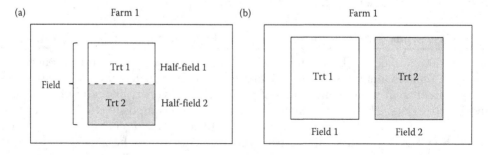

FIGURE 19.1
Schematic representation at one farm of two possible choices of experimental unit for the FSE study (Case Study 19.1). Both choices correspond to a RCBD, but in (a) treatments are applied at random to half-fields (units) within a field (block), and in (b) treatments are applied at random to whole fields (units) within the farm (block).

in Figure 19.2. In this case, the main criterion for the location of the boundary is the presence of a field of beet to the south-east (right of plan) and the presence of nursery buildings to the north-west (left of plan) of the field. A split running from north-east to south-west (vertically on plan) would potentially confound the treatments with this environmental difference. The chosen split running north-west to south-east (horizontally on plan) ensures that each half-field has boundaries including the beet crop

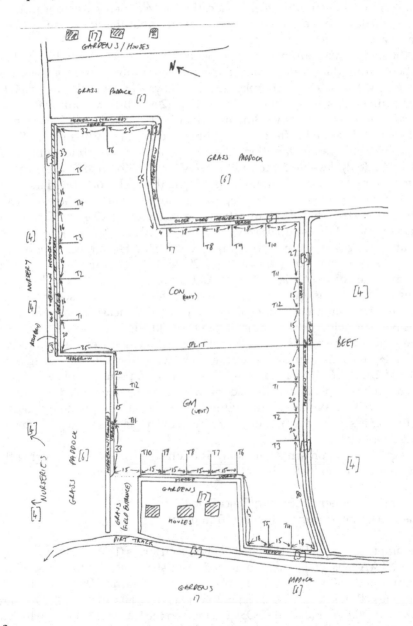

FIGURE 19.2
Detailed field plan showing characteristics of the field to be sampled and other features in the immediate locality that might influence the variability of wildlife within the field, the final choice of boundary for splitting into two half-field units, the final allocation of treatments to half-fields, and locations of within-half-field sampling transects T1–T12 (Case Study 19.1). (Courtesy of Matthew Skellern, Rothamsted Research.)

and nurseries, as well as hedgerows, gardens and grass paddocks. The protocol then labelled the most northerly half-field (or westerly, depending on the overall orientation of the field) as 'A' and the other half as 'B'. An envelope containing a predetermined randomization of the two treatments to halves A and B was then opened to give the final allocation of treatments. For example, in Figure 19.2, half A (west) was sown with the GMHT variety and half B (east) with the conventional variety. This two-stage protocol ensured that the allocation of treatments was not influenced by any of the parties involved in the experiment, and that the final results would have statistical validity.

Next, the sample size (number of whole fields) had to be determined. As noted above, very little existing data were available to inform any power analysis. So, instead, a simulation study was done to complement the analysis of data collected from the pilot study. The full details are given in Rothery et al. (2003). Briefly, count data were simulated according to the Negative Binomial distribution (Section 18.3.4) for a range of scenarios which included different overall mean counts (1, 5, 10 and 50), field effects covering a 100-fold span in variation, sizes of multiplicative treatment effects (1.3-fold, 1.5-fold and 2-fold), levels of variability (%CV = 50%, 80%, 100%) and values of the Negative Binomial exponent parameter (allowing the background variance to be proportional to the expected value or proportional to the square of the expected value). Power was estimated using randomization tests (e.g. Section 5.2.4) and 500 sets of simulated data for each of five sample sizes ($n = 20, 30, 40, 60$ and 90 fields) for each scenario. The results of the power study were complex but the final recommendation was to aim to achieve 60 fields per crop (equivalent to 20 per year over the 3-year period of the experiment). The power of this scheme for detecting 1.5-fold treatment differences and achieving a 50% CV was estimated to exceed 80% for many of the scenarios studied.

Finally, the field-sampling protocols involved taking measurements from up to 12 transects per half-field, each extending from the field edge in towards the centre of the field and spaced as evenly as possible around the three non-treatment-boundary field edges (see Figure 19.2). On each transect, there were five potential sample points at distances of 2, 4, 8, 16 and 32 m into the field. Up to 60 pseudo-replicate observations were therefore made per half-field, and the sub-samples were pooled to give half-field totals for analysis. Nevertheless, the within-half-field sub-sampling gave useful information to later assess and compare variability in the responses at various spatial scales (see Clark et al., 2007).

In Example 16.2, we presented an analysis of one data set collected during the FSEs. There we wrote the model for the data using symbolic notation as

Structural component: Farm/Field/DHalf
Explanatory component: [1] + Year*Treatment

In that example, the factor Field labelled fields within farms (1–3), and the resulting multi-stratum ANOVA table (Table 16.8) contained three strata relating to farms, whole fields within farms and half-fields within whole fields within farms. The first publications of FSE results (e.g. Brooks et al., 2003; Haughton et al., 2003; Hawes et al., 2003; Heard et al., 2003; Roy et al., 2003; Bohan et al., 2005) focussed on the treatment effects, which are estimated within fields. Effects in higher strata were of less interest and so the farm and field effects were combined by specifying the structural component of the model as

Structural component: Farm.Field/DHalf

The Year main effect was also excluded from the explanatory model. As a result of these two changes, the two upper strata in Example 16.2 were combined into a single stratum subsuming the farm, whole field and year main effects. In addition, preliminary analyses showed no evidence of treatment × year interactions (as in Example 16.2), and hence this term was also excluded from the model, with the explanatory structure specified as

Explanatory component: [1] + Treatment

Case Study 19.2: Multi-Phase Experiments – Gene Expression Microarrays for Plant Response to Pathogens

One important application of multi-phase experiments is the study of gene expression responses in plants grown in different environmental conditions, or exposed to different stress treatments. Here, we discuss a microarray study concerned with the measurement of gene expression responses in *Arabidopsis thaliana*, the model plant, over a 24-h time period post-infection with the pathogen *Botrytis cinerea*. Initial interest was in assessing the impact of infection by two contrasting isolates of the pathogen, with a third 'mock inoculation' treatment (inoculation with water) included to provide a baseline response.

The first phase of the study was the production of plant material to be inoculated, from which samples of genetic material (RNA, cDNA) could be obtained for processing before application to the microarrays. The pathogen isolates and mock inoculation were to be applied to detached leaves, with the whole leaf then being processed to generate the genetic sample. Hence, separate leaves were needed at each time point for each inoculation treatment. Different plants would be used for replicates of the treatments, but there were several options for use of plants within replicates. Three options were considered, as illustrated in Figure 19.3:

a. Use a separate plant for each inoculation treatment with leaves within plants allocated to the time points (i.e. plants are the experimental units for inoculation and leaves within plants are the experimental units for time)
b. Use a separate plant for each time point, with leaves within plants allocated to the different inoculation treatments (i.e. plants are the experimental units for time and leaves within plants are the experimental units for inoculation)
c. Use a separate plant for each inoculation × time point combination, but use a specific leaf (e.g. the seventh true leaf) within each plant (i.e. the plant.leaf combinations are the experimental units for both inoculation and time)

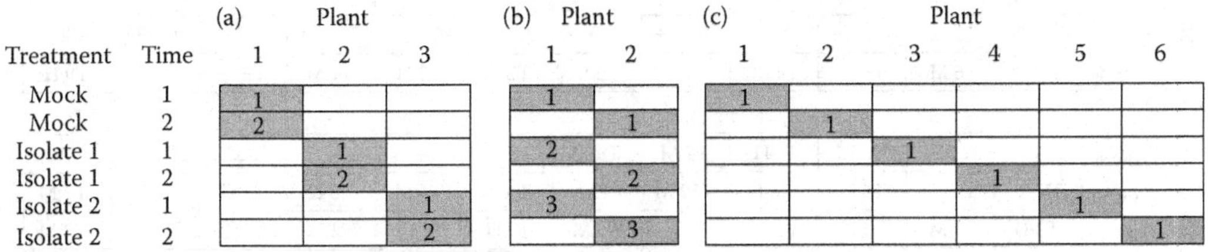

Treatment	Time	(a) Plant 1	2	3	(b) Plant 1	2	(c) Plant 1	2	3	4	5	6
Mock	1	1			1		1					
Mock	2	2				1		1				
Isolate 1	1		1		2				1			
Isolate 1	2		2			2				1		
Isolate 2	1			1	3						1	
Isolate 2	2			2		3						1

FIGURE 19.3
Three options for selecting leaves from plants to be treated with three different inoculation treatments and incubated for two different time periods, using separate plants for each (a) inoculation treatment, (b) time point and (c) inoculation × time point combination (Case Study 19.2). Highlighted boxes represent individual leaves to be sampled; numbers in boxes indicate leaf numbers within plants.

Option (a) would use fewest plants, and within-plant comparisons would avoid genetic variation between plants, therefore potentially providing for a more precise comparison of responses between time points. Option (b) would use more plants (assuming more than three time points are used), but would potentially provide a more precise comparison of responses between inoculation treatments within a time point. However, previous studies had identified substantial variation in gene expression between leaves of different ages, and this variation was often greater than that between plants. Hence option (c), which inoculates leaves of the same age, was preferred.

The total number of plants required then depended on both the number of replicates required for each inoculation × time combination and the number of time points. The researchers expected that there would be subtle changes in gene expression over the first 12 h, though some genes were not expected to show any response until 18–20 h post-infection. With a wide range of potential shapes of expression profiles over time, it was considered best to have the sampling times equally spaced – possibilities included sampling every hour (25 time points, starting immediately after inoculation), every 2 h (13 time points), every 3 h (9 time points) or every 4 h (7 time points). While replicate samples (technical replicates) would be generated during the post-harvest processing, so that the gene expression for each plant sample would be measured on multiple microarrays, it was also important to be able to compare the variation in gene expression due to the different treatment (inoculation × time) combinations with the between-plant (biological) variation. Therefore, it was considered necessary to also include replicate plants for each treatment.

The plants were to be grown in controlled environment cabinets (to minimize variation due to the growing environment), with two separate cabinets available. Each cabinet had two shelves with space for 48 plants to be grown on each shelf in an array of four rows of 12 plants (Figure 19.4), giving an upper limit of 192 plants.

Biological replicates could be processed separately but, within each replicate, leaves for all treatments must be harvested at the same time, inoculated, and then sampled at different times after inoculation. Sampling every hour would require 75 plants per biological replicate (3 inoculations × 25 time points), so that only two replicates would be possible (2 × 75 = 150 plants), while sampling every 4 h would allow up to nine replicates (3 inoculations × 7 time points = 21 treatments, 9 replicates × 21 treatments = 189 plants). The choice here is between improved precision for comparison of

		Plant 1	Plant 2	Plant 3	Plant 1	Plant 2	Plant 3	Plant 1	Plant 2	Plant 3	Plant 1	Plant 2	Plant 3
Shelf 1	Row 1	–	–	–	T7 M	T7 I1	T7 I2	T5 I2	T5 M	T5 I1	T13 I2	T13 M	T13 I1
	Row 2	T4 I2	T4 I1	T4 M	–	–	–	–	–	–	T8 I2	T8 M	T8 I1
	Row 3	T1 I2	T1 I1	T1 M	T10 M	T10 I1	T10 I2	T11 I1	T11 M	T11 I2	T12 I1	T12 I2	T12 M
	Row 4	T6 I1	T6 M	T6 I2	T3 M	T3 I1	T3 I2	T2 I2	T2 I1	T2 M	T9 M	T9 I2	T9 I1

		Plant 1	Plant 2	Plant 3	Plant 1	Plant 2	Plant 3	Plant 1	Plant 2	Plant 3	Plant 1	Plant 2	Plant 3
Shelf 2	Row 1	T11 I2	T11 M	T11 I1	T4 I1	T4 M	T4 I2	–	–	–	T13 I2	T13 I1	T13 M
	Row 2	T9 M	T9 I2	T9 I1	T12 I2	T12 M	T12 I1	T6 I1	T6 M	T6 I2	T7 I1	T7 I2	T7 M
	Row 3	T1 I1	T1 M	T1 I2	T5 I2	T5 I1	T5 M	T2 I1	T2 I2	T2 M	T3 I2	T3 I1	T3 M
	Row 4	T8 M	T8 I1	T8 I2	–	–	–	T10 I1	T10 M	T10 I2	–	–	–

FIGURE 19.4

Arrangement of plants on two shelves within one CE cabinet, with random allocation of sampling times (T1 … T13) to sets of three adjacent plants within rows, and of three inoculation treatments (M = Mock, I1 = Isolate 1, I2 = Isolate 2) to plants within each set (Case Study 19.2). Dashes indicate unused positions.

treatment differences at given time points, and better information about the pattern of changes in gene expression over time. The compromise was to sample every 2 h (13 time points), which allows a reasonable number of time points over which to measure the responses of late-expressing genes. This scheme results in 39 treatments and each replicate could comfortably fit on a single shelf, allowing four complete biological replicates. Any systematic differences between biological replicates introduced at later stages could then be confounded with differences between cabinets and shelves. There was potential variation both along and between rows within each shelf. It was therefore decided to randomly allocate sets of three adjacent plants in a row to a particular time point, with the three inoculation treatments randomly allocated (but not yet applied) to plants within each of these sets. The arrangement for one cabinet is shown in Figure 19.4, and a new randomization was used for the second cabinet. If data were measured at this point of the experiment, the structural component of the model would take the form

Structural component: Cabinet/Shelf/Set/Plant

The four biological replicates were harvested on four separate days. For each plant, the seventh true leaf was excised and the allocated inoculation treatment applied. At each 2-hourly time point, the appropriate set of three leaves (i.e. the set previously allocated to that time point) was freeze-dried to stop any further development prior to processing to obtain the genetic material. Throughout the subsequent processing steps (amplification, labelling), the 39 samples in each biological replicate were processed together where possible. Where this was not possible, samples were processed in batches comprising either the 13 samples for a particular inoculation treatment or the three samples for a particular post-infection sampling time, with the order in which the batches and samples within batches were processed being randomized. As a two-channel microarray system was to be used to assess the gene expression for each sample, the labelling phase required the division of each sample into two sub-samples, one to be labelled with each dye. This step automatically introduces some processing/measurement replication, with the potential for further such technical replication to be introduced during the microarray phase.

The final phase of the study involved the allocation of samples to microarrays to measure relative gene expression levels. Two separate samples can be compared directly on each array (essentially each array is a block of size two), with consistent differences in responses between the two dye labels also expected. The labelling of each sample with both dyes had already satisfied the 'dye-balance' principle, with each treatment measured using both dyes. There were 312 samples (three inoculation treatments × 13 time points × four biological replicates × two sub-samples) to be allocated to arrays, requiring a minimum of 156 two-channel arrays (where the two channels relate to the two wavelengths used to read the expression response for the different dyes). Clearly, it would not be possible to directly compare all pairs of treatments using a reasonable number of arrays and so the strategy must be to directly make the comparisons of most interest. The most important comparisons are between adjacent time points within each inoculation treatment and between the different inoculation treatments at each time point. It is also of interest to allow direct comparison across the different biological replicates so that interactions can be investigated. To allow all of these comparisons to be made, each sub-sample was used on two arrays, resulting in a total of 312 arrays, with each plant sample being measured on four separate arrays, two of these technical replicates being labelled with each dye. The allocation of treatments to microarrays was split into two parts, each using 156 arrays, with each part focussing on a different set of comparisons.

The first part of the design focussed on the comparison of samples along the time course using a 'loop design' (Kerr and Churchill, 2001; Wit et al., 2005). Within each inoculation treatment and biological replicate, each time point appeared on an array with samples from the previous and next time points, and the two occurrences of each time point were labelled with different dyes (as shown in Table 19.1a).

This provided 12 blocks (one for each combination of the three inoculation treatments and four biological replicates) of 13 arrays, and each block was processed on a separate day. Ignoring the controlled environment phase of the design, a model for this part could be written as

Structural component: (Day/Array)*Channel
Explanatory component: [1] + Dye + Inoculation*Time

This is a partially balanced incomplete block design, with the same 13 (out of the 78 possible) time point comparisons appearing together on an array for each biological replicate of each inoculation treatment. Comparisons between inoculation treatments are made between days, and comparisons between time points are made partly within arrays (for adjacent time points), partly within channels and partly between arrays (other comparisons). Comparisons between the two dyes are completely confounded

TABLE 19.1

Allocation of Treatments to Microarrays and Dyes: (a) Allocation of Time Points (1–13) for Each Inoculation Treatment and Biological Replicate in Part 1; (b) Allocation of Inoculation Treatments (M, 1 or 2) and Biological Replicates (a, b, c, d) for Each Time Point in Part 2 (Case Study 19.2)

(a)			(b)		
	Dye			**Dye**	
Array	**Red**	**Green**	**Array**	**Red**	**Green**
1	1	2	1	Ma	1b
2	2	3	2	1b	2c
3	3	4	3	2c	Md
4	4	5	4	Md	1a
5	5	6	5	1a	2b
6	6	7	6	2b	Mc
7	7	8	7	Mc	1d
8	8	9	8	1d	2a
9	9	10	9	2a	Mb
10	10	11	10	Mb	1c
11	11	12	11	1c	2d
12	12	13	12	2d	Ma
13	13	1			

with differences between the two channels. Incorporation of structure from the controlled environment phase can be achieved by adding this directly, so that the structural component becomes

Structural component: (Cabinet/Shelf/Set) + (Day/Array)*Channel

This structure is no longer balanced, and if any of the terms from the different phases are completely confounded, it is important to include only one of them in the structural component. In this example, there is no confounding, but the complexity of the structure is greatly increased, and some terms may be difficult to estimate. If we believe that variation within the CE cabinet shelves is small and can be ignored, then we could simplify matters by creating a new factor to represent the biological replicates, RepCE (equivalent to the Cabinet.Shelf combinations), and use this term in the explanatory component. This uses the ideas of intra-block analysis (Section 11.6) within the multiphase context. The model can then be written as

Structural component: (Day/Array)*Channel
Explanatory component: [1] + RepCE + Dye + Inoculation*Time

The RepCE term is estimated within the Day stratum, using 3 df and resulting in only 6 residual df for that stratum. A dummy ANOVA table for this model is shown in Table 19.2a.

TABLE 19.2

Dummy ANOVA Table for (a) Part 1 and (b) Part 2 of Microarray Design Stage, Adjusting for Controlled Environment Phase (Case Study 19.2)

(a) Part 1		(b) Part 2	
Source of Variation	**df**	**Source of Variation**	**df**
Day stratum		Day stratum	
RepCE	3	Time	12
Inoculation	2		
Residual	6		
Day.Array stratum		Day.Array stratum	
Time	12	RepCE	3
Inoculation.Time	24	Inoculation	2
Residual	108	Inoculation.RepCE	6
		Residual	132
Channel stratum		Channel stratum	
Dye	1	Dye	1
Day.Channel stratum		Day.Channel stratum	
Residual	11	Residual	11
Day.Array.Channel stratum		Day.Array.Channel stratum	
Time	12	RepCE	3
Inoculation.Time	24	Inoculation	2
Residual	108	Inoculation.RepCE	6
		Residual	132
Total	311	Total	311

The second part of the design was used to make direct comparisons between inoculation treatments and biological replicates at each time point. This part of the design consisted of 13 blocks, one for the samples from each time point. Each block was processed on a different day and contained 12 arrays, again using a loop design, with the comparisons shown in Table 19.1b. Each array gave a comparison across inoculation treatments and across biological replicates within a time point, and all combinations of inoculation treatment and biological replicate appeared within each block, labelled by both dyes. Using the same strategy as in Part 1, the structure for this part of the design can be written as

Structural component: (Day/Array)*Channel
Explanatory component: [1] + Time + Dye + Inoculation*RepCE

A dummy ANOVA table for this model is shown in Table 19.2b. If this part of the design is considered alone, then the effects of time are completely aliased with differences between days. But there is information in this part of the design on comparisons between inoculation treatments and it is possible to check for differences in response to inoculation across the biological replicates. Combined analysis of the two parts of the design simultaneously allows these two different aspects to be combined, using the model

Structural component: (Day/Array)*Channel
Explanatory component: [1] + Dye + Inoculation*Time*RepCE

This experiment generated a vast quantity of data, with measurements being made on over 32,000 genes on each array. Initial analysis was performed on a gene-by-gene basis, rather than trying to analyse for treatment effects across all genes simultaneously. The lack of balance in the combined analysis requires the use of linear mixed models (Chapter 16). The careful consideration of the different constraints during each phase of the design, and the careful confounding of sources of variation between different phases are crucial in ensuring that the analysis model is relatively easy to construct and interpret.

Case Study 19.3: Designing a Sampling Scheme to Detect Variation within a Population

---

Mapping populations are derived from inbred parental lines. In the case of recombinant inbred lines (RILs), the two parental lines are crossed and then the individual offspring is self-crossed (usually by a process of single seed descent, see Kearsey and Pooni, 1996) for a number of generations (usually eight or more) to produce a population of genetically stable offspring lines. The resulting population can be used to detect quantitative trait loci (QTLs, see Kearsey and Pooni, 1996), but it is helpful to first identify traits where variation between lines is present.

A new RIL population of 110 oilseed rape lines was grown in a glasshouse, using a RCBD with three replicates, with pots containing single plants as the experimental units. The aim of the study was to identify seed traits showing substantial variation between lines; here, we focus on seed weight (measured in grams). Oilseed rape plants are structured, with pods growing on branches within plants, and branches flowering in succession from the top of the plant, starting from the end of each branch. A pilot study was done to identify structured sources of variation within plants, using all three replicates of 10 lines (chosen at random) and sampling eight pods along each of the first, third and fifth branches. The model for the pilot study was written as

Response variable: *SeedWeight*
Structural component: Rep/Unit/Branch/Pod
Explanatory component: [1] + Line*BranchNo*Position

where Position labels the relative position of a pod within the branch, BranchNo specifies the position of a branch within a plant, Unit labels the experimental units (pots) within replicates and the remaining factor names are self-explanatory. The average seed weight per pod was analysed using ANOVA; the results gave weak evidence for differences between lines and suggested that seed weight might be larger on the first branch, and possibly also for the first pods set within each branch, but there was no evidence of any interactions within the explanatory model. We concluded that a sampling scheme that takes account of the structure within plants should give a more efficient comparison across lines. The next task was to design a more comprehensive study that covered a much larger number of lines.

The constraints on the second study were that a maximum of 600 pods could be processed, with average seed weight per pod derived from the total weight per pod divided by the number of seeds per pod (usually 8–14). Using the methods of Section 16.4.1, we derived estimated variance components for each component of the structural model as shown in Table 19.3. The experimental units for the lines are the individual pots, labelled as Rep.Unit combinations. For any structured sample, we can use the method of Section 16.4.1 to predict the Rep.Unit stratum variance in an ANOVA table, and use this to make power calculations for detecting line differences.

We consider a generic balanced scenario. We could sample n_L lines (chosen at random) using n_R replicates, taking n_P pods from n_B branches. We would specify the branches to be used and the positions within each branch to give reasonable coverage of the plant. We can then use terms in the explanatory model to account for the systematic differences between branches and positions found in the pilot study. Our model for this structure hence takes the form

Structural component: Rep/Unit/Branch/Pod
Explanatory component: [1] + Line + BranchNo + Position

TABLE 19.3

Estimated Variance Components for Each Term in Structural Model for Measurements of Seed Weight (g) (Case Study 19.3)

Term	Variance Component	Estimate
Rep	σ_R^2	$\hat{\sigma}_R^2 = 0.0050$
Rep.Unit	σ_U^2	$\hat{\sigma}_U^2 = 0.0046$
Rep.Unit.Branch	σ_B^2	$\hat{\sigma}_B^2 = 0.0031$
Rep.Unit.Branch.Pod	σ_P^2	$\hat{\sigma}_P^2 = 0.0017$

and a dummy ANOVA table for this structure can be constructed, as shown in Table 19.4a. Using the methods of Section 16.4.1, we can predict the stratum variances (Table 19.4a), and our prediction of the Rep.Unit stratum variance takes the generic form

$$s_U^2 = n_B n_P \hat{\sigma}_U^2 + n_P \hat{\sigma}_B^2 + \hat{\sigma}_P^2 \, ,$$

where $\hat{\sigma}_U^2, \hat{\sigma}_B^2$ and $\hat{\sigma}_P^2$ are the estimated variance components for units (pots), branches and pods, respectively. The variance of a line prediction is based on the Rep.Unit stratum variance divided by the replication calculated as the number of pods sampled per line, i.e. $n_R \times n_B \times n_P$, giving

$$\text{SE} = \sqrt{\frac{s_U^2}{n_R n_B n_P}} = \sqrt{\left(\frac{\hat{\sigma}_U^2}{n_R} + \frac{\hat{\sigma}_B^2}{n_R n_B} + \frac{\hat{\sigma}_P^2}{n_R n_B n_P} \right)} \, .$$

As a starting point, we assess a scenario using all three replicates ($n_R = 3$) of 50 lines (chosen at random from the set not used in the pilot study), taking two pods ($n_P = 2$) from each of two branches ($n_B = 2$) and giving 600 pods in total. In this case, we might use the first and third branches, taking the two pods from the end and middle of each branch to maximize coverage of the plant structure. This structure matches a subset of the data from the pilot study which sampled the same experiment so, as long as the measurement methods have not changed in the interim, we can include the subset from the pilot study to obtain data on 60 lines, giving 720 observations in total and the dummy ANOVA table shown in Table 19.4b. In this case, the background variability for lines is $s_U^2 = 0.0263$ (with 118 df) and so the SE of a line prediction is equal to

$$\text{SE} = \sqrt{\frac{s_U^2}{n_R n_B n_P}} = \sqrt{\frac{0.0263}{3 \times 2 \times 2}} = \sqrt{\frac{0.0263}{12}} = 0.0468 \, ,$$

TABLE 19.4

Dummy ANOVA Table for a Balanced Sample from the Oilseed Rape Study Using a RCBD (a) with n_R Replicates of n_L Lines, Taking n_P Pods from n_B Branches from Each Plant, and (b) with 60 Lines Each with Three Replicates, Taking Two Pods from Each of Two Branches (Case Study 19.3)

	(a)		(b)	
Source of Variation	df	Predicted Stratum Variance	df	Predicted Stratum Variance
Block stratum				
Residual	$n_R - 1$	$n_L n_B n_P \hat{\sigma}_R^2 + n_B n_P \hat{\sigma}_U^2 + n_P \hat{\sigma}_B^2 + \hat{\sigma}_P^2$	2	0.6263
Block.Pot stratum				
Line	$n_L - 1$		59	
Residual	$(n_R - 1)(n_R - 1)$	$s_U^2 = n_B n_P \hat{\sigma}_U^2 + n_P \hat{\sigma}_B^2 + \hat{\sigma}_P^2$	118	0.0263
Block.Pot.Branch stratum				
BranchNo	$n_B - 1$		1	
Residual	$(n_R n_L - 1)(n_B - 1)$	$n_P \hat{\sigma}_B^2 + \hat{\sigma}_P^2$	179	0.0079
Block.Pot.Branch. Pod stratum				
Position	$n_P - 1$		1	
Residual	$(n_R n_L n_B - 1)(n_P - 1)$	$\hat{\sigma}_P^2$	359	0.0017
Total	$n_R n_L n_B n_P - 1$		719	

with SED = 0.0662. In the pilot study, the average seed weight was 0.4 g and the new study is required to detect a change of ± 20%, i.e. differences of the order of 0.16 g. Using the methods of Section 10.3 to calculate the power for a RCBD, we find that the power of this design is 66.9%, i.e. giving a probability of 0.669 of detecting a difference in seed size of 0.16 g between two lines. This does not make any allowance for adjustments required for multiple testing (Section 8.8).

This sampling scheme utilizes only just over half of the available lines, giving a high chance that we might omit some of the more extreme lines, and so we should also assess an alternative scenario that samples a greater proportion of the lines. To do this, we have to sacrifice replication at some other level in the structure, and this is best done where the background variability is relatively low. The pod variance component is the smallest value in Table 19.3, so suppose we sample only one pod from each of two branches (i.e. $n_P = 1$), with the pod position fixed to avoid introducing variability due to pod position. We can then sample the full set of 100 lines not used in the pilot study, using 100 lines × 3 blocks × 2 branches × 1 pod = 600 pods. As we have overlap with the structure of the pilot study, we can again incorporate those data, giving data on the full set of lines. In this case, the bottom stratum is removed from the ANOVA table, as we cannot assess variation between pods within branches when only a single pod is sampled from each branch; however, the form of the rest of the table is unchanged. The Rep.Unit stratum variance is now estimated as

$$s_U^2 = 3\hat{\sigma}_U^2 + \hat{\sigma}_B^2 + \hat{\sigma}_P^2 = (3 \times 0.0046) + 0.0031 + 0.0017 = 0.0186 ,$$

with line predictions having SE = 0.056 and SED = 0.079 (all with 218 df). This design has power of 52.5% for detecting differences of size 0.16 g. For a reasonably small decrease in power, we gain information on our full set of lines. In contrast, if instead we sampled two pods from one branch on each plant, then the Rep.Unit stratum variance would increase to 0.0355 (with 218 df) and the power would fall to 31.0%, which is unacceptably low.

Other options can be investigated in a similar manner, but the second option of sampling one pod from the first and third branches of the 100 lines excluded from the pilot study appears to give the best option for getting information on the full set of lines with reasonable power and precision. This case study gives an example where a pilot study can be used to gain useful information that can also be incorporated into the main study.

19.2 Choosing the Best Analysis Approach

Different traditions for statistical analysis have developed for the analysis of data from designed experiments (where there is careful control of the levels of the explanatory variables used, usually resulting in a balanced structure) and for the analysis of data from observational studies (where the explanatory variables are usually not under the control of the researcher). These differences may seem illogical since both traditions work with the same underlying linear model, but can be understood by considering differences in the aims of the two types of study, and in the approaches used to collect the data. The biggest difference between the two traditions is in the approach to building a model for the response.

In the analysis of a designed experiment, the model will usually have been determined by the design of the experiment. All of the terms incorporated into the design (in both the explanatory and structural components of the model) are fitted and, for orthogonal structures, all terms are retained in the model, whether or not the associated variance ratio is statistically significant (see Section 8.2.4). This approach retains the residual mean square in each stratum as an estimate of pure error, since it is based on variation between units with the same treatment combination applied. It can be argued that model terms that are not statistically significant can legitimately be merged with the residual; however, this may not be true when the term is not significant because the statistical power is low. For non-orthogonal explanatory structures, any non-significant terms would be dropped from the model so that they do not influence the model predictions (Section 11.2.4). In both cases, all significant terms and any terms marginal to them form the model used for prediction.

In the analysis of data from observational studies, data may have been gathered on many different explanatory variables regarded as speculative or exploratory, with the statistical analysis being used to screen for variables that are related to the response. In many cases, there will be strong correlations within the set of explanatory variables and it would be counter-productive to include the full set in the model, and so the subset of variables (and interactions) that gives a good but parsimonious description of the response is selected (see Sections 14.9 and 15.5). Predictions are then made from this selected model.

The differences in procedure arise from differences in the aims and construction of a study. If an experiment has been designed, giving an orthogonal structure to investigate the effect of certain explanatory variables on the response, then the full model is pre-defined and all effects are estimable. In some cases, such as the presence of extraneous covariates, it may be sensible to add unplanned terms to the model, but it will not be necessary to drop terms from the model unless the structure is non-orthogonal. On the other hand, if a study has collected data on a number of uncontrolled explanatory variables, then there is no pre-defined model and it is appropriate to use model selection techniques (adding and/or dropping terms) to find a parsimonious statistical model that gives a good description of the response, and to identify which explanatory variables have some influence on the response.

However, within each type of study, there are further analysis issues that should be considered.

19.2.1 Analysis of Designed Experiments

As discussed above, the analysis of any designed experiment should be defined by the design chosen for the experiment. Certainly, the structural component of the model should be determined by the design, and an initial explanatory model can be identified based on the explanatory variables and structures (e.g. crossed or nested factorial structures) considered during the construction of the design. Where qualitative factors are included, some refinement of the analysis may be possible using contrasts to address questions more specific than 'are the mean responses for the levels of the factor different?'. Of course, these questions (and contrasts) should have been identified and used to select factor levels during the construction of the design, preferably enabling the contrasts identified to be fitted as an orthogonal set. If this is not the case, care must be taken in both specifying these contrasts (ensuring that the statistical package will not orthogonalize the contrasts during the model fitting process – a step that might change the meaning of each fitted contrast) and in interpreting the results of the individual tests (as they may change according to the order in which the contrasts are fitted, see Section 11.2).

With quantitative factors, the fitting of orthogonal polynomial contrasts to explore the shape of the response can be planned during the construction of the design, using the expected maximum order of the polynomial to determine the number and spread of the factor levels included. Polynomial contrasts generally provide a good assessment of whether the response is linear or more complex, but extracting information about the fitted polynomial is quite challenging in most statistical software. Therefore, the fitting of polynomial contrasts within the analysis of a designed experiment is often a precursor to translation of the model into the regression modelling framework. This can be challenging, as it requires that proper account is taken of the structural component of the model (see Sections 11.6 and 15.3), as well as ensuring that all qualitative explanatory factors and interactions are retained in the model (see Chapter 15). But analysis within the regression modelling framework does make it possible to use more complicated models, such as the non-linear models introduced briefly in Section 17.3. As an alternative, linear mixed models (Chapter 16) allow direct incorporation of the structural component of the model in addition to regression relationships within a general explanatory structure.

Finally, it is always important to investigate whether the observed response might have been influenced by any unplanned (extraneous) sources of variation. These might be noticed by the experimenter (e.g. pigeon grazing in one corner of a field) or detected during statistical analysis (e.g. examining residuals according to their physical position in the experimental layout as in Figure 11.4). These effects can be incorporated in the analysis by including a measure of the unplanned quantity for each experimental unit as a factor or covariate (Sections 11.5 and 15.4).

19.2.2 Analysis of Observational Studies

When identifying the aims of any observational study, the primary response variable of interest should be determined, together with the set of potential explanatory variables to be measured. In most cases, some (or even all) of the explanatory variables will be quantitative, and observations should be collected to cover the full range of values normally observed for each variable. With quantitative explanatory variables present, a regression model will usually be the first approach considered, although this approach should still take account of any structure in the study (see Section 15.3). Before analysis, it is important to understand the extent of correlations among the explanatory variables (see Sections 14.1 and 14.7). One possible preliminary step is to fit simple linear regression models to evaluate the relationship of the primary response variable with each individual quantitative explanatory variable, and to fit a simple factor model for any qualitative explanatory variables. With simple data sets, this may be all that is required, but various extensions will usually need to be considered such as those listed below.

- For quantitative variables, is the relationship a straight line or would some form of curved relationship be more appropriate? Choice of the form of curved relationship depends on whether we just want to describe the response (polynomial models may be adequate) or want some deeper understanding of the underlying mechanism (some form of non-linear model may be more appropriate).
- Are there multiple explanatory variables? For a small number of variables, it might be possible to fit and compare models for all combinations of the explanatory

variables (excluding interactions), but for larger numbers some sort of model selection process (such as stepwise regression, Sections 14.9 and 15.5) might be needed as an initial step, before further exploring the models identified in more detail.

- Are there likely to be interactions among the explanatory variables? This depends on the explanatory variables present, and may require an understanding of the underlying biological science. Recall that interactions between quantitative and qualitative explanatory variables (variates and factors) correspond to regression with groups, and these regressions may be parallel or not (Chapter 15).

19.2.3 Different Types of Data

Much of the data that we collect from either designed experiments or observational studies will satisfy the assumptions associated with the linear model – i.e. that the model deviations are homogeneous (constant variability), that the deviations follow a Normal distribution, and that the deviations are mutually independent (see Sections 4.1 and 12.1). We have introduced diagnostic approaches (Chapters 5 and 13) that can be used to check that the first two of these assumptions are satisfied for the statistical analysis approaches covered in this book. The way in which the data are collected usually determines whether deviations can reasonably be expected to be independent (see Section 5.2.2). It should be obvious when planning a study whether this is likely to be an issue, and the data collection procedure can either be modified to ensure independence, or a more complex statistical analysis can be planned; for example, linear mixed models (Chapter 16) can allow for different patterns of correlation between observations.

Where the assumptions of homogeneity of variance or Normality are not met, then one approach is the use of transformation of the data prior to analysis (Chapter 6). This provides some challenges in the presentation of the results (see Section 6.3), but otherwise the analysis proceeds as for an untransformed variable. For some types of discrete data, such as unconstrained counts, or counts expressed as a proportion of some fixed total, transformation is unlikely to be successful and a better alternative is available. These forms of data are likely to follow either Poisson (counts) or Binomial distributions (proportions), and for these types of response we can fit models for both quantitative and qualitative explanatory variables within the GLM framework introduced briefly in Chapter 18.

Finally, studies might involve the collection of multiple response variables as well as multiple explanatory variables. While each response variable might be analysed individually, using the methods identified in this book, it might be more useful to analyse the set of response variables together, taking account of the associations and relationships between them. This requires the application of multivariate statistical methods, which are beyond the scope of this book.

19.3 Presentation of Statistics in Reports, Theses and Papers

Having carefully designed a study, collected the data and performed an appropriate statistical analysis to address the questions and hypotheses that motivated the study, we usually need to summarize the statistical aspects of the study in a report, thesis or paper.

Statistical information should appear in both the Materials and Methods and the Results sections of any publication, and we discuss each of these separately.

19.3.1 Statistical Information in the Materials and Methods

In the Materials and Methods section of a publication, the aim is to provide sufficient statistical information to allow the reader to be able to understand the structure of the study and repeat the statistical analysis.

The first step is to provide details about the design of the study. For a designed experiment this should include information about the treatments and explanatory structures as well as the physical structure of the experiment, identifying any practical constraints associated with performing the experiment. Where a standard form of design has been used (such as a randomized complete block design or a Latin square design), it will usually be adequate to state the form of design by name with the level of replication (e.g. two replicates of a 3×3 Latin square design). Where a more complex or non-standard form of design has been used, it is necessary to provide more detail, and a diagram or table showing the structure of the design, possibly also including the treatment allocation, can be helpful (e.g. see Table 19.1). It is important to identify the form of the experimental unit that was used for each treatment factor, including the dimensions of the unit if this is otherwise unclear. For some studies, these dimensions might vary between experimental units (e.g. field size in Case Study 19.1), in which case it might be helpful to state the range of values. Where different experimental units are used for different explanatory variables, such as in Example 9.2, it is important to indicate the number of smaller experimental units that are combined to produce each larger experimental unit, as well as to clearly identify the explanatory variables that are applied to each type of experimental unit. It is also important to indicate the numbers of levels of each variable (the actual levels may have already been described as part of the biological methods, but might be usefully identified here as well), how variables are combined (e.g. as a crossed or nested factorial structure), whether there are additional control treatments, and, most importantly, the number of replicates of each treatment combination. It should be possible for the reader to relate the choice of experimental treatments directly to the stated aims of the study, but it may sometimes be useful to clarify the precise hypotheses being tested, and to indicate how these relate to the particular treatments or treatment combinations included. Such information may also be used to justify the inclusion of contrasts within a treatment factor.

For observational studies, similar information needs to be provided, but with more focus on how the samples were selected, possibly identifying the sampling frame and any constraints on the selection of individual samples. Any broad differences in the characteristics of the samples (related to spatial location, time of sampling, environmental variables, etc.) should be reported, together with the structure of the sample. For example, in a study that sampled fruit from trees in an orchard for an assessment of pesticide levels, we would need to indicate how the fruit were selected within each tree, possibly taking account of different locations within the tree (e.g. branch number or spatially defined parts of the tree), as well as how the trees were selected within the orchard, and, if the study used fruit from different orchards, how the orchards were selected. Design Case Study 19.1 contains some elements of an observational study, as well as elements of a designed experiment, illustrating how studies can be a hybrid of these two types.

Having described the structure and design of the study, the next part of the statistical methods relates to the data that have been collected. Details should be given of the number of measurements or assessments made on each experimental unit (we might measure each individual plant within a plot of multiple plants, or take several measurements on each experimental unit), and any manipulation of these data prior to analysis. This would include the calculation of any quantity derived from several variables measured on each experimental unit (e.g. harvest index). If some transformation of the data has been performed prior to analysis in order to satisfy the model assumptions (primarily homogeneity of variance and Normality), then the form of data transformation should be stated, together with the reason for the transformation.

The final section of statistical methodology relates to the method(s) of analysis that have been applied to the data to extract answers to the original questions, i.e. to test the statistical hypotheses. Where a standard method has been used, such as simple linear regression to assess the relationship between a response and explanatory variate or ANOVA to summarize data from a designed experiment, it will often be sufficient to just state the method without the need for further referencing. In other cases, it is helpful to state the full model fitted, including both explanatory and structural components; these can be described in the text (e.g. a crossed structure with two factors), but it is often clearer to explicitly give the symbolic form, including details of any contrasts fitted to extract information about particular treatment comparisons. Where model selection approaches have been used, the description should indicate the model terms or sequence of models being considered, as well as the selection strategy and selection criterion used to identify the best model. Where less standard analysis approaches are used, it will usually be more sensible to provide a reference to a good applied statistics text describing the method than to give details within the publication.

In addition, it is useful to state the statistical software used to perform the analysis, including version number and the relevant functions or procedures. However, this information should be given in addition to, not instead of, the information on the methods listed above. It is good practice to keep a safe copy of the analysis program and data file used to produce the results given in the paper, in case of any future revision or queries.

19.3.2 Presentation of Results

The best approach to the presentation of the results of statistical analyses, and the quantity of information required, varies considerably depending on the type of analysis that has been done. For simple statistical hypothesis tests, such as a two-sample t-test (see Section 2.4.2), it is sufficient to present the test statistic, together with the associated degrees of freedom and the observed significance level, within the text. Interpretation of the test result (whether to reject or fail to reject the null hypothesis) might then follow, together with information about the mean values for the different treatments, and hence the direction of the difference and the biological interpretation.

For the analysis of variance of a designed experiment or the fitting of a regression model, more extensive results need to be presented. It is usually sufficient to identify the model terms used to form the predictive model (i.e. terms that are statistically significant and those terms marginal to them), and then to present predictions. At this stage, a table or a graph provides a succinct yet powerful summary of the analysis, and the choice between these two forms usually depends on the complexity and quantity of the information to be presented. Tables may be the better choice for large, more complex sets of means (as might be produced from a multi-factorial designed experiment) and graphs are often better for showing simpler patterns (such as for a simple designed experiment or regression model).

When presenting the results from the analysis of variance of a designed experiment, it is usually not necessary to present the full ANOVA table, although when the model is complex, a table showing the strata, model terms and their associated df can be a useful supplement to the description in the Materials and Methods section. F-tests for the variance ratios associated with each explanatory term can be quoted in the text (test statistic with df and observed significance level), with tables of means or simple graphs used to show the pattern of responses for the predictive model. Where a high-order interaction term has a significant F-test, then it is usual to show predictions for the levels of the associated main effects alongside the predictions for the interaction (i.e. for the combinations of levels of different factors) in a multi-way table. Careful thought is needed about the choice of factors to label the rows and columns of the table, as the human eye is good at seeing patterns down columns of numbers, but less good across rows. For a two-way table, one strategy is to assign columns to the factor which has larger differences between predictions for different levels, with rows assigned to the other factor. The more subtle differences between predictions for the levels of the row factor (within each level of the column factor) are then more easily seen down each column. There is a natural ordering of the levels for quantitative factors, but careful ordering of levels for qualitative factors can make it easier to see patterns.

Graphs can give a good illustration of two-factor interactions, with the levels of one factor labelling positions on the horizontal axis and different colours or symbols used to indicate the points for the levels of the other factor. Point plots are preferable to bar charts, and it can be useful to draw lines between the points for each level of the second factor (as we did from Chapter 8 onwards), although it is important to realize that these lines do not imply that we can interpolate between levels. Again, there will be a natural ordering of factor levels on the horizontal axis for quantitative factors but where the factor is qualitative, careful ordering of the levels can enhance the interpretation of the interaction. The best choice of factor used to label the horizontal axis depends on context, and it is usually worth creating both of the possible graphs to identify the option that provides the clearest interpretation. Graphs can also be used to illustrate three-way interactions by creating separate plots to illustrate the interaction between two of the factors for each level of the third factor; however, if the patterns are complex, then a table may make better use of space.

Whether presenting predictions using tables or graphs, it is vital to also present information about the precision of the predictions or of differences between them. Where interest is in the estimated response for a particular level of a factor (a single prediction), then the estimated SE is the appropriate measure of precision, either as a summary or used to construct a confidence interval (usually at the 95% level). If the predictions are presented graphically, then bars for confidence intervals (or SEs) can be added to the point for each prediction, with the form of error bar and its associated df clearly stated in the figure legend. For a balanced design with equal replication, the SEs will be the same, and so only this common value needs to be presented with a table of predictions. However, most studies are more concerned with assessing differences between pairs of predictions, and so the presentation of the estimated SE of the difference (SED) for each comparison, with its associated df, is more useful. Alternatively, the LSD can be derived for any comparison and presented with its associated df and chosen level of significance stated. Again, for a balanced design with equal replication, there will be only one SED or LSD value to be presented and, in this case, it is best to just add a single SED or LSD bar to a graph of predicted values, suitably positioned to allow easy visual assessment of the comparison(s) of most interest (e.g. Figure 4.6).

More care is required where a transformation has been applied to the data prior to analysis (see Chapter 6). Here, the analysis has been applied to the transformed response, so that assessment of the significance of differences between predictions must be made on that transformed scale using the appropriate SEDs or LSDs. However, we often want to interpret these differences on the scale on which the data were originally measured. The usual approach is to present both the predictions on the transformed scale, with appropriate SEDs/LSDs (plus associated df), and the back-transformed predictions. This is relatively easy when presenting the predictions in a table, as the back-transformed values can be presented in parentheses alongside the predictions. For graphical presentation, one option is to plot the predictions (with SEDs or LSDs) on the transformed scale but with the vertical axis labelled on the back-transformed scale. Where this is not possible, or does not provide a clear representation of the pattern of response, it may be better to present the back-transformed means graphically with the means on the transformed scale (and their associated precision) shown in an accompanying table. Where interest is in single predictions, so that confidence intervals give an appropriate measure of precision, then confidence intervals can be calculated on the transformed scale and both the mean value and the confidence limits back-transformed for graphical display.

For a regression model with a single explanatory variate, a graph of the fitted model imposed on a scatter plot of the observations will usually be helpful to demonstrate the fit of the model. This is a useful approach for a linear (Chapter 12) or non-linear (Chapter 17) model, as well as for models with additional explanatory factor(s) (Chapter 15). The equation of the fitted model should be presented alongside the graph, showing the estimated parameters and their standard errors (with the associated df), and appropriate goodness-of-fit statistics (Section 14.8). A graphical representation of the fitted model is more difficult for multiple regression models, but a plot of the observations against the fitted values can be helpful. For models with a large number of parameters, presentation of the estimated parameters and their SE in a table may be more effective than trying to list them within the text (e.g. Table 14.14). Comparisons between models within a nested sequence (Chapters 14 and 15) can be reported in terms of the F-tests from the sequential ANOVA table, presenting each test statistic with its df and observed significance level.

Issues in the presentation of results from fitting GLMs (Chapter 18) closely mirror those associated with the analyses of transformed data, with parameters being estimated on the scale defined by the link transformation. Results are best presented in terms of predictions on the back-transformed scale for qualitative variables or plots of the fitted model on the back-transformed scale for quantitative variables, with information about the significance of parameter estimates, or comparisons between predictions, presented on the scale of the link transformation.

Finally, we return briefly to the issue of the precision with which numerical values should be presented, initially discussed in Section 2.6. The conventions described there provide a set of guidelines for the presentation of numerical results. We introduced the convention that test statistics, critical values and observed significance levels should be presented to three decimal places, generally providing sufficient detail for interpretation of the tests. We also introduced the concept of identifying the granularity of the observed response, and that other statistics should be presented with a precision defined in terms of this granularity. Predictions and estimated parameters should be presented to one more significant figure than the granularity of the original data, and variances, standard deviations and standard errors (including LSDs) should use two more significant figures. Following these guidelines, with a sprinkling of common sense thrown in, should ensure that you present your numerical results with sufficient detail to allow your reader to understand and appreciate the analysis and interpretation you present.

19.4 And Finally...

We hope that we have provided a clear understanding of a wide range of statistical methods to underpin your scientific research and education, and have provided you with sufficient knowledge to allow you either to design your own studies and analyse the data that you collect, or to be able to instigate a fruitful collaboration with a professional applied statistician. We have all worked in organizations in the latter role and are well aware of the value, enjoyment and success that such collaborations can bring to both sides.

Probably the most valuable single piece of advice we can give is that you should look critically at your statistical analysis, as it is easy to fit statistical models that do not make biological sense and which may therefore give misleading results. Always make sure that you understand the model you have fitted and cross-check that the results are consistent with simple summaries of your study. If your results contradict previous work, then check for a mistake in your data processing or analysis procedures before celebrating your new discovery!

Our intention is to maintain and expand the online software resources associated with this book (www.stats4biol.info), and hope that you continue to find these and the contents of the book useful as you pursue the application of statistical approaches to add value to your research and study.

References

Agresti, A. 2007. *An Introduction to Categorical Data Analysis* (2nd edition). John Wiley & Sons, Hoboken, NJ. 400 pp.

Agresti, A. 2010. *Analysis of Ordinal Categorical Data* (2nd edition). John Wiley & Sons, Hoboken, NJ. 405 pp.

Amoah, S., Wilkinson, M., Dunwell, J. and King, G.J. 2008. Understanding the relationship between DNA methylation and phenotypic plasticity in crop plants. *Comparative Biochemistry and Physiology*, Part A 150 (Supplement 1), S145.

Atkinson, A.C. 1985. *Plots, Transformations and Regression: An Introduction to Graphical Methods of Diagnostic Regression Analysis.* Oxford University Press, Oxford, Great Britain. 282 pp.

Atteia, O., Dubois, J.-P. and Webster, R. 1994. Geostatistical analysis of soil contamination in the Swiss Jura. *Environmental Pollution*, 86, 315–327.

Bailey, R.A. 2008. *Design of Comparative Experiments.* Cambridge Series in Statistical and Probabilistic Mathematics (No. 25). Cambridge University Press, Cambridge, 346 pp.

Baldwin, T.K., Gaffoor, I., Antoniw, J., Andries, C., Guenther, J., Urban, M., Pitkin, J., Hammond-Kosack, K.E. and Trail, F. 2010. A partial chromosomal deletion caused by random plasmid integration resulted in a reduced virulence phenotype in *Fusarium graminearum. Molecular Plant–Microbe Interactions*, 23, 1083–1096.

Barnett, V. and Lewis, T. 1994. *Outliers in Statistical Data* (3rd edition). John Wiley & Sons Ltd, Chichester. 584 pp.

Bartlett, M.S. 1937. Properties of sufficiency and statistical tests. *Proceedings of the Royal Statistical Society, Series A*, 160, 268–282.

Benjamini, Y. and Hochberg, Y. 1995. Controlling the false discovery rate: A practical and powerful approach to multiple testing. *Journal of the Royal Statistical Society, Series B*, 57, 289–300.

Benjamini, Y. and Yekutieli, D. 2001. The control of the false discovery rate in multiple testing under dependency. *Annals of Statistics*, 29, 1165–1188.

Bliss, C.I. 1970. *Statistics in Biology: Statistical Methods for Research in the Natural Sciences, Volume 2.* McGraw-Hill, New York. 639 pp.

Bohan, D.A., Boffey, C.W.H., Brooks, D.R., Clark, S.J., Dewar, A.M., Firbank L.G., Haughton, A.J. et al. 2005. Effects on weed and invertebrate abundance and diversity of herbicide management in genetically modified herbicide-tolerant winter-sown oilseed rape. *Proceedings of the Royal Society B*, 272, 463–474.

Bondari, K. 1999. Interactions in entomology: Multiple comparisons and statistical interactions in entomological experimentation. *Journal of Entomological Science*, 34, 57–71.

Box, G.E.P., Hunter, W.G. and Hunter, J.S. 1978. *Statistics for Experimenters: An Introduction to Design, Data Analysis, and Model Building.* John Wiley & Sons Ltd, New York. 672 pp.

Brien, C.J. and Bailey, R.A. 2006. Multiple randomizations (with discussion). *Journal of the Royal Statistical Society, Series B*, 68, 571–609.

Brooks, D.R., Bohan, D.A., Champion, G.T., Haughton, A.J., Hawes, C., Heard, M.S., Clark, S.J. et al. 2003. Invertebrate responses to the management of genetically modified herbicide-tolerant and conventional spring crops. I. Soil-surface-active invertebrates. *Philosophical Transactions of the Royal Society of London B*, 358, 1847–1862.

Brown, D. and Rothery, P. 1993. *Models in Biology: Mathematics, Statistics and Computing* (1st edition). John Wiley & Sons Ltd, New York. 708 pp.

Carpenter, J.R. and Kenward, M.G. 2013. *Multiple Imputation and Its Applications.* John Wiley & Sons, Chichester. 364 pp.

Carroll, R.J., Ruppert, D., Stefanski, L.A. and Crainiceanu, C.M. 2006. *Measurement Error in Nonlinear Models* (2nd edition). Chapman & Hall/CRC, Boca Raton, FL. 488 pp.

Casella, G. and Berger, R.L. 2002. *Statistical Inference* (2nd edition). Duxbury, Pacific Grove, CA. 660 pp.

Chilès, J.-P. and Delfiner, P. 2012. *Geostatistics: Modelling Spatial Uncertainty*. Wiley, New Jersey. 734 pp.

Clark, S.J., Rothery, P. and Perry, J.N. 2006. Farm Scale Evaluations of spring-sown genetically modified herbicide-tolerant crops: A statistical assessment. *Proceedings of the Royal Society B*, 273, 237–243.

Clark, S.J., Rothery, P., Perry, J.N. and Heard, M.S. 2007. Farm Scale Evaluations of herbicide-tolerant crops: Assessment of within-field variation and sampling methodology for arable weeds. *Weed Research*, 47, 157–163.

Cochran, W.G. and Cox, G.M. 1957. *Experimental Designs* (2nd edition). Wiley, New York. 611 pp.

Collett, D. 2002. *Modelling Binary Data* (2nd edition). Chapman & Hall/CRC Press, Boca Raton, FL. 387 pp.

Cousens, R. 1988. Misinterpretations of results in weed research through inappropriate use of statistics. *Weed Research*, 28, 281–289.

Cox, D.R. 1961. Design of experiments—The control of error. *Journal of the Royal Statistical Society, Series A*, 124, 44–48.

Cox, D.R. 1992. *Planning of Experiments*. John Wiley & Sons, New York. 320 pp.

Diggle, P.J., Heagerty, P., Liang, K.-Y. and Zeger, S.L. 2002. *Analysis of Longitudinal Data* (2nd edition). Oxford University Press, Oxford. 400 pp.

Dobson, A.J. 1990. *An Introduction to Generalized Linear Models*. Chapman & Hall, London. 174 pp.

Draper, N.R. and Smith, H. 1998. *Applied Regression Analysis* (3rd edition). John Wiley & Sons Ltd, New York. 736 pp.

Efron, B. and Tibshirani, R.J. 1993. *An Introduction to the Bootstrap*. Chapman & Hall, London. 436 pp.

Ekesi, S., Shah, P.A., Clark, S.J. and Pell, J.K. 2005. Conservation biological control with the fungal pathogen *Pandora neoaphidis*: Implications of aphid species, host plant and predator foraging. *Agricultural and Forest Entomology*, 7, 21–30.

Farman, J.C., Gardiner, B.G. and Shanklin, J.D. 1985. Large losses of total ozone in Antarctica reveal seasonal ClO_x/NO_x interaction. *Nature*, 315, 207–210.

Fearn, T. 1983. A misuse of ridge regression in the calibration of a near infrared reflectance instrument. *Applied Statistics*, 32, 73–79.

Finney, D.J. 1971. *Probit Analysis*. Cambridge University Press, Cambridge. 350 pp.

Fisher, R.A. and Yates, F. 1963. *Statistical Tables for Biological, Agricultural and Medical Research* (revised 6th edition). Oliver & Boyd, Edinburgh. 146 pp.

Galwey, N.W. 2006. *Introduction to Mixed Modelling: Beyond Regression and Analysis of Variance*. John Wiley & Sons Ltd, Chichester, England. 366 pp.

Gates, C.E. 1991. A users guide to misanalyzing planned experiments. *Hortscience*, 26, 1262–1265.

Gelman, A. 2006. Prior distributions for variance parameters in hierarchical models. *Bayesian Analysis*, 1, 515–534.

Gilligan, C.A. 1986. Use and misuse of the analysis of variance in plant pathology. *Advances in Plant Pathology*, 5, 225–261.

Gilmour, A.R., Cullis, B.R. and Verbyla, A.P. 1997. Accounting for natural and extraneous variation in the analysis of field experiments. *Journal of Agricultural, Biological and Environmental Statistics*, 2, 269–293.

Goovaerts, G. 1997. *Geostatistics for Natural Resources Evaluation*. Oxford University Press, New York. 483 pp.

Haines, S.J. 2000. *Generation of a Zea mays Mutator Grid and Its Use in the Isolation and Partial Characterisation of a Mutator-Tagged Mutant of the Glutamine Synthetase 1-4 Gene*. PhD thesis, University of Bristol.

Hand, D.J., Daly, F., Lunn, A.D., McConway, K.J. and Ostrowski, E. 1994. *A Handbook of Small Data Sets*. Chapman & Hall/CRC Press, Boca Raton, FL. 458 pp.

Hastie, T., Tibshirani, R. and Friedman, J. 2001. *The Elements of Statistical Learning*. Springer, New York. 533 pp.

Haughton, A.J., Champion, G.T., Hawes, C., Heard, M.S., Brooks, D.R., Bohan, D.A., Clark, S.J. et al. 2003. Invertebrate responses to the management of genetically modified herbicide-tolerant and conventional spring crops. II. Within-field epigeal and aerial arthropods. *Philosophical Transactions of the Royal Society London B*, 358, 1863–1877.

Hawes, C., Haughton, A.J., Osborne, J.L., Roy, D.B., Clark, S.J., Perry, J.N., Rothery, P. et al. 2003. Responses of plants and invertebrate trophic groups to contrasting herbicide regimes in the Farm Scale Evaluations of genetically modified herbicide-tolerant crops. *Philosophical Transactions of the Royal Society London B*, 358, 1899–1913.

Healy, M.J.T. and Westmacott, M.H. 1956. Missing values in experiments analysed on automatic computers. *Applied Statistics*, 5, 203–206.

Heard, M.S., Hawes, C., Champion, G.T., Clark, S.J., Firbank, L.G., Haughton, A.J., Parish, A.M. et al. 2003. Weeds in fields with contrasting conventional and genetically modified herbicide-tolerant crops. I. Effects on abundance and diversity. *Philosophical Transactions of the Royal Society London B*, 358, 1819–1832.

Hill, A.B. 1965. The environment and disease: Association or causation? *Proceedings of the Royal Society of Medicine*, 58, 295–300.

Hoel, P. 1984. *Introduction to Mathematical Statistics* (5th edition). John Wiley & Sons Ltd, New York. 409 pp.

Hsu, J.C. 1996. *Multiple Comparisons: Theory and Methods*. Chapman & Hall, Boca Raton, FL. 277 pp.

Jing, H.-C., Kornyukhin, D., Kanyuka, K., Orford, S., Zlatska, A., Mitrofanova, O.P., Koebner, R. and Hammond-Kosack, K. 2007. Identification of variation in adaptively important traits and genome-wide analysis of trait–marker associations in *Triticum monococcum*. *Journal of Experimental Botany*, 58, 3749–3764.

Johnston, A.E., Poulton, P.R. and Syers, J.K. 2001. *Phosphorus, Potassium and Sulphur Cycles in Agricultural Soils*. Proceedings 465, International Fertiliser Society, York, UK. 44 pp.

Kearsey, M.J. and Pooni, H.S. 1996. *The Genetical Analysis of Quantitative Traits*. Chapman & Hall, London. 381 pp.

Kenward, M.G. and Roger, J.H. 1997. Small sample inference for fixed effects estimators from restricted maximum likelihood. *Biometrics*, 53, 983–997.

Kenward, M.G. and Roger, J.H. 2009. An improved approximation to the precision of fixed effects from restricted maximum likelihood. *Computational Statistics and Data Analysis*, 53, 2583–2595.

Kerr, M.K. and Churchill, G.A. 2001. Statistical design and the analysis of gene expression microarray data. *Genetical Research*, 77, 123–128.

Kuehl, R.O. 2000. *Design of Experiments: Statistical Principles of Research Design and Analysis* (2nd edition). Thomson Learning (Duxbury Press), Pacific Grove, CA. 666 pp.

Lane, P.W. and Nelder, J.A. 1982. Analysis of covariance and standardization as instances of prediction. *Biometrics*, 38, 613–621.

Lee, Y., Nelder, J.A. and Pawitan, Y. 2006. *Generalized Linear Models with Random Effects: Unified Analysis via H-Likelihood*. Chapman & Hall, Boca Raton, FL. 416 pp.

Littell, R.C., Milliken, G.A., Stroup, W.W., Wolfinger, R.D. and Schabenberger, O. 2006. *SAS for Mixed Models* (2nd edition). SAS Institute Inc, Cary, NC. 814 pp.

Macdonald, A.J., Poulton, P.R., Powlson, D.S. and Jenkinson, D.S. 1997. Effects of season, soil type and cropping on recoveries, residues and losses of ^{15}N-labelled fertilizer applied to arable crops in spring. *Journal of Agricultural Science*, 129, 125–154.

Madden, L.V. 1982. Considerations for the use of multiple comparison procedures in phytopathological investigations. *Phytopathology*, 72, 1015–1017.

McBride, G.B. 1999. Equivalence tests can enhance environmental science and management. *Australian and New Zealand Journal of Statistics*, 41, 19–29.

McClave, J.T. and Sincich, T. 2012. *Statistics* (12th edition). Prentice Hall, New Jersey. 840 pp.

McCullagh, P. and Nelder, J.A. 1989. *Generalized Linear Models* (2nd edition). Chapman & Hall, London. 511 pp.

McLean, R.A., Sanders, W.L. and Stroup, W.W. 1991. A unified approach to linear mixed models. *American Statistician*, 45, 54–64.

Mead, R., Curnow, R.N. and Hasted, A.M. 2003. *Statistical Methods in Agriculture and Experimental Biology* (3rd edition). Chapman & Hall, London. 488 pp.

Mead, R., Gilmour, S.G. and Mead, A. 2012. *Statistical Principles for the Design of Experiments: Applications to Real Experiments*. Cambridge University Press, Cambridge, UK. 586 pp.

Miller, A. 2002. *Subset Selection in Regression* (2nd edition). CRC Press, Boca Raton, FL. 234 pp.

Miller, R.G. 1981. *Simultaneous Statistical Inference* (2nd edition). Springer-Verlag, Berlin and Heidelberg. 299 pp.

Milliken, G.A. and Johnson, D.E. 2001. *Analysis of Messy Data, Volume III: Analysis of Covariance*. Chapman & Hall/CRC Press, Boca Raton, FL. 624 pp.

Montgomery, D.C. 1997. *Design and Analysis of Experiments* (4th edition). John Wiley & Sons Ltd, New York. 704 pp.

Montgomery, D.C., Peck, E.A. and Viner, G.G. 2012. *Introduction to Linear Regression Analysis* (5th edition). John Wiley & Sons, New York. 672 pp.

Morris, G.E.L. 1985. The presentation of treatment responses from block experiments after analysis of variance of transformed data. *Annals of Applied Biology*, 107, 571–580.

O'Brien, R.M. 2007. A caution regarding rules of thumb for variance inflation factors. *Quality and Quantity*, 41, 673–690.

Patterson, H.D. and Thompson, R. 1971. Recovery of inter-block information when block sizes are unequal. *Biometrika*, 58, 545–554.

Peacock, L., Batley, J., Dungait, J., Barker, J.H.A., Powers, S. and Karp, A. 2004. A comparative study of interspecies mating of *Phratora vulgatissima* and *P. vitellinae* using behavioural tests and molecular markers. *Entomologia Experimentalis et Applicata*, 110, 231–241.

Peacock, L., Carter, P., Powers, S. and Karp, A. 2003. Geographic variation in phenotype traits in *Phratora* spp. and the effects of conditioning on feeding preference. *Entomologia Experimentalis et Applicata*, 109, 31–37.

Pearce, S.C. 1965. *Biological Statistics: An Introduction*. McGraw-Hill, New York. 212 pp.

Pearce, S.C. 1983. *The Agricultural Field Experiment: A Statistical Examination of Theory and Practice*. John Wiley & Sons, Chichester. 335 pp.

Pearce, S.C. 1993. Data-analysis in agricultural experimentation. 3. Multiple comparisons. *Experimental Agriculture*, 29, 1–8.

Perry, J.N. 1986. Multiple-comparison procedures—A dissenting view. *Journal of Economic Entomology*, 79, 1149–1155.

Perry, J.N., Rothery, P., Clark, S.J., Heard, M.S. and Hawes, C. 2003. Design, analysis and statistical power of the Farm-Scale Evaluations of genetically modified herbicide-tolerant crops. *Journal of Applied Ecology*, 40, 17–31.

Pinheiro, J. and Bates, D.M. 2000. *Mixed Effects Models in S and S-Plus*. Springer-Verlag, New York. 530 pp.

Plackett, R.L. and Burman, J.P. 1946. The design of optimum multifactorial experiments. *Biometrika*, 33, 305–325.

Pukelsheim, F. 1990. Robustness of statistical gossip and the Antarctic ozone hole. *The IMS Bulletin*, 19, 540–545.

Rawlings, J.O., Pantula, S.G. and Dickey, D.A. 1998. *Applied Regression Analysis: A Research Tool* (2nd edition). Springer-Verlag New York, Inc., New York. 657 pp.

Ridout, M., Hinde, J. and Demetrio, C.G.B. 2001. A score test for testing a zero-inflated Poisson regression model against zero-inflated negative-binomial alternatives. *Biometrics*, 57, 219–223.

Rothery, P., Clark, S.J. and Perry, J.N. 2003. Design of the farm-scale evaluations of genetically modified herbicide-tolerant crops. *Environmetrics*, 14, 711–717.

Roy, D.B., Bohan, D.A., Haughton, A.J., Hill, M.O., Osborne, J.L., Clark, S.J., Perry, J.N. et al. 2003. Invertebrates and vegetation of field margins adjacent to crops subject to contrasting herbicide regimes in the Farm Scale Evaluations of genetically modified herbicide-tolerant crops. *Philosophical Transactions of the Royal Society London B*, 358, 1879–1898.

Ruppert, D., Wand, M.P. and Carroll, R.J. 2003. *Semi-Parametric Regression*. Cambridge University Press, Cambridge. 404 pp.

Salisbury, A., Clark, S.J., Powell, W. and Hardie, J. 2010. Susceptibility of six *Lilium* to damage by the lily beetle, *Lilioceris lilii* (Coleoptera; Chrysomelidae). *Annals of Applied Biology*, 156, 103–110.

Satterthwaite, F.E. 1946. An approximate distribution of estimates of variance components. *Biometrics Bulletin*, 2, 110–114.

Seber, G.A.F. and Wild, C.J. 1989. *Nonlinear Regression*. Wiley, New York. 768 pp.

Shortall, C.R., Moore, A., Smith, E., Hall, M.J., Woiwod, I.P. and Harrington, R. 2009. Long-term changes in the abundance of flying insects. *Insect Conservation and Diversity*, 2, 251–260.

Smith, A.B., Cullis, B.R. and Thompson, R. 2005. The analysis of crop cultivar breeding and evaluation trials: An overview of current mixed model approaches. *Journal of Agricultural Science*, 143, 449–462.

Smith, V.C. 2007. *Invertebrate Response to Weed Diversity and Spatial Arrangement within Arable Fields*. PhD thesis, The University of Reading, Berkshire.

Snedecor, G.W. and Cochran, W.G. 1989. *Statistical Methods* (8th edition). Iowa State University Press, Ames, IA. 503 pp.

Sokal, R.R. and Rohlf, F.J. 1995. *Biometry: The Principles and Practice of Statistics in Biological Research* (3rd edition). W.H. Freeman & Co., New York. 887 pp.

Stroup, W.W. 2012. *Generalized Linear Mixed Models: Modern Concepts, Methods and Applications*. CRC Press, Boca Raton, FL. 555 pp.

Taylor, J. 1973. The analysis of designed experiments with censored observations. *Biometrics*, 29, 35–43.

Verbeke, G. and Molenberghs, G. 2000. *Linear Mixed Models for Longitudinal Data*. Springer Verlag, New York. 568 pp.

Wackerly, D., Mendenhall, W. and Scheaffer, R.L. 2007. *Mathematical Statistics with Applications* (7th edition). Cengage Learning, Brooks/Cole, Belmont, CA. 944 pp.

Webster, R. 2007. Analysis of variance, inference, multiple comparisons and sampling effects in soil science. *European Journal of Soil Science*, 58, 74–82.

Webster, R., Atteia, O. and Dubois, J.-P. 1994. Coregionalization of trace metals in the soil in the Swiss Jura. *European Journal of Soil Science*, 45, 205–218.

Webster, R. and Oliver, M.A. 2007. *Geostatistics for Environmental Scientists* (2nd edition). Wiley, Chicester. 330 pp.

Welham, S.J., Cullis, B.R., Gogel, B.J., Gilmour, A.R. and Thompson, R. 2004. Prediction in linear mixed models. *Australian and New Zealand Journal of Statistics*, 46, 325–347.

Wilkinson, G.N. and Rogers, C.E. 1973. Symbolic description of factorial models for analysis of variance. *Applied Statistics*, 22, 392–399.

Wit, E., Nobile, A. and Khanin, R. 2005. Near-optimal designs for dual channel microarray studies. *Applied Statistics*, 54, 817–830.

Wood, C.R., Reynolds, D.R., Wells, P.M., Barlow, J.F., Woiwod, I.P. and Chapman, J.W. 2009. Flight periodicity and the vertical distribution of high-altitude moth migration over southern Britain. *Bulletin of Entomological Research*, 99, 525–535.

Wright, E.L. 2013. *The Effect of Pathogens on Honeybee Learning and Foraging Behaviour*. PhD thesis, University of Warwick.

Appendix A: Data Tables

TABLE A.1

Measurements of Weight (W), Length (L), Diameter (D), Moisture Content (M) and Hardness Index (H) for 190 Seeds (Example 12.1A and File TRITICUM.DAT)

Seed	W	L	D	M	H	Seed	W	L	D	M	H
1	30.15	3.27	2.09	10.27	−16.63	39	25.42	3.01	1.93	10.68	−22.04
2	35.51	3.65	2.34	10.61	−8.27	40	30.28	3.45	2.21	10.37	2.25
3	29.16	3.36	2.15	10.27	−21.45	41	27.41	3.30	2.11	10.41	−3.94
4	16.82	2.77	1.79	11.05	4.13	42	38.75	3.84	2.47	10.68	−20.41
5	23.42	2.78	1.80	10.02	−2.05	43	19.69	3.14	2.01	10.28	−11.05
6	31.77	3.37	2.15	10.34	−41.78	44	24.80	3.09	1.98	10.67	−30.84
7	16.45	2.52	1.66	10.64	−5.33	45	33.27	3.39	2.17	10.79	−21.12
8	32.89	3.48	2.23	10.44	−13.91	46	22.43	2.91	1.87	10.47	−28.66
9	22.55	3.17	2.03	10.28	−10.87	47	49.47	4.12	2.66	10.59	−42.47
10	28.03	3.20	2.05	10.22	−16.28	48	22.30	3.07	1.97	10.97	−1.61
11	32.27	3.58	2.29	10.32	−12.81	49	27.29	3.42	2.19	10.37	11.24
12	40.62	3.97	2.56	10.40	10.46	50	34.26	3.63	2.33	10.39	−4.45
13	29.28	3.54	2.27	10.64	−32.43	51	24.30	3.06	1.96	10.85	11.87
14	22.68	3.23	2.07	10.78	−19.04	52	24.55	3.24	2.07	10.30	−21.16
15	29.78	3.53	2.26	10.39	−25.78	53	19.06	2.89	1.86	10.25	−9.72
16	27.16	3.05	1.96	10.49	−34.65	54	27.04	3.18	2.04	10.36	−6.46
17	17.94	2.86	1.85	10.37	−5.24	55	29.03	3.36	2.15	10.67	−8.63
18	20.93	3.08	1.97	10.97	−6.41	56	36.38	3.64	2.33	10.59	−19.23
19	30.78	3.48	2.23	10.83	−4.09	57	24.30	3.22	2.06	10.78	−0.93
20	45.85	3.78	2.43	10.37	−18.00	58	22.68	3.11	1.99	10.38	−22.53
21	30.78	3.27	2.09	10.70	−3.21	59	33.89	3.51	2.25	10.64	−5.04
22	33.64	3.54	2.27	10.52	−21.18	60	33.39	3.45	2.21	10.65	−18.14
23	34.89	3.47	2.22	10.74	−18.36	61	25.54	3.45	2.21	10.28	7.83
24	22.55	3.09	1.98	10.55	−9.35	62	32.52	3.67	2.35	10.51	−13.93
25	28.28	3.38	2.16	10.51	−7.74	63	31.27	3.41	2.18	10.30	−24.13
26	25.04	3.07	1.97	10.06	−7.46	64	31.15	3.40	2.17	10.48	3.35
27	39.25	3.80	2.44	10.76	−28.34	65	19.19	2.93	1.89	10.59	−5.89
28	26.79	3.09	1.98	10.26	6.31	66	23.42	3.22	2.06	10.34	−1.76
29	24.55	3.10	1.99	10.36	4.18	67	25.17	2.87	1.85	10.54	−20.72
30	33.02	3.58	2.29	10.56	−22.34	68	28.91	3.10	1.99	10.56	−11.31
31	24.30	2.74	1.78	10.54	−19.77	69	27.79	3.38	2.16	10.45	−6.73
32	25.92	3.10	1.99	10.67	−28.64	70	34.76	3.45	2.21	10.70	−36.98
33	34.51	3.57	2.29	10.61	−9.14	71	29.53	3.24	2.07	10.05	−20.59
34	28.16	3.00	1.93	10.21	−15.51	72	24.80	3.30	2.11	10.54	−7.83
35	28.16	3.09	1.98	10.52	−11.11	73	29.65	3.31	2.12	10.52	−25.70
36	19.81	2.85	1.84	10.94	−14.08	74	29.90	3.36	2.15	10.42	−34.76
37	27.16	3.39	2.17	10.37	−19.96	75	21.81	2.82	1.82	10.71	−8.90
38	16.07	2.61	1.71	10.31	−5.11	76	25.29	3.12	2.00	10.77	15.07

TABLE A.1 (continued)

Measurements of Weight (W), Length (L), Diameter (D), Moisture Content (M) and Hardness Index (H) for 190 Seeds (Example 12.1A and File TRITICUM.DAT)

Seed	W	L	D	M	H	Seed	W	L	D	M	H
77	33.77	3.43	2.19	10.61	−24.60	121	31.90	3.33	2.13	10.56	−5.52
78	37.13	3.81	2.45	10.70	−15.98	122	24.05	2.73	1.77	10.77	−30.30
79	31.52	3.43	2.19	10.55	−14.59	123	35.88	3.63	2.33	10.65	−17.86
80	32.65	3.48	2.23	10.50	−22.41	124	41.12	3.83	2.46	10.70	−8.30
81	28.16	3.55	2.27	10.79	−42.05	125	30.90	3.71	2.38	10.88	−25.06
82	25.67	3.33	2.13	10.65	8.38	126	23.67	3.18	2.04	10.65	−1.61
83	16.07	2.46	1.63	10.69	−4.47	127	22.43	2.92	1.88	10.54	−11.17
84	31.90	3.34	2.14	10.71	−14.63	128	26.79	3.23	2.07	10.59	−23.49
85	40.99	4.00	2.58	10.68	−2.88	129	43.49	3.96	2.55	10.67	−30.67
86	22.68	3.19	2.04	10.20	6.77	130	27.66	3.42	2.19	10.65	−7.37
87	24.55	3.17	2.03	10.90	−19.53	131	30.28	3.49	2.23	10.70	4.89
88	31.77	3.61	2.31	10.62	−9.93	132	21.81	3.20	2.05	10.91	1.78
89	30.90	3.63	2.33	10.51	−10.35	133	33.52	3.46	2.21	10.30	−21.35
90	29.65	3.56	2.28	10.66	16.00	134	17.19	2.81	1.82	10.72	−17.29
91	33.39	3.34	2.14	10.38	−21.67	135	15.57	2.64	1.72	10.67	−22.46
92	27.54	3.49	2.23	10.29	−10.27	136	23.80	3.10	1.99	10.34	−20.90
93	35.14	3.82	2.45	10.24	4.35	137	35.26	3.60	2.31	10.77	−14.34
94	37.13	3.64	2.33	10.21	−31.31	138	47.47	4.13	2.67	10.72	−27.76
95	19.06	2.94	1.89	10.48	22.05	139	25.17	2.90	1.87	10.28	26.26
96	31.65	3.31	2.12	10.01	−10.07	140	41.99	3.86	2.48	10.38	−20.44
97	25.92	3.01	1.93	10.55	−4.62	141	29.16	3.62	2.32	10.57	−7.22
98	25.17	3.09	1.98	10.77	−19.43	142	18.57	3.08	1.97	10.54	−26.78
99	21.06	2.90	1.87	10.61	−6.36	143	24.17	3.31	2.12	10.98	−24.60
100	27.66	3.12	2.00	10.54	−5.06	144	35.88	3.66	2.34	10.14	−1.33
101	32.27	3.15	2.02	10.70	−34.37	145	20.56	2.90	1.87	10.85	2.23
102	22.93	2.84	1.83	10.21	−6.84	146	30.28	3.44	2.20	10.26	−21.36
103	29.41	3.03	1.94	10.62	−10.74	147	28.41	3.35	2.14	10.39	−16.57
104	31.03	3.54	2.27	10.43	1.11	148	24.80	2.88	1.86	10.44	−14.64
105	25.54	3.23	2.07	10.52	−29.74	149	28.53	3.45	2.21	10.58	−10.61
106	32.77	3.36	2.15	10.49	11.20	150	32.52	3.46	2.21	10.71	−8.90
107	34.89	3.64	2.33	10.90	−14.96	151	26.91	2.97	1.91	10.63	−6.45
108	22.93	2.87	1.85	10.97	−20.46	152	25.42	3.01	1.93	10.57	−38.91
109	35.26	3.54	2.27	10.58	−5.39	153	33.27	3.61	2.31	10.49	−0.79
110	21.81	3.16	2.02	10.53	−14.53	154	29.16	3.34	2.14	10.36	−18.54
111	31.27	3.49	2.23	10.61	−31.34	155	25.17	3.09	1.98	10.74	−28.62
112	32.02	3.50	2.24	10.78	−20.25	156	28.53	3.17	2.03	10.52	−21.63
113	29.41	3.49	2.23	10.25	−9.70	157	20.68	2.92	1.88	10.81	−47.75
114	25.04	3.11	1.99	10.34	−10.69	158	23.42	2.76	1.79	10.42	−7.09
115	37.50	3.83	2.46	10.70	−4.61	159	27.41	3.06	1.96	10.98	−24.94
116	17.94	2.74	1.78	10.84	−30.76	160	22.18	3.06	1.96	10.68	−21.83
117	32.15	3.26	2.09	10.74	−11.28	161	28.03	3.22	2.06	10.29	−19.64
118	26.29	3.33	2.13	10.57	−21.02	162	35.01	3.36	2.15	10.66	−22.98
119	27.29	3.24	2.07	10.59	0.42	163	29.03	3.10	1.99	10.45	−22.79
120	32.15	3.47	2.22	10.34	−19.32	164	23.92	2.95	1.90	10.82	−10.84

TABLE A.1 (continued)

Measurements of Weight (W), Length (L), Diameter (D), Moisture Content (M) and Hardness Index (H) for 190 Seeds (Example 12.1A and File TRITICUM.DAT)

Seed	W	L	D	M	H	Seed	W	L	D	M	H
165	26.54	3.32	2.12	10.88	−15.84	178	37.38	3.67	2.35	10.73	−17.74
166	36.63	3.59	2.30	10.50	−5.22	179	35.01	3.65	2.34	10.34	−15.01
167	31.90	3.51	2.25	10.42	−8.95	180	28.03	3.30	2.11	10.83	−5.60
168	22.80	2.90	1.87	10.85	−5.56	181	34.14	3.51	2.25	10.56	−4.95
169	28.28	3.44	2.20	10.46	−20.97	182	26.79	3.34	2.14	10.55	−19.84
170	39.12	3.69	2.37	10.08	−19.53	183	38.13	3.65	2.34	10.31	−27.10
171	28.16	3.51	2.25	10.19	−27.23	184	31.90	3.37	2.15	10.40	−16.98
172	38.88	3.71	2.38	10.64	−17.95	185	33.64	3.56	2.28	10.73	−34.58
173	23.92	3.07	1.97	10.34	−3.73	186	27.29	3.04	1.95	10.55	−7.29
174	20.81	2.90	1.87	10.54	−1.51	187	27.66	3.60	2.31	10.88	−22.68
175	29.03	3.22	2.06	10.36	−6.98	188	26.54	3.58	2.29	10.49	3.30
176	19.69	3.02	1.94	10.38	3.78	189	30.90	3.17	2.03	10.37	−17.83
177	33.27	3.47	2.22	10.48	−14.60	190	18.94	2.45	1.62	10.08	−7.06

Source: Data from H.-C. Jing and K. Hammond-Kosack, Rothamsted Research.

TABLE A.2

Measurements of Air Temperature (°C) for 100 Days during 2006 from a Standard Glass Mercury Thermometer (M) and a New Electronic Thermistor (ET) (Exercise 12.2 and File AIRTEMP.DAT)

Day	M	ET	Day	M	ET	Day	M	ET
1	5.3	5.3	127	12.1	12.4	244	17.9	17.4
7	6.6	5.5	132	12.6	12.7	246	20.7	19.5
8	8.9	8.7	134	11.7	11.4	251	20.3	19.6
13	6.9	6.7	139	11.3	11.9	253	18.4	18.3
15	8.8	8.6	141	12.0	11.6	258	16.3	16.6
20	2.0	1.8	148	10.0	10.3	260	17.0	17.5
22	0.5	−0.1	153	13.6	13.6	261	21.0	20.5
27	3.9	3.8	155	19.1	19.3	265	16.0	16.1
29	−0.5	−0.7	160	28.4	23.0	267	15.2	15.9
34	4.0	3.8	162	14.3	13.8	272	15.6	14.9
36	5.5	4.6	167	17.5	17.8	274	12.1	12.3
41	8.2	8.0	169	15.0	15.3	279	16.9	15.8
43	9.3	8.9	174	12.7	12.9	281	15.5	15.6
48	4.2	3.4	176	16.4	16.7	286	12.5	12.7
50	2.0	2.0	181	24.1	24.2	288	14.5	14.5
55	3.1	2.7	183	23.0	22.7	293	11.0	10.9
57	0.5	−0.2	188	18.5	19.0	294	11.0	10.9
62	4.0	2.7	190	19.9	20.0	295	11.5	10.9
64	9.4	8.7	195	24.9	23.9	300	12.5	12.0
69	1.2	0.7	197	27.9	26.4	302	4.6	4.3
71	4.1	3.5	202	23.0	22.5	307	6.6	5.8
76	4.3	4.0	204	26.9	25.0	309	9.0	8.1
78	3.6	2.7	209	20.5	20.4	314	14.7	14.3
83	10.9	11.0	211	16.3	17.2	316	12.4	12.2
85	7.9	7.6	216	16.6	16.8	321	11.8	11.2
90	8.4	8.4	217	17.3	18.3	323	5.0	4.8
92	4.5	4.5	218	16.9	17.2	330	6.1	4.7
97	4.8	5.0	223	14.2	14.3	335	8.8	8.1
99	8.9	8.5	225	18.8	17.6	337	8.5	8.6
106	10.1	10.4	230	15.9	16.1	344	10.6	10.4
111	8.4	8.5	232	17.3	17.6	349	4.9	4.0
113	11.9	11.7	239	14.8	14.7	351	−1.5	−2.4
120	13.9	14.2	241	16.3	16.7	353	0.0	−3.2
125	11.8	12.0						

Source: Data from T. Scott and M. Glendining, Rothamsted Research.

TABLE A.3A

Data from 50 Suction Trap Locations (Trap) During 1995: Julian Day When Aphid *Myzus persicae* First Caught in Trap (JDay), Trap Location (Latitude, Longitude and Altitude) and Monthly Rainfall from October 1994 to May 1995 (Example 14.2 and File EXAMINE.DAT)

| Trap | JDay | Location | | | Monthly Rain | | | | | | | |
		Latitude	Longitude	Altitude	October	November	December	January	February	March	April	May
1	181	56.11	13.11	10	62.0	81.2	113.4	89.9	77.5	50.5	71.7	67.3
2	95	47.48	-0.58	20	74.4	77.2	105.4	125.2	100.1	70.2	34.1	54.1
3	170	51.76	11.44	130	57.5	43.0	29.1	45.3	52.2	39.1	42.4	55.4
4	120	47.81	3.58	100	36.7	47.5	46.7	98.7	66.3	58.9	71.8	89.1
5	145	55.45	-4.55	46	85.3	153.1	253.6	175.7	160.8	138.0	31.5	70.8
6	120	52.26	0.57	70	75.3	22.4	53.0	102.2	65.5	53.3	14.3	23.3
7	95	49.17	-0.41	10	82.9	78.9	134.8	124.6	111.6	63.4	20.1	34.9
8	146	49.91	15.39	231	22.6	25.2	37.6	34.5	33.6	40.4	36.5	84.9
9	126	46.40	6.23	430	95.0	59.5	101.0	195.9	131.6	81.4	29.5	136.3
10	144	49.13	16.66	205	39.0	17.5	27.8	27.9	20.2	41.6	30.9	62.9
11	112	48.07	7.35	200	32.5	19.5	49.2	93.3	57.0	81.1	44.6	130.6
12	163	56.45	-3.05	20	55.1	79.7	72.9	96.0	97.1	35.7	30.9	62.8
13	161	55.94	-3.30	60	56.9	77.2	128.0	92.9	91.7	51.0	27.7	54.2
14	123	57.63	-3.33	30	69.2	57.9	70.9	113.5	65.0	50.3	54.4	55.5
15	126	50.57	4.70	165	62.1	31.9	111.7	149.3	100.9	79.6	53.9	37.9
16	93	52.12	-2.64	84	74.2	60.0	124.2	135.4	90.0	41.4	15.3	56.2
17	205	57.12	14.11	10	64.0	57.1	102.2	63.8	84.2	62.3	66.4	58.2
18	104	52.94	-0.07	3	61.6	34.1	60.9	90.3	54.8	41.7	10.2	32.4
19	170	58.81	15.45	90	49.2	25.1	67.9	52.7	58.9	43.1	77.2	53.6
20	37	38.38	23.12	94	133.7	57.1	133.1	150.3	13.7	62.2	12.9	11.9
21	198	58.58	13.24	70	56.9	39.9	75.1	54.5	53.1	49.2	53.8	51.9
22	82	48.11	-1.80	40	71.5	67.1	114.8	131.5	114.1	64.7	26.4	54.1
23	170	49.90	5.38	550	86.2	67.1	142.3	257.7	134.2	127.9	89.5	60.1
24	171	49.55	15.54	505	31.8	36.3	38.6	39.6	37.9	42.6	37.8	77.2
25	105	51.43	-2.67	46	106.4	104.4	160.3	177.4	108.0	56.2	31.2	60.6
26	103	50.45	2.79	40	66.4	33.9	73.6	125.1	78.5	63.0	48.0	43.4

continued

TABLE A.3A (continued)

Data from 50 Suction Trap Locations (Trap) During 1995: Julian Day When Aphid *Myzus persicae* First Caught in Trap (JDay), Trap Location (Latitude, Longitude and Altitude) and Monthly Rainfall from October 1994 to May 1995 (Example 14.2 and File EXAMINE.DAT)

| Trap | JDay | Location | | | Monthly Rain | | | | | | | |
		Latitude	Longitude	Altitude	October	November	December	January	February	March	April	May
27	19	43.58	3.88	30	159.8	73.3	18.2	62.3	23.3	29.7	67.5	23.8
28	139	55.21	-1.68	93	51.4	78.3	49.5	75.8	73.1	33.8	30.5	41.4
29	122	47.84	1.92	120	42.1	69.8	60.5	98.8	70.9	65.7	66.7	63.3
30	92	48.52	-4.32	80	120.0	73.2	185.0	267.9	202.2	86.6	57.9	60.9
31	101	46.56	0.33	80	82.5	90.6	96.9	122.1	112.1	83.2	41.9	55.5
32	1	45.99	13.18	75	116.7	55.8	39.4	59.7	104.9	150.5	44.1	205.2
33	121	53.85	-2.76	15	143.4	109.0	218.1	211.6	153.9	104.8	25.4	61.6
34	122	46.42	15.88	227	120.2	60.4	89.2	55.7	72.4	88.3	61.1	93.1
35	141	49.26	4.03	80	37.8	28.8	63.4	107.4	76.7	61.5	61.5	57.2
36	100	51.81	-0.36	119	86.4	42.8	81.3	129.6	80.2	58.0	15.7	34.3
37	108	46.24	14.41	404	235.4	99.2	100.3	85.5	189.6	201.4	56.7	160.4
38	127	50.63	-3.45	12	103.6	97.7	128.1	178.4	112.0	44.4	22.1	34.0
39	198	57.98	18.41	10	46.3	30.0	80.3	53.7	50.5	38.3	69.4	64.3
40	122	47.16	20.22	88	38.8	18.1	24.9	26.0	45.3	28.6	43.8	49.7
41	123	53.86	-1.34	98	61.0	85.8	106.1	109.5	82.7	48.3	17.8	37.9
42	189	60.39	18.13	20	46.2	18.1	37.9	56.0	43.1	41.7	74.6	41.2
43	85	44.98	4.96	150	103.8	103.1	17.3	83.0	97.9	31.4	131.2	118.8
44	195	57.49	14.69	200	60.9	43.4	91.3	52.3	77.3	53.3	65.1	57.7
45	74	39.73	23.32	85	109.3	53.1	69.2	80.5	16.5	36.6	7.4	39.8
46	145	49.46	17.29	206	50.4	16.5	35.6	31.8	25.3	46.8	37.4	73.9
47	92	48.80	2.11	130	41.4	49.4	54.1	113.8	73.0	62.4	73.5	63.7
48	72	51.73	0.43	38	79.7	25.3	63.9	117.6	66.7	52.1	11.3	21.8
49	123	51.20	0.94	43	107.2	33.3	92.5	137.4	80.1	61.8	17.2	26.8
50	141	50.32	13.55	240	15.3	20.1	39.5	32.6	25.8	34.1	66.1	65.4

Source: Data from R. Harrington, Rothamsted Research.

TABLE A.3B

Data from 50 Suction Trap Locations (Trap) During 1995: Mean Temperature for the Coldest 30-Day Period (C30Day) and the Following 60-Day Period (F60Day), and Proportion of Land in a Circle of Radius 75 km around the Trap Identified as Coniferous, Deciduous or Mixed Forest, Grassland, Arable Crops, Inland Waters, Sea or Urban Use (Example 14.2 and File EXAMINE.DAT)

| | Winter Temperature | | Land Use | | | | | | | |
| | | | Forest | | | | | | | |
Trap	C30Day	F60Day	Coniferous	Deciduous	Mixed	Grass	Arable	Water	Sea	Urban
1	0.09	2.89	0.218	0.035	0.039	0.000	0.378	0.017	0.230	0.043
2	7.11	8.37	0.057	0.019	0.027	0.095	0.784	0.004	0.000	0.013
3	1.65	4.10	0.086	0.080	0.048	0.002	0.738	0.001	0.000	0.045
4	5.53	7.14	0.007	0.329	0.014	0.109	0.531	0.001	0.000	0.008
5	4.05	5.02	0.066	0.054	0.079	0.393	0.022	0.005	0.225	0.025
6	4.80	6.51	0.011	0.008	0.002	0.004	0.875	0.001	0.074	0.026
7	6.80	8.18	0.001	0.033	0.007	0.480	0.196	0.000	0.268	0.014
8	-0.32	4.49	0.109	0.006	0.040	0.014	0.809	0.002	0.000	0.020
9	1.74	3.88	0.074	0.364	0.179	0.104	0.191	0.050	0.000	0.013
10	-0.64	4.43	0.083	0.058	0.063	0.000	0.772	0.002	0.000	0.019
11	3.06	4.88	0.086	0.278	0.250	0.108	0.246	0.000	0.000	0.031
12	3.18	4.75	0.067	0.009	0.030	0.202	0.182	0.003	0.303	0.015
13	3.55	4.93	0.035	0.026	0.040	0.354	0.203	0.001	0.157	0.032
14	2.44	3.67	0.086	0.010	0.040	0.133	0.046	0.001	0.363	0.002
15	4.03	6.35	0.018	0.079	0.072	0.235	0.475	0.001	0.000	0.119
16	5.54	6.64	0.009	0.029	0.019	0.478	0.369	0.000	0.014	0.058
17	-1.44	1.90	0.768	0.001	0.000	0.000	0.084	0.064	0.000	0.013
18	5.05	6.66	0.006	0.002	0.001	0.014	0.762	0.001	0.191	0.015
19	-1.55	1.81	0.436	0.005	0.001	0.003	0.378	0.131	0.009	0.022
20	8.28	10.58	0.028	0.013	0.022	0.044	0.229	0.005	0.335	0.020
21	-1.63	2.19	0.275	0.007	0.002	0.006	0.359	0.320	0.000	0.012
22	7.04	8.78	0.024	0.012	0.017	0.183	0.705	0.000	0.044	0.013
23	7.98	5.82	0.018	0.322	0.141	0.345	0.152	0.000	0.000	0.022
24	-1.55	2.95	0.135	0.009	0.057	0.011	0.775	0.002	0.000	0.010
25	6.34	7.38	0.012	0.026	0.019	0.464	0.342	0.000	0.087	0.033

continued

TABLE A.3B (continued)

Data from 50 Suction Trap Locations (Trap) During 1995: Mean Temperature for the Coldest 30-Day Period (C30Day) and the Following 60-Day Period (F60Day), and Proportion of Land in a Circle of Radius 75 km around the Trap Identified as Coniferous, Deciduous or Mixed Forest, Grassland, Arable Crops, Inland Waters, Sea or Urban Use (Example 14.2 and File EXAMINE.DAT)

| | Winter Temperature | | Land Use | | | | | | | |
| | | | Forest | | | | | | | |
Trap	C30Day	F60Day	Coniferous	Deciduous	Mixed	Grass	Arable	Water	Sea	Urban
26	4.97	7.01	0.004	0.007	0.004	0.048	0.873	0.000	0.001	0.063
27	7.87	10.32	0.153	0.019	0.057	0.012	0.140	0.021	0.330	0.010
28	4.03	5.26	0.038	0.020	0.015	0.267	0.118	0.001	0.450	0.022
29	5.69	7.48	0.151	0.093	0.079	0.013	0.644	0.000	0.000	0.019
30	9.11	9.25	0.008	0.001	0.010	0.041	0.377	0.000	0.553	0.009
31	5.87	7.65	0.038	0.018	0.022	0.162	0.749	0.000	0.000	0.010
32	5.19	6.83	0.144	0.146	0.162	0.024	0.317	0.005	0.152	0.019
33	4.40	5.55	0.007	0.008	0.005	0.401	0.092	0.001	0.285	0.088
34	1.39	5.32	0.092	0.153	0.160	0.021	0.570	0.000	0.000	0.003
35	4.41	7.13	0.020	0.208	0.029	0.080	0.644	0.000	0.000	0.010
36	5.05	6.70	0.008	0.022	0.006	0.088	0.733	0.001	0.006	0.137
37	−0.28	2.40	0.239	0.260	0.308	0.012	0.168	0.002	0.000	0.005
38	7.50	7.95	0.002	0.011	0.004	0.483	0.073	0.000	0.409	0.013
39	1.41	3.07	0.042	0.001	0.021	0.020	0.047	0.000	0.866	0.001
40	0.42	5.97	0.007	0.000	0.001	0.060	0.883	0.002	0.000	0.028
41	4.48	5.71	0.015	0.011	0.007	0.270	0.522	0.000	0.008	0.067
42	−0.39	1.41	0.259	0.002	0.000	0.001	0.179	0.000	0.518	0.010
43	5.71	7.37	0.196	0.153	0.145	0.075	0.401	0.021	0.000	0.021
44	−1.50	1.17	0.760	0.002	0.000	0.001	0.117	0.001	0.000	0.014
45	5.96	9.48	0.015	0.041	0.027	0.020	0.100	0.079	0.000	0.002
46	−0.46	4.28	0.128	0.072	0.125	0.003	0.646	0.000	0.733	0.022
47	5.85	7.40	0.033	0.147	0.044	0.004	0.648	0.000	0.000	0.125
48	5.52	6.81	0.005	0.010	0.006	0.057	0.640	0.000	0.157	0.125
49	5.75	7.06	0.004	0.007	0.006	0.098	0.300	0.000	0.536	0.048
50	−0.95	3.69	0.186	0.026	0.056	0.177	0.515	0.002	0.000	0.038

Source: Data from R. Harrington, Rothamsted Research.

Appendix B: Quantiles of Statistical Distributions

TABLE B.1

95th Percentiles of F-Distribution with N Numerator and D Denominator df

	N									
D	1	2	3	4	5	6	7	8	9	10
1	161.448	199.500	215.707	224.583	230.162	233.986	236.768	238.883	240.543	241.882
2	18.513	19.000	19.164	19.247	19.296	19.330	19.353	19.371	19.385	19.396
3	10.128	9.552	9.277	9.117	9.013	8.941	8.887	8.845	8.812	8.785
4	7.709	6.944	6.591	6.388	6.256	6.163	6.094	6.041	5.999	5.964
5	6.608	5.786	5.409	5.192	5.050	4.950	4.876	4.818	4.772	4.735
6	5.987	5.143	4.757	4.534	4.387	4.284	4.207	4.147	4.099	4.060
7	5.591	4.737	4.347	4.120	3.972	3.866	3.787	3.726	3.677	3.637
8	5.318	4.459	4.066	3.838	3.687	3.581	3.500	3.438	3.388	3.347
9	5.117	4.256	3.863	3.633	3.482	3.374	3.293	3.230	3.179	3.137
10	4.965	4.103	3.708	3.478	3.326	3.217	3.135	3.072	3.020	2.978
11	4.844	3.982	3.587	3.357	3.204	3.095	3.012	2.948	2.896	2.854
12	4.747	3.885	3.490	3.259	3.106	2.996	2.913	2.849	2.796	2.753
13	4.667	3.806	3.411	3.179	3.025	2.915	2.832	2.767	2.714	2.671
14	4.600	3.739	3.344	3.112	2.958	2.848	2.764	2.699	2.646	2.602
15	4.543	3.682	3.287	3.056	2.901	2.790	2.707	2.641	2.588	2.544
16	4.494	3.634	3.239	3.007	2.852	2.741	2.657	2.591	2.538	2.494
17	4.451	3.592	3.197	2.965	2.810	2.699	2.614	2.548	2.494	2.450
18	4.414	3.555	3.160	2.928	2.773	2.661	2.577	2.510	2.456	2.412
19	4.381	3.522	3.127	2.895	2.740	2.628	2.544	2.477	2.423	2.378
20	4.351	3.493	3.098	2.866	2.711	2.599	2.514	2.447	2.393	2.348
22	4.301	3.443	3.049	2.817	2.661	2.549	2.464	2.397	2.342	2.297
24	4.260	3.403	3.009	2.776	2.621	2.508	2.423	2.355	2.300	2.255
26	4.225	3.369	2.975	2.743	2.587	2.474	2.388	2.321	2.265	2.220
28	4.196	3.340	2.947	2.714	2.558	2.445	2.359	2.291	2.236	2.190
30	4.171	3.316	2.922	2.690	2.534	2.421	2.334	2.266	2.211	2.165
32	4.149	3.295	2.901	2.668	2.512	2.399	2.313	2.244	2.189	2.142
34	4.130	3.276	2.883	2.650	2.494	2.380	2.294	2.225	2.170	2.123
36	4.113	3.259	2.866	2.634	2.477	2.364	2.277	2.209	2.153	2.106
38	4.098	3.245	2.852	2.619	2.463	2.349	2.262	2.194	2.138	2.091
40	4.085	3.232	2.839	2.606	2.449	2.336	2.249	2.180	2.124	2.077
45	4.057	3.204	2.812	2.579	2.422	2.308	2.221	2.152	2.096	2.049
50	4.034	3.183	2.790	2.557	2.400	2.286	2.199	2.130	2.073	2.026
55	4.016	3.165	2.773	2.540	2.383	2.269	2.181	2.112	2.055	2.008
60	4.001	3.150	2.758	2.525	2.368	2.254	2.167	2.097	2.040	1.993
70	3.978	3.128	2.736	2.503	2.346	2.231	2.143	2.074	2.017	1.969
80	3.960	3.111	2.719	2.486	2.329	2.214	2.126	2.056	1.999	1.951
100	3.936	3.087	2.696	2.463	2.305	2.191	2.103	2.032	1.975	1.927

continued

TABLE B.1 (continued)

95th Percentiles of F-Distribution with N Numerator and D Denominator df

					N					
D	11	12	13	14	15	16	18	20	22	24
1	242.984	243.906	244.690	245.364	245.950	246.464	247.323	248.013	248.579	249.052
2	19.405	19.413	19.419	19.438	19.443	19.447	19.454	19.460	19.464	19.468
3	8.763	8.745	8.729	8.715	8.703	8.692	8.674	8.660	8.648	8.638
4	5.936	5.912	5.891	5.873	5.858	5.844	5.821	5.802	5.787	5.774
5	4.704	4.678	4.655	4.636	4.619	4.604	4.578	4.558	4.541	4.527
6	4.027	4.000	3.976	3.956	3.938	3.922	3.896	3.874	3.856	3.841
7	3.603	3.575	3.550	3.529	3.511	3.494	3.467	3.444	3.426	3.410
8	3.313	3.284	3.259	3.237	3.218	3.202	3.173	3.150	3.131	3.115
9	3.102	3.073	3.048	3.025	3.006	2.989	2.960	2.936	2.917	2.900
10	2.943	2.913	2.887	2.865	2.845	2.828	2.798	2.774	2.754	2.737
11	2.818	2.788	2.761	2.739	2.719	2.701	2.671	2.646	2.626	2.609
12	2.717	2.687	2.660	2.637	2.617	2.599	2.568	2.544	2.523	2.505
13	2.635	2.604	2.577	2.554	2.533	2.515	2.484	2.459	2.438	2.420
14	2.565	2.534	2.507	2.484	2.463	2.445	2.413	2.388	2.367	2.349
15	2.507	2.475	2.448	2.424	2.403	2.385	2.353	2.328	2.306	2.288
16	2.456	2.425	2.397	2.373	2.352	2.333	2.302	2.276	2.254	2.235
17	2.413	2.381	2.353	2.329	2.308	2.289	2.257	2.230	2.208	2.190
18	2.374	2.342	2.314	2.290	2.269	2.250	2.217	2.191	2.168	2.150
19	2.340	2.308	2.280	2.256	2.234	2.215	2.182	2.155	2.133	2.114
20	2.310	2.278	2.250	2.225	2.203	2.184	2.151	2.124	2.102	2.082
22	2.259	2.226	2.198	2.173	2.151	2.131	2.098	2.071	2.048	2.028
24	2.216	2.183	2.155	2.130	2.108	2.088	2.054	2.027	2.003	1.984
26	2.181	2.148	2.119	2.094	2.072	2.052	2.018	1.990	1.966	1.946
28	2.151	2.118	2.089	2.064	2.041	2.021	1.987	1.959	1.935	1.915
30	2.126	2.092	2.063	2.037	2.015	1.995	1.960	1.932	1.908	1.887
32	2.103	2.070	2.040	2.015	1.992	1.972	1.937	1.908	1.884	1.864
34	2.084	2.050	2.021	1.995	1.972	1.952	1.917	1.888	1.863	1.843
36	2.067	2.033	2.003	1.977	1.954	1.934	1.899	1.870	1.845	1.824
38	2.051	2.017	1.988	1.962	1.939	1.918	1.883	1.853	1.829	1.808
40	2.038	2.003	1.974	1.948	1.924	1.904	1.868	1.839	1.814	1.793
45	2.009	1.974	1.945	1.918	1.895	1.874	1.838	1.808	1.783	1.762
50	1.986	1.952	1.921	1.895	1.871	1.850	1.814	1.784	1.759	1.737
55	1.968	1.933	1.903	1.876	1.852	1.831	1.795	1.764	1.739	1.717
60	1.952	1.917	1.887	1.860	1.836	1.815	1.778	1.748	1.722	1.700
70	1.928	1.893	1.863	1.836	1.812	1.790	1.753	1.722	1.696	1.674
80	1.910	1.875	1.845	1.817	1.793	1.772	1.734	1.703	1.677	1.654
100	1.886	1.850	1.819	1.792	1.768	1.746	1.708	1.676	1.650	1.627

TABLE B.1 (continued)

95th Percentiles of F-Distribution with N Numerator and D Denominator df

	N									
D	**26**	**28**	**30**	**35**	**40**	**45**	**50**	**60**	**80**	**100**
1	249.453	249.797	250.095	250.693	251.143	251.494	251.774	252.196	252.724	253.041
2	19.472	19.474	19.477	19.482	19.485	19.488	19.491	19.494	19.498	19.501
3	8.630	8.623	8.617	8.604	8.594	8.587	8.581	8.572	8.561	8.554
4	5.763	5.754	5.746	5.729	5.717	5.707	5.699	5.688	5.673	5.664
5	4.515	4.505	4.496	4.477	4.464	4.453	4.444	4.431	4.415	4.405
6	3.829	3.818	3.808	3.789	3.774	3.763	3.754	3.740	3.722	3.712
7	3.397	3.386	3.376	3.356	3.340	3.328	3.319	3.304	3.286	3.275
8	3.101	3.090	3.079	3.058	3.043	3.030	3.020	3.005	2.986	2.975
9	2.886	2.874	2.864	2.842	2.826	2.813	2.803	2.787	2.767	2.755
10	2.723	2.710	2.700	2.678	2.661	2.648	2.637	2.621	2.601	2.588
11	2.594	2.582	2.570	2.548	2.531	2.517	2.506	2.490	2.469	2.456
12	2.491	2.478	2.466	2.443	2.426	2.412	2.401	2.384	2.363	2.350
13	2.405	2.392	2.380	2.357	2.339	2.325	2.314	2.297	2.275	2.261
14	2.333	2.320	2.308	2.284	2.266	2.252	2.241	2.223	2.200	2.187
15	2.272	2.259	2.247	2.223	2.204	2.190	2.178	2.160	2.137	2.123
16	2.220	2.206	2.194	2.169	2.151	2.136	2.124	2.106	2.083	2.068
17	2.174	2.160	2.148	2.123	2.104	2.089	2.077	2.058	2.035	2.020
18	2.134	2.119	2.107	2.082	2.063	2.048	2.035	2.017	1.993	1.978
19	2.098	2.084	2.071	2.046	2.026	2.011	1.999	1.980	1.955	1.940
20	2.066	2.052	2.039	2.013	1.994	1.978	1.966	1.946	1.922	1.907
22	2.012	1.997	1.984	1.958	1.938	1.922	1.909	1.889	1.864	1.849
24	1.967	1.952	1.939	1.912	1.892	1.876	1.863	1.842	1.816	1.800
26	1.929	1.914	1.901	1.874	1.853	1.837	1.823	1.803	1.776	1.760
28	1.897	1.882	1.869	1.841	1.820	1.803	1.790	1.769	1.742	1.725
30	1.870	1.854	1.841	1.813	1.792	1.775	1.761	1.740	1.712	1.695
32	1.846	1.830	1.817	1.789	1.767	1.750	1.736	1.714	1.686	1.669
34	1.825	1.809	1.795	1.767	1.745	1.728	1.713	1.691	1.663	1.645
36	1.806	1.790	1.776	1.748	1.726	1.708	1.694	1.671	1.643	1.625
38	1.790	1.774	1.760	1.731	1.708	1.691	1.676	1.653	1.624	1.606
40	1.775	1.759	1.744	1.715	1.693	1.675	1.660	1.637	1.608	1.589
45	1.743	1.727	1.713	1.683	1.660	1.642	1.626	1.603	1.573	1.554
50	1.718	1.702	1.687	1.657	1.634	1.615	1.599	1.576	1.544	1.525
55	1.698	1.681	1.666	1.636	1.612	1.593	1.577	1.553	1.521	1.501
60	1.681	1.664	1.649	1.618	1.594	1.575	1.559	1.534	1.502	1.481
70	1.654	1.637	1.622	1.591	1.566	1.546	1.530	1.505	1.471	1.450
80	1.634	1.617	1.602	1.570	1.545	1.525	1.508	1.482	1.448	1.426
100	1.607	1.589	1.573	1.541	1.515	1.494	1.477	1.450	1.415	1.392

TABLE B.2

Percentiles of t- and Chi-Squared Distributions with D df

D	t-Distribution				Chi-Squared Distribution			
	95th	97.5th	99th	99.5th	95th	97.5th	99th	99.5th
1	6.314	12.706	31.821	63.657	3.841	5.024	6.635	7.879
2	2.920	4.303	6.965	9.925	5.991	7.378	9.210	10.597
3	2.353	3.182	4.541	5.841	7.815	9.348	11.345	12.838
4	2.132	2.776	3.747	4.604	9.488	11.143	13.277	14.860
5	2.015	2.571	3.365	4.032	11.070	12.833	15.086	16.750
6	1.943	2.447	3.143	3.707	12.592	14.449	16.812	18.548
7	1.895	2.365	2.998	3.499	14.067	16.013	18.475	20.278
8	1.860	2.306	2.896	3.355	15.507	17.535	20.090	21.955
9	1.833	2.262	2.821	3.250	16.919	19.023	21.666	23.589
10	1.812	2.228	2.764	3.169	18.307	20.483	23.209	25.188
11	1.796	2.201	2.718	3.106	19.675	21.920	24.725	26.757
12	1.782	2.179	2.681	3.055	21.026	23.337	26.217	28.300
13	1.771	2.160	2.650	3.012	22.362	24.736	27.688	29.819
14	1.761	2.145	2.624	2.977	23.685	26.119	29.141	31.319
15	1.753	2.131	2.602	2.947	24.996	27.488	30.578	32.801
16	1.746	2.120	2.583	2.921	26.296	28.845	32.000	34.267
17	1.740	2.110	2.567	2.898	27.587	30.191	33.409	35.718
18	1.734	2.101	2.552	2.878	28.869	31.526	34.805	37.156
19	1.729	2.093	2.539	2.861	30.144	32.852	36.191	38.582
20	1.725	2.086	2.528	2.845	31.410	34.170	37.566	39.997
22	1.717	2.074	2.508	2.819	33.924	36.781	40.289	42.796
24	1.711	2.064	2.492	2.797	36.415	39.364	42.980	45.559
26	1.706	2.056	2.479	2.779	38.885	41.923	45.642	48.290
28	1.701	2.048	2.467	2.763	41.337	44.461	48.278	50.993
30	1.697	2.042	2.457	2.750	43.773	46.979	50.892	53.672
32	1.694	2.037	2.449	2.738	46.194	49.480	53.486	56.328
34	1.691	2.032	2.441	2.728	48.602	51.966	56.061	58.964
36	1.688	2.028	2.434	2.719	50.998	54.437	58.619	61.581
38	1.686	2.024	2.429	2.712	53.384	56.896	61.162	64.181
40	1.684	2.021	2.423	2.704	55.758	59.342	63.691	66.766
42	1.682	2.018	2.418	2.698	58.124	61.777	66.206	69.336
44	1.680	2.015	2.414	2.692	60.481	64.201	68.710	71.893
46	1.679	2.013	2.410	2.687	62.830	66.617	71.201	74.437
48	1.677	2.011	2.407	2.682	65.171	69.023	73.683	76.969
50	1.676	2.009	2.403	2.678	67.505	71.420	76.154	79.490
55	1.673	2.004	2.396	2.668	73.311	77.380	82.292	85.749
60	1.671	2.000	2.390	2.660	79.082	83.298	88.379	91.952
65	1.669	1.997	2.385	2.654	84.821	89.177	94.422	98.105
70	1.667	1.994	2.381	2.648	90.531	95.023	100.425	104.215
75	1.665	1.992	2.377	2.643	96.217	100.839	106.393	110.286
80	1.664	1.990	2.374	2.639	101.879	106.629	112.329	116.321
85	1.663	1.988	2.371	2.635	107.522	112.393	118.236	122.325
90	1.662	1.987	2.368	2.632	113.145	118.136	124.116	128.299
100	1.660	1.984	2.364	2.626	124.342	129.561	135.807	140.169

Appendix C: Statistical and Mathematical Results

C.1 Derivation of Least Squares Estimates for a Model with a Single Factor

For a set of N observations with a single explanatory factor, we represent the data as y_{jk}, $j = 1 \ldots t$, $k = 1 \ldots n_j$, where y_{jk} is the kth observation for the jth treatment, t is the number of treatments and n_j is the number of replicates of the jth treatment, with $N = n_1 + n_2 + \ldots + n_t$. The model (Equation 4.1) is written as

$$y_{jk} = \mu_j + e_{jk} \, ,$$

where μ_j is the unknown population mean for the jth treatment, and e_{jk} is the deviation from that population mean for the kth observation on that treatment. We can write the residual sum of squares (Section 4.2) as a function of the estimated population means:

$$\mathrm{ResSS}(\hat{\mu}_1 \ldots \hat{\mu}_t) = \sum_{j=1}^{t} \sum_{k=1}^{n_j} (y_{jk} - \hat{\mu}_j)^2 \, .$$

We use a standard mathematical approach to find the estimates that minimize this function. At any local minimum of a continuous function, its first derivative will be equal to zero and its second derivative will be positive. We therefore take the first derivative of the ResSS function with respect to each of the estimates, set the resulting equations equal to zero, and solve them to obtain estimates that minimize the ResSS. We can verify that we have found a minimum by calculating the second derivative of the ResSS function at these estimates.

The first derivative of the ResSS function with respect to $\hat{\mu}_j$ is

$$\frac{\partial \mathrm{ResSS}}{\partial \hat{\mu}_j} = -2 \sum_{k=1}^{n_j} (y_{jk} - \hat{\mu}_j) \, .$$

If we set this equation equal to zero and solve for $\hat{\mu}_j$, we find

$$0 = -2 \sum_{k=1}^{n_j} (y_{jk} - \hat{\mu}_j) = -2 \sum_{k=1}^{n_j} y_{jk} + 2 \sum_{k=1}^{n_j} \hat{\mu}_j = -2 \sum_{k=1}^{n_j} y_{jk} + 2 n_j \hat{\mu}_j \, ,$$

which can be rearranged to give a unique solution as

$$\hat{\mu}_j = \frac{1}{n_j} \sum_{k=1}^{n_j} y_{jk} = \bar{y}_{j\cdot} \, .$$

To check that we have found a minimum, we calculate the second derivative as

$$\frac{\partial^2 \text{ResSS}}{\partial \hat{\mu}_j^2} = 2n_j \ ,$$

which is positive as required, and in fact is constant (and hence positive everywhere). The set of estimates that minimize the ResSS are hence the set of treatment sample means. For further details, Kuehl (2000) presents a simple demonstration for the CRD and Searle (1982) shows a complete derivation for any linear model using matrix notation (matrix notation is introduced in Section 15.6.1).

C.2 Partitioning the Total Sum of Squares for a Model with a Single Factor

In Section 4.3.1, we saw that the total sum of squares takes the form

$$\text{TotSS} = \sum_{j=1}^{t} \sum_{k=1}^{n_j} (y_{jk} - \bar{y})^2 \ .$$

This formula can be expanded, without any change in its value, by subtraction and then addition of each group mean to give

$$\text{TotSS} = \sum_{j=1}^{t} \sum_{k=1}^{n_j} (y_{jk} - \bar{y}_{j.} + \bar{y}_{j.} - \bar{y})^2 = \sum_{j=1}^{t} \sum_{k=1}^{n_j} [(y_{jk} - \bar{y}_{j.}) + (\bar{y}_{j.} - \bar{y})]^2 .$$

We then make use of the following relationship for two quantities A and B:

$$(A + B)^2 = (A + B)(A + B) = A^2 + 2AB + B^2 \ .$$

We now substitute $A = y_{jk} - \bar{y}_{j.}$ and $B = \bar{y}_{j.} - \bar{y}$ into this expression to get

$$\text{TotSS} = \sum_{j=1}^{t} \sum_{k=1}^{n_j} [(y_{jk} - \bar{y}_{j.}) + (\bar{y}_{j.} - \bar{y})]^2$$

$$= \sum_{j=1}^{t} \sum_{k=1}^{n_j} [(y_{jk} - \bar{y}_{j.})^2 + 2(y_{jk} - \bar{y}_{j.})(\bar{y}_{j.} - \bar{y}) + (\bar{y}_{j.} - \bar{y})^2]$$

$$= \sum_{j=1}^{t} \sum_{k=1}^{n_j} (y_{jk} - \bar{y}_{j.})^2 + 2 \sum_{j=1}^{t} \sum_{k=1}^{n_j} (y_{jk} - \bar{y}_{j.})(\bar{y}_{j.} - \bar{y}) + \sum_{j=1}^{t} \sum_{k=1}^{n_j} (\bar{y}_{j.} - \bar{y})^2 . \quad \text{(C.1)}$$

We will consider each of the components of Equation C.1 in turn. The first component is equal to the residual sum of squares (ResSS, Section 4.3.1). The third component is the treatment sum of squares (TrtSS, Section 4.3.1), which can be rewritten as

$$\text{TrtSS} = \sum_{j=1}^{t} \sum_{k=1}^{n_j} (\bar{y}_{j\cdot} - \bar{y})^2 = \sum_{j=1}^{t} n_j (\bar{y}_{j\cdot} - \bar{y})^2 .$$

We now look at the second component of Equation C.1, which can be rewritten as

$$2 \sum_{j=1}^{t} \sum_{k=1}^{n_j} (y_{jk} - \bar{y}_{j\cdot})(\bar{y}_{j\cdot} - \bar{y}) = 2 \sum_{j=1}^{t} \left\{ (\bar{y}_{j\cdot} - \bar{y}) \sum_{k=1}^{n_j} (y_{jk} - \bar{y}_{j\cdot}) \right\} .$$

We can perform summation over the k index first (for each value of j), to give

$$\sum_{k=1}^{n_j} (y_{jk} - \bar{y}_{j\cdot}) = \sum_{k=1}^{n_j} y_{jk} - \sum_{k=1}^{n_j} \bar{y}_{j\cdot} = y_{j\cdot} - n_j \bar{y}_{j\cdot} = y_{j\cdot} - y_{j\cdot} = 0 .$$

The second component of Equation C.1 is therefore also equal to zero, leaving the result as required, i.e.

$$\sum_{j=1}^{t} \sum_{k=1}^{n_j} (y_{jk} - \bar{y})^2 = \sum_{j=1}^{t} \sum_{k=1}^{n_j} (\bar{y}_{j\cdot} - \bar{y})^2 + \sum_{j=1}^{t} \sum_{k=1}^{n_j} (y_{jk} - \bar{y}_{j\cdot})^2 ,$$

which is equivalent to

$$\text{TotSS} = \text{TrtSS} + \text{ResSS} .$$

The same result holds for any linear model, and the principle of this proof still holds although the details become more complicated when there are more terms in the model.

C.3 Derivation of Least Squares Estimates for a Model with a Single Variate

For a set of N observations with a single explanatory variate, we represent the data as y_i, for $i = 1 \ldots N$, where y_i is the ith observation. The model (Equation 12.1) is written as

$$y_i = \alpha + \beta x_i + e_i ,$$

where α is the intercept and β is the slope of the straight line relationship, x_i is the value of the explanatory variate and e_i is the deviation from the straight line for the ith observation.

We can write the residual sum of squares (Section 12.2) as a function of parameter estimates in the form

$$\text{ResSS}(\hat{\alpha}, \hat{\beta}) = \sum_{i=1}^{N} (y_i - \hat{\alpha} - \hat{\beta}x_i)^2 .$$

We use the same approach as Section C.1. We take the first derivative of the ResSS function with respect to each of the estimates, set the resulting equations equal to zero, and solve them to obtain estimates that minimize the ResSS. We can verify that we have found a minimum by calculating the second derivative at these estimates.

The first derivative of the ResSS function with respect to $\hat{\alpha}$ is

$$\frac{\partial \text{ResSS}}{\partial \hat{\alpha}} = -2 \sum_{i=1}^{N} (y_i - \hat{\alpha} - \hat{\beta}x_i) = -2 \sum_{i=1}^{N} y_i + 2N\hat{\alpha} + 2\hat{\beta} \sum_{i=1}^{N} x_i .$$

If we set this equation equal to zero and solve for $\hat{\alpha}$, we find

$$N\hat{\alpha} = \sum_{i=1}^{N} y_i - \hat{\beta} \sum_{i=1}^{N} x_i ,$$

which can be rearranged to give a unique solution as

$$\hat{\alpha} = \bar{y} - \hat{\beta}\bar{x} .$$

The first derivative of the ResSS function with respect to $\hat{\beta}$ is

$$\frac{\partial \text{ResSS}}{\partial \hat{\beta}} = -2 \sum_{i=1}^{N} x_i(y_i - \hat{\alpha} - \hat{\beta}x_i) = -2 \sum_{i=1}^{N} x_i y_i + 2\hat{\alpha} \sum_{i=1}^{N} x_i + 2\hat{\beta} \sum_{i=1}^{N} x_i^2 .$$

If we set this equation equal to zero and solve for $\hat{\beta}$, we find

$$\hat{\beta} \sum_{i=1}^{N} x_i^2 = \sum_{i=1}^{N} x_i y_i - \hat{\alpha} \sum_{i=1}^{N} x_i .$$

At this point, we need to substitute for our estimate $\hat{\alpha}$, as we cannot have both estimates defined in terms of the other. This gives

$$\hat{\beta} \sum_{i=1}^{N} x_i^2 = \sum_{i=1}^{N} x_i y_i - (\bar{y} - \hat{\beta}\bar{x}) \sum_{i=1}^{N} x_i = \sum_{i=1}^{N} x_i y_i - N\bar{x}\bar{y} + N\hat{\beta}\bar{x}^2 .$$

We need to group terms with $\hat{\beta}$ together, to get the revised form

$$\hat{\beta} \left(\sum_{i=1}^{N} x_i^2 - N\bar{x}^2 \right) = \sum_{i=1}^{N} x_i y_i - N\bar{x}\bar{y} .$$

We can then use the following identity for sums of squares to simplify the expressions:

$$\sum_{i=1}^{N} (y_i - \bar{y})(x_i - \bar{x}) = \sum_{i=1}^{N} y_i x_i - N\bar{y}\bar{x} .$$

This relationship holds for any two variables, giving

$$\hat{\beta} = \frac{\sum x_i y_i - N\bar{x}\bar{y}}{\sum x_i^2 - N\bar{x}^2} = \frac{\sum (x_i - \bar{x})(y_i - \bar{y})}{\sum (x_i - \bar{x})(x_i - \bar{x})} = \frac{SS_{xy}}{SS_{xx}} ,$$

as in Section 12.2. In both cases, the second derivative is positive as required.

C.4 Variances and Standard Errors of Linear Combinations of Random Variables

For a set of m random variables $Y_1 \ldots Y_m$, the variance of a linear combination Z, where

$$Z = \sum_{i=1}^{m} a_i Y_i = a_1 Y_1 + \ldots + a_m Y_m ,$$

is calculated as

$$\mathrm{Var}(Z) = \sum_{i=1}^{m} a_i^2 \mathrm{Var}(Y_i) + 2\sum_{i=1}^{m}\sum_{j=i+1}^{m} a_i a_j \mathrm{Cov}(Y_i, Y_j) .$$

The following results can be derived directly:

$$\mathrm{Var}(Y_1 + Y_2) = \mathrm{Var}(Y_1) + \mathrm{Var}(Y_2) + 2\mathrm{Cov}(Y_1, Y_2)$$
$$\mathrm{Var}(Y_1 - Y_2) = \mathrm{Var}(Y_1) + \mathrm{Var}(Y_2) - 2\mathrm{Cov}(Y_1, Y_2)$$
$$\mathrm{Var}(Y_1 + Y_2 + Y_3) = \mathrm{Var}(Y_1) + \mathrm{Var}(Y_2) + \mathrm{Var}(Y_3) + 2\mathrm{Cov}(Y_1, Y_2)$$
$$+ 2\mathrm{Cov}(Y_1, Y_3) + 2\mathrm{Cov}(Y_2, Y_3)$$

For the mean of m variables, $\bar{Y} = \dfrac{1}{m}\sum_{i=1}^{m} Y_i$, we have $a_i = \dfrac{1}{m}$ for $i = 1 \ldots m$. Then

$$\mathrm{Var}(\bar{Y}) = \frac{1}{m^2}\sum_{i=1}^{m} \mathrm{Var}(Y_i) + \frac{2}{m^2}\sum_{i=1}^{m}\sum_{j=i+1}^{m} \mathrm{Cov}(Y_i, Y_j) .$$

If the m variables are independent (zero covariance) with variance equal to σ^2, this gives

$$\mathrm{Var}(\bar{Y}) \; = \; \frac{1}{m^2} \sum_{i=1}^{m} \sigma^2 \; = \; \frac{\sigma^2}{m} \; .$$

C.5 Matrix Addition and Multiplication

A matrix $\mathbf{A}$ of size $p \times q$ is an array of numbers with p rows and q columns. We write the elements of matrix $\mathbf{A}$ as A_{ij} for $i = 1 \dots p$ and $j = 1 \dots q$, for example a 2×3 matrix $\mathbf{A}$ takes the form

$$\mathbf{A} \; = \; \begin{bmatrix} A_{11} & A_{12} & A_{13} \\ A_{21} & A_{22} & A_{23} \end{bmatrix} .$$

A vector is a special case of a matrix with one column ($q = 1$). Two matrices with the same dimensions (same number of rows and same number of columns) can be added together via addition of their corresponding elements. So, for matrices $\mathbf{A}$ and $\mathbf{B}$ of the same size, matrix $\mathbf{C} = \mathbf{A} + \mathbf{B}$ means that $C_{ij} = A_{ij} + B_{ij}$. The matrix product operation is more complex, and we denote it with the compound symbol '*+'. The product of two matrices $\mathbf{A}$ and $\mathbf{B}$ can be formed as $\mathbf{AB} = \mathbf{A} \; {}^*{+} \; \mathbf{B}$ only if the number of columns of $\mathbf{A}$ is equal to the number of rows in $\mathbf{B}$. For a $p \times q$ matrix $\mathbf{A}$ and a $q \times r$ matrix $\mathbf{B}$, we can form their product as $\mathbf{C} = \mathbf{A} \; {}^*{+} \; \mathbf{B}$ with matrix $\mathbf{C}$, of size $p \times r$, having elements defined as

$$C_{ij} \; = \; \sum_{k=1}^{q} A_{ik} B_{kj} \quad \text{for } i = 1 \dots p, \, j = 1 \dots r \, .$$

This is sometimes described as taking the vector product of a row of matrix $\mathbf{A}$ with a column of matrix $\mathbf{B}$.

Index

A

Absolute residuals plot, 96, 116
Added variable plot, 332, 361
Additive model, 152–153, 155, 171, 173, 260–262
Additive structure of linear model, 7, 125, 127
Adjusted coefficient of determination, *see* Adjusted R^2 statistic
Adjusted R^2 statistic, 306, 365, 367, 373, 437, 452, 475
AIC, *see* Akaike information criterion
Akaike information criterion, 365, 367, 373, 475
Aliasing, 177
Alternative hypothesis (H$_1$), 29–30; *see also* Hypothesis testing; Null hypothesis
  completely randomized design, 75
  equivalence testing, 253
  one-sample t-test, 30
  one-sided test, 29–30
  power calculations, 246–248
  simple linear regression, 298
  single factor model, 75
  two-sample t-test, 32
  two-sided test, 29–30
Analysis of covariance, 381, 409–413, 415
Analysis of deviance, 487–488, 490, 506
  sequential, 493, 496, 506
Analysis of variance; *see also* Multi-stratum analysis of variance
  balanced incomplete block design, 236–237
  completely randomized design, 75
  Latin square design, 212–213
  multiple linear regression, 352–353
  multi-way factorial, 165
  one-way, 75
  one-way with blocks, 134
  power calculations for, 248–249
  principles, 9–10, 74–75
  randomized complete block design, 134
  regression in designed experiments, 406–409
  sequential, 265–267, 353–355, 391–393
  simple linear regression, 296–298
  simple linear regression with groups, 388, 391–393
  single factor model, 75
  split-plot design, 223–224
  two-way, 150, 158–159

Analysis of variance table; *see also* Multi-stratum analysis of variance table; Sequential analysis of variance table
  completely randomized design, 82–83
  including contrasts, 183, 184, 186
  with missing data, 275
  model selection from, 166, 171
  nested treatment structure, 174
  presentation in publications, 541
  randomized complete block design, 137
  regression through the origin, 315
  simple linear regression, 298
  single factor model, 82–83
  two treatment factors, 158–159
ANCOVA, *see* Analysis of covariance
Angular transformation, 122
ANODEV, *see* Analysis of deviance
ANOVA, *see* Analysis of variance
Arcsine transformation, 122
Assumptions about deviations
  analysis of designed experiments, 70–71, 83, 130
  checking assumptions, 96–102, 330–332, 437, 538
  data transformation to satisfy, 113–114, 171, 451, 540
  failure of, 462, 479
  linear mixed model, 429, 437
  model selection, 366, 375
  non-linear regression, 473
  regression analysis, 291–292, 306–307, 315, 348, 383, 395
Asymptotic approximation, 432, 488
Attenuation, 307, 321
Automatic model selection, 367–378, 414–415
  all subsets selection, 367
  backward elimination, 270, 369, 372, 414
  backward stepwise selection, 370, 372
  cross-validation, 377
  forward selection, 270, 369, 372, 376
  forward stepwise selection, 370, 372
  stepwise selection, 270

B

Background variation, 9–10
  balanced incomplete block design, 236

Printed in the United States
by Baker & Taylor Publisher Services